JavaTech

An Introduction to Scientific and Technical Computing with Java

JavaTech is a practical introduction to the Java programming language with an emphasis on the features that benefit technical computing, such as platform independence, extensive graphics capabilities, multi-threading, and tools to develop network and distributed computing software and embedded processor applications.

The book is divided into three parts. The first presents the basics of object-oriented programming in Java and then examines topics such as graphical interfaces, thread processes, I/O, and image processing. The second part begins with a review of network programming and develops Web client-server examples for tasks such as monitoring of remote devices. The focus then shifts to distributed computing with RMI, which allows programs on different platforms to exchange objects and call each other's methods. CORBA is also discussed and a survey of web services is presented. The final part examines how Java programs can access the local platform and interact with hardware. Topics include combining native code with Java, communication via serial lines, and programming embedded processors.

JavaTech demonstrates the ease with which Java can be used to create powerful network applications and distributed computing applications. It can be used as a textbook for introductory or intermediate level programming courses, and for more advanced students and researchers who need to learn Java for a particular task. *JavaTech* is up to date with Java 5.0.

CLARK S. LINDSEY received his Ph.D. in physics from the University of California at Riverside and has held research positions at Iowa State University, Fermilab, and the Royal Institute of Technology, Sweden. This book grew out of a course in Java programming he developed with Professor Lindblad. He now runs his own company that develops Java applications, Web publications, and educational tools and materials.

JOHNNY S. TOLLIVER holds a Ph.D. in Computational Plasma Physics and has worked in fusion energy research, computer security, and trusted operating systems. He is a Sun Certified Java Programmer and has been actively using Java since 1997. He is currently at Oak Ridge National Laboratory, developing Web services software and a GPS vehicle tracking application using GPS-enabled wireless phone handsets and other GPS devices.

THOMAS LINDBLAD received his Ph.D. in physics at the University of Stockholm in 1972 and became associate professor two years later. He is currently a professor in the Department of Physics at the Royal Institute of Technology, Stockholm, and also serves part time as Director of Undergraduate Studies. His research currently concentrates on techniques in image and data analysis in high data rate systems.

JavaTech

*An Introduction to Scientific and Technical
Computing with Java*

**Clark S. Lindsey, Johnny S. Tolliver
and Thomas Lindblad**

CAMBRIDGE
UNIVERSITY PRESS

CAMBRIDGE UNIVERSITY PRESS
Cambridge, New York, Melbourne, Madrid, Cape Town, Singapore, São Paulo

Cambridge University Press
The Edinburgh Building, Cambridge CB2 2RU, UK

Published in the United States of America by Cambridge University Press, New York

www.cambridge.org
Information on this title: www.cambridge.org/9780521821131

First published 2005

Printed in the United Kingdom at the University Press, Cambridge

A catalog record for this publication is available from the British Library

Library of Congress Cataloging in Publication data

ISBN-13 978-0-521-82113-1 hardback
ISBN-10 0-521-82113-4 hardback

Contents

Preface

Java is a serious language suitable for demanding applications in science and engineering. Really, we promise! Java offers a lot more than just those little applets in your Web browser.

In *JavaTech* we focus on how Java can perform useful tasks in technical computing. These tasks might involve an animated simulation to demonstrate a scientific principle, a graphical user interface for an existing C or C++ computational engine, a distributed computing project, controlling and monitoring an experiment remotely via the Internet, or programming an embedded Java hardware processor in a device such as a remote sensor. While other Java books intended for the science and engineering audience concentrate primarily on numerical programming, we take a much broader approach and examine ways that Java can benefit programmers working on many different types of technical applications.

This project grew out of a course given by two of us (C.S.L. and Th.L.) at the Royal Institute of Technology in Stockholm, Sweden in which students of diverse backgrounds followed the class via the Internet. For this type of *distance learning* situation, we developed hypertext instructional material for delivery via the Web browser that allows for a high degree of self-study. This approach works especially well with Java since many of the demonstration programs run as applets within the browser.

This book provides a handy print companion to this hypertext course, which is available online at www.javatechbook.com. The book includes additional material that deals with distributed computing techniques based on work done by one of us (J.T.) at Oak Ridge National Laboratory in the USA. Throughout the book we refer to the hypertext materials as the *Web Course*.

Who should use this book

JavaTech targets primarily those who want to learn the Java programming language so as to apply it to practical applications in science and engineering. From the freshman science major to the experienced programmer in a technical field, we believe this book and the Web Course will be helpful.

For those unfamiliar with the language and with object-oriented programming, we begin with a compact introduction to Java. Since Java has grown into a very big field we only touch on the essential elements needed to begin doing useful programming. We include examples of how Java can apply to technical tasks such as histogramming of data and image analysis.

While familiarity with C and C++ will hasten a reader's understanding of Java programming, we do not assume the reader knows these languages.

After the Java introduction we discuss network programming, which we consider to be one of Java's strongest features. We focus particularly on how to build client/server systems for distributed computing applications. If you have a network application, such as the need to monitor remote devices or to give distant users access to a complex simulation running on a central server, the survey here should help you get started. Our aim is to show that you can create powerful network software with Java without needing first to become an authority on all the arcane intricacies of network systems. Java's networking tools and platform portability allow you to focus more on your application than on the underlying mechanisms.

The final part of the book looks at how Java can interact with the local platform, with code in other languages, and with embedded processors. For example, perhaps you have a legacy program in C that represents many years of development and tuning, but it lacks a graphical interface to make it interactive and flexible. We discuss the Java Native Interface (JNI) that allows you to connect your program to Java and to take advantage of the extensive graphical tools available in Java to build an interface around your computational engine in C (or in Fortran via intermediate C code as discussed in the Web Course). You can also add the networking capabilities of Java discussed above. For example, remote clients could connect with your legacy program that runs on a central server.

Hardware microprocessors designed especially to run Java are now widely available. Those who work on embedded processor applications will be interested in our survey of the field of Java processors. In a demonstration program, for example, we show how to connect via a serial port to a microcontroller that is programmed with Java and used to read a sensor.

We look at compact, low-cost platforms that contain Java processors, Ethernet connectors, analog-to-digital inputs, digital-to-analog outputs and other useful features. With such systems you can run servers that allow remote clients to monitor, control, and diagnose an instrument of some kind. This offers the opportunity to those who work with large complex installations, such as an elaborate scientific apparatus or a power plant, to access and control a system at a fine-grained level. We provide a demonstration of a server on such a Java processor platform in which the server responds to a Web browser with an HTML file containing a voltage reading.

Organization and topics

We attempted with this book and Web Course combination to create an innovative and highly flexible approach that allows readers with a diverse range of interests and backgrounds to find and use effectively the materials for their particular needs. The Web Course includes hypertext tutorial materials, many demonstration programs, and exercises. The book compliments the Web Course with more extensive discussions on a range of topics and with tables and diagrams for quick reference.

We follow an example-based teaching approach, using lots of applets and application programs to demonstrate the concepts and techniques described. In addition, we supply a large selection of *starter* programs that provide templates with which readers can quickly begin to develop their own programs.

The chapters in the book correspond directly to those in the Web Course. Note that while one of Java's strongest features is its extensive graphics capability, we do not discuss graphics programming in the first five chapters. Instead we focus on the components and structure of the language. We demonstrate techniques with stand-alone programs (referred to in Java as *applications*) that print to the console and applets that send output to the web browser's Java console window.

The book and Web Course are divided into three parts plus appendices.

Part I Introduction to Java

The 12 chapters in Part I provide an introduction to the Java language. These chapters focus on the Java language but also discuss various topics relevant to applying Java to technical areas. The Web Course expands the introductory material into three tracks:

The *Java Track* provides an introduction to Java programming. The reader can follow this track alone for a quick course in the basics of Java programming. *Supplements* provide additional information on both basic and advanced topics.

The *Tech Track* focuses on topics relevant to general math, science, and engineering applications of Java such as floating-point numbers, random number generators, and image processing.

The *Physics Track* provides an example of how to apply Java to a particular technical subject. The track corresponds to a short course for undergraduate students on the use of numerical computing, simulations, and data analysis in experimental physics.

Part II Java and the network

This part focuses on the application of Java to network programming and distributed computing. It begins with an introduction to TCP/IP programming and then looks at several topics including socket based client/server demonstration programs and distributed computing with RMI, CORBA, and other techniques. An introduction is given to Unified Modeling Language (UML), which leads to better object oriented code design and analysis. A brief overview of web services and XML is also provided.

Part III Out of the sandbox

This part deals with how Java programs can access information and resources on the underlying platforms on which the Java Virtual Machine (JVM) is installed and how the JVM can interact with its local environment. It also reviews implementations of Java in hardware rather than in a virtual machine. Topics include interfacing Java programs to C/C++ and Fortran codes with the Java Native Interface (JNI), communicating with devices via serial/parallel ports, and working with embedded Java processors.

Appendices

Appendices 1 and 2 provide tables of Java language elements and operators, respectively. Appendix 3 gives additional information about floating-point numbers in Java.

Topics not discussed

Java has grown into an enormous industry since it first appeared in the mid-1990s. No single book could possibly do justice to all of the Java classes, packages, tools, techniques, and applications of the language. In fact, there exist many books devoted to individual topics such as Java I/O, graphics, and multithreading. The Java industry expands further every day.

For this book we have chosen what we consider to be an important subset of Java topics relevant to technical applications. Some important topics not treated include:

- Java Enterprise techniques, such as database access and Java application servers
- Security topics such as the Java Cryptography Extension (JCE)
- Java 3D graphics

We do provide in the Web Course a large set of links to references and resources for these and other Java subjects. We also believe that this book provides the reader with a solid base of understanding on which to pursue further learning. All Java programmers must deal with the need to continually learn new classes and APIs (Application Program Interfaces). As we go to press, Sun is about to release Java 2 Standard Edition version 5.0, which contains significant additions to the language. We discuss the most important of these but some are beyond the scope of this book.

We emphasize the use of the web for access to language specifications, online tutorials, and other resources needed to tackle new Java techniques. We include references and web links in each chapter and in the Web Course. You can also find many online resources at `http://java.sun.com`, `java.net`, and `www.ibm.com/developerworks/java/`.

As mentioned in the introduction, we do not delve into numerical programming with Java. We only touch on this subject here while the Web Course *Tech* and

Physics tracks contain several introductory level sections. See the reference list at the end of Chapter 1 for a list of several books that deal extensively with numerical programming in Java.

How to use this book and Web Course

We designed the book and Web Course in a way that lets readers follow individualized paths through the materials. Part I, in particular, allows for a variety of different approaches. You could, for example, study only the Java sections of each chapter and get a fast introduction to the basics of Java programming. You could also study the sections with particular relevance to technical applications (the Web Course expands on these in its *Tech Track*) or, alternatively, you could skip these tech topics in a first pass and return to them later. Those already familiar with Java basics could focus just on the tech-related topics.

You can proceed through the book and Web Course at your own pace and experiment with the many applets and application demonstration programs. There is an emphasis on coding by the reader since ultimately you can only learn Java or any other language by writing lots of programs yourself.

Part II and Part III deal with specialized topics. If you are already familiar with the basics of Java programming, you could proceed directly to the chapter or sub-section of interest in those parts.

One of the most important features of Java is its extensive network programming capability. So we designed the course around the assumption that the reader has easy access to the Internet. Most of the Web Course pages include links to reference and resource materials, especially the tutorials and language specifications on the `http://java.sun.com` website. Rather than reinvent the wheel we try to incorporate resources such as the Sun tutorials in a way that takes best advantage of what is already available.

The Web Course hypertext materials and demonstration codes, along with updates and corrections to the book, are available at the website `www.JavaTechBook.com`. (A mirror site is available at `www.particle.kth.se/~lindsey/JavaCourse/Book/`.)

Note that if we included in the book the source codes for all the demonstration programs, it would be a very long book indeed. Since the source codes are easily available from the Web Course, we often print only "code snippets" rather than entire classes or programs.

Conventions

`Fixed width` style indicates:

- code samples such as: `for (i=0; i < 4; i++) j++;`
- Java class names, variable names, and other code-related terms
- console commands such as: `c:\> java HelloWorld`
- web addresses such as `http://java.sun.com`

In code listings, italicized *fix width* indicates that the text is not actually in the code but included to emphasize some aspect of the code or to summarize code that was skipped. We also put the class name in **bold** in the code listings. (Coding style conventions are discussed in Section 5.9.) When discussing a method in the text we may often ignore the argument list for the sake of brevity. So `aMethod (int x, float y, double z)` is abbreviated as `aMethod()`.

In the main text, new terms of particular importance are italicized. The book name and *Web Course* sections are also italicized.

In Chapter 22 on the Java Native Interface, we use the notation `Xxx` and `xxx` as placeholders to represent the many possible names that can replace the `Xxx` or `xxx`. For example, JNI has a `GetIntField()` method. It also has `GetFloatField()`, `GetDoubleField()`, etc. methods. We refer to these as a group with the `GetXxxField()` notation. Similarly, the `xxx` in `jxxxArray` can be replaced with `int`, `float`, `double`, etc. to produce `jintArray`, `jfloat-Array`, `jdoubleArray`, etc.

Java version

The code in *JavaTech* primarily follows that of Java version 1.4 released in 2002, but we discuss the significant enhancements available in the Java 5.0 release where relevant. (This release was under development for at least two years and became available in beta form near the end of the writing of this book.) Since many web browsers currently in use only run Java 1.1 applets and also since some small platforms (e.g. embedded processors) with limited resources only run Java 1.1, we also include in the *Web Course* some discussion of programming techniques for this version and provide sample codes.

The programs do not usually assume a particular platform and should run on MS Windows, Mac OS X, Linux, as well as Solaris and most Unix platforms.

Acknowledgements

We would like to thank our editors Simon Capelin and Vince Higgs for their help and patience. We thank Roger Sundman and Carl Wedlin of Imsys Technologies for their review of the Java hardware discussion and helpful suggestions. Thanks also to Michele Cianciulli for his comments on the manuscript.

We thank the many students who took our Web Course over the years and gave us a great amount of useful feedback. We especially want to thank one of our first students, Conny Carlberg, who encouraged us by quickly applying Java to his research. Several students at the Royal Institute of Technology (KTH) have used parts of the manuscript of this book, and many of them have come with interesting and useful comments, especially Bruno Janvier and Jaakko Pajunen. At a very early stage, when the Java course was introduced, and when Java was not generally too well known, we received encouraging support from many professors at KTH, the University of Stockholm, and the Manne Siegbahn Institute of Physics.

One of us (C. S. L.) would like to dedicate this book to his wife Kerima who provided great support and encouragement.

One of us (J. S. T.) would like to thank his wife Janey and children Kevin and Chelsea for their enduring patience with a too-often absent or preoccupied husband and father during many months on a project that grew to be longer and more difficult than anyone expected. Thank you.

One of us (Th. L.) makes a dedication to whoever said "do not write any more books, it is a much bigger undertaking than you recall from writing the previous one."

Part I
Introduction to Java

Chapter 1
Introduction

1.1 What is Java?

The term Java refers to more than just a computer language like C or Pascal. Java encompasses several distinct components:

- **A high-level language** – Java is an object-oriented language whose source code at a glance looks very similar to C and C++ but is unique in many ways.
- **Java bytecode** – A compiler transforms the Java language source code to files of binary instructions and data called bytecode that run in the Java Virtual Machine.
- **Java Virtual Machine (JVM)** – A JVM program takes bytecode as input and interprets the instructions just as if it were a physical processor executing machine code. (We discuss actual hardware implementations of the Java interpreter in Chapter 24.)

Sun Microsystems owns the Java trademark (see the next section on the history of Java) and provides a set of programming tools and class libraries in bundles called Java Software Development Kits (SDKs). The tools include `javac`, which compiles Java source code into bytecode, and `java`, the executable program that creates a JVM that executes the bytecode. Sun provides SDKs for Windows, Linux, and Solaris. Other vendors provide SDKs for their own platforms (IBM AIX and Apple Mac OS X, for example). Sun also provides a runtime bundle with just the JVM and a few tools for users who want to run Java programs on their machines but have no intention of creating Java programs. This runtime bundle is called the Java Runtime Environment (JRE).

In hope of making Java a widely used standard, Sun placed minimal restrictions on Java and gave substantial control of the development of the language over to a broadly based Java community organization (see Section 1.4 "Java: open or closed?"). So as long as other implementations obey the official Java specifications, any or all of the Java components can be replaced by non-Sun components. For example, just as compilers for different languages can create machine code for the same processor, there are programs for compiling source code written in other languages, such as Pascal and C, into Java bytecode. There are even Java bytecode assembler programs. Many JVMs have been written by independent sources.

Java might be said more accurately to refer to a set of programming and computing specifications. However, in this book and the Web Course, unless otherwise indicated, we follow the common use of the term Java to refer to the high-level language that follows the official specifications and the virtual machine platform on which the compiled language runs. The usage is normally clear from the context.

Finally, many people know Java only from the applets that run in their web browsers. Java programs, however, can also run as standalone programs just like any other language. Such standalone programs are referred to as "applications" to distinguish them from applets.

1.2   History of Java

During 1990, James Gosling, Bill Joy and others at Sun Microsystems began developing a language called Oak. They primarily intended it as a language for microprocessors embedded in consumer devices such as cable set-top boxes, VCRs, and handheld computers (now known as personal data assistants or PDAs).

To serve these goals, Oak needed the following features:

- platform independence, since it must run on devices from multiple manufacturers
- extreme reliability (can't expect consumers to reboot their VCRs!)
- compactness, since embedded processors typically have limited memory

They also wanted a next-generation language that built on the strengths and avoided the weaknesses of earlier languages. Such features would help the new language provide more rapid software development and faster debugging.

By 1993 the interactive TV and PDA markets had failed to take off, but internet and web activity began its upward zoom. So Sun shifted the target market to internet applications and changed the name of the project to Java.

The portability of Java made it ideal for the Web, and in 1995 Sun's *HotJava* browser appeared. Written in Java in only a few months, it illustrated the power of applets – programs that run within a browser – and the ability of Java to accelerate program development.

Riding atop the tidal wave of interest and publicity in the Internet, Java quickly gained widespread recognition (some would say *hype*), and expectations grew for it to become the dominant software for browsers and perhaps even for desktop programs. However, the early versions of Java did not possess the breadth and depth of capabilities needed for desktop applications. For example, the graphics in Java 1.0 appeared crude and clumsy compared with the graphics features available in software written in C and other languages and targeted at a single operating system.

Applets did in fact become popular and remain a common component of web page design. However, they do not dominate interactive or multimedia displays in the browser as expected. Many other "plug-in" programs also run within the browser environment.

Though Java's capabilities grew enormously with the release of several expanded versions (see Section 1.3), Java has not yet found wide success in desktop client applications. Instead Java gained widespread acceptance at the two opposite ends of the platform spectrum: large business systems and very small systems.

Java is used extensively in business to develop enterprise, or middleware, applications such as on-line stores, transactions processing, dynamic web page generation, and database interactions. Java has also returned to its Oak roots and become very common on small platforms such as smart cards, cell phones, and PDAs. For example, as of mid-2004 there are over 350 different Java-enabled mobile phone handsets available across the world, and over 600 million Java Cards have been distributed

1.3 Versions of Java

Since its introduction, Sun has released new versions of the Java core language with significant enhancements about every two years or so. Until recently, Sun denoted the versions with a 1.x number, where x reached up to 4. (Less drastic releases with bug fixes were indicated with a third number as in 1.4.2.) The next version, however, will be called Java 5.0. Furthermore, Sun has split its development kits into so-called editions, each aimed towards a platform with different capabilities. Here we try to clarify all of this.

1.3.1 Standard Edition

Below is a time line for the different versions of the *Standard Edition* (SE) of Java, which offers the core language libraries (called *packages* in Java) and is aimed at desktop platforms. We include a sampling of the new features that came with each release.

- **1995** – Version 1.0. The Java Development Kit (JDK) included:
  - 8 packages with 212 classes.
  - Netscape 2.0–4.0 included Java 1.0.
  - Microsoft and other companies licensed Java.
- **1997** – Version 1.1:
  - 23 packages, 504 classes.
  - Improvements included better event handling, inner classes, improved VM.

- Microsoft developed its own 1.1-compatible Java Virtual Machine for the Internet Explorer.
- Many browsers in use are still compatible only with 1.1.
- Swing packages with greatly improved graphics became available during this time but were not included with the core language.
- **1999** – Version 1.2. Sun began referring to the 1.2 and above versions as the Java 2 Platform. The Software Development Kit (SDK) included:
  - 59 packages, 1520 classes.
  - Java Foundation Classes (JFC), based on Swing, for improved graphics and user interfaces, now included with the core language.
  - Collections Framework API included support for various lists, sets, and hash maps.
- **2000** – Version 1.3:
  - 76 packages, 1842 classes.
  - Performance enhancements including the Hotspot virtual machine.
- **2002** – Version 1.4:
  - 135 packages, 2991 classes.
  - Improved IO, XML support, etc.
- **2004** – Version 5.0 (previously known as 1.5). This version was available only in beta release as this book went to press. See Section 1.9 for an overview of what is one of the most extensive updates of Java since version 1.0. It includes a number of tools and additions to the language to enhance program development, such as:
  - faster startup and smaller memory footprint
  - metadata
  - formatted output
  - generics
  - improved multithreading features
  - 165 packages, over 3000 classes

During the early years, versions for Windows platforms and Solaris (Sun's version of Unix) typically became available before Linux and the Apple Macintosh. Over the last few years, Sun has fully supported Linux and has released Linux, Solaris, and Windows versions of Java simultaneously. Apple Computer releases its own version for the Mac OS X operating system, typically a few months after the official Sun release of a new version. In addition, in the year 1999, Sun split off two separate editions of Java 2 software under the general categories of *Micro* and *Enterprise* editions, which we discuss next.

1.3.2   Other editions

Embedded processor systems, such as cell phones and PDAs, typically offer very limited resources as compared to desktop PCs. This means small amounts of RAM and very little disk space or non-volatile memory. It also usually means small, low-resolution displays, if any at all. So Sun offers slimmed-down versions of Java for such applications. Until recently this involved three separate bundles of Java 1.1-based packages, organized according to the size of the platform. *Java Card* is intended for extremely limited Java for systems with only 16 KB non-volatile memory and 512 bytes volatile. The *EmbeddedJava* and *PersonalJava* bundles are intended for systems with memory resources in the 512 KB and 2 MB ranges, respectively.

To provide a more unified structure to programming for small platforms, Sun has replaced EmbeddedJava and PersonalJava (but not JavaCard) with the Java 2 Micro Edition (J2ME). The developer chooses from different subsets of the packages to suit the capacity of a given system. (We briefly review J2ME in Chapter 24.)

At the other extreme are high-end platforms, often involving multiple processors, that carry out large-scale computing tasks such as online stores, interactions with massive databases, etc. With the Java 2 Platform came a separate set of libraries called the Java 2 *Enterprise Edition* (J2EE) with enhanced resources targeted at these types of applications. Built around the same core as Standard Edition packages, it provides an additional array of tools for building these so-called middleware products.

1.3.3   Naming conventions

We note that the naming and version numbering scheme in Java can be rather confusing. As we see from the time line above, the original Java release included the Java Development Kit (JDK) and was referred to as either Java 1.0 or JDK 1.0. Then came JDK 1.1 with a number of significant changes. The name Java 2 first appeared with what would have been JDK 1.2. At that time the JDK moniker was dropped in favor of SDK (for Software Development Kit). Thus the official name of the first Java 2 development kit was something like Java 2 Platform Standard Edition (J2SE) SDK Version 1.2. Versions 1.3 and 1.4 continued the same naming/numbering scheme.

Meanwhile many people continue to use the JDK terminology – thus JDK 1.4 when referring to J2SE SDK Version 1.4. Another common usage is the simpler but less specific Java Version 1.x, or even just Java 1.x to mean J2SE Version 1.x. Both of these usages are imprecise because there is also a Java 2 Enterprise Edition (J2EE) Version 1.4. To make it clear what you mean, you should probably either use J2SE or J2EE rather than just Java when mentioning a version number unless the meaning is clear from context. This book is not about J2EE, though we do touch on Java Servlet technology in Chapters 14 and 21 and on web services in general in Chapter 21. Since we never need to refer to the Enterprise Edition, we use the terms Java 1.x, SDK 1.x, and J2SE 1.x interchangeably.

By the time this book is in your hands, Sun will have released the Java 2 Standard Edition Version 5.0. The version number 5.0 replaces what would have been version number 1.5. Undoubtedly many people will continue to use the 1.5 terminology. In fact, the Beta 2 release of J2SE 5.0 (the latest available at the time of this writing) continues to use the value 1.5 in some places. You may also come across the code name Tiger for the 5.0 release; however, we expect that usage to fade away just like previous code names Kestrel and Merlin have all but disappeared from the scene. This book uses the notation J2SE 5.0 or Release 5.0 or Version 5.0 or Java 5.0 or sometimes just 5.0 when referring to this very significant new release of Java.

We provide a brief overview of Java 5.0 in Section 1.9 and examine a number of 5.0 topics in some detail throughout the book.

1.4   Java – open or closed?

Java is not a true open language but not quite a proprietary one either. All the core language components – compiler, virtual machines, core language class packages, and many other tools – are free from Sun. Furthermore, Sun makes detailed specifications and source code openly available for the core language. Another company can legally create a so-called *clean room* compiler and/or a Java Virtual Machine as long as it follows the detailed publicly available specifications and agrees to the trademark and licensing terms. Microsoft, for example, created its own version 1.1 JVM for the Internet Explorer browser. See the Web Course for a listing of other independent Java compilers and virtual machines.

Sun, other companies, and independent programmers participate in the Java Community Process (JCP) organization whose charter is "to develop and revise Java technology specifications, reference implementations, and technology compatibility kits." Proposals for new APIs, classes, and other changes to the language now follow a formal process in the JCP to achieve acceptance.

Sun, however, does assert final say on the specifications and maintains various restrictions and trademarks (such as the *Java* name). For example, Microsoft's JVM differed in some significant details from the specifications and Sun filed a lawsuit (later settled out of court) that claimed Microsoft attempted to weaken Java's "Write Once, Run Anywhere" capabilities.

1.5   Java features and benefits

Before we examine how Java can benefit technical applications, we look at the features that make Java a powerful and popular general programming language. These features include:

- **Compiler/interpreter combination**
  - Code is compiled to bytecode, which is interpreted by a Java Virtual Machine (JVM).

- This provides portability to any base operating system platform for which a virtual machine has been written.
- The two-step procedure of compilation and interpretation allows for extensive code checking and improved security.
- **Object-oriented**
  - Object-oriented programming (OOP) throughout – no coding outside of class definitions.
  - The bytecode retains an object structure.
  - An extensive class library available in the core language packages.
- **Automatic memory management**
  - A *garbage collector* in the JVM takes care of allocating and reclaiming memory.
- **Several drawbacks of C and C++ eliminated**
  - No accessible memory pointers.
  - No preprocessor.
  - Array limits automatically checked.
- **Robust**
  - Exception handling built-in, strong type checking (that is, all variables must be assigned an explicit data type), local variables must be initialized.
- **Platform independence**
  - The bytecode runs on any platform with a compatible JVM.
  - The "Write Once Run Anywhere" ideal has not been achieved (tuning for different platforms usually required), but is closer than with other languages.
- **Security**
  - The elimination of direct memory pointers and automatic array limit checking prevents rogue programs from reaching into sections of memory where they shouldn't.
  - Untrusted programs are restricted to run inside the virtual machine sandbox. Access to the platform can be strictly controlled by a Security Manager.
  - Code is checked for pathologies by a class loader and a bytecode verifier.
  - Core language includes many security related tools, classes, etc.
- **Dynamic binding**
  - Classes, methods, and variables are linked at runtime.
- **Good performance**
  - Interpretation of bytecodes slowed performance in early versions, but advanced virtual machines with adaptive and just-in-time compilation and other techniques now typically provide performance up to 50% to 100% the speed of C++ programs.
- **Threading**
  - Lightweight processes, called threads, can easily be spun off to perform multiprocessing.
- **Built-in networking**
  - Java was designed with networking in mind. The core language packages come with many classes to program Internet communications.
  - The Enterprise Edition provides an extensive set of tools for building middleware systems for advanced network applications.

These features provide a number of benefits compared to program development in other languages. For example, C/C++ programs are beset by bugs resulting from direct memory pointers, which are eliminated in Java. Similarly, the array limit checking prevents another common source of bugs. The garbage collector relieves the programmer of the big job of memory management. It's often said that these features lead to a significant speedup in program development and debugging compared to working with C/C++.

1.5.1   Java features and benefits for technical programming

The above features benefit all types of programming. For science and engineering specifically, Java provides a number of advantages:

- **Platform independence** – Engineers and scientists, particularly experimentalists, probably use more types of computers and operating systems that any other group. Code that can run on different machines without rewrites and recompilation saves time and effort.
- **Object-oriented** – Besides the usual benefits from OOP, scientific programming can often benefit from thinking in terms of objects. For example, atomic particles in a scattering simulation are naturally self-contained objects.
- **Threading** – Multiprocessing is very useful for many scientific tasks such as simulations of phenomena where many processes occur simultaneously. This can be quite useful in the conceptual design of a program even when it will run on a single-processor machine. However, Java Virtual Machines on multiprocessor platforms also can distribute threads to the different processors to obtain true parallel performance.
- **Simulation tools** – The extensive graphics resources and multithreading in the core Java language provide for depicting and animating engineering and scientific devices and phenomena.
- **Networking** – Java comes with many networking capabilities that allow one to build distributed systems. Such capabilities can be applied, for example, to data collection from remote sensors.
- **Interfacing and enhancing legacy code** – Java's strong graphics and networking capabilities can be applied to existing C and Fortran programs. A Java graphical user interface (GUI) can bring enhanced ease of use to a Fortran or C program, which then acts as a computational engine behind the GUI.

1.5.2   Java shortcomings for technical programming

Several features of Java that make it a powerful and highly secure language, such as array limit checking and the absence of direct memory pointers, can also slow it down, especially for large-scale intensive mathematical calculations. Furthermore, the interpretation of bytecode that makes Java programs so easily portable can cause a big reduction in performance as compared to running a program in local machine code. This was a particular problem in the early days of Java, and its slow performance led many who experimented with Java to drop it.

Fortunately, more sophisticated JVMs have greatly improved Java performance. So called *Just-in-Time* compilers, for example, convert bytecode to local machine code on the fly. This is especially effective for repetitive sections of code. During the first pass through a loop the code is interpreted and converted to native code so that in subsequent passes the code will run at full native speeds.

Another approach involves adaptive interpretation, such as in Sun's Hotspot, in which the JVM dynamically determines where performance in a program is slow and then optimizes that section in local machine code. Such an approach can actually lead to faster performance than C/C++ programs in some cases.

Here are some other problems in applying Java to technical programming:

- **No rectangular arrays** – Java 2D arrays are actually 1D arrays of references to other 1D arrays. For example, a 4×4 sized 2D array in Java behaves internally like a single 1D array whose elements point to four other 1D arrays:

```
A[0] ==> B[0] B[1] B[2] B[3] B[4]
A[1] ==> C[0] C[1] C[2] C[3] C[4]
A[2] ==> D[0] D[1] D[2] D[3] D[4]
A[3] ==> E[0] E[1] E[2] E[3] E[4]
```

The B, C, D, and E arrays could be in widely separated locations in memory. This differs from Fortran or C in which 2D array elements are contiguous in memory as in this 4×4 array:

```
A(0,0) A(0,1) A(0,2) A(0,3) A(0,4)
A(1,0) A(1,1) A(1,2) A(1,3) A(1,4)
A(2,0) A(2,1) A(2,2) A(2,3) A(2,4)
A(3,0) A(3,1) A(3,2) A(3,3) A(3,4)
```

Therefore, with the Java arrays, moving from one element to the next requires extra memory operations as compared to simply incrementing a pointer as in C/C++. This slows the processing when the calculations require multiple operations on large arrays.

- **No complex primitive type** – Numerical and scientific calculations often require imaginary numbers but Java does not include a complex primitive data type (we discuss the primitive data types in Chapter 2). You can easily create a complex number class, but the processing is slower than if a primitive type had been available.

- **No operator overloading** – For coding of mathematical equations and algorithms it would be very convenient to have the option of redefining (or "overloading") the definition of operators such as "+" and "–". In C++, for example, the addition of two complex number objects could use `c1 + c2`, where you define the + operator to properly add such objects. In Java, you must use method invocations instead. This lengthens the code significantly for nontrivial equations and also makes it more susceptible to bugs. (The `String` class in Java does provide a "+" operator for appending two strings.)

- **Floating-point limitations** – The early versions of Java provided only 32- and 64-bit floating-point regardless of whether the host machine provided for greater precision. This insured that for the same program the JVMs produced exactly the same result on all machines. However, in Java 1.2 it became possible to use a wider exponent (but not significand) representation if available on the platform. The keyword modifier `strictfp` forces the floating-point calculations to follow the previous lower precision mode to ensure exactly the same results on all machines. We discuss floating-point representations and operations in Chapter 2.

So for some types of intensive mathematical processing, such as those using very large arrays, it may be difficult to achieve the performance levels of C or Fortran, especially when that code has been optimized over many years of use. In such cases, it may be advantageous to let Java provide a graphical user interface and networking tools, but keep the C/Fortran program as a calculation engine inside the Java body. Chapter 22 discusses how to link Java with native code.

1.6   Real-world Java applications in science and engineering

Java has been used extensively for several years now to solve real-world programming challenges in numerous areas of endeavor. While applications in science and engineering may be less well known, there certainly are many and we present a few examples here.

A Java-based system called Maestro provided for data visualization, collaboration, and command and control services for the NASA JPL team in charge of the Mars rovers that landed on the Red Planet in January 2004. James Gosling called it the "the coolest Java app ever" [1]. The system provides an elaborate set of tools for analyzing images, 3D modeling of the terrain around a rover, and collaborative planning for rover maneuvers and experimental operations. Figure 1.1 shows a display of a special version of Maestro made available to the public for personal use. You can use it "to create your own driving and science activities, using all of the rover's instruments to enact your own day of mission operations." Large sets of actual data and imagery can be downloaded for different periods of the mission [2].

One of us is part of team working on a project known as SensorNet that is using Java-based web services to enable the collection and archiving of data from sensors that are distributed nationwide [3]. Web services involve the exchange of data in XML (Extensible Markup Language) format via web client/server systems. (We give an introduction to Web services in Chapter 21.) SensorNet uses web services and open standards so that sensor information is available to a wide variety of users in a standard format. Java was chosen as the implementation

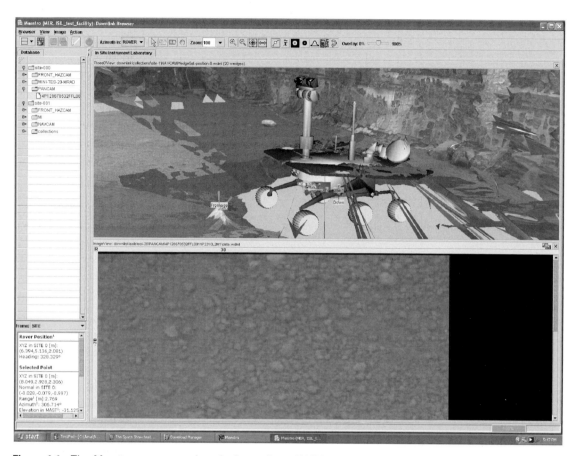

Figure 1.1 The *Maestro* program, written in Java, allows NASA operators of the Mars rovers to analyze images, create stereo views and 3D terrain models, and plan maneuvers and experiments. The figure shows a screen capture of the public version of the program.

language because of the portability and other features mentioned above that make it an excellent software platform in general and for network programming in particular. Furthermore, Sun provides a free Java Web Services Developer Pack (JWSDP) that includes extensive tools and documentation for creating such services.

The Aviation Digital Data Service (ADDS), created by a consortium that includes the National Oceanic and Atmospheric Administration (NOAA), offers several Java tools to provide graphical displays of an assortment of meteorological data of importance to the aviation community [4]. Another aviation-related Java program is AirportMonitor™. This commercial product is used by a number of airports to provide near-live displays of air traffic arriving and departing from their facilities. The data also includes flight ID, aircraft type, altitude, origin, and destination [5].

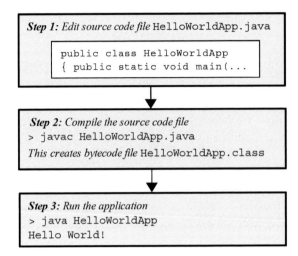

Figure 1.2   The steps to create and run a Java standalone application program.

The Swedish Institute of Space Physics uses a Java program to collect and view data from infrasound (acoustic waves in the 0.1–25 Hz range) detectors distributed in the north of the country. The infrasound system can detect distant events such as Shuttle launches and meteors [6].

The open source program BioJava, developed at the Sanger Institute in Great Britain, provides researchers in the field of bioinformatics with an extensive toolkit for manipulating genomic sequences [7]. The large program (over 1200 classes and interfaces) is now used at major laboratories around the world [8].

The Web Course provides a resource section with links to sites that describe many other applications of Java in science and engineering. There are also links to Java programming tools in numerical computing, data analysis, image processing, and other areas. In addition, we link to a small sample of the thousands of applets available on the Web that use Java graphics and animation techniques to demonstrate scientific principles and complex systems. Such simulations provide powerful tools for teaching technical subjects.

1.7   The Java programming procedure

Figure 1.2 illustrates the basic steps to create and run a Java *application*, as standalone Java programs are called. You first create the source code with a text editor and save it to a file with a " .java" file extension appended to the name. Here the file HelloWorldApp.java is created. The file name must exactly match the class name in both spelling and case. (We define *class* in Chapter 3.)

With the `javac` program, you compile this file to create a bytecode file (or files if the source code file included more than one class). The bytecode file ends with the "`.class`" extension appended. Here the output is `HelloWorldApp.class`. The bytecode consists of the instructions for the Java Virtual Machine (JVM).

With the `java` program you run your class file. The JVM interprets the byte-code instructions just as if it were a processor executing native machine code instructions. (In Chapter 24 we discuss hardware processors that directly execute Java bytecode.)

The platform independence of Java thus depends on the prior creation of a JVM for different platforms. The Java bytecode can then run on any platform on which a JVM is available and should perform the same regardless of platform differences. This "Write Once, Run Anywhere" ideal is a key goal of the Java language. For applets the browser runs a JVM in the same manner that it runs other plug-in programs.

1.7.1   Java tools

Programming in Java typically involves two alternative approaches that we will call *manual* and *graphical*:

- **Manual** – Use your favorite text editor to create the java source code files (`*.java`) and use the command line tools in the Software Development Kit (SDK) to compile and run the programs. The SDK is provided by Sun and includes several tools, the most important of which are:
  - `javac` – compiles Java language source files to bytecode files
  - `java` – the JVM that executes java applications
  - `appletviewer` – tests applets independently of a web browser.
- **Graphical** – Graphical user interface programming environments (or *GUI builders*) are elaborate programs that allow you to edit Java programs and interactively build graphical interfaces in a WYSIWYG (*What You See Is What You Get*) manner. They also include a compiler and a JVM so that you can both create and run Java applets and applications all within one environment. These are also known as Integrated Development Environments, or IDEs.

During this course we recommend the manual approach so that you will learn first hand the essential details of Java graphics programming. While it is fine to use the editor in the GUI builder or IDE, you should write all of the code yourself rather than use the interactive user interface building tools. Otherwise, the GUI builder does most of the work and you learn very little.

When you later begin developing programs for your own tasks that require graphical interfaces with many components, you may want to use a GUI builder. However, even then you will occasionally want to modify the graphics code by hand to debug a problem or tweak the appearance of a layout. You then obviously need to understand the details of graphics programming.

For editing the source code, we recommend a language sensitive editor that color codes the Java text so that the language keywords and symbols are high-lighted. This helps to avoid simple spelling mistakes and assists with debugging. Chapter 1 in the Web Course provides links to several freeware and commercial editors.

The `appletviewer` program that comes with the SDK provides an ideal tool for applet debugging. As we discuss later, browsers do not provide a good environment in which to debug applet code except near the end of the development process.

1.7.2   Documentation

The Java language elements available from Sun currently fall into these two broad categories:

- **Core language** – This refers to the set of packages and classes that must be available in the JVM for a particular edition (see Section 1.3) regardless of the platform.
- **Optional APIs** – A number of useful *Application Programming Interfaces* (APIs) are available from Sun for audio and video applications, 3D graphics, and several other areas. However, they are not part of the core language so may not be available for all platforms.

These elements now involve an enormous number of packages, classes and meth-ods. So access to documentation is essential. Sun provides a large set of docu-mentation freely available on its site at `http://java.sun.com`.

If you do not have continual online web access, we recommend that you download from the `http://java.sun.com` site a copy of the Java 2 Platform, Standard Edition API specifications for the latest version of Java. This set of web pages provides detailed descriptions of all the packages, classes, and methods in the core language. You will frequently refer to this documentation to find out what exactly a given method does, what methods are available for a given class, and to obtain other useful information.

Note that the online documentation indicates what version of Java a class or method appeared in the core language (where this is not explicit, assume it came with Version 1.0). This is important if you want to write a program that is consistent with, for example, version 1.1.

1.7.3   Code compatibility

In this course we primarily use code consistent with version 1.4 or higher. We note those places where we do not. Currently (circa 2004) many browsers in use still only run applets with Java 1.1 compatibility. For such browsers, Sun provides a Java plug-in that implements Java 2 features. Applet tags in the web page can initiate the downloading of the plug-in, but the plug-in file is large and users on slow dial-up lines may refuse the download.

If your goal is to write applets for the broadest audience, then you need to write code limited to version 1.1 classes and methods. The *Supplements* sections in the Web Course provide alternative instruction in the 1.1 graphics techniques.

So far, Java maintains strict *backwards compatibility*. The bytecode from a Java 1.0 compiler still runs in a Java 1.4 virtual machine. A program written according to the Java 1.0 specification still compiles in a Java 1.4 compiler, though the compiler will flag "deprecated" elements that are no longer considered recommended practice.

Note, however, that you can in some cases run into problems if you mix code from different versions within the same program. For example, the handling of *events*, such as mouse clicks, changed significantly from version 1.0 to 1.1. A program can use either event-handling approach but it cannot contain both.

1.8   Getting started

You can quickly begin creating simple Java applets for the browser and applications for the console by following the code in the examples given here. Initially you do not need to understand all the elements of the language. We discuss the meaning of terms such as `class` and `extends` in the following chapters. For now it is important just to get the basic programming tools installed and learn how to run them. The details will come later.

1.8.1   Setup for Java programming

To begin developing Java programs, follow these steps:

1. Obtain the Java 2 Software Development Kit (SDK) for your platform:
   - The SDK is available free from Sun Microsystems for Solaris, Linux, and Windows platforms. See the Resources page in the Web Course for links to sites that provide kits for alternate platforms as well as kits from non-Sun sources.
   - The SDK contains the essentials needed to create Java programs.
   - The SDK includes:
     - Java executables such as `javac` (compiler) and `java` (the JVM)
     - Class packages, which correspond to code libraries.
2. Install the SDK:
   - The Sun SDKs now come with an installer application that does most of the work for you. Just run this program and follow its instructions.
3. Install the documentation:
   - It is recommended that you have ready access to the Java 2 API Specification since it will be very useful during program development. If you use the Sun SDK, you should

download the current documentation set and install it following the directions available at http://java.sun.com. If you use a third-party SDK, the documentation should be available from the same source as the SDK itself.

The following two sections give an example of a simple applet and a simple application program. These illustrate the basics for creating such programs and provide some initial experience with the SDK programming tools. The general workings of the code should be readily apparent so we wait until later chapters to discuss the exact meaning and function of the various language keywords and structures.

1.8.2   First application

Standalone programs in Java are referred to as applications. As mentioned previously, Figure 1.2 shows the steps needed to create and run an application. The code for the standard "Hello World" application that prints a string to the console is shown here:

```java
public class HelloWorldApp
{
   public static void main (String[] args)
   {
      System.out.println ("Hello World!");
   }
}
```

Put this code into a file called HelloWorldApp.java. (A Java file name must match exactly the class name, including the capitalization; this is true for all platforms.) Compile the application as follows:

```
> javac HelloWorldApp.java
```

This creates the class file HelloWorldApp.class. Use the java command to run the program:

```
> java HelloWorldApp
```

This produces the output

```
Hello World!
```

in the console window.

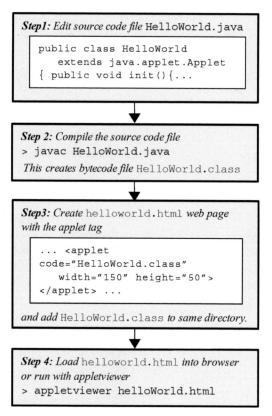

Step1: *Edit source code file* `HelloWorld.java`

```
public class HelloWorld
    extends java.applet.Applet
{ public void init(){...
```

Step 2: *Compile the source code file*
> `javac HelloWorld.java`
This creates bytecode file `HelloWorld.class`

Step3: *Create* `helloworld.html` *web page*
with the applet tag

```
... <applet
code="HelloWorld.class"
    width="150" height="50">
</applet> ...
```

and add `HelloWorld.class` *to same directory.*

Step 4: *Load* `helloworld.html` *into browser*
or run with appletviewer
> `appletviewer helloworld.html`

Figure 1.3 The steps to create a Java applet and a web page to hold it. Loading the web page in a browser or running it with the `appletviewer` program will display the applet as shown in Figure 1.4.

1.8.3 First applet

Figure 1.3 shows the steps to create an applet. The following applet code displays the string "Hello World!" in a window on a browser page:

```
public class HelloWorld extends java.applet.Applet
{
   public void paint (java.awt.Graphics g)
   {
     g.drawString ("Hello World!", 45, 30);
   }
}
```

Put this code into a file called `HelloWorld.java`. (Again, the name of the file, including the cases of the letters, must match exactly with the class name

Hello World!

Figure 1.4 Display of a very basic applet.

HelloWorld.) Then compile the code with

```
> javac HelloWorld.java
```

This creates the class file HelloWorld.class.

Applets are intended for display in browsers so you need to create a Web page file with the proper *tags* that indicate where the browser should find the class code and how big a window it should provide to display the applet's graphical output. (A brief tutorial on HTML coding for Web pages is available in the Web Course Chapter 1: *Supplements* section.)

Put the following code into a file called HelloWorld.html:

```
<html>
  <head>
    <title> A Simple Applet</TITLE>
  </head>
  <body>
    <applet code = "HelloWorld.class" width = ";150" height = "50">
    </applet>
  </body>
</html>
```

With this <applet> tag the file must reside in the same directory as the HelloWorld.class file. Open the file HelloWorld.html in your browser and you should see something like that in Figure 1.4.

1.8.4 Starter programs

We provide with the Web Course the codes for a number of programs that act as templates into which you insert code to make them do something useful. Since the term "template" implies something else in object-oriented programming, we call these *starter* programs. You can use them to begin a program without starting from scratch every time.

A very simple application program, StartApp1.java, is shown here. You can insert code snippets, such as those to be discussed in Chapter 2, into the area indicated.

```
public class StartApp1
{
  public static void main (String[] args)
  {
    // Put code between this line

    Insert the example codes or your exercise code
    between these lines.

    // and this line.
  }
}
```

Follow the instructions above for compiling and running an application. Rename the class for each new program. For example, in the above starter replace StartApp1 with MyClassApp1 or just MyClass. (We indicate at the top of each demonstration program file on the Web Course the starter program that we began with for that demo.) Remember that in the Java scheme the file name and the class name **must match exactly** in both characters and case. Also, the name cannot begin with a number or punctuation character ("." or "?", etc.).

We also provide a starter for applets as shown below with StartApplet1. java. Applets are intended to present a graphics display of some sort in a browser window. However, at this early stage we do not discuss Java graphics so that we can focus initially on the basics of the language. (We begin discussing graphics in Chapter 6.) Other than "painting" a short message to the applet display window, we send output via print statements to the browser's Java console rather than paint text to the browser window. (A browser's Java console can be activated via the browser's preferences or options settings.) This applet includes an init() method, which is called by the browser when the applet first begins. The init() is a convenient place to put simple example code with print output.

```
public class StartApplet1 extends          We learn later how applets
                  java.applet.Applet        inherit the Java applet class.
{

  public void init ()                       The browser always invokes
  {                                         this initialization method
                                            when the applet begins running

    // Put code between this line

                                            Insert example codes or your
                                            exercise code between these
                                            lines.

    // and this line.
  }
```

```
// Paint a message in the applet window.    This line is a comment
public void paint (java.awt.Graphics g)
{                                           This paints the string "Test"
   g.drawString ("Test", 20, 20);           in the applet window.
}
}
```

After inserting code into your applet, follow the same procedure as we discussed above for the `HelloWorld` applet: compile the program, create a web page with the appropriate applet tags, and place the web page and class file into the same directory.

You can test the applet by loading the web page into a browser or you can use the `appletviewer` program that comes with the SDK tools. If, for example, you put your applet into a web page named `MyFirstApplet.html`, you run `appletviewer` on it from the command line as follows:

```
> appletviewer MyFirstApplet.html
```

It is usually more convenient to use the `appletviewer` during debugging and tuning of your applets. Then once they work satisfactorily with `appletviewer`, you can examine how they appear in the browser. With the older JVMs that came installed in the browsers, it could be difficult to force the browser to load a new class file rather than use the old one in the cache. With Sun's Java Plug-in, which should have been installed into your browsers when you installed the SDK and is also available for downloading from `www.java.com`, the console window allows for various commands to the JVM including the "x" command that clears the cache.

We note here that we focus on applications in much of the course and even show later how you can add a `main()` method to an applet so that it can also run as an application. An applet has the inconvenient requirement that you need an HTML file containing the applet tags if you want to test it in a browser or `appletviewer`. We like applets, however, since they are pedagogically very useful for the Web Course. You can immediately see the applet running in a web page and can compare its output to the source code, which is displayed for every applet. You can download the code, make modifications, and test it. To avoid the need to make a web page to test the applet, we use the trick of putting the tags within a comment block in all of our applet source code files as shown here for the `HelloWorld` example:

```
/*<applet code = "HelloWorld.class" width = "150"
   height = "50">
   </applet> */
public class HelloWorld {
   . . .
```

The `appletviewer` program doesn't require that a file be a legitimate HTML file, only that it contains the applet tags. You can then run the applet with

```
> appletviewer HelloWorld.java
```

To reduce space, we don't show the applet tags section in the applet code listings in the book but it appears in all the downloadable code at the Web Course.

1.9   Changes in Java 2 Standard Edition 5.0

We mentioned in Section 1.3 that Sun is introducing a new version of the language called Java 5.0. Most of the changes fall into the *ease of development* category. With a few important exceptions, the changes do not add new functionality but rather provide an easier way of doing the same things you could do before but with less code and better compiler-time error detection. We provide here a brief rundown of the most important changes to the platform.

1.9.1   Major themes in 5.0 Development

Release 5.0 revolves around five major themes:

- **Quality, stability, and compatibility** – The designers of J2SE 5.0 considered quality, stability, and compatibility to be the most important aspect of the new release. Release 5.0 is the most tested release ever. Great efforts were made to ensure compatibility with previous versions of Java. The Sun engineers have made a public plea for users world-wide to test their code with the 5.0 Beta releases and report any problems that appear, especially any code that works with earlier versions of Java but fails under 5.0.
- **Performance and scalability** – Faster JVM startup time and smaller memory footprint were important goals. These have been achieved through careful tuning of the software and use of class data sharing. (Refer to `http://java.sun.com/j2se/1.5.0/docs/guide/vm/class-data-sharing.html` for more information about class data sharing and why it helps.)
- **Ease of Development** – Most of us take the first two themes for granted. We *expect* quality, stability, and compatibility, and the performance and scalability enhancements are very nice to have. We applaud the extensive efforts Sun has made in those areas. However, the Ease of Development (EoD) theme and the two that follow are likely to be the most important changes noticed by developers.

  It is in the EoD area that the most significant changes appear. In most cases, no new functionality was added in the sense that almost anything you can do with 5.0 you could do with 1.4, it just sometimes took a lot more boilerplate code (i.e. code that is repeated frequently) to do it. The exception to this general statement has to do with the new multithreading and concurrency features that provide capabilities previously unavailable.

In many cases, the new EoD features are all about syntax shortcuts that greatly reduce the amount of code that must be entered, making coding faster and more error-free. Some features enable improved compile-time type checking, thus producing fewer runtime errors.

- **Monitoring and manageability** – The 5.0 release includes the ability to remotely monitor and even manage a running Java application. For example, it is now much easier to watch memory usage and detect and respond to a low-memory condition. Many of these features are built right in to the system, and you can add additional monitoring and managing features to your own code.

- **Improved desktop client** – The last great theme of the 5.0 release was an improved experience on the desktop client. In addition to better performance because of a faster startup time and smaller memory footprint, there is a new, improved Swing (see Chapter 6) look and feel called Ocean, and a new easy-to-customize skinnable look and feel called Synth in which you can use XML configuration files to specify the appearance of every visual component in the system. In addition, the GTK and XP look and feels introduced in J2SE 1.4.2 have received further improvements. There is support for OpenGL and better performance on Unix X11 platforms. The Java Web Start and Java Plug-In technologies (both used to run Java applications downloaded over the Web) have been improved.

Other new features in J2SE 5.0 include core XML support, improvements to Unicode, improvements to Java's database connectivity package known as JDBC, and an improved, high-compression format for JAR files that can greatly reduce download times for applets and other networked applications.

1.9.2   Major Ease of Development changes in J2SE 5.0

We list below the most important of the many EoD improvements in Java 5.0, roughly in the order in which they are encountered in the rest of this book, not in the order of importance. Most of these enhancements to the language can only be appreciated after having had experience with programming in Java. If you are completely new to Java, you might want to skip this section and come back to it after you complete Part I.

- **Autoboxing and unboxing** – Chapter 2 explains that Java has primitive types like `int` for integers, and Chapter 3 explains "object" types like `Integer`. The difference between the two types is very important as we will see. Previous versions of Java made it necessary to explicitly convert between the primitive types and the object types. In Chapter 3 we examine the so-called autoboxing and unboxing feature added with J2SE 5.0 that removes the need for explicit conversions in most cases, and thus improves code readability and removes boilerplate code and sources of errors.

- **Enhanced `for` loop** – Chapter 2 looks at the several types of looping structures in the Java language, one of which is the `for` loop (quite similar to the C/C++ `for` loop). Version 5.0 includes an enhanced `for` loop syntax that reduces code complexity and improves readability. We introduce the enhanced `for` loop in Chapter 2 and explain it in more detail in Chapter 10 after we have described the object types with which the enhanced `for` loop works.

- **Metadata** – In Chapter 4 we encounter the `@Override` annotation. It falls under the category of *metadata* or "data about data." In this case, it is better thought of as "data about code". The new metadata facility, also called the annotation facility, is designed to use an "annotation" in your code that greatly reduces much of the boilerplate code that would be required in previous versions of Java.

  An annotation is a new 5.0 language element that begins with "`@`". Some annotations are processed by the `javac` compiler and some require the new annotation processing tool `apt`. There are currently only three annotations in the beta release of version 5.0. However, now that the metadata framework is available, we anticipate the appearance of many useful annotations and annotation processors in the future.

- **Formatted input and output and varargs** – In Chapter 5 we discuss how to format numerical output with Java. Version 5.0 finally adds the oft-requested ability to produce formatted output easily in the form of a `printf()` method that behaves very similarly to the `printf()` function in the C/C++ `stdio` library. There is also a new formatted input feature that is described in Chapter 9.

  Both these features rely on another new feature known as "varargs", which stands for "variable argument list'" in which the number of parameters passed to a Java method is not known when the source is constructed. Varargs is a useful new feature that can be of value in your own code, not just in the new `printf()` feature. Chapter 10 presents another EoD enhancement that provides for automatic format and output of the elements of an array (see the `java.util.Arrays` class). This feature really has nothing to do with `printf` or `varargs`, but we mention it here because it eases the amount of work that was necessary in pre-5.0 releases to output all the elements in an array in a nicely formatted style.

- **Static import** – Release 5.0 includes a new technique for accessing Java static methods and constants in another class without the need to include the full package and class name every time they are used. (We explain what the terms *class*, *package*, *static*, *import*, etc., mean in Chapters 3–5.) This new "static import" facility makes your code easier to write and, since there's less of it, less error-prone. We discuss static import in more detail in Chapter 5 after discussing `import` in general.

- **New `pack200` hyper-compression JAR format** – Chapter 5 discusses JAR (Java Archive) files used to combine and compress Java class files. We also look at the new `pack200` format that compresses JAR files very tightly, reducing bandwidth and saving download time. (This is not really an EoD change, but more of an "ease of deployment" change.)

- **Graphics system improvements** – Release 5.0 includes numerous bug fixes and minor tweaks to Java's graphics subsystems known as AWT and Swing (see Chapters 6 and 7), including reduced memory usage. In the EoD area, perhaps the biggest improvement is that it is no longer necessary to call `getContentPane()` when using Swing components (see Chapter 6 for details). Other enhancements include improved popup menu support, improved printing support for some graphics components, and the ability to query for the mouse location on the desktop.

- **New concurrency features** – Chapter 8 discusses Java's multithreading support, which has been present since version 1.0. Release 5.0 adds new capabilities that greatly enhance the multithreading features of Java. Some of these additions depend upon the generics concept (see next item), so we wait until Chapter 10 to introduce these important new capabilities.
- **Generics** – In Chapter 10 we introduce the new *generics* feature, a large and important subject that we do not have space to cover in detail in this book. Java is already a very type-safe language, which simply means that every variable has a specific type and that only compatible types can be assigned to each other. However, the use of generics brings an even greater amount of type safety to the Java language. Java includes a number of "object containers," such as the `ArrayList` class, that can contain "objects" of many different types. When retrieving an object from one of these containers, it must be retrieved as a very basic type and then converted back to the original type. If, however, an incorrect type is added to the container, then an error occurs at runtime during the conversion attempt. The use of generics makes it possible for the object containers to require that only certain types can be placed into them, else a compile time error occurs. Since mistakes are found at compile time, runtime safety and correctness is improved. In addition, since the specialized containers only contain items of the desired type, retrieval of items from the containers is easier since no explicit conversion to the desired type is necessary.

    You are not required to use the generics approach but see the note at the end of Section 10.7 about using the 5.0 compiler with code that uses the older containers.
- **Enumerated types** – Chapter 10 presents a feature of C/C++ that many programmers have missed in Java. Version 5.0 adds an *enumerated* type using the `enum` keyword. The new Java enumerated type includes all the features of C/C++ `enum` and more, including type safety.
- **New `StringBuilder` class** – We discuss this new class in Chapter 10, along with the older `StringBuffer` class. Both are used in the building, concatenating, and appending of string types, but the new class has improved performance.
- **Changes to ease RMI development** – Chapter 18 explains the Remote Method Invocation (RMI) techniques, including a simple but important change in J2SE 5.0 that makes RMI development simpler.

We do not go into great depth in this book on these changes. To do so would require an entire book just for those changes. In fact, at least one book devoted entirely to Release 5.0 is available at the time of this writing [9]. There is also much documentation available online at `http://java.sun.com`, though reading that documentation is sometimes difficult. By the time this book is in your hands, there are sure to be more books devoted to Java Version 5.0 that explain all the new features in more detail than we can provide here. We expect one of those books to be the consistently good *Java in a Nutshell* series [10], the 5th edition of which should include coverage of J2SE 5.0.

1.10 Web Course materials

The Web Course Chapter 1: *Supplements* section covers a number of topics related to those mentioned here including the issue of interpretation versus compilation, the design of JVMs like those with Just-In-Time compilation, and options for deploying your Java programs to users. It provides a list of web links to JVMs and compilers produced by independent sources. It also provides further information on J2SE 5.0.

Some practical programming topics are also presented such as how to create web pages with applets. For applets compatible only with version 1.2 or later, the Web Course gives instructions on how to set up the web page so that a Java plug-in will automatically download if needed.

The Web Course Chapter 1: *Tech* section discusses the benefits and shortcomings of Java for technical applications. The Web Course Chapter 1: *Physics* section looks similarly at the benefits and shortcomings of Java for physics computation and simulation. It gives examples of different kinds of simulations of physics phenomena. It also emphasizes the benefits of *learning by coding* in which the process of converting a physics theory into a simulation will help to deepen your understanding of the phenomena.

References

[1] James Gosling, *Java Technology and the Mission to Mars*, Sun News, January 16, 2004, www.sun.com/aboutsun/media/features/mars.html.

[2] Maestro and the Mars rover data sets are available at http://mars.telascience.org.

[3] SensorNet Project, www.sensornet.gov.

[4] Aviation Digital Data Service, http://adds.aviationweather.gov/java/.

[5] AirportMonitor™, www.passur.com/am_airport.htm.

[6] Infrasound viewer, Swedish Institute of Space Physics, Umeå, Sweden, www.umea.irf.se/ilfil/.

[7] BioJava, www.biojava.org.

[8] Steve Meloan, *BioJava – Java Technology Powers Toolkit for Deciphering Genomic Codes*, Sun Developer Network, June 2004, http://java.sun.com/developer/technicalArticles/javaopensource/biojava/.

[9] Brett McLaughlin and David Flanagan, *Java 1.5 Tiger, A Developer's Notebook*, O'Reilly, 2004.

[10] David Flanagan, *Java in a Nutshell*, 4th edn, O'Reilly, 2002. (Note: we expect that the 5th edition, covering J2SE 5.0, will be published by the time you read this.)

Resources

Calvin Austin, *J2SE 1.5 in a Nutshell*, May 2004, `http://java.sun.com/developer/technicalArticles/releases/j2se15/`.

Judith Bishop and Nigel Bishop, *Java Gently for Scientists and Engineers*, Addison-Wesley, 2000.

Stephen J. Chapman, *Java for Engineers and Scientists*, Prentice-Hall, 2000.

Richard Davies, *Java for Scientists and Engineers*, Addison-Wesley, 1999.

Ronald Mak, *Java Number Cruncher: the Java Programmer's Guide to Numerical Computing*, Prentice-Hall, 2003.

Grant Palmer, *Technical Java: Applications for Science and Engineering*, Prentice-Hall, 2003.

Java Community Process Program – `www.jcp.com`.

Java resources at IBM Developerworks, `www.ibm.com/developerworks/java`.

Java software development on the Apple Mac, `http://developer.apple.com/java/`.

Java at Sun Microsystems, Inc., `http://java.sun.com`.

Java 2 Standard Edition API Specification, `http://java.sun.com/j2se/1.5.0/docs/api/`.

The Java Tutorial at Sun Microsystems, `http://java.sun.com/docs/books/tutorial/`.

Chapter 2
Language basics

2.1 Introduction

If you buy one of those do-it-yourself furniture kits, the best way to start is to just dump all of those screws, nuts, planks, tools and other odd looking widgets on the floor, group them into piles of similar looking items, and then go read the instructions. Even if you don't know what all of those widgets are for, it helps to pick them up and look them over so that you become familiar with them and can recognize them in the instructions.

So rather than dribbling them out over several chapters, here we dump out most of the basic widgets needed to construct Java programs. The goal is to start to become familiar with Java's symbols, keywords, operators, expressions, and other building blocks of the language with which to construct programs. We provide examples and starter programs (on the Web Course) that allow you to begin to write programs without needing to understand yet all of these language elements at a deep level. You should refer back to this chapter as you proceed and as your understanding of the language increases.

Note that in this chapter we occasionally mention the terms *class*, *method*, and *object*. If you are new to object-oriented programming, do not worry about these terms for now. We discuss them in detail in the following chapters.

We begin with a listing of the basic elements and then outline the structure of a generic program. We proceed through the individual elements of the language beginning with the Java reserved words, or *keywords*. We then discuss the basic data types in Java called *primitives*. These are used in expressions and with various operators to create statements.

With regard to technical programming, we look at the floating-point representations in Java and the various issues regarding them. We also discuss the math functions available in the core language.

2.2 Language elements and structures

Like any computer language, Java consists of a set of basic elements that the programmer arranges according to the syntax of the language to build a program.

Characters and symbols make up the keywords and the punctuation with which to build expressions, statements, and other higher-level structures. We first briefly list these basic elements and structures here and then examine them in more detail in the rest of the chapter.

2.2.1 Basic elements

Java codes consist of the following:

- **Keywords and symbols** – The programmer has wide latitude for the names of variables and classes but Java reserves some words for itself. These keywords include, for example,

```
class, float, return, if, else
```

Symbols in java include, for example,

```
{}; () //
```

See Table A.1.1 and Table A.1.2 in Appendix 1 for listings of the Java keywords and Java reserved symbols, respectively.
- **Data types** – Eight keywords name the eight kinds of basic data types in Java:
  - `byte, short, int, long` – four integer types, all signed
  - `float, double` – two floating-point types
  - `boolean` – logical (true/false)
  - `char` – 2-byte Unicode character encoding (can be used for 2-byte unsigned integers)
  These data types do not use object representations but simply hold a data value in one to eight bytes of memory. They are referred to as *primitives*.
- **Operators** – Java operators include the arithmetic operators (+, -, /, *, %), Boolean operators (&&, ||, etc.), comparisons (==, <=, etc.) and others. See the tables of operators in Appendix 2.
- **Identifiers** – The programmer chooses the names of variables, classes, and methods but these names cannot begin with a number and cannot contain a punctuation character (though underscore _ and dollar sign $ are allowed). Java is a strongly typed language, which means that all variables must be explicitly declared as a particular type and that strict rules apply when assigning variables of one type to another. The types of variables include:
  - **Primitive** type variables:

```
int n = 5; // n is a variable containing a primitive
           // int type
double x; // x contains a double type
```

- **Reference** variables refer to an object (see Chapter 3):

```
Vector vec = new Vector (); // vec is a reference to a
                           // Vector object
```

- **Array** variables (see Chapter 3):

```
x = b * c[5]; // c[5] refers to the element at index 5
              // of array c
```

- **Literals** – A specific value in a line of code is called a literal:

```
double x = 4.0;          // 4.0 is the double floating-point
                         // literal
String str = "A string"; // "A string" is a string literal
char c = 'c';            // 'c' is a character literal
```

2.2.2 Language structures

The basic language elements above combine to create higher-level structures that perform the operations of a program. These structures include:

- **Expressions** – An expression contains an operation on one or more operands and returns a value:

```
    x = 5       – assignment operation
    x < 5       – comparison operation
    14. * y     – multiplication operation
```

- **Statements** – A statement, which can hold one or more expressions, presents a complete action or set of actions to perform:

```
x = 5;
if (y < 5) x = 3;
return 3.0 * (14. * y);
```

Statements end with a semicolon.

Finally, these structures in turn make up the structure of a class and its fields and methods (methods resemble subroutines or functions in other languages). We fully develop the definition of Java classes in later chapters.

2.3 A simple application

The following example code illustrates the bare essentials of a Java application program that prints a short text string to the console. You can see that the code must always occur within the framework of a class. Follow the instructions in Chapter 1 for compiling and executing this application.

```
/*                                      Begin with comments describing
 * A simple application.                the class.
 */
public class SimpleApp1                 The class signature.
{

    public static void main (String[] args)     Application processing always
    {                                           begins in this method.

        // Print a string to the console.       Single line comment

        System.out.println ("An application");  Print to console.

    }                                       Braces span the code for a class
}                                           and the code for each method.
```

The code includes the keywords `public`, `class`, `static`, `void`, and `main`. The brace symbols { } enclose the contents of a class, which in this case has the name (i.e. the identifier) `SimpleApp1`. Similarly, matching braces enclose the contents of the method named `main`. Parentheses enclose a method's parameter list, which in this case consists of an array of strings named `args`. (Programmers in C/C++ will note the resemblance to the `main` method in those languages, though its argument is not a string array.)

Methods, like subroutines and functions in other languages, carry out specific tasks. This program has only the single method `main()`. This method is required for all application programs and is the first method invoked when the program starts. This `main()` method only holds a single statement, which invokes another method to print a string literal to the console. The `println()` method comes with the core language.

2.4 Comments

Java denotes comments in three ways. Double slashes precede a single line comment as in

```
// Print a string to the console.
```

As seen in some of the examples above, double slashes can also be used at the end of a statement, after the semicolon, to add a comment on the rest of the line.

Alternatively, you can bracket one- or multiple-line comments with matching slash-asterisk and asterisk-slash, as in

```
/*
   Print a string
   to the console.
*/
```

The `javadoc` tool, which comes with the SDK, automatically generates hypertext documentation when operated on Java source code files with special tags. Starting a comment with a slash and two asterisks indicates comments for `javadoc`, as in

```
/**
   This applet tests graphics.
*/
public class TestApplet extends applet { . . .
```

The command

```
> javadoc TestApplet.java
```

will include this comment in the hypertext description of the class. Several special comment tags are also available to describe method parameters, return values, and other items. (Note that asterisks within the comments are ignored by `javadoc` so we often use them to mark the left side of a block of `javadoc` comments.) See Chapter 6 in the Web Course for more about `javadoc`.

2.5 Data types and Java primitives

How numbers and other data, such as characters, are represented in memory is of great practical importance. Ideally a single memory representation, or type, could represent all numerical data. However, computer memory and transfer rates are not infinite and designers must strike a compromise among the demands for numerical values to span the widest possible range of values while conserving memory and maximizing program performance.

Very large numbers and fractional numbers require a floating-point type. Floating-point operations involve the extra complexity of dealing with both a significand and an exponent. Therefore, integer data types normally provide faster performance for operations where floating-point is not required. (Some processors don't do any floating-point arithmetic at all, only integer arithmetic.) A single universal numerical data type will not be efficient or sufficient for all situations.

Table 2.1 lists the eight primitive data types and shows the features of each. Four integer types provide for efficient use of memory for a given task. The `byte` type (8-bit) is often useful for IO tasks while the `long` type (64-bit) is needed for representing very large integer values. In between are the 16-bit `short` and the 32-bit `int` types. The `int` type is the default for most integer operations in Java. The integer types use two's complement signed representations.

Table 2.1 *Java primitive data types.*

Type	Values	Default	Size	Range
byte	signed integer	0	8-bit	-128 to 127
short	signed integer	0	16-bit	-32768 to 32767
int	signed integer	0	32-bit	-2147483648 to 2147483647
long	signed integer	0	64-bit	-9223372036854775808 to 9223372036854775807
float	IEEE 754 floating-point	0.0	32-bit	±1.4012985E-45 to ±3.4028235E+38, ±infinity, ±0, NaN
double	IEEE 754 floating-point	0.0	64-bit	±4.9E-324 to ±1.7976931348623157E+308, ±infinity, ±0, NaN
char	Unicode character or Unsigned integer	\u0000 or 0	16-bit	\u0000 to \uFFFF or 0 to 65535
boolean	true, false	false	1 bit in 32 bit integer	Not applicable

The IEEE 754 floating-point standard is used for the 32-bit float and 64-bit double types. We discuss floating-point in more detail later in Section 2.11 and in Appendix 3.

Historically, standard characters were typically represented with the 7-bit ASCII encoding. Various other 8-bit encoding schemes exist to provide an additional 128 characters for symbols or for the characters needed for particular languages. To provide internationalization with a single encoding requires more than the 256 characters available with 1 byte. The Java designers decided to use the 2-byte character type encoding called the Unicode system for the char type. (J2SE 5.0 adds support for an even larger 4-byte system known as Unicode 4.0.) The char type can also act as a 16-bit unsigned integer in some cases. (See Section 9.7 for more about character encoding.)

The boolean (true/false) type provides for the many kinds of logical operations carried out in almost any program. Though internally a boolean value is either a 1 or 0 integer, it cannot be used in arithmetic operations. (Representations of Boolean arrays are not specified for the JVM. Some implementations use bit arrays.)

We discuss classes and objects later, but we note here that primitive types in Java are not instances of classes. The decision to use simple data types broke the symmetry and elegance of the language to some degree but vastly reduced overhead and execution times compared to the case where all data types are objects.

2.6   Strings

As in C/C++, an array of Java char primitive type values can represent a string
of characters. (We discuss arrays in Chapter 3.) However, you will find it much
more convenient and common to use the String class that comes with the core
language. Strings were deliberately made very easy to use in Java, and they behave
almost like a primitive type. Though we haven't discussed classes yet, we present
some basics about Java strings here since they are essential to begin programming
anything of interest.

You can create a string variable by enclosing the string literal in quotations:

```
String str1 = "A string ";
String str2 = "and another string";
```

(This is the only class where the "new" operator isn't required to create an object
of that class.) You can append one string to another with a simple "+" operation.
(This is also the only case in Java of overloading an operator, i.e. redefining its
operation according to context.) For example,

```
String str3 = str1 + str2;
```

results in str3 holding:

```
"A string and another string";
```

A very useful capability of the "+" operator allows for default conversion of
primitive types to strings. Whenever you "add" a primitive type value, such as an
int value, to a string, the primitive value converts to a String type and appends
to the string. For example, this code:

```
String str1 = "x = ";
int i = 5;
String str2 = str1 + i;
```

results in str2 holding "x = 5". This also works with boolean type values,
which result in "true" and "false" strings.

We discuss the String class and other string handling classes in more detail
in Chapters 3 and 10. Note that whenever we use the term string we refer to a
String class object unless otherwise indicated.

2.7   Expressions

An expression performs an operation and returns a value. Here are some
examples:

- i = 2   – assignment puts 2 into the i variable and returns the value 2.
- ++k   – the k operand is incremented by 1 and the new value returned
- x < y   – logical "less than" comparison, returns a Boolean true or false value.

- `i | j`   – returns the value of a bitwise OR operation on bits in the two variables.
- `4.0 * Math.sin (i * Math.PI)`   – combines several operations in this expression, including multiplication and a method call.

Expressions involve at least one operator. A value is returned from the expression and, in some cases, an operand is also modified (such as ++k).

2.8   Operators

In Java an expression carries out some operation or operations that act according to the particular operator in use. An operator acts upon one, two, or three operands. Here we discuss some general properties of operators and their operands. Refer to Appendix 2 for tables of the allowed operators in Java.

2.8.1   Operands

An operand can be:

- a numeric variable – integer, floating-point, or character
- any primitive type variable – numeric and boolean
- a reference variable to an object
- a literal numeric value, boolean value, or string
- an array element – a[2]
- char primitive, which in a numeric operation is treated as an unsigned 2-byte integer.

The operator is unary if it acts on a single operand; binary if it requires two operands. The conditional operator, to be discussed later, is the only ternary operator in Java.

Each operator places specific requirements on the operand types allowed. For example, the subtraction operator "–" in x = a - b; requires that a and b variables be numeric types. The assignment operator "=" in that same expression requires that x also be a numeric type. (If a and b were wider types than x, a casting operation, see Section 2.10, would also be required.)

2.8.2   Returned value

A value is "returned" at the completion of an operation. The following statements use the assignment operator "=" and the addition operator "+":

```
int x = 3;
int y = x + 5;
```

These statements result in x holding the value 3 and y holding the value 8. The entire expression y = x + 5 could be used in another expression:

```
int x = 3;
int y;
int z = (y = x + 5) * 4;
```

This results in y holding 8 and z holding 32. The assignment operator "=" in the expression

```
(y = x + 5)
```

produces a new value for y and also returns the value of y to be used in the expression for z.

2.8.3 Effects on operands

In most of the operations, the operands themselves are not changed. However, for some operators, the operand(s) do undergo a change:

- **Assignment operators** – "x = y" replaces the value of the first operand with that of the second. The other assignment operators, "*=, +=, -=, /=", also replace the value of the first operand but only after using its initial value in the operation indicated by the symbol before the equals sign. For example,

```
x *= y
```

  results in x being replaced by x * y. Also, this is the value returned from the operation.
- **Increment and decrement operators**
(++x)	– x is incremented before its value is returned.
(--x)	– x is decremented before its value is returned.
(x++)	– the initial value of x is returned and then x is incremented.
(x--)	– the initial value of x is returned and then x is decremented.

For the increment and decrement operations, note that in a standalone expression such as

```
x++;
```

there is no effective difference between x++ and ++x. Both expressions increment the value stored in the variable x. However, in expressions such as

```
y = x++;
```

and

```
z = ++i;
```

the order of the appearance of the increment operator is important. In the former case, y takes on the value of x *before* the increment occurs. If x is initially 3, then y becomes 3 and x becomes 4. In the latter case the increment occurs before the value is used. So an initial value of 3 for i leads to i incrementing to 4 and then z taking on the new value, 4.

Remember that if an operand is changed by the operation and the statement holding that expression is processed again, as in a loop, the operand's value will be different for each pass.

2.8.4 Expression evaluation

The operands of an operator are always evaluated from left to right. For example, in

```
x = a + b;
```

the "+" operator will determine the value of expression a and then expression b. Do not get this rule confused with the precedence and associativity rules, discussed next.

Precedence determines the order in which operators act in an expression with more than one operator. Table A.2.9 gives the precedence rating for each operator, the higher number indicating higher precedence.

Associativity rules determine how the compiler groups the operands and operators in an expression with more than one operator of the same precedence. For example, in the expression

```
x = a + b * c;
```

the evaluation begins with a and then the "+" operator determines its right operand. But in this case the right operand consists of the expression "b * c". The multiplication operator "*" has a higher precedence than the additive operator "+" so b multiplies c rather than sums with a.

Precedence can be overridden with parentheses, as in

```
x = (a + b) * c;
```

The parentheses force the addition of b to a, and then c multiplies this sum.

Although the precedence ratings, which are similar to those in C/C++, were chosen for the most "natural" ordering of the operator evaluations, it never hurts to use the parentheses if you are unsure of the precedence and to make the code more readable.

When the operations in an expression all have the same precedence rating, the associativity rules determine the order of the operations. For most operators, the evaluation is done from left to right, as in

```
x = a - b + c;
```

Here, addition and subtraction have the same precedence rating, so a and b are subtracted and then c added to the difference. Again, parentheses can be used to overrule the default associativity, as in

```
x = a - (b + c);
```

The assignment and unary operators, on the other hand, are associated right to left. For example, the statement

```
x += y -= -~4;
```

is equivalent to

```
x += (y -= (-(~4)));
```

or, in long hand,

```
int a = ~4;
a = -a;
y = y - a;
x = x + y;
```

2.8.5 More operator tricks

Finally, here are some additional notes about operators. As indicated in the last example above, assignment operations can be chained:

```
x = y = z = 4;
```

with the evaluations proceeding from right to left.

Assignments can be combined into other operations, to compact the code, as in

```
if ((x = 5) == b) y = 10;
```

which first sets x to 5, then returns that value for the test of b. This technique should generally be avoided as it makes the code difficult to read.

2.9 Statements

A statement in Java, like most computer languages, corresponds to a complete sentence in a spoken language. It performs some complete action or set of actions. It can encompass multiple operators and operands, as well as multiple sub-statements. A single statement ends with a semi-colon and a set of multiple sub-statements is enclosed in braces.

The statement,

```
x = 5.3 *(4.1 / Math.cos (0.2*y));
```

consists of several expression – multiplication, division, and an invocation of a method – but is still considered a single statement.

A group of statements enclosed in braces – called a code block or compound statement – acts as a single statement. For example, a simple if test looks like this

```
if (test) statement;
```

and causes the `statement` to be executed if the `test` is true. For the case where a series of statements are needed after an `if` test, then they should be enclosed in braces as in

```
if (test)
{
    statement 1;
    statement 2;
        . . .
}
```

where all the statements within the braces will execute if a is less than b. Note that a semi-colon is not necessary after the final brace. For these *compound* statements we follow the common practice of putting the first brace after the control part of the statement as in

```
if (test) {
    statement 1;
    statement 2;
        . . .
}
```

We discuss below several important kinds of statements: declaration, conditional, and flow control.

2.9.1 Declarations

A declaration gives both an identifier and a data type to a variable and provides memory space for it. Java is a strongly typed language so every variable must be explicitly declared a particular type. An initializer in the declaration can assign a value to the variable:

```
int x;
int x, y, z; // multiple declaration
double y = 1.0; // Declaration and initializer
double x = 5.0;
```

Local variables in methods must be assigned a specific value either in the declaration, or afterwards, before they are used, else the compiler complains. Class and instance variables (see Chapter 3) will be given default values by the compiler: zero for numerical types, false for `boolean`, empty (zero bits) for `char`, and `null` for object reference types.

2.9.2 Conditional statements

The conditional `if` statement evaluates a `boolean` expression to decide whether to execute a statement, as in

```
if (test) statement;
```

where `test` indicates a `boolean` expression. If `test` evaluates to `true`, then `statement` is executed.

Similarly, the conditional `if-else` statement evaluates a `boolean` expression to decide which of two statements to evaluate, as in

```
if (test)
    statement1;
else
    statement2;
```

The statement(s) to evaluate can consist of single, one-line statements or of code blocks of multiple statements enclosed in braces.

A sequence of conditions can be tested with multiple `if` tests, as in

```
if (test1)
    statement1;
else if (test2)
    statement2;
else if (test3)
    statement3;
else
    statement4;
```

2.9.3 Flow control statements

Several types of statements affect the flow or sequence of processing such as repeating a section of code or jumping over a section of code. Such flow control statements are essential tools for any type of programming. The following loop statements repeat the processing of a statement or code block for a number of times as set by a logic test.

2.9.3.1 The `for` loop

The `for` loop goes as follows:

```
for (start; test; action) statement;
```

Here `statement` repeatedly executes until the expression `test` returns `false`. The loop begins with an evaluation of the `start` and `test` expressions and, after each loop, the `action` expression is evaluated before `test` is evaluated again. A typical example goes as

```
for (int i=0; i < 10; i++) j = i * j;
```

This begins with the declaration of the integer `i` initialized to 0 followed by the evaluation of the test `i < 10`. Since the variable `i` is less than 10, the evaluation of the `j = i * j` statement proceeds. The processing "loops back" and evaluates the `i++` expression and then the test is evaluated again. The looping continues until the `test` expression returns `false`.

2.9.3.2   The `for` loop

J2SE 5.0 adds a new looping feature called the "enhanced `for` loop," also sometimes called the "for/in" loop or the "for each" loop. For compatibility reasons, no
new keyword was added for the enhanced `for`. Instead a syntax change inside the
parentheses indicates the use of the new loop. Unfortunately, since the enhanced
for loop works on classes and objects or arrays that we don't describe until Chapter 3, we cannot explain the loop's behavior in detail here. Instead, we give a brief
description and defer details until Chapter 10 where we discuss the `Iterator`
class. Briefly, the new loop appears as follows:

```
for (type value: container)

    statement
```

The colon is normally read as "in," thus the name "for/in" loop. The whole
expression can be read "for each value in container, do the statement". (Of course,
the `statement` can be a block of statements enclosed in braces.) The "container"
here is a Java object that contains other objects. For example, an array (Section
3.8) is a kind of container. Some arrays can hold object types, others hold primitive
types. If we have an array of `int` types, we can loop through all elements in the
array as follows:

```
for (int i: array) statement;
```

This loops through each element in the array and performs the statement using
each element, one at a time. The real power of the enhanced `for` loop becomes
evident after we learn about classes and objects (Chapter 3) and, in particular, the
`Iterator` class (Chapter 10).

2.9.3.3   The `while` and `do-while` loops

The `while` loop goes as follows:

```
while (test) statement;
```

It repeatedly executes `statement` until the `test` expression returns `false`.
Here is an example:

```
while (i < 5) {
   i = a.func ();
}
```

The `a.func()` method is invoked as long as the variable `i` is less than 5. The
test is done before the first evaluation of `statement`, so that if the initial test
fails, nothing inside the code block is executed even once. As an alternative, the
`do-while` statement evaluates the loop code before the test is evaluated. This
ensures that the loop code is executed at least once.

```
do {
   i = a.func ();
} while (i < 5)
```

Sometimes it is necessary to instantly break out of a `while` or `do-while` loop without waiting on the test at either the beginning or end of the loop. The `break` statement provides that functionality. The processing will jump from a loop via a `break` statement and continue to the statements following the loop, as in this example:

```
while (i < 5) {
   i++;
   x = a.func ();
   if (x < 0) break; // jump out of the loop
      b.func ();
}
c.func ();
```

The `break` statement in the loop causes processing to jump immediately to the following `c.func()` statement. The `b.func()` statement is ignored and no further loop processing is done, regardless of the value of the text expression.

Sometimes it is necessary to begin a loop, perform only the first portion, skip the rest, but continue in the loop for further processing. The `continue` statement does just that, causing the processing in a `for` or `while` loop to skip the rest of the loop and return back to the start of the loop. An example is

```
while (i < 5) {
   if (a.func ())
      continue;
   b.func ();
}
```

Here the processing jumps back to the start of the `while` loop and checks the test expression if `a.func()` is true. Otherwise, it executes the `b.func()` statement. In a `do-while` loop a `continue` will cause the processing to skip down to the `while(test)` and execute the test expression.

2.9.3.4   The `switch` *statement*
The final flow control structure is the `switch` statement:

```
switch (int expression) {
   case 1: statement1;
   case 2: statement2;
   default: statement3;
}
```

If the integer expression in the `switch` parameter returns the value 1, then the processing jumps to `statement1`, which has the label `case 1`. If it returns 2, then processing starts with `statement2`. If the value does not match any label value, then the processing goes to `statement3` with the `default` label.

Note that processing continues on to statements that follow it in subsequent "`case`" labeled sections unless a break statement causes the process to jump out of the `switch` area. In this example,

```
switch (i) {
   case 1: m = 5;
   case 2: j = 5;
      break;
   default: j = 2;
}
```

If `i` equals 1 the `m = 5` statement is evaluated and then the `j = 5` statement is evaluated as well. The `break` sends the processing out of the switch section. Most of the time, there are `break` statements at the end of each case label. Also, most `case` sections consist of multiple lines of code. It is not necessary to enclose those lines in braces. For example, the following is perfectly legal and works as expected:

```
switch (i) {
   case 1:
      m = 5;
      q = 6;
      break;

   case 2:
      j = 5;
      k = 6;
      break;
}
```

Note that the `default` label is not required, though it is good practice to use one in every `switch` statement.

The type of the variable `i` in the `switch` statement must be `byte`, `short`, `int`, or `char` type. The long type, `boolean`, any floating-point type, and object references are not permitted (but see the autoboxing discussion in Chapter 3 for J2SE 5.0).

2.10   Casts and mixing

Now that we have presented an overview of the basic elements and structures of the language we can look a bit more closely at how the data types work. Here we discuss how one type converts into another and what happens when statements include a mix of variables of different types.

Converting one type of data into another must follow the rules of *casting*. If a conversion would result in the loss of precision, as in an int type value converted to a short, then the compiler issues an error message unless an explicit cast is made.

To cast type A data into type B data, put the type B name in parentheses in front of the type A data:

```
A a = a_type_data;
B b = (B) a; // cast a, which is originally type A,
             // to type B
```

Of course, it must be "legal" to convert type A data into type B. For the primitive types, most, but not all, conversions are legal. An example of an illegal cast would be an attempt to cast a String object into an int. The rules for what is legal and what is not are described below and generally follow common sense. It is nonsensical, for example, to convert a string into an integer since the integer primitive type cannot "hold" a string, which is not a primitive.

An example of a sensible cast is the conversion of double data to an int:

```
double d = 1.234;
int i = (int) d; // Cast double to int
```

Expressions can promote to a wider (higher precision) type without an explicit cast. For example, an int type can convert to long without a cast, but the reverse requires a cast:

```
int i = 3;
long j = i; // no cast needed
i = (int) j; // cast required
```

Note that a char type value can be cast to other integer types but as an unsigned value. The boolean type is singular; it cannot be cast to anything, and nothing can be cast to a boolean. If a boolean value is needed, it must be the result of a logical expression. For example, if an integer i > 0 is considered to be true, then a boolean value can be obtained as follows:

```
boolean b = (i > 0);
```

2.10.1 Cast rules

Table 2.2 shows to what other primitive types a given primitive data type can be cast. The symbol C indicates that an explicit cast is required since the precision decreases. The symbol A indicates that the precision increases so an automatic conversion occurs without the need for an explicit cast. N indicates that the

Table 2.2 *Converting between primitive data types.*

	int	long	float	double	char	byte	short	boolean
int	–	A	A*	A	C	C	C	N
long	C	–	A*	A*	C	C	C	N
float	C	C	–	A	C	C	C	N
double	C	C	C	–	C	C	C	N
char	A	A	A	A	–	C	C	N
byte	A	A	A	A	C	–	A	N
short	A	A	A	A	C	C	–	N
boolean	N	N	N	N	N	N	N	–

* Indicates that the least significant digits may be lost in the conversion even though the target type allows for larger values. For example, a value in an `int` type that uses all 32 bits will lose some of the lower bits when converted to a `float` since the exponent uses 8 of the 32 bits.

conversion is not allowed. Object types can also be involved in casts, but we defer that discussion until Chapter 3.

When data of one primitive type is cast to another type, the effect on the numerical values must be taken into account.

- **Narrowing conversions** – When an integer type is cast to another integer type of a smaller number n of bits, all but the n lowest-order bits are discarded. Depending on the initial value, the result can have a different value and/or a different sign than the input value.
- **Integer to floating-point conversions** – These are widening conversions so explicit casts are not required. However, note that the precision for large numbers can actually decrease since the exponent occupies part of the four bytes allocated for a `float` and part of the eight bytes for a `double`. So the mantissa decreases accordingly. A large `int` value converted to a `float` can lose low-order bits and similarly for a `long` to a `double`.

2.10.2 Mixed types in expressions

If an expression holds a mix of types, the lower precision or narrower value operand is converted to the higher precision or wider type. This result then must be cast if it goes to a lower precision type:

```
double x, y = 3.0;
int j, i = 3;
x = i * y;        // OK since i is promoted to double
j = i * y;        // Error since result is a double value
j = (int)(i * y)  // OK because of the explicit cast
```

The process of converting a value to a wider or higher precision integer or floating-point type is called "numeric promotion". The Java language specification states the following rules for promotion in an expression of two operands, as in x + i:

- If either operand is of type `double`, the other is converted to `double`.
- Otherwise, if either operand is of type `float`, the other is converted to `float`.
- Otherwise, if either operand is of type `long`, the other is converted to `long`.
- Otherwise, both operands are converted to type `int`.

This last rule can lead to confusion if adding two byte values. For example, the code

```
byte a = 3, b = 6, c;
c = a + b;
```

does not compile because the `a` and `b` operands are automatically promoted to type `int` before the addition occurs, meaning that the result is also an `int`. The code then attempts to place the resulting `int` back into a `byte` type, resulting in a "possible loss of precision" compiler error. The solution is to make an explicit cast to `byte`:

```
c = (byte)(a + b);
```

2.11 Floating-point

Floating-point representation is obviously an important aspect of numerical computation and how Java handles it while maintaining platform portability should be understood. We first look at floating-point in general and then in Java. Appendix 3 gives further details about Java floating-point representation and operations.

2.11.1 Floating-point basics

A floating-point number is represented in binary as

$$\pm b_0.b_1b_2b_3 \quad . \quad . \quad . \quad b_{n-1} \quad * \quad 2^{exponent}$$

where b_i represents the i bit in the n bits of the significand (also called the mantissa). There is also a bit to indicate the sign. A floating-point value is calculated as

$$(-1)^s \cdot (b_0 + b_1 \cdot 2^{-1} + b_2 \cdot 2^{-2} + b_3 \cdot 2^{-3} + \ . \ . \ . \ + b_{n-1} \cdot 2^{-(n-1)}) \cdot 2^{exponent}$$

where s is a bit for the sign. Floating-point numbers involve a number of complications with which the processor designers must deal. These complications include:

- **Approximations** – The limited number of places in the significand means that only a finite number of fractional values can be represented exactly. Similarly, the finite width of the exponents limits the upper and lower size of the numbers.

Table 2.3 *Bit layout of the floating point primitives.*

Type	Sign	Exponent	Significand
float	**1 bit**	8 bits	23 bits
double	**1 bit**	11 bits	52 bits

- **Round-off** – Arithmetic operations often result in the need to round off between the exact value and the value that can be represented by the floating-point type. A round-off (or truncation) algorithm must be chosen by the designer of the language. Round-offs can have significant impact on calculated values, especially during intermediate operations where errors can build up.
- **Overflows/underflows** – Similarly, a calculation may result in a number that is smaller or larger than the floating-point type can represent. Again, the language designer must select a strategy for how to handle such situations.
- **Decimal-binary conversion** – The computer represents numbers in base 2. This can result in loss of precision since many finite decimal fractions (0.1 for example) cannot be represented exactly by binary fractions. (All finite binary fractions, however, can be converted to finite decimal fractions.)

These complications can mean even simple calculations with floating-point give surprising results. For example, the following code:

```
double d = 0.0;
for (int i = 1; i <= 10; i++) {
   d += 0.1;
}
```

does not result in d = 1.0 since 0.1 is not exact in binary format. For this reason, it is best to avoid equality tests between floating-point values, as in

```
if (a == b) statement;
```

Instead you should normally test floating values with <, <=, >=, and >. However, it may be sensible to test for equality to 0.0 if a divide by zero could occur.

2.11.2 Java floating-point

The bit allocations for the floating-point representations of the float and double types in Java are shown in Table 2.3. For each type there is one bit for the sign. The exponents contain 8 and 11 bits and the fractions contain 23 and 52 bits, respectively.

The exponent values 0 and 255 for float are reserved for the special cases discussed below. Otherwise, a bias of 127 is subtracted, giving an effective exponent range of -126 to $+127$. Similarly, for double the exponent values 0 and 2047 are reserved and subtracting a bias of 1023 gives an effective exponent range of -1022 to $+1023$.

For `float` numbers with exponent values in the range -126 to $+127$, a 1 bit to the left of the binary point is assumed (i.e. b_0 is fixed at 1 in the above formula) thereby increasing the number of effective fractional bits by one. This holds similarly for `double` numbers in the -1022 to -1023 range. This provides an effective significand of 24 bits for `float`, 53 bits for `double`. Such numbers are referred to as *normalized*. This scheme for `float` provides at least six digits of decimal precision while `double` provides at least 15 digits of decimal precision.

The special floating-point cases include the following:

- **Denormalized** – If the bits in the exponent all equal 0 and the significand bits do not all equal 0, then the exponent is treated as -126 for `float` and -1022 for `double` (i.e. the binary point moves left by 126 places and 1022 places, respectively) and the implied bit to the left of the binary points is 0. These *denormalized* numbers allow for smaller values (i.e. closer to zero) than normalized alone.
- **± Zero** – If the bits in the exponent and the significand all equal 0, then the floating-point value is -0 or $+0$ depending on the sign bit. (See Appendix 3 for definitions of $+0$ and -0.)
- **± Infinity** – If all the bits in the exponent equal 1 and all the bits in the significand equal 0, then the floating-point value is plus or minus infinity according to the sign.
- **Not-a-Number (NaN)** – If all the bits in the exponent equal 1 and any of the bits in the significand equal 1, then the floating-point value is Not-a-Number (NaN) and the sign value is ignored. Not-a-Number occurs when an operation has no mathematical meaning, such as 0.0/0.0, or when any operation is done with an existing Not-a-Number.

Overflows, underflows, and divide by zero in Java floating-point operations do not lead to error states (Java *Exceptions* are discussed in Section 3.9). A division by zero leads to the plus or minus infinity value unless the numerator is also zero, in which case the Not-a-Number value results. You can test for Not-a-Number values using methods from the floating-point wrapper classes (see Chapter 3) such as `Double.isNaN (double x)`. Also, the Not-a-Number value can be checked for with the test

```
if (x!= x) statement;
```

which always returns `true` for Not-a-Number values. Numerical comparisons such as

```
if (x < y) statement;
```

always return `false` if either or both values are Not-a-Number.

Round-off takes the binary value nearest to the exact (or higher precision intermediate) value. If two binary values are equally close, then the even value (the one with its last bit equal to 0) is chosen.

In general, it is far safer to do floating-point calculations in `double` type. This helps to reduce round-off errors that can reduce precision during intermediate calculations. (You can always cast the final value to `float` if that is a more convenient size for I/O or storage.) There can be some performance tradeoff, since `double` operations involve more data transfer, but the size of the tradeoff depends on the JVM and the platform. (In Chapter 12 we discuss techniques for measuring code performance.)

The representations of the primitives are the same on all machines to ensure the portability of the code. However, during calculations involving floating-point values, intermediate values can exceed the standard exponent ranges if allowed by the particular processor (see Table A.3.1). The `strictfp` modifier of classes or methods requires that the values remain within the range allowed by the Java specifications throughout the calculation to ensure the same results on all platforms.

2.12   Programming

As seen in Chapter 1, you must code within an object-oriented framework. However, you can apply the Java elements, structures, and techniques discussed so far in this chapter to straightforward procedural programming and ignore object-oriented issues until later. You can simply insert code into the `main()` method of an application or in the `init()` method of an applet, as shown in the starter programs of Chapter 1, and not yet deal with the workings of class and object concepts.

To program something interesting you need the ability to display output in some way. Since we postpone graphics interface programming until Chapter 6, we use methods from the core Java library to print to the console in our demonstration programs through Chapter 5. Below, we use this technique to look at an example of code that illustrates various aspects of floating-point operations.

2.12.1   Print to console

Java possesses a very extensive set of Input/Output (I/O) tools and capabilities but Java Input/Output is somewhat complicated and is best introduced after gaining more experience with the objected oriented aspects of the language. So we introduce Java I/O techniques over several chapters and devote all of Chapter 9 to I/O.

For now we give just some simple techniques for printing to the console, which refers to the command line window where you run applications or the Java console window in the browser holding an applet. (The Java environment on the Apple Macintosh also provides a console window for application output.)

Until we discuss graphics and graphical user interfaces in Chapter 6, we rely heavily on the following methods to see the results of a program:

```
System.out.print (string) // no line return
System.out.println (string) // includes line return
```

where `string` denotes any `String` object that you create as explained in Section 2.6.

For example, if you insert the following code snippet into the `main` method of `StarterApp1.java` (see Chapter 1):

```
int i = 5;
int j = 3;
System.out.println ("i * j =" + (i * j));
```

the output to the console looks like

```
i * j = 15
```

The (`i * j`) expression inside the `println` parameter results in an integer value, which the "+" append operator converts automatically to a string and attaches to the preceding string.

Note that the parentheses around the `i * j` term are not necessary according to the higher precedence of the multiplication operator compared to the "+" append operator (see Appendix 2 for a table listing the precedence rules). However, with addition, as in

```
System.out.println ("i + j =" + i + j);
```

you must be careful. Without parentheses, the compiler will treat the two + operands as equal precedence and perform a string concatenation of the `i` value (5) and the `j` value (3) resulting in

```
i + j = 53
```

if you instead desired the numerical sum, as in

```
i + j = 8
```

you must use the following:

```
System.out.println ("i + j =" + (i + j));
```

In general, for the sake of clarity, it is good practice to use parentheses whenever a numerical expression appears inside a `print` or `println` parameter.

For floating-point values, Java automatically adjusts the output based on the number of digits in the fractional part of the value. For this code snippet,

```
double = 5.0;
int y = 3.0;
System.out.println ("x * y =" + (x * y));
System.out.println ("x / y =" + (x / y));
```

the output to the console looks like

```
x * y = 15.0
x / y = 1.6666666666666667
```

Note the variation in the number of digits in the fraction. The basic `println` method does not provide a way to specify the formatting of numerical values.

Until J2SE 5.0, Java separated formatting from input and output operations. You first formatted numbers into a string and then printed the string or displayed it graphically. In Section 5.11 we discuss formatting with the new tools in Java 5.0 that provide combined formatting/output capabilities similar to those of the `printf()` function in C.

2.12.2   Floating-point demo

As discussed in Section 2.11, you must deal with several aspects of floating-point representations and operations when doing numerical computations. The following code illustrates some of these floating-point issues:

```
// FP literals are double type by default.
// Append F or f to make float, or cast to float
float x = 5.1f;
float y = 0.0f;
float z = (float) 1.0;

float div_by_zero = x/y;
System.out.println ("Divide By Zero = x/y = " +
                    div_by_zero);

x = -1.0f;
div_by_zero = x/y;
System.out.println ("Divide negative value by zero = x/y = "
                    + div_by_zero);
x = 2.0e-45f;
y = 1.0e-10f;
float positive_underflow = x*y;
System.out.println ("Positive underflow = "
                    + positive_underflow);
x = -2.0e-45f;
y = 1.0e-10f;
float negative_underflow = x*y;
System.out.println ("Negative underflow = "
                    + negative_underflow);
x = 1.0f;
y = negative_underflow;
float div_by_neg_zero = x/y;
System.out.println("Divide 1 by negative zero = " +
                    div_by_neg_zero + "\n");
x = 0.0f;
y = 0.0f;
float div_zero_by_zero = x/y;
System.out.println ("Divide zero by zero = " +
                    div_zero_by_zero);
```

If we insert this code into the `main` method of the `StartApplet1.java` or `StartApp1.java` programs described in Chapter 1, the output looks as follows:

```
Divide By Zero = x/y = Infinity
Divide negative value by zero = x/y = -Infinity
Positive underflow = 0.0
Negative underflow = -0.0
Divide 1 by negative zero = -Infinity
Divide zero by zero = NaN
```

On some platforms, the symbols on the console for Infinity and Not-a-Number differ from these.

2.13   Basic math in Java

The core Java language comes with some basic mathematical tools built into it. In this section we look at simple arithmetic operations and the mathematical functions provided by the `Math` class.

2.13.1   Arithmetic operations

The Java core language includes the simple arithmetic operators:

- + addition
- – subtraction
- * multiplication
- / division
- % modulo (remainder)

The addition operator "a + b" both adds numerical type values and also performs string concatenation as discussed in Section 2.6. This is the only case in Java of operator overloading. The subtraction operator "a - b" subtracts b from a, but the minus sign can also act as the unary minus operator that performs negation on a single number (a = -b).

The "a/b" division operator divides a by b according to these rules:

- If both a and b are integers, the result is an integer with the remainder truncated.
- If either a or b is a floating-point type, the result is floating-point.
- If a and b are integers and b is zero, an exception is thrown (error and exception handling are discussed in Chapter 3).
- If either a or b is a floating-point type and b is zero, the result is
  - +infinity if a is a non-zero positive value
  - -infinity if a is a non-zero negative value
  - NaN if a is also zero.

The "a % b" modulo operator returns the remainder of a divided by b. For example,

```
5 % 3
```

returns a value of 2. The modulo operator also works with floating-point values. If either operand is floating-point, the remainder is a floating-point type.

Note that the Math class (see below) includes the method

```
Math.IEEEremainder (double a, double b)
```

which computes the remainder of a/b for two double type values according to the specific rules of the IEEE 754 standard (see the Math class in the Java 2 API Specifications).

Java does not include an exponentiation operator, such as x**b for x to the power of b. Instead, you must use the Math class method

```
Math.pow (double a, double b)
```

which computes a to the power of b and returns a double type.

2.13.2 Math functions

Beyond the basic arithmetic operations, a number of mathematical functions and constants are available in the core Java language via the Math class. We discuss exactly what a class is in the next chapter so for now just accept that the function name must be preceded by "Math.".

Math is a class that comes as part of the core language. It offers many useful features:

- Constants:

```
Math.PI
Math.E
```

- Trigonometric functions (radian units):

```
double x =.5;
double y = Math.sin (x * Math.PI);
x = Math.asin (y);
```

- Absolute values:

```
int i = Math.abs (j);
```

- Random number generators:

```
double x = Math.random (); // In the range:
                           // 0.0 <= x < 1.0
```

- Other:

```
double y = Math.sqrt (x);
double x = Math.exp (y);
```

See the Web Course Chapter 2: *Tech* section for a table listing all of the `Math` methods with brief descriptions of each. The Java 2 API Specifications (see Resources) provide a detailed description of the `Math` class.

Java version 1.3 included a new class called `StrictMath` that holds the same methods as the `Math` class but must, according to the class specifications, always implement the same algorithms from Sun's "Freely Distributable Math Library" (`fdlibm` in C) and give the exact same results regardless of the platform. This differs from the `Math` class for which JVM designers have more latitude in its implementation. This means that calculations on one platform may give slightly different results with the use of the functions in Math in some cases than those on another platform.

2.14 Web Course materials

The Web Course Chapter 2: *Supplements* section gives more information and examples dealing with the language structures such as repetition statements and flow control statements. It also examines:

- differences in the language elements between Java and C/C++
- the `javap` tool in the JDK that allows one to look at the bytecode in an assembler style format
- the bytecode instruction set

The Chapter 2: *Tech* section gives more details about floating-point in Java, the `Math` class, and casting and mixing among primitive types. The Chapter 2: *Physics* section demonstrates some basic numerical methods with Java such as example programs using Euler and Predictor-Corrector methods for solving first-order differential equations.

Resources

Joseph D. Darcy, *What Everybody Using the Java™ Programming Language Should Know About Floating-Point Arithmetic*, Sun Microsystems, JavaOne Conference, 2002, `http://servlet.java.sun.com/javaone/sf2002/conf/ sessions/display-1079.en.jsp`.

David Flanagan, *Java in a Nutshell*, 4th edn, O'Reilly, 2002.

David Goldberg, *What Every Computer Scientist Should Know About Floating-point Arithmetic*, Computing Surveys, March 1991, `http://docs.sun.com/source/806-3568/ncg_goldberg.html`.

James Gosling, Bill Joy, Guy Steele and Gilad Bracha, *The Java Language Specification*, 2nd edn, Addison-Wesley, 2000. Online version at `http://java.sun.com/docs/books/ jls/second_edition/html/j.title.doc.html`.

Java 2 Platform, Standard Edition, API Specification, `http://java.sun.com/j2se/1.5/api/`.

Ronald Mak, *Java Number Cruncher: The Java Programmer's Guide to Numerical Computing*, Prentice Hall, 2003.

Glen McCluskey, *Some Things You Should Know about Floating-Point Arithmetic*, Java Tech Tips, February 4, 2003,
`http://java.sun.com/developer/JDCTechTips/2003/tt0204.html#2`.

See also Appendices 1 and 2 for tables of language elements and operators. See Appendix 3 for more about floating-point.

Chapter 3
Classes and objects in Java

3.1 Introduction

In Java the "class" is paramount. Essentially all coding resides within class definitions. Here and in the following chapters we develop the concepts and coding for classes and objects in Java.

For those new to object-oriented programming (OOP) the learning curve can be rather steep because several concepts must be understood before the overall picture comes into focus. The Web Course *Supplements* section for Chapter 3 offers additional introductory material to help get these concepts across.

Note that throughout the book we use the terms *object* and *instance* interchangeably.

3.2 Custom data types

In Chapter 2 we discussed Java primitive data types such as `int`, `float`, and `boolean`. Data of a given type means that memory is reserved for a value of that type and that only operations specific to that type can act upon the data. So, for example, a `float` value has 4 bytes of memory allocated for it with the sign, exponent, and significand bits arranged according to the representation discussed in Chapter 2. When an operation such as an addition or multiplication occurs upon a `float` value, the JVM executes floating-point operations that carry out the proper procedures for addition and multiplication with significands and exponents. For integer addition and multiplication the JVM executes a different set of operations unique to integer type data.

In conventional languages, you are stuck with only the data types that come with the language. Languages like C, C++, and even modern versions of Fortran permit the definition of data types called structures, but these are just convenient ways to group related pieces of data together. Data structures have no innate "behavior" associated with them. You cannot "add" two data structures together.

Object-oriented programming, however, lets the programmer define new data types that include data along with operations (or behavior) unique to that custom type. For example, Java does not offer a complex number primitive type. A programmer, however, can create a complex number type with a class definition. The definition would include data – two floating-point values for the real and imaginary parts of the complex number – and operations performed with that data. That is, the addition, multiplication, conjugation, and other operations for complex numbers would be defined by the *methods* of the class definition.

We see that the class definition uses primitives to hold the data and that the operations are written at the source code level; you do not define new machine level (or virtual machine level in the case of Java) instructions for the class. The principle, however, is the same. The class definition creates a new custom data type with its own operational capabilities.

With the primitives you create instances of a type with a declaration such as

```
float x = 5.0f;
```

This allocates 4 bytes of memory to hold the value 5.0 in memory and the data is labeled as belonging to the floating-point type so that only legal `float` operations can act upon it.

Similarly, once a class definition is available, instances of the class can be created. This means that memory is allocated for all the data fields in the definition. (The code for the operations defined in the class is not duplicated with each instance since Java knows how to find the code when needed.) A special method called the *constructor* initializes the data values when the instance is created. Just as you can create multiple variables to hold `float` values, you can create multiple instances of a class such as the complex number class. Such instances of the class are called *objects*.

With object-oriented programming, we go beyond just thinking in terms of data-type representation and operations and instead look at using classes to represent any sort of self-contained entity. In a physics code we might, for example, define a class that represents a particle. The data would include fields for the name, mass, charge, and other qualities that define a particle. The methods would implement the particle's behavior such as its response to electric and gravitational fields and its interactions with other particles.

3.3 Class definition

In Chapter 1 we presented class definitions of very simple classes in the code for `HelloWorld.java` and `HelloWorldApp.java`. With only `init()` and `paint()` methods in the former and the `main()` method in the latter, the capabilities of those classes were very limited. A class typically contains:

- **Data fields** – Declare the data values belonging to the class or object.
- **Methods** – Functions to carry out tasks for the objects.

- **Constructor** – A special kind of method called when an object is created. Its job is to carry out initialization tasks.

The class `GenericClass` below illustrates these features:

```
public class GenericClass
{
   int i;                          This field declares the
                                   property i as an integer
                                   type. (By default its
                                   value will be 0.)

   public GenericClass (int j) {   A constructor is called
      i = j;                       when an instance of this
   }                               class is first created.
                                   It can be used to
                                   initialize properties.

   public void set (int j) (       This method assigns a
                                   value to the field i.

      i = j;
   }

   public int get () {             This method returns the
                                   value of i.

      return i;
   }
}
```

We can now create an instance of our new data type and invoke its methods:

```
void aMethodSomewhere () {
   // Create an instance of this data type.
   GenericClass g = new GenericClass (5);
   int k = g.get ();
   . . .
```

In the following pages we discuss the main features of a class definition, beginning with data fields.

3.3.1   Data fields

A class definition typically includes one or more fields that declare data of various types. Data fields are also called "member variables." For example, this code shows a class with a single primitive `int` data value:

```
public class GenericClass
{
    int i = 12;                    Field with a declaration of
                                   an integer data variable.


    public int get () {            A method to obtain the value
                                   of the i variable.

      return i;
    }
}
```

Methods (discussed later) can access the data in the fields. Here, for example, the get() method returns the value in the i field.

When a data field declaration does not assign an explicit value to the data, default values are assigned:

- int, byte, short, char – default value   0
- float, double – default value   0.0
- boolean – default value   false

In the following example, we let one variable take a default value and set explicit values for two of the variables. Setting an explicit value, even if it is the same as the default, can be a good practice just to confirm that every field has the initial value that you intended for it.

```
public class GenericClass       Fields can use either the
{                               default values or explicit
    int i;                      initialization.
    double d = 1.3;             Here i will hold a 0 value.
    boolean b = true;


    . . .
}
```

The fields can reside anywhere in the class definition (outside of methods) but putting them all at the top (before the methods) is a popular coding style that we follow.

3.3.2 Methods

A class definition typically includes one or more *methods* that carry out some action with the data and may or may not return a value. The following code shows a class with two methods – get() and triple() – that return the value of an integer datum and one method – set() – that does not return a value (void):

```
public class GenericClass
{
   int i;                          Field with a declaration of an integer
                                   data variable.

   public int get () {             A method to obtain the value of the i
      return i;                    variable.
   }

   public void set (int j)         A method to set the value of the i
      i = j;                       variable. Parameter defines type for
   }                               value passed.

   public int triple (int j) {     A method with a local double variable f. This
      double f = 3.0;              local variable is valid only within the
      i = f * j;                   method. It must be assigned a value before
      return i;                    it is used.
   }
}
```

Methods inside a class can access and modify the data in the class fields. Omitting a few aspects that we discuss later, the structure of a method is:

```
access modifier return type method name (list of parameters)
{
   statements, including local variable declarations
}
```

where:

- *access modifier* – determines what other classes and subclasses can invoke this method. The access modifier may be omitted, in which case the default access rules apply. We discuss access modifiers in Chapter 5.
- *return type* – what primitive or class type value will return from the invocation of the method. In the above get() method, for example, the return type is int. The return type may not be omitted. If there is no value returned, use void for the return type as in the set() method above.
- *method name* – follows the same identifier rules as for data names. Customarily, a method name begins with a lowercase letter.
- *list of method parameters* – the parameters passed to the method. Listed with type and name as in the set (int j) method in the code above.
- *local variables* – data variables can be declared and used within the method. Local variables must be assigned a value before they are used. The variables are discarded when the process returns from the method.
- *statements* – the code to carry out the task for the particular method.

So far we have discussed three locations where a method can store and access data:

- member variables – the data defined in the data fields of the class definition
- local variables – data declared within a method and only valid there
- method parameters – data passed in the method parameter list

For example, the above `triple()` method includes all three types of data: a member variable, local variables, and the parameter variables. Note that you can also access data in another object if the access rule for that data allows it.

3.3.3 Method overloading

Perhaps you create a class that holds a method with an integer parameter:

```
void aMethod (int k) {. . .}
```

You decide later that you need a method that accomplishes essentially the same task but requires a `float` parameter. In some procedural languages you would create a new method with a slightly different name:

```
void aMethod_f (float x) {. . .}
```

A more elegant solution would allow you to use the exact *same* name for the new method and have the compiler determine from the parameter type which method to use:

```
void aMethod (int k) {. . .}
void aMethod (float x) {. . .}
```

This very valuable feature is called *overloading*, which Java permits for methods as well as constructors. Another example of overloading is when the number of parameters changes. For example, there can be yet another method named `aMethod()` that takes two `int` parameters instead of just one:

```
void aMethod (int k, int q) {. . .}
```

Constructors with different numbers of parameters are common in the standard Java class libraries in which none, one, two, or more class properties can be initialized using the same constructor name but with different parameter lists.

Another common example is the `println()` method that we've already used to print messages to the Java console:

```
System.out.println (String)
```

This is actually just one of several `println()` methods in the `PrintStream` class (we discuss Java I/O and stream classes in chapter 9). The variable named `System.out` references an instance of the `PrintStream` class. You can take advantage of the other overloaded versions of `println()`, which include:

```
println () <- prints out just a line separator
println (boolean)
println (char)
println (char[])
println (double)
println (float)
println (int)
println (long)
println (java.lang.Object) - invokes the toString() method
        of the object
```

In general, any change to the type or number of parameters is legal overloading. Note, however, that a change in the return type alone does *not* produce an overloaded method. The compiler does not permit the following:

```
void aMethod (int k) {. . .}
int aMethod (int k) {. . .}
```

because only the return type was changed. Neither does changing the parameter names produce an overloaded method. For example,

```
void aMethod (int x, int y, int z) {. . .}
void aMethod (int i, int j, int k) {. . .}
```

is not legal. Only the parameter types are examined, not the names.

3.3.4 Constructors

The `new` operator creates an instance of a class as in

```
. . .
int i = 4;
Test test = new Test (i);
. . .
```

The statement declares a variable named `test` of the `Test` type and creates an instance of the `Test` class with the `new` operator. The argument of `new` must correspond to a special method in the class called a constructor.

The constructor looks much like a regular method in that it has an access modifier and name and holds a list of parameters in parentheses. However, a constructor has no return type. Instead, an instance of the class type itself is returned. Except for some special situations discussed later, the only way to invoke a constructor is with the `new` operator. The constructor's name must exactly match the class name, including case.

Constructors are useful for initializing variables and invoking any methods needed for initialization. The code here shows a simple class with a constructor and one method.

```
class Test
{
      int i;

      Test (int j) {          A constructor is called when an
          i = j;              instance of this class is first
      }                       created. Here it is used to
      int get () {            initialize a property variable.
          return i;
      }
}
```

The above code for the constructor,

```
Test (int j) {
    i = j;
}
```

shows that in the process of creating an instance of the class, an initial value for the member variable `i` is passed as a parameter to the constructor.

Java does not actually require an explicit constructor in the class description. If you do not include a constructor, the Java compiler creates a default constructor in the bytecode with an empty parameter list. The default constructor for a class `Test` with no constructor is equivalent to explicitly writing

```
Test () {/* do nothing */}
```

In the discussion of data fields, we noted that the data can receive explicit initial values or default values. You might wonder when this initialization actually occurs. The `javac` compiler, in fact, inserts the initialization of the data into the bytecode for the constructor. So, for instance, if the `Test` class had no explicit constructor, the bytecode would be equivalent to that shown below where a constructor explicitly sets the `int` variable to 0:

```
class Test
{
    int i;

    Test () {                  This constructor illustrates
       i = 0;                  explicitly the initialization of
    }                          property values to their default
                               values as would occur if we had used
                               no constructor or included Test() {}


    int get () {
       return i;
    }
}
```

As with methods, you can define multiple overloaded constructors to provide optional ways to create and initialize instances of the class. (We discuss overloading of constructors and methods in more detail in Chapter 4.)

3.4 Class instantiation

Let's use the following class for our explanation of instantiation:

```
class Test
{
    int i;
    double x;

    Test (int j, double y) {
       i = j;
       double x = y;
    }

    int getInt () {
       return i;
    }

    double getDouble () {
       return x;
    }
       double calculate () {
          return i*x;
       }
    }
}
```

The class itself is a somewhat abstract concept. As explained so far, it has little value until an instance of the class is created. When this class is instantiated with the `new` operator, such as in

```
Test g1 = new Test (4, 5.1);
```

then the variable `g1` holds a pointer to an "instance" of the class `Test`. In Java a pointer to an object is called a *reference*. As laid out in memory, `g1` is just a 32-bit value that tells the JVM where to find the data for that class instance, whereas that data itself is considerably larger than 32 bits. Thus, `g1` is (a reference to) an object or an instance of the class `Test`. You can't do much with the class itself but you can use the reference `g1`. (More about references in the following sections.)

During the `new` operation, the JVM allocates memory for, among other things, the data fields `i` and `x` of the class. This data is stored, along with other aspects of the object, somewhere in memory. You can imagine that the JVM creates and keeps track of a unique ID just for that instance of the class.

If we create another instance of the class

```
Test g2 = new Test (53, 34.3);
```

then another set of data is stored under a different unique ID. So there will be two blocks of memory, one for each instance. One block of memory contains the values 4 and 5.1 for the `i` and `x` variables, and the other block contains the values 53 and 34.3, respectively. Since the JVM keeps track of the unique IDs, the JVM always knows which block of memory to look in to find the correct values for a particular instance of `Test`.

When a program invokes the methods of an object, the JVM loads the unique data for that object into the fields, and these values are used in the code for the methods of that class. When it invokes the same methods for a different instance of the same class, then that object's data is used in the code. We often refer to objects in rather abstract or pictorial metaphors as if both the data and methods were contained within each object. However, at the processor level it just comes down to sets of data, unique to each object, shifting in and out of the method codes.

3.4.1 Object references

Java references differ considerably from C and C++ memory pointers where the programmer can access and manipulate pointers directly.

Pointers in C and C++:

- hold the actual addresses of data in memory
- can be cast to different data types
- can be altered to point to other memory locations

A Java reference holds an indirect address of an object in the JVM and:

- the actual memory value of the reference is hidden
- reference values cannot be altered
- references can only be recast to a superclass or subclass of that object, never to other data types (see Chapter 4 for a discussion of superclasses and subclasses)

So references in Java act in some ways like memory pointers but are much safer with regard to creating program errors and security problems.

3.4.2 Accessing methods and fields

To create a useful program in Java, an object obviously needs to invoke methods and to access data in other objects. This is done with a reference and the " . " or *dot* operator. For example, we invoke the `calculate()` method for an instance of the `Test` class above as follows:

```
Test g1 = new Test (4, 5.1);
double z = g1.calculate ();
```

The invocation of `calculate()` with the `g1` reference will cause the data for the `g1` object to be used with the bytecodes for that method.

You can also directly access the data fields as in

```
double y = g1.x;
int j = g1.i;
```

In some situations you may not want one class to have access to certain data (e.g. a password value) and methods. We will discuss access settings in Chapter 5 that allow you to determine what classes can access the fields and methods of a class.

3.5 Static (or class) members

The `Test` class above has essentially no value outside of instantiations of the class. In the course of program development a need often arises for utility methods that might be needed for all instances of the class or for constants that are useful to all instances of the class. Since object creation uses up memory resources it is desirable to have access to such common resources without requiring them to appear in each instance of the class.

However, we have seen that Java does everything within a class framework. There are no global variables, for example, as in C/C++. Instead, Java has the concept of *class* data and *class* methods that are defined in a class definition and apply to all instances of the class. In fact, class data and class methods can be accessed even without creating any instance of the class. Class data and methods are also referred to as "static" members since the keyword `static` is used to identify such members.

Static data are created when the class is loaded, even before any instantiation of the class, and exists in just one place so that no matter how many instances of that class are created, there remains only one copy of the static data. Instance data, on the other hand, belongs solely to an instance of the class and new memory is allocated for the data for each object.

Static values are declared with the `static` keyword. For example, in this class the variable `pi` is declared `static`:

```
public class ConstHolder
{
    public static double pi = 3.14;
}
```

When the JVM loads the byte code for the class description of `ConstHolder`, it creates a single memory location for the `pi` variable and loads it with the value 3.14. All instances of the `ConstHolder` class can access the exact same value of `pi`. The `pi` data exists and we can access it even when no instance of `ConstHolder` is created. Since `pi` is a member variable of the class, though a special one because it is static, it can be accessed just like any other member variable. For example, if `g1` is an instance of `Test`, then we can access the variable `pi` as follows:

```
ConstHolder c = new ConstHolder();
double twopi = 2.0 * c.pi;
```

If there is no instance of `ConstHolder`, we can access the data directly using the name of the class:

```
double twopi = 2.0 * ConstHolder.pi;
```

In practice, it is a good habit to always use the latter syntax, even when a class instance exists, since it makes clear to the reader that the referenced variable is a static member variable of the `Test` class.

In addition to static member variables, we can also define static methods, producing methods that can be called without instantiating the class. For example,

```
public class ConstHolder {
    public static double pi = 3.14;

    public static double getPI () {
        return pi;
    }
}
```

We could, in this case, use the `getPI()` method to obtain the value of `pi`:

```
double x = 2.0 * ConstHolder.getPI ();
```

A static variable or method is also called a class variable or method, since it belongs to the class itself rather than to an instance of that class. We've already seen examples of static methods in the Math class. In fact, Math is a class of nothing but static member variables and static methods. The Math class itself cannot even be instantiated. The only way to access the various constants and static methods in the Math class is with the Math.constant and Math.method() syntax.

If a class property is also declared final, then it becomes a constant value that cannot be altered. In the following example, the static data field PI is now a fixed constant. By convention, constants are normally written with all uppercase letters.

```
class ConstHolder {
   public final static double PI = 3.14;
}
```

3.6   More about primitive and reference variables

Primitive data type operations deal only with value. For example, the following code shows that assigning primitive data variable i to another primitive named j simply passes a copy of the value in i into j. That is, a datum's value is passed, not a reference or pointer to a data location.

```
int i = 1; // Variable i holds the value 1.
int j = i; // Now j holds 1 also, that is, i's value is
           // copied to j

   i = 2; // Now i is 2, but j still holds the value 1
```

Similarly, in method parameters, primitive variables are passed by value. That is, a copy is made and passed to the method. Changes to the passed value inside the method cannot affect the value in the calling code.

The following snippet creates an instance of AClass and then invokes the change() method with an int parameter.

```
. . . a method in some class . . .
int i = 2;
AClass a1 = new AClass ();
a1.change (i);
int m = i; // m = 2, no matter what happens inside
           // a1.change()
System.out.println ("i = " + i);
. . .
```

The `AClass` definition is shown below. The `change()` method might seem to change the value of the variable passed into it.

```java
class AClass {
    void change (int k)
    {
        System.out.println ("k = " + k);
        k = 5;
        System.out.println ("k = " + k);
    }
}
```

However, only the *value* of variable `i` in the calling code is passed to variable `k` in the `change()` parameter list. When that method assigns a value of 5 to `k`, it has no effect whatsoever on the original variable `i` in the calling code, as the output of this program demonstrates:

```
k = 2
k = 5
i = 2
```

Dealing with references and the data they point to is a bit more complicated than primitive variables and so we discuss them in the following two sections.

3.6.1 Modifying reference variables

You can always assign a new object to a reference variable. For example, `AClass` is defined as follows:

```java
class AClass
{
    int j = 1;
    void aMethod (int k) {
        int i = 10 * k;
        k = 5;
        j = k * 10;
    }
}
```

In the following snippet, we first create two instances of `AClass` and reference them with `ac` and `bc`:

```
. . . a method in some class . . .
AClass ac, bc;
ac = new AClass ();
bc = new AClass ();

ac.j = 4;    // Set j = 4 in the ac instance
bc.j = 40;   // Set j = 40 in the bc instance

ac = bc;     // Make ac a reference to the
             // same object that bc references.
bc.j = 400;  // Reset j in bc to 400.

int m = ac.j; // m now holds 400 since ac references the
              // same object as bc and bc's j value has
              // been changed to 400.
. . .
```

The code first assigns values to the j variable in each object. Next we assign ac to the same object that bc references with the ac = bc statement. So both the ac and bc variables now reference the same object. Therefore, when we set bc.j to the value 400 and then obtain the value of ac.j we get back 400 since ac and bc both reference the same object in memory.

The object that was referenced by the original value of ac is no longer available. The memory allocated for that object most likely still exists somewhere within the JVM's data buffers, but our code no longer has any way to access that object. By assigning a new object to the ac variable we effectively "orphaned" the original object that ac pointed to previously. When an object is no longer referenced by any variables, the Java garbage collector eventually reclaims the memory allocated for that object. See the Web Course Chapter 3: *Supplements* section and also Chapter 24 for discussions of garbage collection in Java.

3.6.2 Object references in method parameters

The argument, or parameter, list of a method can include object references. The methods and data of that object can be accessed and modified by the method. However, the reference itself passes by value. That is, the JVM makes a copy of the internal memory reference and that copy goes into the parameter value. If inside the method you set the reference variable to a new object, this will not affect the reference variable in the calling method just like you cannot change primitive values passed from calling methods. On the other hand, if you change the values of the variables "inside" an object, then those changes apply to the object in the calling method since both the calling method's reference variable and the called method's reference variable refer to the exact same object.

For example, we define two classes, `AClass` and `BClass`:

```
class AClass {
    void aMethod (BClass bb) {
       bb.j = 20;
    }
    void anotherMethod (BClass bb) {
       bb = new BClass ();
       bb.j = 100;
    }
}

class BClass {
    int j = 0;
}
```

In the following snippet, a reference to a `BClass` object is passed to `aMethod()` of an `AClass` object. In that method, the value of `bb.j` is changed to 20. Since `bb` inside `aMethod()` refers to the same object as does `b` in the calling code, then `b.j` in the calling code takes on the value 20, losing the value 5 it originally had.

```
    . . .
int i = 2;
AClass a;
BClass b;

a = new AClass ();
b = new BClass ();
b.j = 5;
a.aMethod (b);
i = b.j; // i now holds 20 not 5


    . . .
```

Alternatively, look what happens when we call `anotherMethod()` in which the local variable that originally held the passed reference to `BClass` is reassigned to a brand new instance of `BClass`:

```
    . . .
AClass a;
BClass b;

a = new AClass ();
b = new BClass ();
```

```
b.j = 5;
a.anotherMethod (b);
int i = b.j; // i still holds 5, not 100

. . .
```

The j member of that new BClass is assigned the value 100. Back in the calling code, the original b.j still holds the original value 5. The reason is that the
assignment of 100 to bb.j applied to a completely new instance of BClass,
and, since the original reference was passed by value, that reference was not
changed. The calling code's b variable still references the original BClass
instance.

Comparison of reference variables only tests if the two variables refer to the
same object. That is, the test

```
if (a == b) statement;
```

simply checks whether a and b refer to the same object, not whether the referenced
objects have equal data values.

If you need to test whether two objects hold the same data values, many classes
provide an equals() method like this:

```
a.equals (b)
```

It returns true if the data in object b matches that in object a. Comparing
all the data inside an object for equality with an equals() method is called
a "deep" comparison, whereas the "==" comparison is "shallow." In fact, all
classes implicitly include a shallow equals() method inherited from the top-
level class known as java.lang.Object (we discuss inheritance in Chapter 4).
However, unless the author of a class has explicitly written the equals() method
to do a deep comparison, the results may not be as expected. All classes in the
core Java libraries can be expected to have a properly written equals() method.
However, no such guarantees exist for third-party classes or classes you write
yourself unless you are careful to write your equals() method correctly.

3.7 Wrappers

We noted earlier that Java primitive types are not class objects. The language
designers decided that the higher processing speed and memory efficiency of
simple, non-class structures for such heavily used data types overruled the elegance of a purely object only language. They decided instead that for each primitive type there would be a corresponding *wrapper* class that provides various
useful tools such as methods for converting a given numerical type to a string or
a string to a number.

Table 3.1 *Primitives and their wrapper classes.*

Primitive	Wrapper
`boolean`	`java.lang.Boolean`
`byte`	`java.lang.Byte`
`char`	`java.lang.Character`
`short`	`java.lang.Short`
`int`	`java.lang.Integer`
`long`	`java.lang.Long`
`float`	`java.lang.Float`
`double`	`java.lang.Double`

Table 3.1 lists the primitive types and the corresponding wrapper classes. See the Java 2 Platform API Specification for detailed listings and descriptions of the methods and constants provided with each of the wrapper classes.

With the exception of `java.lang.Character` and `java.lang.Integer`, the wrappers have the exact same name as the corresponding primitive type but with the first letter capitalized. The wrappers are normal classes that inherit from the `Object` superclass like all Java classes (see Chapter 4).

The wrapper constructors create class objects from the primitive types. For example, for a double floating-point number `d`:

```
double d = 5.0;
Double wrapper_double = new Double (d);
```

Here a `Double` wrapper object is created by passing the `double` value to the `Double` constructor.

In turn, the wrapper provides a method to return the primitive value

```
double r = wrapper_double.doubleValue ();
```

Each wrapper class has a similar method to access the corresponding primitive value: `intValue()` for `Integer`, `booleanValue()` for `Boolean`, etc. There are additional convenience methods to return other types to which the wrapped primitive could be converted. For example, `Integer.longValue()` returns the value of the `Integer` as a `long` type, `Integer.byteValue()` returns the value as a `byte` type, etc. Since nothing casts to or from a `boolean` primitive type, the `Boolean` wrapper has only the `booleanValue()` method.

3.7.1 Strings, wrappers and primitives

The wrappers for primitive types also provide a number of useful static methods including some to convert numbers to strings and vice versa. These are very useful when reading textual input that needs to be converted to a primitive type. A common situation where these come in handy involves passing numbers to applets via the applet tag parameters. An applet hypertext tag includes the `param` sub-tag:

```
<applet . . .>
   <param name = "string1" value = "string2" >
</applet>
```

The `name` parameter provides an identifier for the string passed in the `value` parameter to the applet program. For example, we could pass two numbers as follows:

```
<applet code = "MyApplet" width = "100" height = "50">
   <param name = "fpNumber" value ="12.45">
   <param name = "intNumber" value = "10">
</applet>
```

To obtain the parameter values the Applet class provides this method:

```
String getParameter (String paramName);
```

The parameter returns as a string from this method so we need to convert it to a numerical primitive type value. The wrapper classes provide tools for this in the form of static methods. (Since they are static, they can be called without having an instance of the wrapper class.) This is illustrated in the following example code:

```
public void init () {

    string fpStr = getParameter ("fpNumber");
    double fpNum = Double.parseDouble (fpStr);

    String intStr = getParameter ("intNumber");
    int intNum = Integer.parseInt (intStr);
    . . . .
```

Here the `getParameter (String)` method returns the string value for the fpNumber parameter, and the static method `Double.parseDouble (String)` from the `Double` wrapper class converts it to a `double` value. Similarly, we get the integer parameter using the `parseInt (String)` static method from the `Integer` class.

The `parseInt()` method has been available since Java 1.0, but `parseDouble()` only appeared with Java 1.2. Previously, the `valueOf()` method was used to return a `Double` value, which in turn could provide the `double` primitive value using the `doubleValue()` method:

```
double fpNum = Double.valueOf(fpStr).doubleValue();
```

Note that this code demonstrates the common Java technique of executing several commands in a single line. The line executes from left to right. First, the `valueOf()` method returns a `Double` object, which then has its `doubleValue()` method called.

We saw here how to convert a string representation of a number to a primitive type value. Going in the other direction, you can convert a primitive type to a string in several ways. The `String` class provides several overloaded static `valueOf()` methods as well as the overloaded "+" operator. For example, in the following code we first convert numerical values to strings using the `String` class's `valueOf()` methods (there is one for each primitive type) and then using the "+" operator:

```
double d = 5.0;
int i = 1;
String dStr = String.valueOf (d);
String iStr = String.valueOf (i);

String aStr = "d = " + dStr;
String bStr = "i = " + iStr;
```

Now the `dStr` and `iStr` variables reference the strings "5.0" and "1", respectively, while `aStr` references "d = 5.0" and `bStr` references "i = 1". We discuss more about strings and string helper classes in Chapter 10.

3.7.2 Autoboxing and unboxing

In all versions of Java prior to J2SE 5.0, conversions between wrapper classes and the corresponding primitive types (and vice versa) are somewhat messy, as seen above. As another example, creating a `Float` object from a `float` primitive is straightforward:

```
float primitive_float = 3.0f;
Float wrapper_float = new Float (primitive_float);
```

Going in the other direction, however, is not quite as simple. It requires explicitly calling the `floatValue()` method on the `Float` object:

```
float primitive_float = wrapper_float.floatValue ();
```

In J2SE 5.0, the code to create the wrapper object can be simplified to

```
Float wrapper_float = primitive_float;
```

Here, the "wrapping" is done automatically! There is no need to explicitly call the `Float` constructor. This "wrapping" is called "autoboxing" in the sense that the primitive value is automatically "boxed up" into the wrapper object. Autoboxing is available for all the primitive/wrapper types.

Going the other way, from object type to primitive, is just as simple:

```
Integer wrapper_integer = 5; // primitive 5 autoboxed into
                    // an Integer

int primitive_int = wrapper_integer; // automatic unboxing
                    // Integer into int
```

These shortcuts simplify coding and reduce errors in J2SE 5.0, but you might not be too impressed since the simplification is only minor. Note, though, that autoboxing and unboxing can be used just about anywhere. They can even be used in loop control and incrementing and decrementing operations. For a contrived example, consider adding all the integers from 1 to 100:

```
public class Box {
    public static void main (String[] args) {
        int MAX = 100;         // a primitive int type
        Integer counter = 1;  // an Integer type
        Integer sum = 0;      // ditto
        while (true) {
            sum += counter;
            if (counter == MAX) break;
            counter++;
        }
        System.out.println ("counter is now " + counter);
        System.out.println ("sum is now " + sum);
        System.out.println ("MAX*(MAX+1)/2 is " +
                            MAX*(MAX+1)/2);
    }
} // class Box
```

There is a lot of hidden autoboxing and unboxing going on in this simple-looking code. First, the `Integer` types `counter` and `sum` are autoboxed from the primitive values `1` and `0`. Then, in the loop, they are unboxed to primitive values so the `+=` operation can be applied and then reboxed to their "native" `Integer` types. To do the `==` comparison `counter` is unboxed so it can be compared with the `int` type `MAX`. If the `break` does not apply, then `counter` is unboxed, operated on with `++`, and then reboxed.

Autoboxing and unboxing work in a `for` loop as well:

```
Integer sum = 0;
for (Integer counter=1; counter < MAX; counter++) {
    sum += counter;
}
```

Note that both of these loops are likely to perform abysmally because of all the autoboxing and unboxing operations that must occur. Even though the conversions do not appear explicitly in the source code, they still must be done. An optimizing compiler might be able to avoid some of the autoboxing and unboxing operations, but in general looping operations should be done with primitive types unless there is a *very* good reason to use a wrapper type.

Where autoboxing and unboxing *should* be used is whenever an explicit conversion to or from a wrapper type would be required if you were using J2SE 1.4. In 5.0 and above, just write the code "naturally" and let the compiler handle the conversions automatically. But don't go out of your way to demonstrate the automatic conversion feature as we did in the examples above.

Autoboxing and unboxing also work with `Boolean` and `boolean` types. For example,

```
boolean one = true;  // nothing new here
Boolean two = true;  // autoboxing of boolean literal "true"
                     // to Boolean type
if (one && two)      // auto unboxing
    do_something ();
```

Before 5.0, the `if`, `while`, and `do-while` statements (see Chapter 2) all expected `boolean` expressions. Through the use of unboxing, those flow control statements now also accept expressions that evaluate to `Boolean` types. Similarly, the old `switch` statement expects a `byte`, `short`, `int`, or `char` type in Java 1.4 and below. With the addition of autoboxing in 5.0, `switch` now also accepts `Byte`, `Short`, `Integer`, and `Character` types.

Where autoboxing and unboxing become particularly useful is with the insertion and retrieval of primitive values into and out of object containers like `Vector` and `ArrayList` (see Chapter 10). Since the container classes only accept objects, not primitives, prior to J2SE 5.0 it was necessary to convert primitives to wrapper object types for insertion and to convert wrapper objects back to primitives after retrieval. Autoboxing and unboxing now make these operations almost transparent. With the additional new "generics" feature in 5.0, using primitives with container objects is even simpler. We postpone that discussion until we discuss the generics feature in Chapter 10.

3.7.3   Autoboxing and overloading

Autoboxing and unboxing can make method overloading interesting. Consider the two overloaded methods shown here:

```
long method1 (long l) {return l+1;}
long method1 (Integer i) {return i+2;}
```

If you call `method1()` with a primitive `long` parameter, then the first `method1()` is used. If you call `method1()` with an `Integer` object parameter, then the second `method1()` is used. There is nothing new there. But what happens if you call `method1()` with an `int` parameter? In J2SE 1.4 and below, the `int` is promoted to a `long` and the first `method1()` is used. With autoboxing, it is conceivable that the `int` could be boxed into an `Integer` type and the second `method1()` used. That might even be what you want to happen – it might make more sense to convert an `int` to an `Integer` than to promote it to a `long`. While arguably reasonable, that is *not* what happens. The general rule is, for compatibility reasons, the same behavior that applied in pre-5.0 versions must continue to hold. The reason is that existing code cannot suddenly start behaving differently when compiled and run under 5.0.

We would suggest that an even better rule is to realize that using overloads like this is confusing and potentially asking for trouble. There is little reason to write such obfuscated code.

3.8 Arrays

An array provides an ordered, sequential collection of elements. These elements consist of either primitive values or object references. Here we concentrate on primitive array types. We discuss arrays in more detail in the next chapter. However, arrays are very useful for the demonstration programs and exercises, so we learn enough with this brief introduction to implement arrays of primitive type values.

You can declare a one-dimensional array of primitive type values in two ways:

```
int iArray[];
float[] fArray;
```

You can put the brackets after the array name as shown in the first line or after the type declaration as shown in the second line. As a matter of style, the second method is preferred; you are creating a `float` array, so the array symbols go with the `float` keyword.

You create an array of a given size and with default values for the elements using the `new` operator and putting the array size in brackets. For example,

```
int[] iArray = new int[10];
```

creates an `int` array that is ten elements long. Here are some other examples:

```
long[] lArray = new long[20];
int n = 15;
short[] sArray = new short[n];
```

For arrays of primitive type values, the declaration creates the array object (Java arrays are objects) and allocates memory for each primitive element and sets each element to a default value for that type. For numeric types, the default values equal zero (0 for integers, 0.0 for floating-point). For `boolean` arrays the default value for each element is `false`. For `char` the default is the Unicode value " \ u0000" (all bits equal 0).

The size of the array in the declaration must be an `int` integer, so arrays are limited to the maximum value of an `int`, which equals 2 147 483 647.

You can also create and initialize an array to particular values in the declaration by putting the elements between braces and separated by commas, as in

```
int[] iArray = {1, 3, 5, 6};
long[] lArray = {10, 5, 3};
double[] dArray = {1.0, 343.33, -13.1};
```

Here the compiler counts the number of elements provided and automatically sizes the array to that number. Array elements are accessed with an `int` value inside brackets with a value between 0 and n-1, where n equals the size of the array. For example,

```
int ii = iArray[3] * 5;
double d = dArray[0] / 3.2;
```

You can find the size of an array via the `length` property:

```
int arraySize = iArray.length;
```

Note again that the above declarations work only for arrays of primitive type values. For arrays of objects, an object must be created separately for each element in the array. In other words, the array is really just a list of references to other objects that must be created in a separate step. We discuss arrays of objects in the next chapter.

3.9 Exceptions

If a Java program attempts to carry out an illegal operation, the JVM does not necessarily halt processing at that point. In most cases, the JVM allows for the possibility of detecting the problem and recovering from it. To emphasize that such problems are not fatal, the term exception is used rather than error.

For example, what if by accident we put a non-numeric string into an applet tag parameter and it is passed to the wrapper method `Integer.parseInt()` as shown below:

```
class MyApplet extends Applet
{
    . . .
  init () {

      // If the parameter returned is not
      // numeric, there will be a problem:
      int data = Integer.parseInt (getParameter ("intNumber"));

      aMethod (data);
  }
}
```

If `getParameter()` returns, say, the non-numeric string "abc", then the `Integer.parseInt()` method cannot possibly parse the string into a numeric value. Obviously this would cause a serious problem. Java exception handling provides a systematic way for the programmer to respond to such errors and to decide on an appropriate response rather than simply letting the program fail, perhaps ungracefully. Without an exception-handling system built into the language, you would have to write your own routine to test the string for numeric numbers. For example, in the code below we create a special method to test if a `String` holds a valid integer value before the string goes to the `parseInt` method:

```
. . .
String str = getParameter ("dat");
if (testIfInt (str))
   Integer.parseInt (str);
else {
   ErrorFlag = MY_ERROR_BAD_FORMAT;
   return −1;
}

boolean testIfInt (String str) {
   . . . messy code to test characters for numbers. .
}
. . .
```

Thankfully, we can instead use Java exception handling rather than having to write custom test code for all possibilities, or worse, letting a program fail ungracefully. In Java exception handling, an exception is said to be "thrown" when the JVM detects that something is awry. In the example here, the attempt to parse a non-numeric string throws a particular type of exception called a `NumberFormatException`.

Whenever an exception is possible, our code can "catch" the exception, should one occur, using the `try-catch` syntax as follows:

```
try {
   code that can throw an exception
}
catch (Exception e) {
   code to handle the exception
}
```

In the following code segment we surround the `parseInt()` method invocation with a `try-catch` pair:

```
try {
   int data = Integer.parseInt (getParameter ("dat"));
   aFunctionSetup (data);
   . . .
}
catch (NumberFormatException e) {
   data = -1;
}
```

Here, the `Integer` class method `parseInt()` throws an instance of the `NumberFormatException` class if the string passed does not represent a valid integer number. The program processing jumps from the line that causes the exception (the `parseInt()` method call) to the code in the `catch` block. All the code from the line that caused the exception to the end of the `try` block is permanently skipped.

The `parseInt()` method in `Integer` is written something like the following:

```
public static int parseInt (String s)
   throws NumberFormatException {
   . . .

   . . . code to check if the string is a number
   . . . if it isn't then:
   throw new NumberFormatException ("some error message");
   . . .

}
```

The `throws NumberFormatException` phrase in the method signature indicates that the method includes a `throw` statement of that type somewhere in the method code.

We see that the `throw` statement actually creates an instance of the exception (like everything else in Java, exceptions are class types) and causes the routine to return with the exception thrown. The constructors for an exception may include parameters with which you can pass useful information about the circumstances of the exception. The catch code can then examine this information using methods on the `Exception` class.

Java divides exceptions into two categories:

• general exceptions
• run-time exceptions

General exceptions *must* be handled in source code. They are also called "checked" exceptions, meaning they must be checked for. For any method that can throw a general exception, you must either place a `try-catch` around the invocation of that method or arrange to have the exception propagate up to the calling method that invoked your method. If you want the exception to propagate, then you must first declare that your method might throw the exception. This declaration is made with the `throws` clause as the following code shows:

```
public void myMethod () throws NumberFormatException {
    . . .
    Integer.parseInt (str)
    . . .
}
```

Here the `myMethod()` invokes the `parseInt()` method and, since it does not use the `try-catch` to handle the exception, the method declaration includes `throws NumberFormatException` in the method signature. You need do nothing special when you call `Integer.parseInt()`. The JVM automatically causes the exception to be propagated should one occur anywhere in your method. Then the caller of your method must handle the exception.

In summary, whenever you call any method that might throw a checked exception, you must either put that method call into a `try-catch` block and handle the exception yourself or declare that your method could throw the same exception. (You can also declare that your method throws a superclass of the exception; we discuss super- and subclasses in the next chapter.) If you do not use one of these techniques to handle checked exceptions, the `javac` compiler returns errors during compilation.

Run-time exceptions, unlike checked exceptions, do not have to be explicitly handled in the source code. This avoids requiring that a `try-catch` be placed around every integer divide operation, for example, to catch a possible divide by zero or around every array variable to avoid indices going out of bounds. However, you can handle possible run-time exceptions with `try-catch` if you think there is a reasonable chance of one occurring.

(We discuss class inheritance in the next chapter but we note here that all exceptions are instances of either the `Exception` class or its many subclasses.)

Note that you can use multiple `catch` clauses if the code can throw different types of exceptions. For example, in this code we provide for two possible `Exception` subclasses:

```
. . .
try {
    . . . code . . .
    } catch (NumberFormatException e) {
    . . .
    } catch (IOException e) {
```

```
    . . .
} catch (Exception e) {
    . . .
} finally {// optional
    . . . this code always executed even if
        no exception occurs . . .
}
. . .
```

The code bracketed by the `try-catch` may throw the specific exceptions `NumberFormatException` or `IOException`, which are caught in the corresponding `catch` sections. Any other type of exception thrown will be caught by the third `catch` since it catches `Exception`, which is the root of all exception types. The `finally` keyword indicates that, regardless of whether an exception is thrown or not, the code block following `finally` is always executed. Sometimes it is useful to use a `try-finally` block in which no exceptions appear but one wants to be certain that the code in the `finally` block always executes.

3.10   OOP in engineering and science

The same reasons that object-oriented programming benefits general programming also apply to science and engineering applications. These benefits include:

- enhanced reusability of the code
- modularity makes the code structure easier to understand and maintain
- encapsulation helps to reduce the breaking of other codes when a change is made
- enhancements of the code via inheritance (see next chapter) are made in a systematic manner

OOP can also provide additional benefits to science and engineering. For example, in the Web Course Chapter 3: *Physics* section we discuss how complicated physical systems are reduced to essential parts and how these parts naturally fall into object-type descriptions. Science, in general, usually seeks to segment complex systems into simpler components. These components can easily correspond to objects. Then these simpler objects can be grouped into composite objects that correspond to the higher-order complex systems.

For example, one could easily imagine a set of classes providing the properties and structures of a group of proteins. These could be useful to both chemists and microbiologists, who would use them in their own programs for different applications.

In a similar way, engineers could use objects to represent the different components that make up a complex machine or a group of machines like a power plant. Each object representing a component of a system can act and interact with the other objects of the system. A modified or enhanced component can then correspond to an extended class (see Chapter 4).

3.10.1 Complex number class

Fortran, the long time programming language in science, includes a complex number type, but Java, unfortunately, does not. As mentioned earlier we can, however, create a complex class of our own. A complex number needs two memory locations reserved for the real and imaginary parts and it needs methods to carry out operations such as addition and subtraction.

Below we show code for a complex number class that possesses two floating-point fields for the real and imaginary values plus two methods to carry out operations on these values:

```
/** A very limited complex class. **/

public class BasicComplex
{
   double real;

   double img;

   /** Constructor initializes the values. **/
   BasicComplex (double r, double i)
   {real = r; img = i;}

   /** Define a complex add method. **/
   public void add (BasicComplex cvalue) {
      real = real + cvalue.real;
      img = img + cvalue.img;
   }

   /** Define a complex subtract method. **/
   public void subtract (BasicComplex cvalue) {
      real = real - cvalue.real;
      img = img - cvalue.img;
   }
}
```

Then in another program we could create two instances of our complex class and add one to the other, as shown below:

```
public class ComplexTest {
  static public void main (String[] args) {

     // Create two complex objects
     BasicComplex ac = new BasicComplex (1,2);
     BasicComplex bc = new BasicComplex (3,1);

     // Add ac and bc. The ac object will hold the sum.
     ac.add (bc);
     . . .

  }
}
```

This example illustrates some of the basic concepts of classes, how they relate to data types, and how they could be used in mathematical applications. This class is very limited, though. In Chapter 4 we create a complex class with more features and take advantage of class techniques such as overloading of methods.

3.10.2   Histogram class

A histogram provides a frequency distribution for the range of values that can be taken by a parameter of interest. These types of distributions are often made in science and engineering studies. For example, say that an experiment measures the voltage output of a sensor with a range of $-2\,V$ to $+2\,V$. We create a histogram with, say, ten "bins" in which the first bin holds the number of hits measured between -2.0 and -1.6, the second bin for between -1.6 and -1.2, and so forth, up to the last bin, which counts the number of times the hits were between 1.6 and 2.0.

The following `BasicHist.java` class provides some essential histogram features. The class properties include instance variables for the number of bins, an array of bins, over and underflow counts, and the range over which to bin the values. The constructor creates an instance of the class for a given set of bins and for a lower and upper parameter range. The three methods provide for adding an entry to the histogram, clearing the histogram, and for obtaining the values in the bins (including the overflow and underflow counts).

```
/** A simple histogram class to record the frequency
  * of values of a parameter of interest.
  **/
public class BasicHist
{
    int[] bins;
    int numBins;
```

```
   int underflows;
   int overflows;

   double lo;
   double hi;
   double range;

   /** The constructor will create an array of a given
     * number of bins. The range of the histogram is given
     * by the upper and lower limit values.
     **/
   public BasicHist (int numBins, double lo, double hi)
   {
      this.numBins = numBins;
      bins = new int[numBins];
      this.lo = lo;
      this.hi = hi;
      range = hi − lo;
   }

   /** Add an entry to a bin. Include if value is in the
     * range lo <= x < hi
     **/
   public void add (double x) {
      if (x >= hi) overflows++;
      else if (x < lo) underflows++;
      else {
         double val = x − lo;

         // Casting to int will round off to lower integer
         // value.
         int bin = (int) (numBins * (val/range));

         // Increment the corresponding bin.
         bins[bin]++;
      }
   }

   /** Clear the histogram bins. **/
   public void clear () {
     for (int i=0; i < numBins; i++) {
        bins[i] = 0;
     }
     overflows = 0;
     underflows= 0;
   }
```

```
    /** Provide access to the bin values. **/
    public int getValue (int bin) {
      if (bin < 0)
          return underflows;
        else if (bin >= numBins)
          return overflows;
      else
          return bins[bin];
    }
}
```

The applet below creates an instance of `BasicHist` and uses it to provide a histogram of the distribution of values generated by a Gaussian random number generator. (We discuss details about random number generation in Java in Chapter 4.) We give the histogram ten bins and set the limits from −2.0 to 2.0.

```
/** Class built from StarterApplet1. **/
public class BasicHistApplet1 extends java.applet.Applet
{
  public void init () {
      // Create an instance of the Random class for
      // producing our random values.
      java.util.Random r = new java.util.Random ();

      // Create an instance of our basic histogram class.
      // Make it wide enough enough to include most of the
      // gaussian values.
      BasicHist bh = new BasicHist (10, −2.0, 2.0);

      // The method nextGaussian () in the class Random
      // produces values centered at 0.0 and with a standard
      // deviation of 1.0. Use it to fill the histogram
      for (int i=0; i < 100; i++) {
        double val = r.nextGaussian ();
        bh.add (val);
      }

      // Print out the frequency values in each bin.
      for (int i=0; i < 10; i++) {
        System.out.println ("Bin " + i + " = " +
                            bh.getValue (i));
      }
```

```
        // Negative bin values gives the underflows
        System.out.println ("Underflows = "+ bh.getValue
                     (-1));

        // Bin values above the range give the overflows.
        System.out.println ("Overflows = "+ bh.getValue (10));
    }
}
```

When we run the above applet, an output similar to the following is produced. We see that the distribution does in fact roughly follow the general shape of a Gaussian centered in the middle bins.

```
Bin 0 = 3
Bin 1 = 8
Bin 2 = 12
Bin 3 = 14
Bin 4 = 15
Bin 5 = 17
Bin 6 = 9
Bin 7 = 9
Bin 8 = 7
Bin 9 = 3
Underflows = 1
Overflows = 2
```

In the coming chapters and in the Web Course we develop a more capable histogram class and use it for many examples.

3.10.3 Object-oriented vs. procedural programming

If your program had 20 parameters to examine, you can simply create 20 instances of BasicHist, each with its own number of bins and range limits relevant to that parameter. If at some later point, we add new methods and instance variables to the BasicHist, you don't need to modify the code in your program as long as the changes are internal to BasicHist and don't affect the parameter lists in the methods that you invoke.

If you think about how to do histogramming in a traditional procedural code approach, you should start to appreciate the elegance of the object-oriented approach. In a procedural program, you would need to create arrays to hold the histogram values. You might create a number of 1D histograms or, for the case of histograms with the same number of bins, you could use a 2D array with the first index indicating the histogram and the second index corresponding to the bins.

Similarly, you could use arrays to hold the parameters of the histograms, such as the number of bins, the lower and upper ranges, and so forth. Functions to add entries to the histograms would require a lot of bookkeeping to determine which histogram was needed. With objects, we just create 20 instances of the `BasicHist` type and each instance knows which histogram it is, and you don't have to worry about keeping track of the histogramming details in your code.

Furthermore, if you wanted to use the histogram code in another program, or let someone else use your histograms, it would be messy to extract just that code from the program and move it into the new one. The encapsulation aspect inherent to the object approach makes reusability far easier than with procedural code.

3.11   Web Course materials

The Chapter 3 Web Course: *Supplements* section provides additional introductory material about class definitions and objects. It also lists differences between object-oriented programming in Java and C++, and reviews memory management in Java (Garbage Collection) and the internals of the JVM.

The Chapter 3: *Tech* section looks further at OOP in science and engineering applications. It also provides complex number and histogram codes and demonstration programs. The Chapter 3: *Physics* section gives a tutorial on OOP in physics and continues with more examples of numerical computing techniques with Java.

Resources

Didier H. Besset, *Object-Oriented Implementation of Numerical Methods: An Introduction with Java and Smalltalk*, Morgan Kaufmann, 2001.

Cay S. Horstmann and Gary Cornell, *Core Java 2: Vol. 1 – Fundamentals*, 6th edn, Sun Microsystems, 2002.

Patrick Niemeyer and Joshua Peck, *Learning Java*, 2nd edn, O'Reilly, 2002.

Monica Pawlan, *Essentials of the Java Programming Language: A Hands-on Guide*, March 1999, http://developer.java.sun.com/developer/onlineTraining/Programming/BasicJava1/.

The Java Tutorial, Sun Microsystems, http://java.sun.com/docs/books/tutorial/.

Chapter 4
More about objects in Java

4.1 Introduction

Chapter 3 introduced the basic concepts of classes and objects in Java such as the class definition, instantiation, and object reference. We emphasized the analogy of classes with data types, but the class approach allows for more than just defining a new data type. Java allows you to build upon, or inherit from, a class to create a new child class, or *subclass*, with additional capabilities. In this chapter we introduce *class inheritance* in Java. Inheritance involves the *overriding* (not overloading) of constructors and methods, abstract classes and interfaces, polymorphism, the Object class, and the casting of object references to sub- or superclass types. We discuss each of these concepts in detail.

This chapter also includes additional discussion of arrays and how to use them for vectors and matrices in mathematical operations. The chapter ends with a couple of examples of classes for technical applications. We create an improved complex number class and also an enhanced Histogram class.

4.2 Class inheritance

A key feature of object-oriented programming concerns the ability of a class to inherit from an existing class, retaining all the features of the base class but adding new features, thus creating a subclass with increased capabilities. Here class B inherits from class A, also known as "extending" class A (thus the Java keyword extends):

```
                         public class A {
                             int i = 0;
                             void doSomething () {
                                 i = 5;
                             }
Class  A                 }

                         class B extends A {
Class  B                     int j = 0;
                             void doSomethingMore () {
                                 j = 10;
                                 i += j;
                             }
                         }
```

The diagram on the left indicates the class hierarchy. By convention the *superclass* is on top, subclasses are below, and the arrow points upwards from the subclass to the superclass The subclass B has all the data and methods from class A plus the new data and methods added by B. We can think of class B as having the data and methods equivalent to an imaginary class (let's call it "BA") shown here:

```
class BA {
   int i = 0;
   int j = 0;

   void doSomething () {
       i = 5;
   }
   void doSomethingMore () {
       j = 10;
       i += j;
   }
}
```

By using inheritance we get the features of the imaginary class BA without having to duplicate the code from the base class A. We can now create instances of class B and access methods and data in both class B and class A:

```
   . . .
   B b = new B ();          // Create an instance of class B
   b.doSomething ();        // Access a method defined in class A
   b.doSomethingMore ();    // And a method defined in class B
   . . .
```

Another class can, in turn, inherit from class B, as shown here with class C:

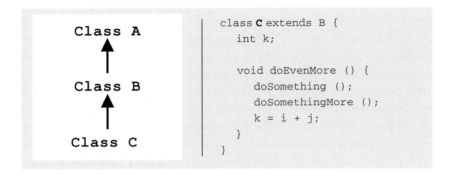

```
class C extends B {
    int k;

    void doEvenMore () {
        doSomething ();
        doSomethingMore ();
        k = i + j;
    }
}
```

Here the `doEvenMore()` method internally calls the `doSomething()` method from class `A` and the `doSomethingMore()` method from class `B`. An instance of class `C` can use the class `C` data and methods and also those of both classes `A` and `B`.

Inheritance does more than just reduce the size of the class definitions. We see shortly that the inheritance mechanism offers several new capabilities including the ability to redefine, or *override*, a method in the superclass with a new one. (The terms *superclass*, *base class*, and *parent class* all mean the same thing and are used interchangeably, as are the terms *subclass* and *child class*.)

Class inheritance in Java is strictly linear. A subclass may extend only one direct superclass, though all of that parent's superclasses get inherited as well in a chaining fashion, as shown in the class `C` example above. Unlike C++, Java does not permit multiple class inheritance, which is inheriting from more than one direct parent class. That is, given two classes `X` and `Y`, it is not possible in Java to create a class `Z` that extends both `X` and `Y`.

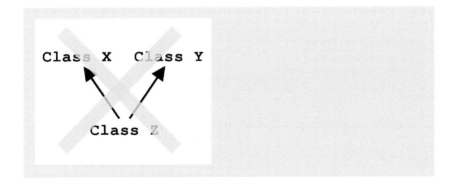

There are times that multiple class inheritance could be useful, but it was intentionally omitted by the Java designers because correctly implementing and using multiple class inheritance is fraught with difficulty. Java interfaces, to be discussed later, do permit multiple inheritance, providing many of the benefits of multiple class inheritance without the drawbacks.

4.2.1 Overriding

A common situation is when a class is needed that provides most of the functionality of a potential superclass except one of the superclass methods doesn't do quite the right thing. Adding a new method with a different name in a subclass doesn't really solve the problem because the original superclass method remains accessible to users of the subclass, thereby resulting in a source of errors should a user inadvertently use the original name instead of the new name. What is really needed is a way to change the behavior of that one superclass method without having to rewrite the superclass. Often we may not even have the superclass source code, making rewriting it impossible. Even if we do have the source code, rewriting it would be the wrong approach. That method in the superclass is assumed to be completely appropriate for the superclass and should not be changed. We wish to change the behavior of the method only for instances of our subclass, retaining the existing behavior for instances of the superclass and other subclasses that expect the original behavior of the method.

Java provides just this capability in a technique known as *overriding*. Overriding permits a subclass to provide a new version of a method already defined in a superclass. Instances of the original superclass (and other subclasses) see the original method. Instances of the overriding subclass see the new (overridden) method. In fact, overriding is often the whole reason to create a subclass.

Overriding occurs when a subclass method *exactly* matches the signature (the method name, return type, and parameter types) of a method in a superclass. If the return type is different, a compile-time error occurs. If the parameter list is different, then *overloading* occurs (already discussed in Chapter 3), not overriding. In the next section we discuss the differences, which are very important, but first we give an example of overriding. In the code below, we see that subclass Child overrides the method doSomething() in class Parent:

```
public class Parent {
   int parent_int = 0;
   void doSomething (int i) {
      parent_int = i;
   }
}


class Child extends Parent {
   int child_int = 0;
   void doSomething (int i) {
      child_int = 10;
      parent_int = 2 * i;
   }
}
```

When we have an instance of class `Child`, an invocation of the method `doSomething()` results in a call to the overridden `doSomething()` code in class `Child` rather than `Parent`:

```
. . .
Parent p = new Parent (); // Create instance of class Parent
Child c = new Child ();   // Create instance of class Child
c.doSomething (5); // The method in class Child is invoked.
p.doSomething (3); // The method in class Parent is invoked.
```

On the other hand, if we call the `doSomething()` method on a `Parent` instance, then the original `doSomething()` code from class `Parent` is invoked. Java automatically invokes the correct method based on the type of the object reference.

The real power of overriding, however, is illustrated by this code:

```
. . .
Parent p = new Child (); // Create an instance of Child
                         // but use a Parent type reference.

p.doSomething (); // Though the Parent type reference
                  // is used, the Child class's doSomething()
                  // is executed.
. . .
```

This code has created an instance of class `Child` but declared it to be of type `Parent`. Doing so is legal when `Child` is a subclass of `Parent`, since `Child` has all the methods and data of type `Parent`. Even though the variable `p` is declared to be the superclass type, it actually references the subclass object. So the subclass method is executed rather than the method in the superclass. This happens because the instance `p` really is of type `Child`, not type `Parent`. The actual type of the object referred to by an object reference is the type that it is "born as," not the type of variable that holds the object reference.

This feature is very useful when, for example, the elements of an array of the base class type contain references to instances of various subclasses. Looping through the array and calling a method that is overridden will result in the method in the subclass being called rather than the method in the base class.

The following code illustrates this so-called *polymorphic* feature of object-oriented languages. We begin with a superclass named A and three subclasses B, C, and D, all of which override the `doSomething()` method from A (classes C and D could be direct subclasses of A or they could be indirect subclasses of A by subclassing B).

```
A[] a = new A[3]; // Class A type array with three elements

a[0] = new B (); // Create an instance of class B but use
                 // an A reference since the array is
                 // type A.
a[1] = new C (); // Ditto for C
a[2] = new D (); // And D

for (int i=0; i < 3; i++) {// Call doSomething() for each
                           // element of the A array.
a[i].doSomething (); // Though the A type reference is used,
                     // the overriding doSomething() method
                     // of the actual referenced object is
                     // invoked.
}
```

It is important to understand that even though the array type is that of the super-class A, the code used for the doSomething() methods is that of the actual object that is referenced in each array element, not the code for the method in the A base class.

4.2.2   Overriding versus overloading

It is important to note how *overriding* differs from *overloading*. The latter refers to reusing the same method name but with a different parameter list and was explained in Chapter 3. Briefly, if a class contains two (or more) methods of the same name but with different parameter lists, all those methods are said to be overloaded. The compiler automatically decides which method to call based on the parameters used when the method is invoked. What was not mentioned in Chapter 3 is that overloading can occur *across* inherited classes. If a subclass reuses a method name from a parent class but changes the parameter list, then the method is still overloaded, just as if both methods appeared in the same class. (Note that via inheritance both methods really do appear in the subclass; the fact that the source code appears in two different places makes no difference.) In over*loading*, the new method does not replace the superclass method; it just reuses the name with a different parameter list. Calling the method with the original parameter list invokes the original method; calling it with the new parameter list invokes the new method.

Confusing overriding and overloading is a vexing error, both for novices and experienced Java developers. If a subclass attempts to override a method in a superclass but doesn't use the exact same parameter list, then the method is really *overloaded*, not over*ridden*. We illustrate this with the following example:

```
public class Parent {
   int i = 0;
   void doSomething (int k) {
      i = k;
   }
}

class Child extends Parent {
   void doSomething (long k) {
      i = 2 * k;
   }
}
```

Here we created class Child with the intention of overriding the doSomething (int k) method in class Parent but we mistakenly changed the int parameter to a long parameter as shown. Then the Child version of doSomething() has *overloaded* the Parent version, not overridden it. Look what happens when we attempt to call doSomething() from an instance of Child:

```
. . .
Parent p = new Parent (); // Create a Parnet instance.
Child c = new Child ();    // Create a Child instance.
p.doSomething (5); // The method in Parent is invoked,
                   // as expected.
c.doSomething (3); // The method in Parent, not Child, is
                   // invoked, probably not as expected.
```

The call to c.doSomething(3) passes an int parameter, not a long (a literal 3 is an int; to make it a long, an l or L must be appended, as in 3L). Therefore the overloaded method that takes an int is invoked, not the Child version expected. Even though we have explicitly asked for c.doSomething(), the int version of the method named doSomething() gets invoked – again, the fact that the source code happens to appear in the superclass makes no difference.

This error is often difficult to uncover. It occurs most often when an overridden superclass method is changed while forgetting to make the same change in the corresponding overriding subclass methods at the same time.

4.2.3 The `@Override` annotation in J2SE 5.0

One of the annotations available with the addition of the metadata facility in Java Version 5.0 (see Chapter 1) greatly reduces the chance of accidentally overloading when you really want to override. The `@Override` annotation tells the compiler that you intend to override a method from a superclass. If you don't get the parameter list quite right so that you're really overloading the method name, the compiler emits a compile-time error. This annotation is used as follows:

```
public class Parent {
   int i = 0;
   void doSomething (int k) {
      i = k;
   }
}

class Child extends Parent {
   @Override
   void doSomething (long k) {
      i = 2 * k;
   }
}
```

The metadata facility in Java 5.0 supports simple and complex annotation types, which are closely related to Java interfaces (discussed in Section 4.5). Some annotation types define member methods and member variables and require parameters when used. However, the `@Override` annotation is just a *marker* interface (see Section 4.5.3). It has no members, and thus accepts no parameters when used, as shown above. It must appear on a line by itself and indicates that the method name on the next line should override a method from a superclass. If the method signature on the next line isn't really an overriding signature, then the compiler complains as follows:

```
Parent.java:10: method does not override a method from its
superclass
   @Override
   ∧
1 error
```

By using `@Override` each time you intend to override a method from a superclass, you are safe from accidentally overloading instead of overriding.

4.2.4 The `this` and `super` reference operators

Perhaps you need to create a subclass that overrides a method in the base class. However, you want to take advantage of code already in the overridden method rather than rewriting it in the overriding method. That is, you want to do everything that the original method did but add some extra functionality to it for the subclass.

When in a subclass, the special reference variable `super` always refers to the superclass object. Therefore, you can obtain access to overridden methods and data with the `super` reference. In the following code class `Child` overrides the `doSomething()` method in class `Parent` but calls the overridden method by using `super.doSomething()`:

```
public class Parent {
   int i = 0;
   void doSomething () {
      i =5;
   }
}

class Child extends Parent {
   int j=0;
   void doSomething () {
      j = 10;
      // Call the overridden method
      super.doSomething ();
      j += i; // then do something more
   }
}
```

You cannot cascade `super` references to access methods more than one class deep as in

```
j = super.super.doSomething(); // Error!! Not a valid use of
                               //super
```

This usage would seem logical but it is not allowed. You can only access the overridden method in the immediate superclass with the `super` reference.

Note that you can also "override" data fields by declaring a field in a subclass with the same name as used for a field in its superclass. This technique is seldom useful and is very likely to be confusing to anyone using your code. Its use is not recommended.

A related concept is known as *shadowing* in which a local variable has the same name as a member variable. For example,

```
public class Shadow {
    int x = 1;
    void someMethod () {
        int x = 2;
        . . .
    }
}
```

Here the x inside someMethod() *shadows* the member variable x in the class definition. The local value 2 is used inside someMethod() while the member variable value 1 is used elsewhere. Such usage is often a mistake, and can certainly lead to hard-to-find bugs. This technique is not recommended. In fact, the variable naming conventions explained in Chapter 5 are designed to prevent accidental shadowing of member variables.

We can also explicitly reference instance variables in the current object with the this reference. The code below illustrates a common technique to distinguish parameter variables from instance or class variables:

```
public class A {
    int x;
    void doSomething (int x) {
        // x holds the value passed in the parameter list.
        // To access the instance variable x we must
        // specify it with 'this'.
        this.x = x;
    }
}
```

Here the local parameter variable shadows the instance variable with the same name. However, the this reference in this.x explicitly indicates that the left-hand side of the equation refers to the instance variable x instead of the local variable x from the parameter list.

4.3 More about constructors

In Chapter 3 we discussed the basics of constructors, including the overloading of constructors. Here we discuss some additional aspects of constructors.

4.3.1 this()

In addition to the this reference, there is also a special method named this() which invokes constructors from within other constructors. When a class holds overloaded constructors, typically they include one constructor that carries out basic initialization tasks and then each of the other constructors does optional tasks. Rather than repeating the initialization code in each constructor, an overloaded constructor can invoke `this()` to call another constructor to carry out the initialization tasks.

For example, the following code shows a class with two constructors:

```
class Test {
    int x,y;
    int i,k;

    Test (int a, int b) {
        x = a;
        y = b;
    }

    Test (int a, int b, int c, int d) {
        this (a,b);// Must be in first line
        i = c;
        k = d;
    }
}
```

The first constructor explicitly initializes the values of two of the data variables (the other two variables receive the default 0 value for integers). The second constructor needs to initialize the same two variables plus two more. Rather than include redundant code, the second constructor first invokes `this (a, b)`, which executes the first constructor, and then initializes the other two variables.

The parameter list in the invocation of `this()` must match that of the desired constructor (every constructor must have a unique parameter list in number and types). In this case, `this (a, b)` matches that of the first constructor with two `int` arguments. The invocation of `this()` must be the first executable statement in a constructor and cannot be used in a regular method.

4.3.2 `super()`

There is another special method named `super()`. When we create an instance of a subclass, its constructor plus a constructor in each of its superclasses are invoked (we discuss below the invocation sequence of the constructors). If there are multiple overloaded constructors somewhere in the chain, we might care which constructor gets used. We choose which overloaded superclass constructor we want with `super()`.

For example, in the following code, class `Test2` extends class `Test1`, class `Test1` has a one-argument constructor and a two-argument constructor while the constructor in class `Test2` takes three parameters. Which constructor in the superclass should be invoked? It is unwise to leave it to the compiler to "guess." (Actually, the compiler does not guess; it follows specific rules, which we discuss later.) Let's suppose that our design requires that the two-argument constructor in `Test1` be called. Therefore, the `Test2` constructor invokes the second constructor in class `Test1` by using `super(a, b)`. Had we wanted the one-argument constructor, we would use `super (a)` or `super (b)`.

```java
class Test1 {
    int i;
    int j;

    Test1(int i)
    {this.i = i;}

    Test1 (int i, int j) {
        this.i = i;
        this.j = j;
    }
}

class Test2 extends Test1 {
    float x;

    Test2 (int a, int b, float c) {
        super (a, b); // Must be first statement
        x = c;
    }
}
```

As with `this()`, the parameter list identifies which of the overloaded constructors in the superclass to invoke. And as with `this()`, the `super()` invocation must occur as the first statement of the constructor and cannot appear in regular methods.

Do not confuse the `this` and `super` references with the `this()` and `super()` constructor operators. The `this` and `super` references are used to gain access to data and methods in a class and superclass, respectively, while the `this()` and `super()` constructor operators indicate which constructors in the class and superclass to invoke.

4.3.3 Construction sequence

When you instantiate a subclass, the object construction begins with an invocation of the constructor in the topmost base class and *initializes downward* through the constructors in each subclass until it reaches the final subclass constructor. The question then arises: if one or more of the superclasses have multiple constructors, which constructor does the JVM invoke? The answer is that, unless told otherwise with `super()`, the JVM will always choose the zero-argument constructor.

Let's begin with the simplest case of a superclass definition without any constructors. In this case, as we learned in Chapter 3, the compiler automatically generates a zero-argument constructor that does nothing. Almost as simple is the case of a superclass with an explicit zero-argument constructor and no other constructors. In both of these cases, the subclass constructor does not need to explicitly invoke `super()` because the JVM automatically invokes the zero-argument constructor in the superclass – either the zero-argument constructor provided in the superclass source code if there is one, or the default do-nothing "free" constructor if no explicit constructor is provided.

If the superclass contains one or more explicit constructors, then the compiler does *not* generate a free zero-argument constructor. A subclass that does not utilize `super()` to choose one of the existing constructors fails to compile since there is no zero-argument superclass constructor to use. Therefore, the subclass must employ a `super()` with a parameter list matching one of the superclass constructors.

If the subclass also holds several constructors, each must invoke a `super()` to one of the superclass constructors (or perhaps use `this()` to refer to a subclass constructor that does use `super()`). The compiler and JVM figure out the proper sequence of constructors to call as the subclass instance is being built according to which constructor is used with the `new` operator.

The example code here shows two different sequences of constructors invoked for the case of a base class and two subclasses, all with overloaded constructors:

```java
public class ConstructApp3 {
    public static void main (String[] args) {

        // Create two instances of Test2
        // using two different constructors.

        System.out.println ("First test2 object");
        Test2 test2 = new Test2 (1.2, 1.3);

        System.out.println ("\nSecond test2 object");
        test2 = new Test2 (true, 1.2, 1.3);
    }
}
```

```java
class Test {
    int i;
    double d;
    boolean flag;

    // No-arg constructor
    Test () {
        d = 1.1;
        flag = true;
        System.out.println ("In Test()");
    }

    // One-arg constructor
    Test (int j) {
        this ();
        i = j;
        System.out.println ("In Test(int j)");
    }
}
```

```java
/** Test1 is a subclass of Test **/
class Test1 extends Test {
    int k;
    // One-arg constructor
    Test1 (boolean b) {
        super (3);
        flag = b;
        System.out.println ("In Test1(boolean b)");
    }
    // Two-arg constructor
    Test1 (boolean b, int j) {
        this (b);
        k = j;
        System.out.println ("In Test1(boolean b, int j)");
    }
}
```

```java
/** Test2 is a subclass of Test1. **/
class Test2 extends Test1 {
    double x,y;
    // Two-arg constructor
    Test2 (double x, double y) {
        super (false);
```

```
        this.x = x;
        this.y = y;
        System.out.println ("In Test2(double x, double y)");
    }

    // Three-arg constructor
    Test2 (boolean b, double x, double y) {
      super (b, 5);
      flag = b;
      System.out.println (
         "In Test2(boolean b, double x, double y)");
    }
  }
}
```

The output of `ConstructApp3` goes as:

```
First test2 object
In Test()
In Test(int j)
In Test1(boolean b)
In Test2(double x, double y)

Second test2 object
In Test()
In Test(int j)
In Test1(boolean b)
In Test1(boolean b, int j)
In Test2(boolean b, double x, double y)
```

This illustrates the different sequence of constructors invoked according to which of the `Test2` constructors we choose.

4.4 Abstract methods and classes

For some applications we might need a generic base class that we never actually instantiate. Instead, we want always to use subclasses of that base class. That is, the base class handles behavior that is common to all the subclasses but does not contain enough data or behavior to be useful on its own. In a sense, the common behavior has been "factored out" of the subclasses and moved to the common base class.

In the following standard example, we create a base class `Shape`, which provides a method that calculates the area of some 2D shape:

```
public class Shape {
   double getArea () {
      return 0.0;
   }
}
```

The `Shape` class itself does almost nothing. To be useful, there must be subclasses of `Shape` defined for each desired 2D shape, and each subclass should override `getArea()` to perform the proper area calculation for that particular shape. We illustrate with two shapes – a rectangle and a circle.

```
public class Rectangle extends Shape {
   double ht = 0.0;
   double wd = 0.0;

   public double getArea () {
      return ht*wd;
   }
   public void setHeight (double ht) {
      this.ht = ht;
   }
   public void setWidth (double wd) {
      this.wd = wd;
   }
}

public class Circle extends Shape {
   double r =0.0;
   public double getArea () {
      return Math.PI * r * r;
   }
   public void setRadius (double r) {
      this.r = r;
   }
}
```

The subclasses `Rectangle` and `Circle` extend `Shape` and each overrides the `getArea()` method. We could define similar subclasses for other shapes as well. Each shape subclass requires a unique area calculation and returns a `double` value. The default area calculation in the base class does essentially nothing but it must be declared to return a `double` for the benefit of the subclass methods that do return values. Since its signature requires that it return something, it was defined to return 0.0. In practice, since the superclass Shape should never be instantiated, only the subclasses, then the superclass getArea() will never be called anyway.

The capability to reference instances of Rectangle and Circle as Shape types uses the advantage of polymorphism (see Section 4.2.1) in which a set of different types of shapes can be treated as one common type. For example, in the following code, a Shape array passed in the parameter list contains references to different types of subclass instances:

```
void double aMethod (Shape[] shapes) {
    areaSum = 0.0;
    for (int i=0; i < shapes.length; i++) {
        areaSum += shapes[i].getArea ();
    }
}
```

This method calculates the sum of all the areas in the array with a simple loop that calls the getArea() method for each instance. The polymorphic feature means that the subclass-overriding version of getArea() executes, not that of the base class.

The careful reader will have observed that the technique used above is messy and error-prone. There is no way, for instance, to *require* that subclasses override getArea(). And there is no way to ensure that the base class is never instantiated. The above scheme works only if the subclasses and the users of the Shape class follow the rules. Suppose someone does instantiate a Shape base class and then uses its getArea() method to calculate pressure, as in the force per unit area. Since the area is 0.0, the pressure will be infinite (or NaN). The Java language can do much better than that.

A much better way to create such a generic base class is to declare a method abstract. This makes it explicit that the method is intended to be overridden. In fact, all abstract methods *must* be overridden in some subclass or the compiler will emit errors. No code body is provided for an abstract method. It is just a marker for a method signature, including return type, that must be overridden and given a *concrete* implementation in some subclass.

In the above case, we add the abstract modifier to the getArea() method declaration in our Shape class and remove the spurious code body as shown here:

```
public abstract class Shape {
    abstract double getArea ();
}
```

Note that if any method is declared abstract, the class must be declared abstract as well or the compiler will give an error message. The compiler will not permit an abstract class to be instantiated. An abstract class need not include only abstract methods. It can also include concrete methods as well, in case there is common behavior that should apply to all subclasses. In fact, a class marked abstract is not required to include *any* abstract methods. In that case, the abstract modifier simply prevents the class from being instantiated on its own. Abstract classes, unlike interfaces (see next section), can also declare instance variables. As an example, our abstract Shape class might declare an instance variable name:

```
public abstract class Shape {
   String name;
   abstract double getArea ();
   String getName () {
     return name;
   }
}
```

Here each subclass inherits the name instance variable. Each subclass also inherits the concrete method getName() that returns the value of the name instance variable.

When an abstract class does declare an abstract method, then that method must be made concrete in some subclass. For example, let's suppose that class A is abstract and defines method doSomething(). Then class B extends A but does not provide a doSomething() method:

```
abstract class A {
   abstract void doSomething ();
}
class B extends A {
   // Fails to provide a concrete implementation
   // of doSomething ()
   void doSomethingElse () {. . .}
}
```

In this case, the compiler complains as follows:

```
B is not abstract and does not override abstract method
doSomething() in A class B extends A {
 ^
```

This message indicates that not overriding doSomething() in class B is okay if B is declared to be abstract too. In fact, that is true. If we don't want B to provide doSomething(), then we can declare B abstract as well:

```
abstract class A {
   abstract void doSomething ();
}
abstract class B extends A {
   // Does not provide a concrete implementation
   // of doSomething ()
   void doSomethingElse () {. . .}
}
```

This code compiles without errors. Of course, classes A and B may never be instantiated directly (since they are abstract). Eventually, there must be some subclass of A or B that provides a concrete implementation of all the abstract methods:

```
class C extends B {
   // Provides a concrete implementation of doSomething()
   void doSomething () {. . .}
}
```

4.5 Interfaces

As discussed in Section 4.2, Java does not allow a class to inherit directly from more than one class. That is,

```
class Test extends AClass, BClass // Error!!
```

There are situations where multiple inheritance could be useful, but it can also lead to problems; an example is dealing with the ambiguity when the inherited classes include methods and fields with the same identifiers (i.e. the names and parameter lists).

Interfaces provide most of the advantages of multiple inheritance with fewer problems. An interface is basically an abstract class but with *all* methods abstract. The methods in an interface do not need an explicit abstract modifier since they are abstract by definition. A concrete class implements an interface rather than extends it, and a class can implement more than one interface. Any class that implements an interface must provide an implementation of each interface method (or be declared abstract).

In the example below, Runnable is an interface with a single method: run(). Any class that implements Runnable must provide an implementation of run().

```
class Test extends Applet implements Runnable {
   . . .
   public void run () {
      . . .
   }
}
public interface Runnable {
   public void run ();
}
```

To implement multiple interfaces, just separate the interface names with a comma:

```
class Test extends Applet implements Runnable, AnotherInterface
{
   . . .
}
```

If two interfaces each define a method with the same name and parameter list, this presents no ambiguity since both methods are abstract and carry no code body. In a sense, both are overridden by the single method with that signature in the implementing class.

Any class that implements an interface can be referenced as a type of that interface, as illustrated by this code:

```
class User implements Runnable {
   public void run () {
      . . .
   }
}

class Test {
   public static void main (String[] args) {
      Runnable r = new User ();
      . . .
   }
}
```

Here the class User implements Runnable, so it can be referenced in a variable of type User or in a variable of type Runnable as shown. The value of using the type Runnable instead of User is illustrated in the next section.

4.5.1 Interfacing classes

The term *interface* is a very suitable name for these kinds of abstract classes because they can provide a systematic approach to adding access to a class. That is, they can provide a common *interface*.

For example, say that we have classes Relay and Valve that are completely independent, perhaps written by two different programmers. The class Test could communicate easily with both of these classes if they were modified to implement the same interface. Let's define an interface called Switchable, which holds a single method called getState(), as in

```
public interface Switchable {
   public boolean getState ();
}
```

We want both the Relay and Valve classes to implement Switchable and provide a getState() method that returns a value true or false that indicates whether a relay or a valve is in the on or off state.

In the code below we show the class Test that references instances of Relay and Valve as Switchable types. Test can then invoke their respective getState() methods to communicate with them.

```
class Test {
   public static void main (String[] args) {
      Switchable[] switches = new Switchable[2];
      switches[0] = new Relay ();
      switches[1] = new Valve ();

      for (int i=0; i < 2; i++) {
         if (switches[i].getState ()) doSomething (i);
      }
   }
}

class Relay implements Switchable {
   boolean setting = false;
   // Implement the interface method getState()
   boolean getState () {
      return setting;
   }

   . . other code . .
}

class Valve implements Switchable {
   boolean valveOpen = false;
```

```
    // Implement the interface method getState()
    boolean getState () {
       return valveOpen;
    }
    . . other code . .
}

interface Switchable {
   boolean getState ();
}
```

So we see that an interface can serve literally to *interface* otherwise incompatible classes together. The modifications required for the classes `Relay` and `Valve` involve only the implementation of the interface `Switchable`. Class `Test` illustrates how we can treat instances of `Relay` and `Valve` both as the type `Switchable` and invoke `getState()` to find the desired information for the particular class. If additional classes that represent other components with on/off states are created for our system simulation, we can ask that they also implement `Switchable`.

   Note that if we don't have the source code for `Valve` and `Relay`, we could still create subclasses of them and have those subclasses implement `Switchable`. For example,

```
class SwitchableValve extends Valve implements Switchable {
   boolean getState () {
      . . .
   }
}
```

4.5.2   Interfaces for callbacks

With the C language, programmers often use pointers to functions for tasks such as passing a pointer in an argument list. The receiving function can use the pointer to invoke the passed function. This technique is referred to as a "callback" and is very useful in situations where you want to invoke different functions without needing to know which particular one is being invoked or when library code needs to invoke a function that is supplied by a programmer using the library.

For example, a plotting function could receive a pointer to a function that takes an x axis value as an argument and returns a value for the y axis (*sin(x), cos(x)*, for example). The plotting function could then plot any such function whose pointer is passed to it without knowing explicitly the name of the function.

Java, however, does not provide pointers (actual memory addresses), only object references. A reference cannot refer to a method. Instead, Java provides interfaces for callbacks. In this case, a library method holds a reference to an interface in its parameter list and then invokes a method declared in that interface. The programmer provides a class that implements the required interface and provides an object reference to the library method. When the library method invokes the required interface method, the concrete implementation in the provided object is invoked.

In the following code we see that the `aFunc(Switchable sw)` method invokes the `getState()` method of the `Switchable` interface. An instance of any class that implements the `Switchable` interface can thus be passed to `aFunc()`. This technique provides the same generality as pointer callbacks in C. The only drawback is that a class must implement the interface.

```
public class TestCallBack {
    public static void main(String [] args){
        Switchable[] switches = new Switchable[3];
        switches[0] = new Relay();
        switches[1] = new Relay();
        switches[2] = new Valve();

        // Pass Switchable objects to aFunc ()
        for (int i=0; i < 3; i++) {
            aFunc (switches[i]);
        }
    }
    // Receive Switchable objects and call their getState ()
    void aFunc (Switchable sw) {
        if (sw.getState ()) doSomething ();
    }
}
    . . . See previous example for Relay and Valve definitions.
```

4.5.3   More about interfaces

Interfaces can extend other interfaces, much like class inheritance. All the methods declared in the super-interface are effectively present in the sub-interface. Unlike classes, however, interfaces can participate in multiple inheritance. The following code shows an interface extending two interfaces at once using a comma in the `extends` clause:

```
public interface A {. . .}

public interface B {. . .}

public interface C extends A, B {. . .}
```

An interface can also contain data fields, and those fields can be seen by implementing classes. Any data fields in an interface are implicitly `static` and `final` though those qualifiers need not appear. Thus data fields in interfaces are effectively constants and, by convention, are best declared using all uppercase characters.

Placing constants in an interface is a common, though not recommended, practice. As an illustration of the convenience of this technique, consider the `MyConstants` interface shown here:

```
public interface MyConstants {
    final static double G = 9.8;
    final static double C = 2.99792458e10;
}
```

The following `Calculations` class `implements` `MyConstants` and so can refer to the constants directly:

```
class Calculations implements MyConstants {
    // Can directly use the constants defined
    // in the MyConstants interface
    public double calc (double t) {
        double y = 0.5*G*t*t;
        return y;
    }
}
```

If we instead made `MyConstants` a class, we would need to reference the constants with a class name prefix as follows:

```
double y = 0.5 * MyConstants.G * t * t;
```

This obviously becomes awkward if you have a long equation with lots of constants taken from other classes.

However, despite its usefulness, using an interface just to hold constants is not recommended since it really is an abuse of the interface concept. An interface full of nothing but constants does not define a type, as a proper interface is expected to do. And a class that "implements" such an interface isn't really *implementing* anything – it is just *using* the constants in the interface (perhaps a uses keyword would be more appropriate). Seeing the implements keyword should imply that the class actually implements something.

For these reasons, the use of a class instead of an interface to define constants is the recommended practice, accepting the need for the more verbose syntax to refer to those constants. We note that J2SE 5.0, in fact, solves this problem with the "static import" facility, which we explain in Chapter 5 after discussing the import keyword.

Another interesting feature of interfaces is that an interface need not contain either method declarations or data. It can be completely empty. Such an interface can be useful as a "marker" of classes. That is, you can use the instanceof operator to determine if a class is of the particular marker type, which then can imply some quality of the class. One can use the empty Cloneable interface to indicate whether a class overrides the clone() method from Object (see Section 4.6.3) to make copies of instances of the class.

We discuss access rules and modifiers in the next chapter but here we note that interface methods and constants are implicitly public. This means that any class can access the methods and constants in the interface. The concrete implementations of interface methods in classes that implement the interface must also be public otherwise the compiler will complain.

4.6 More about classes

In this section we continue our introduction to the basics of class definitions and objects with an examination of casting, the Object class, and the toString() method.

4.6.1 Converting and casting object references

In Chapter 2 we discussed the topic of mixing different primitive types in the same operation and the need in some cases to explicitly cast one type into another. The same concepts apply when dealing with objects instead of primitives. Sometimes, as with primitives, the type conversion is automatically handled by the compiler. Consider a superclass Fruit with a subclass Pineapple:

```
class Fruit {. . .}
class Pineapple extends Fruit {. . .}
```

Let f be a variable of type Fruit and p be of type Pineapple. Then we can assign the Pineapple reference to the Fruit variable:

```
class Conversion {
    Fruit f;
    Pineapple p;
    public void convert () {
        p = new Pineapple ();
        f = p;
    }
}
```

The compiler automatically handles the assignment since the types are compatible. That is, the type `Fruit` can "hold" the type `Pineapple` since a `Pineapple` "is a" `Fruit`. Such automatic cases are called *conversions*.

A related automatic conversion is with interfaces. Let the class `Fruit` implement the `Sweet` interface:

```
interface Sweet {. . .}
class Fruit implements Sweet {. . .}
```

Then we see that a variable of type `Fruit` can be automatically converted to a variable of type `Sweet`. This makes perfect sense since a `Fruit` "is" `Sweet`.

```
Fruit f;
Sweet s;
public void good_convert () {
    s = f; // legal conversion from class type to interface type
}
```

However, an attempt to convert from the interface type to the class type does not compile:

```
public void bad_convert () {
    f = s; // illegal conversion from interface type to class type
}
```

As with primitives, if the compiler cannot perform an automatic conversion, an explicit cast is required. In most cases you can force the compiler to permit the desired type conversion by using a cast. Like with primitive types, the class type that an object is being cast to is enclosed in parentheses in front of the object reference. Doing so essentially tells the compiler to ignore the apparent type incompatibility and proceed anyway. If the types really are incompatible then runtime errors will ensue.

For example, let `BClass` be a subclass of `AClass`. Let `AClass` hold aMethod(), which, of course, is inherited by `BClass`. In addition, bMethod() is a new method in `BClass`.

```
class AClass {
    void aMethod () {. . .}
}

class BClass extends AClass {
    void bMethod () {. . .}
}
```

In the following code, `miscMethod()` is declared to receive an `AClass` object as a parameter. When used, the actual object passed in might in fact be an instance of `BClass`, which is perfectly legal since `BClass` is a subclasses of `AClass`.

```
public void miscMethod (AClass obj) {
    obj.aMethod ();
    if (obj instanceof BClass) ((BClass)obj).bMethod ();
}
```

We see that we can invoke `aMethod()` on the object received in the parameter list whether that object is an `AClass` or a `BClass` since both types have this method. However, to invoke the `bMethod()`, we need first to check the object type. We use the `instanceof` operator to find out if the object really is a `BClass` and then cast to `BClass` if appropriate. Without the cast the compiler complains that it cannot find `bMethod()` in the `AClass` definition.

4.6.2 Casting rules

The casting rules can be confusing, but in most cases common sense applies. There are compile-time rules and runtime rules. The compile-time rules are there to catch attempted casts in cases that are simply not possible. For instance, suppose we have classes A and B that are completely unrelated – i.e. neither inherits from the other and neither implements the same interface as the other, if any. It is nonsensical to attempt to cast a B object to an A object, and the compiler does not permit it even with an explicit cast. Instead, the compiler issues an "inconvertible types" error message.

Casts that are permitted at compile-time include casting any object to its own class or to one of its sub- or superclass types or interfaces. Almost anything can be cast to almost any interface, and an interface can be cast to almost any class type. There are some obscure cases (see the Java Language Specification for the details), but these common sense rules cover most situations.

The compile-time rules cannot catch every invalid cast attempt. If the compile-time rules permit a cast, then additional, more stringent rules apply at runtime. These runtime rules basically require that the object being cast is compatible with the new type it is being cast to. Else, a `ClassCastException` is thrown at runtime.

4.6.3 The `Object` class

All classes in Java implicitly extend the class `Object`. That is,

```
public class Test
{. . .}
```

is equivalent to

```
public class Test extends Object
{. . .}
```

So, all Java objects are instances of `Object`. This ability to treat all objects as one type provides the ultimate in polymorphism. An example of this usage is the `ArrayList` class, which is a part of the `java.util` package (we discuss Java packages in Chapter 5). The `ArrayList` class can hold any object type. The `ArrayList.add()` method is used to input objects into the `ArrayList`. The parameter list for the `add()` method is declared to receive an `Object` parameter. That way, any object type can be added, since all object types always inherit from the `Object` base class. When an element is retrieved from the `ArrayList`, it is of type `Object` and should be cast to the type needed.

A simpler example is the following case, where the parameter type of `miscMethod()` is `Object` so any class whatsoever can be provided in a method call to `miscMethod()`. Inside `miscMethod()` we decide what type the received object reference really is and call appropriate methods based on that type. Except for the case where we want to invoke a method belonging to the `Object` class, we need to cast the object to one of the classes that we expect as a parameter before we can invoke a method or access a field in that class.

```
public void miscMethod (Object obj) {
    if (obj instanceof AClass) ((AClass)obj).aMethod ();
    if (obj instanceof BClass) ((BClass)obj).bMethod ();
    if (obj instanceof CClass) ((CClass)obj).cMethod ();
}
```

The `Object` class provides several methods that are useful to all of its subclasses. A subclass can also override these methods to provide behavior unique to the particular subclass. These methods include:

- `clone ()` – produces copies of an object. (See Web Course Supplements.)
- `equals (Object obj)` – tests whether an object is equal to the object `obj`. The default is to test simply whether `obj` references the same object (i.e. a shallow equals), not whether two independent objects contain identical properties. This method is often overridden to perform a deep equals as in the `String` class, which tests whether the strings actually match.

- `toString ()` – provides a string representation of this object. The default for a class consists of a string constructed from the name of the class plus the character "@" plus a hash code of the object in hex format. This method is often overridden to provide more illuminating information. (See the next section.)
- `finalize ()` – called by the garbage collector when there are no more references to this object. You can override this method to take care of any housecleaning operations needed before the object disappears.
- `getClass ()` – gets the runtime class of the object, returned as a `Class` type (see the Web Course Chapter 5: *Supplements* section for a discussion of the `Class` class).
- `hashCode ()` – generates a hash code value unique for this object.

The following methods involve thread synchronization that we introduce in Chapter 8. They can only be called from within a synchronized method or code block:

- `notify ()` – called by a thread that owns an object's lock to tell a waiting thread, as chosen by the JVM, that the lock is now available.
- `notifyAll ()` – similar to `notify()` but wakes all waiting threads and then they compete for the lock.
- `wait ()` – the thread that owns the lock on this object releases the lock and then waits for a `notify()` or `notifyAll()` to get the lock back.
- `wait (long msecs)` – same as `wait()` but if a notify fails to arrive within the specified time, it wakes up and starts competing for the lock on this object anyway.
- `wait (long msecs, int nanosecs)` – same as `wait (long msecs)` but specified to the nanosecond.

(We note that most operating systems do not provide a clock that is accurate to a nanosecond and some not even to a few milliseconds.)

4.6.4 Objects to strings

We discussed in Chapter 3 how to convert primitive types to and from strings. You can also convert any Java object to a string. If you just print any object, as in

```
System.out.println (someObjectReference);
```

then that object's `toString()` method is called automatically to produce string output. All objects inherit the `toString()` method from the `Object` class. This default version of `toString()` from `Object` produces a string beginning with the class name with certain data values appended to it.

However, the `toString()` method typically is overridden by most classes to provide output in a more readable format customized for that class. Most of the classes in the Java core class libraries provide sensible `toString()` methods, and classes that you write should too for convenience when printing.

You can call the `toString()` method directly, or, alternatively, the "+" operator calls the `toString()` method whenever the variable refers to an object. For example, consider

```
Double aDouble = 5.0;
String aDoubleString = "aDouble = " + aDouble;
```

The plus operator in the second line invokes the `toString()` method of the `Double` object `aDouble`. This results in `aDoubleString` referencing the string "aDouble = 5.0".

4.7   More about arrays

Here we look at other aspects of Java arrays and at tools to use with them. Note that like much of Java syntax, arrays at first glance seem very similar to those in C/C++. However, there are several differences from these languages in how Java arrays are built and how they work.

4.7.1   Object arrays

In the previous chapter we introduced arrays of primitive types, which generally behave in the manner that is expected of such arrays. For example, to create an array of ten integers we could use the following:

```
int[] iArray = new int[10];
```

This sets aside ten `int` type memory locations, each containing the value 0.

For arrays of objects, however, the array declaration only creates an array of references for that particular object type. It does not create the actual objects of that type. Creating the objects themselves requires an additional step. For example, let's say we want to create an array of five `String` objects. We first create a `String` type array:

```
String[] strArray = new String[5];
```

When the array is created, five memory locations are set aside to contain object references of the `String` type with the expectation that each reference will eventually "point" to a `String` object. But initially, each element contains the special `null` reference value; that is, it points *nowhere*. So if we followed the above declaration with an attempt to use a `String` method, as in

```
int numChars = strArray[0].length ();
```

an error message results:

```
Exception in thread "main" java.lang.NullPointerException at
ArrayTest.main (ArrayTest.java:8)
```

Before using the array elements, we must first create an object for each array element to reference. For example,

```
strArray[0] = new String ("Alice");
strArray[1] = new String ("Bob");
strArray[2] = new String ("Cindy");
strArray[3] = new String ("Dan");
strArray[4] = new String ("Ed");
```

This code sets each element to reference a particular string.

Note that there is an alternative declaration that only works for `String` objects:

```
strArray[0] = "Alice";
```

That is, the string literal `"Alice"` is equivalent to `new String ("Alice")`.

4.7.2 Array copying

A copy of an array can be made with the static method `System.arrayCopy()` as shown here:

```
System.arraycopy (Object src, int src_position,
                  Object dst, int dst_position, int length)
```

Here `src` is the array to be copied and `dst` is the destination array (of the same type). The copy begins from the array element at the index value of `src_position` and starts in destination at `dst_position` for `length` number of elements. If the value of the `length` parameter is too long, or if any situation occurs such that either the source or destination arrays are accessed beyond their actual array length, then an `IndexOutOfBoundsException` is thrown at runtime. This optimized method works for primitive arrays as well as object arrays. It even handles the case where the destination array overlaps the source array.

4.7.3 Multi-dimensional arrays

In Java, multi-dimensional arrays are arrays of arrays. That is, each element is a reference to an array object. For example, we could declare a two-dimensional array as follows:

```
String[][] str = new String[3][2];
```

This is equivalent to

```
String [][] str = new String[3][];
str[0] = new String[2];
str[1] = new String[2];
str[2] = new String[2];
```

However, we don't need to keep the sub-array lengths the same. This also works:

```
str[0] = new String[2];
str[1] = new String[33];
str[2] = new String[444];
```

We can combine the string array declaration and initialization, as in

```
str[0] = new String[]{"alice", "bob"};
str[1] = new String[]{"cathy", "don", "ed"};
str[2] = new String[]{"fay", "grant", "hedwig", "ward"};

System.out.println ("str[1][2],str[2][3] = " +
str[1][2] + str[2][3]);
```

The print statement would show

```
str[1][1],str[2][3] = edward
```

4.7.4 More about arrays as objects

As mentioned earlier, arrays in Java are objects. An array inherits `Object` and possesses an accessible property – `length` – that gives the number of elements in the array. For example, if a method uses `Object` as a parameter, as in

```
void aMethod (Object obj) {. . .}
```

then an array can be passed as the actual parameter since an array is a subclass of `Object`:

```
. . .
int[] i_array = new int[10];
aMethod (i_array);
. . .
```

To make arrays appear in a convenient and familiar form (as in C, for example), the language designers provided brackets as the means of accessing the array elements as already seen above. Without brackets, an array class would have to provide a method such as `getElementAtIndex()` to access array elements. For example,

```
String string_one = str_array.getElementAtIndex (1);
```

Fortunately, the simpler syntax using brackets was chosen instead:

```
String string_one = strArray[1];
```

Since arrays are objects, arrays are somewhat more complicated in Java than in other languages, but the class structure also provides important benefits. For example, each use of an array element results in a check on the element number, and if the element exceeds the declared length of the array, an out of bounds run-time exception is thrown.

Thus, unlike in C or C++, a program cannot run off the end of an array and write to places in memory where it should not. This avoids a very common program bug and source of security attacks that can be difficult to track down since the problem may not show up until well after the write occurs. On the other hand, there is some performance penalty in the bounds checking that can show up when doing intensive processing with arrays.

4.7.5 Mathematical vectors and matrices

Vector and matrix operations are obviously standard tools throughout science and engineering. Here we look at some ways to use Java arrays to represent and carry out operations for vectors and matrices.

Note that the Java core language includes a class called Vector in the java.util package (see Chapter 10). Vector is similar to the ArrayList discussed above (see Section 4.6.3); both provide a dynamic list that allows for both adding and removing elements. ArrayList and Vector are often quite useful, but they are slow and not intended for mathematical operations.

4.7.5.1 Mathematical vectors

The elements of a floating-point array can represent the component values of a vector, as in

```
double[] vec1 = {0.5,0.5,0.5};
double[] vec2 = {1.0,0.0,0.2};
```

We then need methods to carry out various vector operations such as the dot product:

```
double dot (double[] a, double[] b) {
   double dot_prod = 0.0;
   for (int i=0; i < a.length; i++) {
     dot_prod += a[i]*b[i];
   }
   return dot_prod;
}
```

Note that a more robust method would check that the vector arguments are not null and that the array lengths are equal.

Several numerical libraries are available that provide classes with methods to carry out vector operations. The Web Course Chapter 4 provides links to several of these.

4.7.5.2 Matrices

The obvious approach for matrices is to use arrays with two indices:

```
double[][] dMatrix = new double[n][m];
```

However, as indicated by the discussion in Section 4.7.2, this does not produce a true two dimensional array in memory but is actually a one-dimensional array of references to other one-dimensional arrays, each of which can be located in a different area of memory.

In the C language, moving from one element to the next in a 2D array requires only incrementing a memory pointer. This does not apply for Java, which uses an indirect referencing approach that causes a performance penalty, especially if the matrix is used in intensive calculations.

One approach to ameliorate this problem to some extent is to use a 1D array. The code below shows how one might develop a matrix class to use a 1D array for 2D operations. A sophisticated compiler can optimize such a class and in some cases provide better performance than a standard Java two-dimensional array.

```java
public class Matrix2D {
   private final double[] fMat;

   private final int fCols;
   private final int fRows;
   private final int fCol;
   private final int fRow;

   public Matrix2D (int rows, int cols) {
     fCols = cols;
     fRows = rows;
     fMat= new double[rows * cols];
   }

   /** r = row number, c = column number **/
   public double get (int r, int c) {
     return fMat[r * fCols + c];
   }

   /** r = row number, c = column number **/
   public double set (int r, int c, double val) {
     fMat[r * fCols + c] = val;
   }

   . . . other methods, e.g. to fill the array, access a
   subset of elements, etc.
}
```

4.8   Improved complex number class

In the Chapter 3 we created a class with the bare essentials needed to represent complex numbers. Here we expand on that class. For example, we would often like to add two complex numbers and put the sum into another complex number rather than modify one of the current complex objects. Because of overloading we can still use the `add()` method name. A new, improved version of our complex number class appears here:

```
public class Complex {

  double real;
  double imag;
  /** Constructor that initializes the real & imag values
   **/
  Complex (double r, double i) {
    real = r; imag = i;
  }


  /** Getter methods for real & imaginary parts **/
  public double getReal ()
  {return real;}
  public double getImag ()
  {return imag;}

  /** Define an add method **/
  public void add (Complex cvalue) {
    real = real + cvalue.real;
    imag = imag + cvalue.imag;
  }

  /** Define a subtract method. **/
  public void subtract (Complex cvalue) {
    real = real - cvalue.real;
    imag = imag - cvalue.imag;
  }

  /** Define a static add method that returns a
   *  a new Complex object with the sum.
   **/
  public static Complex add (Complex cvalue1,
                             Complex cvalue2) {
    double r = cvalue1.real + cvalue2.real;
    double i = cvalue1.imag + cvalue2.imag;
    return new Complex (r, i);
  }
```

```
/** Define a static subtract method that returns a
 *  a new Complex object with the difference.
 **/
public static Complex subtract (Complex cvalue1,
                                Complex cvalue2) {
  double r = cvalue1.real - cvalue2.real;
  double i = cvalue1.imag - cvalue2.imag;
  return new Complex (r, i);
}
} // class Complex
```

Here the new static add() and subtract() methods each create a new complex object to hold the sum and difference, respectively, of the two input complex objects. The operand objects are unchanged by the method.

As we discussed in Chapter 3, a static method is invoked by giving the name of the class and the dot operator. Unfortunately, in Java, unlike C++, we cannot override the + operator and create a special + operator for complex addition. The following code shows how to add two complex numbers together using our static add() method:

```
public class ComplexTest {
public static void main (String[] args) {
  // Create complex objects
  Complex a = new Complex (1.0, 2.1);
  Complex b = new Complex (3.3, 1.2);

  Complex c = Complex.add (a, b); // c now holds a + b

  . . . other code . . .
  }
}
```

The Web Course Chapter 4 gives a more complete version of the class (e.g. it includes modulus, multiplication, etc.).

4.9   Random number generation

Random values can be obtained from the Math class using the method

```
public static double random ()
```

This method produces pseudo-random double values in the range

```
0.0 <= r < 1.0
```

The first time it is invoked, it initializes the seed with a value derived from the current time.

The `java.util.Random` class provides a more extensive set of random generators. Two constructors – `Random()` and `Random (long seed)` – offer the options of initialization from the current time or from a specific `seed` value.

The methods in the `Random` class include:

- `nextInt ()` – integers in the range $0 <= r < 2**32$
- `nextInt (int n)` – integers in the range $0 <= r < n$
- `nextBoolean (int n)` – randomly chosen true/false
- `nextGaussian ()` – random double values with mean 0.0 and sigma of 1.0

The last three methods first became available with Java 1.2.

4.9.1   Random number algorithm

The `Random` class uses a *linear congruential algorithm* [1,2] with a 48-bit seed. If the constructor `Random (long)` or the `setSeed (long)` method is invoked, the algorithm uses only the lower 48 bits of the seed value.

Random number generator formulas actually produce a sequence of numbers that eventually repeat. For the same seed value a formula always produces the same sequence. A seed simply selects where in the sequence to start. The generator will eventually repeat that seed value and start the same sequence again. Compared to the randomness of physical fluctuations, such as in radio noise, these formulas are said to produce *pseudo-random* numbers.

To insure that applications ported to different platforms give the same results, all implementations of Java must use the same algorithm so that the same seed returns the same sequence regardless of the platform.

The linear congruential formula in Java goes as

```
x_i+1 = (a * x_i + c) mod m
```

As discussed in the references, you should use such formulas with care. They can produce random number sequences of a length up to `m` but not necessarily that long. The length depends on the set of `a`, `c`, and `m` values chosen.

Also, if you grab consecutive sequences of numbers of K length, and plot them as points in K-dimensional space, they do not fully populate the volume randomly but instead lie on K-1dimensional planes. There are no more than $m^{1/K}$ planes and possibly less. If you need to create points in a space this way, you should shuffle the values obtained from the generator. [2]

In Java the values in the linear congruential formula in `Random` are

```
a = 0x5DEECE66DL
c = 11
m = 2^48 - 1.
```

The actual code in `next (int bits)` goes as

```
synchronized protected int next (int bits) {
   seed = (seed * 0x5DEECE66DL + 0xBL) & ((1L << 48) - 1);
   return (int)(seed >>> (48 - bits));
}
```

Here the mod operation comes via the AND operation since m in this case has all
47 bits set to 1.

This method is protected (see Section 5.3.3, Access Rules). The public
random number methods accessible by all classes use the next() method. For
example, nextInt() simply includes the statement

```
return next (32);
```

The nextLong() method invokes next(32), shifts the result by 32 bits to the
left, invokes next(32) again and then ORs the two values together to obtain a
64-bit random number:

```
return ((long)next (32) << 32) + next (32);
```

The nextFloat() method provides values in the range 0.0f <= x < 1.0f:

```
return next (24) / ((float)(1 << 24));
```

The nextDouble() method provides values in the range 0.0d <= x < 1.0d
using the statement

```
return (((long)next (26) << 27) + next (27))/(double)(1L << 53)
```

The nextBoolean() method uses the statement

```
return next (1)!= 0;
```

See the java.util.Random class specification for more detailed descriptions
of the algorithms used for these and the other nextXxx() methods.

4.10   Improved histogram class

Here we make a subclass of the BasicHist class discussed in Chapter 3.
The class definition below shows that BetterHist inherits from BasicHist,
obtaining the properties of the latter while providing new capabilities.

Note how the constructor invokes super() to select a constructor in the base
class. Also, we see how the new methods in the subclass can access the data
variables in the base class. (In the next chapter we discuss access modifiers such
as private, which prevents subclasses from accessing a field or method.)

We add several methods to our histogram that provide various parameters
specifying the histogram. Also, a calculation of the mean and standard deviation
of the distribution in the histogram is included.

```
/** A simple histogram class to count the frequency of
 * values of a parameter of interest. **/
class BetterHist extends BasicHist
{
  /** This constructor initializes the basic elements of
   * the histogram.
   **/
  public BetterHist (int numBins, double lo, double hi) {
    super (numBins, lo, hi);
  }

  /** Get the low end of the range. **/
  public double getLo ()
  {return lo;}

  /** Get the high end of the range. **/
  public double getHi ()
  {return hi;}

  /** Get the number of entries in the largest bin. **/
  public int getMax () {
    int max = 0;

    for (int i=0; i < numBins; i++)
      if (max < bins[i]) max = bins[i];
    return max;
  }
  /** Get the number of entries in the smallest bin. **/
  public int getMin () {
    int min = getMax ();

    for (int i=0; i < numBins; i++)
      if (min > bins[i]) min = bins[i];
    return min;
  }

  /** Get the total number of entries **/
  public int getTotal () {
    int total = 0;
    for (int i=0; i < numBins; i++)
      total += bins[i];
    return total;
  }
  /** Get the average and std. dev. of the distribution. **/
  public double [] getStats () {
    int total = 0;
```

```
        double wtTotal = 0;
        double wtTotal2 = 0;
        double [] stat = new double[2];
        double binWidth = range/numBins;

        for (int i=0; i < numBins; i++) {
            total += bins[i];
            double binMid = (i - 0.5) * binWidth + lo;
            wtTotal += bins[i] * binMid;
            wtTotal2 += bins[i] * binMid * binMid;
        }

        if (total > 0) {
            stat[0] = wtTotal/total;
            double av2 = wtTotal2/total;
            stat[1] = Math.sqrt (av2 - stat[0]*stat[0]);
        }
        else {
            stat[0] = 0.0;
            stat[1] = -1.0;
        }
        return stat;
    } // getStats
} // class BetterHist
```

4.11   Understanding OOP

Chapters 3 and 4 present the fundamentals of class definitions and objects. In Chapter 5 we look at how classes are organized into files and directories and how the JVM locates classes. If you find that object-oriented programming (OOP) remains somewhat vague, your understanding of the concepts involved will deepen as you see OOP techniques applied to graphics, threading, I/O, and other areas in subsequent chapters. We return to class structure, design, and analysis in Chapter 16 where we give a brief overview of the Unified Modeling Language (UML). UML provides a systematic approach to the design of classes and to analysis of the interactions among objects. We then use UML to design a set of classes for a distributed computing example.

4.12   Web Course materials

The Web Course Chapter 4: *Supplements* section provides more discussion of inheritance and the overriding and overloading features. There is also discussion of security aspects of Java including the checking of code by the JVM during class loading. It also gives a brief overview of the security manager.

The Chapter 4: *Tech* section provides additional discussion and demonstration programs for the vector/matrices in Java, the complex number class, random number generation, and the improved histogram class. The Chapter 4: *Physics* section continues with a tutorial on numerical computing techniques with Java.

References

[1] Donald Knuth, *The Art of Computer Programming: Semi-numerical Algorithms Volume 2*, 3rd edn, Addison-Wesley, 1997.
[2] W. H. Press, B. P. Flannery, S. A. Teukolsky and W. T. Vetterling, *Numerical Recipes in C: The Art of Scientific Computing*, Cambridge, 1992. (Subsequent versions are available for Fortran 90 and C++.)

Resources

Ronald Mak, *Java Number Cruncher: The Java Programmer's Guide to Numerical Computing*, Prentice Hall, 2003.

Chapter 5
Organizing Java files and other practicalities

5.1 Introduction

In this chapter we look first at several topics related to the organization of Java files. In fact, a scheme for organizing Java files and classes comes built into the language. When a class is used, the name of the class includes, either explicitly or implicitly (via the `import` directive), its location in a particular *package*. The Java package resembles the code libraries of other languages and provides a name space that successfully avoids name collisions. In practice, a large class library will contain many packages – the J2SE 1.4 class library contains over 100 separate packages – arranged in some sensible order. A very small, single-purpose library might reside entirely in just one package.

In some of the examples in previous chapters the code included the `public` modifier. We finally explain in this chapter exactly what that modifier does. It and the other access modifiers determine what classes, methods and fields can be used by methods in other classes and subclasses, in the same and in other packages.

For faster downloading, you can pack your Java packages, classes, images, audio files and other program resources into a single file called a JAR (Java Archive) file. JAR files use the ZIP format and compression system (a variation of Lempel-Ziv) to hold files and to maintain internally a hierarchical directory system like that on disk. We show how to create JAR files and how to extract files from them. We then discuss the `pack200` tool in J2SE 5.0 for compressing JAR files even further.

Other topics presented in this chapter, include the `javadoc` tool for automatic documentation of packages and classes, distributing applet files into subdirectories, declaring constants, and coding style conventions. We also look at string formatting of numerical values with the tools available in J2SE 1.4 and with the new `tools` added in J2SE 5.0.

5.2 Class definition files

We first summarize the format of Java files. A file containing a Java class definition must follow these guidelines:

- A Java class definition must fit into one file; it cannot be split among multiple files.
- Only one `public` class allowed per file (we discuss access modifiers in Section 5.3.3).
- The file name must match exactly, including usage of upper and lower case characters, the name of the `public` class definition in the file plus the extension "`.java`".

It is also recommended that each class definition appear in its own file except for small helper classes used only by the primary class. The compilation of a Java file results in separate "`.class`" files for each class definition in the file.

5.3 Packages

A package is the Java version of a library. It allows for the grouping of "`.class`" files that share a common purpose. The package organization follows that of a hierarchical file directory system. To create a package you place a group of Java source files into the same directory and in each source file include a `package` directive with the name of the directory at the top of the file.

For example, we put the files `TestA.java` and `TestB.java` into the directory `mypack`, which is a subdirectory of `myApps`. On a Windows platform the file paths might look like

```
C:\myApps\mypack\TestA.java
```

and

```
C:\myApps\mypack\TestB.java
```

At the top of each file we insert the statement

```
package mypack;
```

as shown in the following code:

```
package mypack;                    package mypack;
public class TestA {               public class TestB {
  public int a;                      public double x;
  public TestA (int arg1) {          public TestB (double y) {
    a = arg1;                          x = y;
  }                                  }
}                                  }
```

A program that needs to use these classes must include, either explicitly or implicitly (we discuss the `import` directive below), the package name wherever it gives

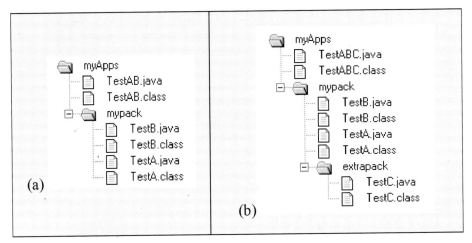

Figure 5.1 (a) Directory hierarchy for the package mypack. (b) Directory hierarchy for the package myPack and its sub-package mypack.extrapack.

the class type name. For example, we illustrate how to use package mypack with the program TestAB.java, shown below.

```
public class TestAB {
   public static void main (String[] args) {
      mypack.TestA testa = new mypack.TestA (4);
      mypack.TestB testb = new mypack.TestB (31.3);
      System.out.println ("Prod = " + (testa.a * testb.x));
   }
}
```

The package names are relative to the current directory or to a *classpath* setting. (We discuss the CLASSPATH setting in Section 5.7.) If the program TestAB.class resides in the c:\myApps directory, then by default the compiler and JVM look for the packages relative to this directory. The term mypack.TestA tells the compiler to look for the TestA.class in the subdirectory mypack relative to the directory where TestAB.class resides. Figure 5.1(a) shows the directory hierarchy for the files.

Move to the directory myApps and compile the file:

```
C:\myApps> javac TestAB.java
```

The compiler looks for the TestA.class and TestB.class files in the mypack subdirectory relative to the current directory. They are compiled if the compiler does not find class files newer than the last modification date of the source files. The compiler sees the combination of the package and class names for mypack.TestA and mypack.TestB and treats the " . " as if it were the directory name separator.

You can also compile the package files directly, but if the compilation occurs from the higher directory, you must specify the proper directory name:

```
C:\myApps> javac mypack/TestAB.java
```

(The `javac` compiler interprets the forward slash "`/`" as a directory separator. The backslash "`\`" also works on Windows.) Similarly, the JVM by default looks for class files in subdirectories as specified by the package names.

Just as a subdirectory can itself contain lower levels of subdirectories, packages can be nested to lower levels as well. For example, the class definition for `TestC` shown below includes the directive "`package mypack.extrapack;`". The `TestC.java` file must then go into the subdirectory `myApps/mypack/extrapack` (or `myApps\mypack\extrapack` on a MS Windows platform) as shown in Figure 5.1(b).

```
package mypack.extrapack;

public class TestC {
   public boolean flag;
   public TestC (boolean b) {
      flag = b;
   }
}
```

When the compiler acts on the `TestABC` class shown below, it uses the class specification `mypack.extrapack.TestC` to look in subdirectory `mypack/extrapack/` relative to the directory of `TestABC`. It compiles `TestC.java` if there is no class file or if the class file is older than the `TestC.java` file.

```
public class TestABC
{
   public static void main (String[] args) {

      mypack.TestA testa = new mypack.TestA (4);
      mypack.TestB testb = new mypack.TestB (1.3);

      mypack.extrapack.TestC testc =
         new mypack.extrapack.TestC (true);

      if (testc.flag)
         System.out.println ("testa.a = " + testa.a);
      else
         System.out.println ("testb.x = " + testb.x);
   }
}
```

All classes belong to some package. If a class definition does not include the `package` directive, then it goes into the default *unnamed* package maintained internally by the JVM. Note that by convention package names begin with a lower-case letter while class names begin with an upper-case letter. In fact, package names are often completely lower case, as is true of all but 28 of the 135 packages in J2SE 1.4.2. (Those 28 packages with upper-case characters are all part of the CORBA system in Java, which we discuss in Chapter 19, and which use upper-case names for consistency with the CORBA naming conventions.)

5.3.1 Namespaces

Packages were also designed to provide namespace protection – that is, in order to avoid name collisions. In our examples above, we used the not uncommon names `TestA`, `TestB`, etc. If we kept all of our source code in one large directory, we would have to think of more creative names to avoid confusing the various test classes we might write. By organizing our classes in packages, which means they must be organized on disk in the same directory hierarchy as used for the package names, we have a way to keep like classes together and unlike classes somewhere else. That way, we could create a package named `basic` for the simple test classes from the first five chapters of this book and a package named `graphics` for our graphics demos, etc.

Now consider the commercial Java packages one might purchase. Several different competitive companies might offer Java graphics software, for example, each with similar types of graphing classes, and each with similar or perhaps even identical names for some of the classes. If you have two or more such software libraries installed, how do you distinguish between them? The answer is with packages. Each vendor should create a separate package name for that vendor's offerings. Still, there could be naming conflicts if two vendors choose identical package names.

Suppose vendors ABC and XYZ both create a `PieChart` graphing class. To keep the two different `PieChart` classes separate, we would hope that one might be in a package named, for example, `abc`, while the other `PieChart` class would be in a package named `xyz`. Thus you would refer to either `abc.PieChart` or `xyz.PieChart`, depending on which version you wanted.

An ingenious naming convention promulgated by Sun from the earliest days of Java is to name packages based on a company's internet domain name with the order of the names reversed. Since domain names are unique, package names based on domain names are completely under the control of the company that owns that domain name. Thus our companies ABC and XYZ would probably name their packages `com.abc` and `com.xyz`, respectively.

In practice, company ABC might have several products – say a scientific graphics package and a business graphics package, along with other non-graphics classes. Thus, it might offer several packages – `com.abc.graphics.scientific`, `com.abc.graphics.business`, and `com.abc.utils`, etc.

This package-naming convention is so widely used in the Java industry as to be completely standardized.

Packages can be nested to any depth. In most situations the first two or more names reflect the domain name of the vendor or creator of the package. Many open source packages exist as well, with package names like `org.apache.ant`, for example. Another example is the CORBA package (see Chapter 19). CORBA is maintained by the Open Management Group, a standards body with the internet domain name `omg.org` (see `www.omg.org` for much more information about the OMG). Thus all the CORBA classes live in package structures that begin with `org.omg.CORBA`, `org.omg.Messaging`, etc. As mentioned above, these are the only packages in the entire Java class library that use uppercase letters anywhere in the package names.

5.3.2 Import

Unless special arrangements are made, all classes must be referred to by their complete name, which includes the full package name (such names are called "fully qualified" names). Therefore, in `TestABC.java` above we specified the full package name of the `TestC` class in the declaration and `new` expression:

```
mypack.extrapack.TestC testc = new mypack.extrapack.TestC ();
```

An exception to this rule holds for all the classes in the standard `java.lang` package. Thus, whenever we use the `Float` class, for instance, we merely use something like

```
Float someFloat = new Float (3.14);
```

There are nearly 3000 classes in the 135 packages in the J2SE 1.4.2 standard class library (over 3000 classes in 165 packages in J2SE 5.0). Commercial class libraries that one might use add many more packages. Using fully qualified class names obviously becomes unwieldy in any non-trivial program since such a program will involve many different classes from many different packages.

Thankfully, the Java compiler provides the special arrangements necessary to abbreviate most fully qualified class names with just the short class name by using the `import` directive. That is, the compiler understands a shortcut that allows us to refer to something like `java.awt.event.ActionEvent` (see Chapter 7) as simply `ActionEvent`. The appearance of an `import` directive tells the compiler where to look for class definitions when it comes upon a short class name that is not fully qualified. Every class in the `java.lang` package is imported automatically. That's why we can refer to the `java.lang.Float` class as just `Float`. To use this abbreviation technique for classes in other packages requires that the class name or package name appear in an `import` statement.

As an example in a class definition for an applet class, this `import` directive:

```
import java.applet.Applet;
```

allows the source code to refer to `Applet` rather than `java.applet.Applet`. That is, instead of

```
public class MyApplet extends java.applet.Applet {. . .}
```

we use

```
import java.applet.Applet;
public class MyApplet extends Applet {. . .}
```

Any appearance of the name `Applet` is automatically interpreted as `java.applet.Applet`. There can be multiple `import` statements to import fully qualified class names from multiple packages. All `import` statements appear at the top of a class definition file, immediately after the `package` statement and before the opening line of the class body itself. Of course, valid comments may appear anywhere.

The `java.applet` package is small, containing only the `Applet` class. A large package like `java.io` contains (in Java 1.4.2) 50 classes plus 16 exception classes and 10 interfaces (see Chapter 9 for a description of I/O in Java). To use the `import` notation for each of these classes could become unwieldy even though a typical program would be highly unlikely to use more than just a few of the 50 classes. For example, a typical program might use the following classes from the `java.io` package: `BufferedWriter`, `FileInputStream`, `FileOutputStream`, `FileReader`, `FileWriter`, etc. That program would also likely check for at least `java.io.IOException` if not some of the sub-exceptions like `java.io.EOFException` and `java.io.FileNotFoundException`.

In this case the list of imports at the top of the file would look like this:

```
import java.io.BufferedWriter;
import java.io.FileInputStream;
import java.io.FileOutoutStream;
import java.io.FileReader;
import java.io.FileWriter;
import java.io.IOException;
import java.io.EOFException;
import java.io.FileNotFoundException;
. . .
public class MyClass {. . .}
```

Using classes from several other packages, as most non-trivial programs do, would lead to a very long list of `import` statements. To reduce the size of the list, the compiler also supports the * notation to import an entire package at once rather than specifying each imported class name separately. Therefore, the above example can be reduced to

```
import java.io.*;
. . .
public class MyClass {. . .}
```

With this notation, our class can use any of the 50 classes, 16 exceptions, and 10 interfaces in the java.io package by referring simply to the short names.

What the import statement with the * notation really does is instruct the compiler into which package(s) to look whenever a short unqualified class name is encountered. Therefore, in the example above, if an unqualified FileWriter class name appears, the compiler is able to find that class in the java.io package. It is not an error to use the fully qualified name even when an import statement appears that would permit the short unqualified name.

Note that the * notation does not look into sub-packages. For example, importing java.awt.* does not also import java.awt.event.*.

As a complete but simple example, consider our TestABCApplet class, which uses the import directives

```
import mypack.*;
import mypack.extrapack.*;
```

These instruct the compiler to look in the mypack package and the mypack.extrapack package for any class references that cannot be found in the default package (java.lang).

```
import java.applet.Applet;
import mypack.*;
import mypack.extrapack.*;

public class TestABCApplet extends Applet
{
  public static void main (String[] args) {
    TestA testa = new TestA (4);
    TestB testb = new TestB (31.3);
    TestC testc = new TestC (true);

    if (testc.flag)
      System.out.println ("testa.i = " + testa.i);
    else
      System.out.println ("testa.x = " + testb.x);
  }

  // Paint message in Applet window.
  public void paint (java.awt.Graphics g){
    g.drawString ("TestABCApplet", 20, 20);
  }
}
```

In rare circumstances, two different packages might use the same class name. In this case the compiler cannot know which class you really want. For example, java.lang contains the Object class, and the org.omg.CORBA package contains an interface named Object. These two names collide if you use import org.omg.CORBA.*, in which case the compiler issues a warning. The solution is to use the fully qualified names whenever a name collision occurs. In other words, if you import org.omg.CORBA.* and need to refer to a plain java.lang.Object, then refer to it as java.lang.Object and refer to CORBA objects as org.omg.CORBA.Object.

C/C++ programmers new to Java usually assume that an import statement actually brings code into the class bytecode files in a manner similar to that of include files in C/C++. However, that is not the case. An import statement simply provides an address where the compiler can look for class definitions.

5.3.3 Access rules

Access or visibility rules determine whether a method or a variable can be accessed by another method in another class or subclass. We have used the public modifier frequently in class definitions. It makes classes, methods, or data available to any other method in any other class or subclass in any package.

Java provides four access levels:

- public – access by any other class, anywhere.
- protected – accessible by classes in the same package and by any subclasses of those classes whether in the same package or in other packages.
- Default (also known as "package private") – accessible to classes in the same package but not by classes in other packages, even if they are subclasses of classes in the package.
- private – accessible only within the class. Even methods in subclasses in the same package do not have access.

These access rules allow one to control the degree of encapsulation of classes. For example, you may distribute your class files to other users who can make subclasses of them. By making some of your data and methods private, you can prevent the subclass code from interfering with the internal workings of your classes. Also, if you distribute new versions of your classes, you can change or eliminate private fields and methods without affecting the subclasses made from your superclasses.

If you do not put your class into a package, it is placed into the "default unnamed package." For simple demonstration programs this usually suffices. However, with more serious programs you should use packages since otherwise any other class in the unnamed package has access to your class's fields and methods that are not set to private.

5.4 The `final` modifier and constants

We already briefly mentioned the final modifier in Chapter 3. It indicates that a data field cannot be modified. In the declaration

```
final double PI = 3.14;
```

any attempt to assign a new value to the PI variable results in a compiler error. Therefore final is useful to guarantee that constants are truly constant.

Constants should also be declared static, which means there will be only one copy for the entire class (there's no point in wasting space with multiple copies of constants). The following class defines the constant TWO_PI:

```
public class MyMath {
    public final static double TWO_PI = 6.28;
    . . .
}
```

Then other classes can reference the constant, as in

```
    . . .
    double y = theta / MyMath.TWO_PI;
    . . .
```

The final modifier is also used with methods to indicate that they cannot be overridden by subclasses:

```
public class MyMath {
    . . .
    public final double myFormula() {. . .}
}
```

This ensures that any subclass of MyMath does not override the myFormula() method. It also helps to improve performance since the JVM does not need to check for overriding versions each time the method is invoked.

5.5   Static import in J2SE 5.0

Many classes, including many in the Java core libraries, contain static constants that are used within the class and are also useful outside the class. For example, the java.lang.Math class contains the constants PI and E for π and e, respectively. As another example, we see in Chapter 7 that the BorderLayout class contains constants such as NORTH, CENTER, etc. that are useful for laying out graphics components.

Prior to Java version 5.0, the only way to access those constants was by fully spelling out the names Math.PI, Math.E, BorderLayout.NORTH, etc. in your code. With static imports, you can use just PI, E, and NORTH without all the extra typing. To use static imports, add the static keyword to the import statement as follows:

```
import static java.lang.Math.PI;
```

Then in the code body use

```
double area = PI * radius * radius;
```

instead of

```
double area = Math.PI * radius * radius;
```

Wildcards also work in the `import static` statement:

```
import static java.lang.Math.*;
```

For infrequent use, typing the complete name `Math.PI` is probably easier, but if constants from another class are used extensively, then the static import feature saves quite a bit of typing. In addition to the constants, all the many static methods in the `Math` class are also available when you use the wildcard static import line above. For example,

```
double logarithm = log (number);
```

instead of

```
double logarithm = Math.log (number);
```

For extensive use of the mathematical functions, the static import feature is a great addition to the Java language.

As explained in Chapter 4, constants can also be defined in interfaces, where they are automatically `static`, even if the `static` keyword is not used. An example from Chapter 4 was

```
public interface MyConstants {
   double G = 9.8;
   double C = 2.99792458e10;
}
```

Prior to Java version 5.0, these constants could be accessed with either the `MyConstants.G` notation as used above with the `Math` constants or by implementing `MyConstants` in the class definition. However, static import also works fine with a package, as shown here:

```
import static somepackage.MyConstants;
public class UsesMyConstants {
   /** Calculate E = m * c-squared. **/
   double calculateEnergy (double mass) {
     return = mass * C * C;
   }
}
```

We note that this technique works only if the constants interface is in a package. That is, classes or interfaces using the default package, which is discouraged anyway, cannot be statically imported.

It was pointed out in Chapter 4 that using an interface just to access a set of constants is ill advised and considered bad programming style since nothing is really being implemented. Use a class instead and then statically import the class.

5.6 JAR files

A useful technique for organizing classes and packages is to combine them into a file called a JAR or Java Archive file. JAR files (often referred to as "jar" files but we use the uppercase JAR in this book) are based on the ZIP archiving and compression system. JARs provide a number of advantages:

- Loading a single large JAR file over the network instead of several small class files reduces the overhead that occurs for each network transfer.
- Compression, which is the default but may be turned off, also helps for faster loading since a compressed JAR file is usually significantly smaller than the sum of all the class files in the JAR.
- Internally the ZIP format maintains the directory structure of the packages.

The applet `TestABCApplet` discussed above (section 5.3.2) needed classes from two packages. Here we use the `jar` tool (provided with the SDK) to create a JAR file that holds the `TestABCApplet` class file and the `mypack` and `mypack.extrapack` packages. The command line and output here shows the `jar` tool in action:

```
c:\. . .\myApps>jar -cvf TestABCApplet.jar TestABCApplet.class
mypack
added manifest
adding: TestABCApplet.class(in = 1109) (out= 648) (deflated 41%)
adding: mypack/(in = 0) (out= 0) (stored 0%)
adding: mypack/extrapack/(in = 0) (out= 0) (stored 0%)
adding: mypack/extrapack/TestC.class(in = 250) (out= 203)
(deflated 18%)
adding: mypack/extrapack/TestC.java(in = 129) (out= 93)
(deflated 27%)
adding: mypack/TestA.class(in = 237) (out= 194) (deflated 18%)
adding: mypack/TestA.java(in = 111) (out= 81) (deflated 27%)
adding: mypack/TestB.class(in = 237) (out= 192) (deflated 18%)
adding: mypack/TestB.java(in = 110) (out= 79) (deflated 28%)

C:\. . .\myApps>
```

The `-c` option means to "create" the JAR file; the `v` option means to use "verbose" output; and the `f` option specifies the file name of the created file. JAR files should use the ".jar" filename extension. We see that the command above adds the applet class file `TestABCApplet.class` and the classes in the packages to the archive, compresses them, and maintains the package directory structure.

for running the program. In general, setting the CLASSPATH on the command line rather than in an environment variable is preferred. That way, different classpaths can be used for different programs. Interestingly, in J2SE 1.4.2 and below, the javac tool only recognizes the -classpath option while the java tool recognizes both -classpath and -cp. In J2SE 5.0, both tools recognize both options.

5.8   Applet directories

For organizational purposes it is often convenient to put applet files into different directories from the one holding the web page that includes the applet. For example, suppose that we want the simpleApplet.html web page, located in the myApplets directory to use an applet class file located in

```
myApplets/Beginners/HelloWorld/simpleApplet.class
```

Then the applet tag in the web page should be

```
<applet
  code = "simpleApplet.class"
  codebase = "Beginners/HelloWorld/"
  width = "100" height = "100">
</applet>
```

The code attribute can only include a class file name, not its directory path. So the following tag does not work:

```
<applet
  code = "basic/demo/simpleApplet.class" *** ERROR ***
  width = "100" height = "100">
</applet>
```

The directory location of the applet classes can also go above these directories as long as they don't go outside limits imposed by the web server and the browser security manager (security managers are discussed in Chapter 14).

That is, suppose that the class file is in

```
Course/Code/Java/Beginners/HelloWorld/
```

and the web page file is in

```
Course/Java/Chapter01/
```

Then you can use:

```
<applet
  code = "simpleApplet.class"
  codebase = ". ./. ./Code/Java/Beginners/HelloWorld/"
  width = "100" height = "100">
</applet>
```

Finally, it is possible for the code to reside at a completely different URL address, such as in this example:

```
<applet
  code = "HelloWorld.class"
  codebase = "http://xyz.edu/Course/Code/Java/Beginners/HelloWorld/"
  width = "100" height = "100">
</applet>
```

The applet tag also accepts JAR files in the codebase. In fact, use of the JAR format is recommended because all the required class files can be obtained in one download operation rather than in several downloads, one for each class file. In addition the JAR format is compressed, further saving download time.

For J2SE 5.0 applets, the highly compressed `pack200` format (see Section 5.6.3) can also be used. Most browsers only support Java version 1.1, so the `pack200` format is of little use for those browsers. However, the Java Plug-in is available for the Java 2 Platform for versions 1.2 and up. The Java Plug-in is a browser plug-in that works with Microsoft Internet Explorer and Mozilla browsers to enable Java 2 support in those browsers. (See Web Course Chapter 1: *Supplements* and reference [1]). The Java Plug-in has been updated in J2SE 5.0 to support the `pack200` format.

5.9 Javadoc

The `javadoc` tool provided with the SDK provides for automatic organization of package and class documentation. Applying the tool to a package results in a set of web pages that list the fields and methods of each class. The Javadoc API is quite powerful. The entire voluminous Java API Specification documentation is created with `javadoc`.

You can add comments and other information describing the class in the source files with the use of special tags, and these then appear in the `javadoc` web pages. We noted in Chapter 2 that Java recognizes a block of comments bracketed with `/**. . .*/` in addition to the usual `/*. . .*/`. The double asterisks tell `javadoc` to include the comments in its documentation output. For example, here is a `HelloWorld.java` file commented in the Javadoc style:

```
import java.applet.Applet;
import java.awt.Graphics;

/** This applet tests graphics.
  * (This comment block that describes a class must be
  * placed immediately before the class line.)
  */
public class HelloWorld extends Applet {
```

```
/** This method paints the text to the screen.
 * (The comment block that describes a method must be
 * placed immediately before the method line.)
 */
public void paint (Graphics g) {
   g.drawString ("Hello World!", 50, 25);
}
}
```

Note that the `/**. . .*/` comments should be placed immediately before the class, method, or variable they are describing. Running `javadoc` on this file creates several hypertext documentation files such as `HelloWorld.html` whose contents are shown in the Web Course Chapter 5. Since the output files are in hypertext format, you can also use hypertext tags in the `javadoc` comments, such as `<br>` for line breaks and list tags such as `<ul><li>abc</li></ul>`.

Special Javadoc tags that begin with the "`@`" symbol allow you to put specific information into the documentation output. Examples include `@author`, `@param`, `@throws`, and `@return`, which specify the author of the class, a description of a method parameter, the exceptions that might be thrown, and a method's return value, respectively.

Over subsequent versions, `javadoc` has grown increasingly sophisticated and now provides for extensive customization. The `doclets` API (via the `com.sun.javadoc` package) allows for Java classes that can modify the output from `javadoc`. The `taglets` API allows for Java classes that create custom tags. See the Javadoc reference [2] for documentation on `doclets` and `taglets`.

5.10 Coding conventions

As with any programming language, you should strive to make your Java programs as readable and understandable as possible, both for yourself as well as for others who might use your code. This means clear, concise, and plentiful comments and descriptive, informative names for classes, method, and variables. Furthermore, a consistency in the style of the code helps both in reading and debugging your programs.

There are a number of Java coding standards that have been proposed. For our demonstration programs in the following chapters, we follow the guidelines set out by noted Java guru Doug Lea. The full standard is too lengthy to include here (see Lea [3] for the complete standard) but we give the essentials that suffice for our short demos.

Our naming conventions go as follows:

- `OurClass` – class names use concatenated words with the first letter of each word upper case.
- `ourMethod ()` – non-private methods use concatenated words with first letter lower case and the first letter of each subsequent word upper case.
- `a_priv_method ()` – private methods use lower-case words separated by "_".
- `fOurInstanceVar` – member variables start with "f" (for field) and consist of concatenated words with the first letter of each word uppercase.
- `our_local_var` – local variables are in lower case with the words separated by "_".
- `fOurStaticVar_` – static variables begin with "f" like member variables but end with two "_" characters.
- `javatech.pack` – package names use lower case.
- `A_CONSTANT_VAL` – constants are upper-case words separated by "_".

Our interfaces are distinguished from classes by appending "Ifc" or "able". We use `javadoc` comments before each class and method (unless they are short and obvious). We terminate them with "**/" for symmetry with the "/**" that must start a `javadoc` comment.

The sample code shown below illustrates the code styling we use for the remainder of this book. Each distinct code section, such as a method or an `if-else` statement, is indented to the right from the section in which it is nested. All primitive variables are initialized explicitly. (Default values would be assigned automatically but by initializing them you make clear that the values given are what you intended for those variables.) For the class definition and for long methods we put the name in a comment after the final brace. (We use "// ctor" for long constructors.)

```
// Non javadoc comments, authorship, and
// class development history.

package javatech.xxx.yyy.zzz;
import java.io.*;

/** javadoc comments about the class. **/
public class SomeClassName
{
    int fInstanceVal = 1;
    double fVal = 0;
    Integer fInstanceRef;

public SomeClassName {
    . . . constructor code. . .
} // ctor — for longer constructors
```

```java
/** Describe the method. **/
  public void methodName (. . .) {
    int x = 5;
    some_method (x);
    if (test) {
      do_something ();
    }
    else {
      . . .
    }
    . . .
    try {
      xxx ();
    }
    catch (Exception e) {
      handle_it ();
    }
  } // methodName - for longer methods

  /** Longer comment.
    * -- this asterisk is ignored by javadoc
    **/
  public void someMethodWithLotsOfParameters (
    int param1,
    int param2,
    etc.
  ) throws SomeException {
    . . .
  } // someMethodWithLotsOfParameters

  /** Private method names with underscores. **/
  private void some_method (int i) {
    . . .
  } // some_method

  /** Getter method. **/
  double getVal () {
    return fVal;
  }

  /** Setter method. **/
  void setVel (double val) {
    fVal = val;
  }

} // class SomeClassName
```

The getVal() and setVal()are examples of the common *getter* and *setter* methods used to get and set the value of a particular variable whose name is included in the method name. In this way we follow the important JavaBeans standard although the use of JavaBeans is not important to this book. We also note that the consistent use of the fMemberVariable notation for member variables and local_variable_name notation for local variables inside a method makes it impossible to accidentally shadow a member variable with a local variable (see brief discussion on shadowing in Chapter 4).

5.11 Formatting numbers

Until recently Java separated I/O from the formatting. By formatting we mean converting a number to a string in a specific form. A typical example is to specify a fixed number of decimal places to a floating-point value. Since strings were usually intended for graphical displays such as labels and text fields (we discuss Java graphics in Chapters 6 and 7), the procedure was to format a string in one operation and then in a second operation send the formatted string to where it was to be displayed. However, J2SE 5.0 provides the means both to format and send data to a file or stream at the same time. Here we give a brief introduction first to the older formatting approach and then to the new 5.0 techniques.

5.11.1 Format, NumberFormat, and DecimalFormat

Number formatting techniques available as of Java 1.4 and earlier are still useful and commonly found in Java programs. In many of the previous demonstration programs, we often found that a simple floating-point operation such as division could result in a long fraction when we used the default conversion of a double value to a string. Control of the format for a numerical value to string conversion was added in Java 1.1 with the java.text package. This package includes the abstract class java.text.Format, which is subclassed by java.text.NumberFormat. This class is in turn subclassed by java.text.DecimalFormat.

These classes focus primarily on internationalization of numerical output such as using either a period or comma for the decimal point according to the current locale setting. (See the tutorial on internationalization of formatting given in the reference [4].) However, scientific notation was added to DecimalFormat with Java 1.2.

The approach of DecimalFormat is to specify the format with a pattern of symbols. For example, the pattern "0.000" indicates that the number should begin with 0 if it is less than 1.0 and that the number of decimal places should be three and padded with 0 if necessary. The pattern "0.###" indicates that the number should begin with 0 if it is less than 1.0 and that the number of decimal places should be up to three, padded with blanks if the fraction needs less than three places.

An exponential pattern appends "E0" to the right side. So "0.000E0" results in three decimal places in the significand and then an exponent, such as "1.234E10". See the Java 2 API specifications for DecimalFormat to find the complete set of formatting pattern symbols.

The recipe for formatting a floating-point value goes as follows. First specify a string pattern such as

```
String fmt = "0.00#";
```

Then create an instance of the DecimalFormat class with this pattern:

```
DecimalFormat df = new DecimalFormat (fmt);
```

Invoke the format() method to create a formatted string:

```
double q = 10.0/4.0;
String str = df.format (q);
System.out.println ("q = " + str);
```

In this case the resulting string is:

```
q = 2.50
```

The DecimalFormat object can be reused for new format patterns by invoking the applyPattern (String format) method. The following snippet from the program DecimalFormatDemo illustrates several format patterns for floating-point output:

```
. . . code segment in DecimalFormatDemo . . .
    double q = 1.0/3.0;

    // Define the format pattern in a string
    String fmt = "0.000";

    // Create a DecimalFormat instance for this format
    DecimalFormat df = new DecimalFormat (fmt);

    // Create a string from the double according to the
    // format
    valStr = df.format (q);

    System.out.println ("1.0/3.0 = " + valStr);

    // Can change the format pattern:
    df.applyPattern ("0.00000");
```

```
        valStr = df.format (q);
        System.out.println ("1.0/3.0 = " + valStr);

        // The# symbol indicates trailing blanks
        df.applyPattern ("0.#####");
        valStr = df.format (1.0/2.0);
        System.out.println ("1.0/2.0 = " + valStr);

        // Fix the number of places in the fraction.
        df.applyPattern ("0.00E0");
        valStr = df.format (1000.0/3.0);
        System.out.println ("1000.0/3.0 = " + valStr);

        // Scientific notation
        df.applyPattern ("0.00E0");
        valStr = df.format (3.0/4567.0);
        System.out.println ("3.0/4567.0 = " + valStr);

        // Negative infinity
        df.applyPattern ("0.000E0");
        valStr = df.format (-1.0/0.0);
        System.out.println ("-1.0/0.0 = " + valStr);

        // NaN
        df.applyPattern ("0.000E0");
        valStr = df.format (0.0/0.0);
        System.out.println ("0.0/0.0 = " + valStr);
. . . .
```

The following shows the output of `DecimalFormatDemo`:

```
1.0/3.0 = 0.333
1.0/3.0 = 0.33333
1.0/2.0 = 0.5
1000.0/3.0 = 3.33E2
3.0/4567.0 = 6.57E-4
-1.0/0.0 = - ?
0.0/0.0 =?
```

Note that the `format()` method returns the infinity and Not-a-Number (NaN) values (see Section 2.12) as a "?" character.

5.11.2   The `printf()` method

The `Format` subclasses discussed above clearly have limited formatting capabilities and are rather clumsy to implement. Also, the two-step process of formatting a string and then sending it to where it is needed can be rather tedious, especially for output to the console or to a file. Those with experience in the C language have missed the `printf()` function, which offers a wide range of formatting options and also combines formatting and output.

To satisfy the demands of such programmers and to facilitate the porting of C programs to Java, J2SE 5.0 comes with the class `java.util.Formatter`, which can both format numerical output into a string and send the string to a file or other destination (we discuss I/O in Chapter 9). Numerical values are formatted according to format *specifiers* like those for the `printf()` function in C.

In fact, J2SE 5.0 went a step further and actually added a `printf()` method to the `PrintStream` class (see Chapter 9), of which `System.out` and `System.err` are instances. So now you can use `System.out.printf()` to send formatted numerical output to the console. It uses a `java.util.Formatter` object internally and closely emulates the `printf()` function in C.

In Section 9.4.2 we return to the `java.util.Formatter` class so that we can discuss it in the context of Java I/O. Here we introduce the `printf()` method so that you can begin to take advantage of it.

The simplest of the overloaded versions of the method is

```
printf (String format, Object . . . args)
```

The " . . . " indicates the varargs functionality, which we noted in Chapter 1 was introduced with J2SE 5.0. It allows a method to accept a variable number of arguments. We note that the arguments can be primitives as well as objects, e.g. the wrappers for the primitives.

The `format` argument is a string in which you embed specifier substrings that indicate how the arguments appear in the output. For example,

```
double pi = Math.PI;
System.out.printf ("pi = %5.3f%n", pi);
```

results in the console output

```
pi = 3.142
```

The format string includes the specifier "`%5.3f`" that is applied to the argument. The `%` sign signals a specifier. The *width* value 5 requires at least five characters for the number, the *precision* value 3 requires three places in the fraction, and the *conversion* symbol `f` indicates a decimal representation of a floating-point number.

A specifier needs at least the conversion character, of which there are several besides `f`. Some of the other conversions include

```
d - decimal integer
o - octal integer
e - floating-point in scientific notation
```

There are also special conversions for dates and times (calendar and time classes are discussed in Chapter 10). The general form of the specifier includes several optional terms:

```
%[argument_index$][flags][width][.precision]conversion
```

The `argument_index` indicates to which argument the specifier applies. For example, `%2$` indicates the second argument in the list. A `flag` indicates an option for the format. For example, "+" requires that a sign be included and "0" requires padding with zeros. The `width` indicates the minimum number of characters and the precision is the number of places for the fraction.

There is also one specifier that doesn't correspond to an argument. It is "`%n`" and outputs a line break. A "`\n`" can also be used in some cases, but since `%n` always outputs the correct platform-specific line separator, it is portable across platforms whereas "`\n`" is not.

We don't have space here for more than this brief overview of the new format tools. The Java API Specifications for J2SE 5.0 provides a lengthy description of the `java.util.Formatter` class and all of the specifiers and their options. The following program provides several examples of `printf()`:

```
. . . code segment in PrintfDemo . . .
    double q = 1.0/3.0;

    // Print the number with 3 decimal places.
    System.out.printf ("1.0/3.0 = % 5.3f %n", q);

    // Increase the number of decimal places
    System.out.printf ("1.0/3.0 = %7.5f %n", q);

    q = 1.0/2.0;
    // Pad with zeros.
    System.out.printf ("1.0/2.0 = %09.3f %n", q);

    q = 1000.0/3.0;
    // Scientific notation
    System.out.printf ("1000/3.0 = %7.2e %n", q);

    q = 3.0/4567.0;
    // More scientific notation
    System.out.printf ("3.0/4567.0 = %7.2e %n", q);
```

Chapter 6
Java graphics

6.1 Introduction

Java's graphics capability has always been a leading feature of the language. The Java designers clearly expected the graphical user interface (GUI) to dominate interactions with Java programs on all but the smallest platforms. Java appeared at the start of the Internet boom and applets were expected to bring interactivity to the browser. Many thought that Java would also quickly become popular for standalone client applications on platforms with graphical operating systems.

In Java 1.0, however, the graphical elements provided for workable interfaces but they appeared crude compared to platform-specific graphics developed with other languages. The goal of portability had led to a lowest common denominator approach that was not very pretty. This became one of the main stumbling blocks that prevented Java from becoming a popular language for desktop applications.

However, with the inclusion of the *Swing* packages in version 1.2, Java graphics took a huge leap forward in visual appeal and in the breadth and depth of its features. With subsequent versions, Java graphics continued to improve and now compares quite well with that available with any other programming language and still provides for relatively easy portability.

In this chapter we introduce Java graphics starting with a quick overview of the Abstract Windowing Toolkit from Java 1.0. We then look at the Java Foundation Classes system, also known as "Swing," in some detail. We wait until Chapter 7 to discuss how to bring interactivity to the user interface. This chapter focuses on using Swing components, drawing on Swing panels, and displaying text and images. We use the display of histograms as an example of a technical application of Java graphics.

We do not have space here to provide in-depth descriptions of all the capabilities of the many graphics related classes. We focus on the fundamental features of the visual components and illustrate their use in demonstration programs. See the Web Course Chapter 6 and the Java 2 API specifications and the references for more information about Java graphics resources.

160

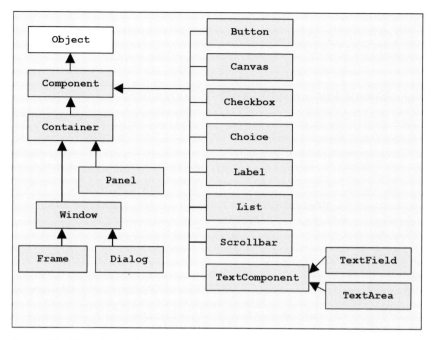

Figure 6.1   The primary classes in the AWT hierarchy.

6.2 AWT

The AWT (Abstract Windowing Toolkit) package (`java.awt`) and sub-packages came with Java 1.0 and still provide the essential set of classes for all Java graphics. Even Swing builds on the graphics base in AWT. The diagram in Figure 6.1 shows the primary members of the AWT class hierarchy.

The key `Component` class provides the base class for all the AWT visual components and also for the Swing components. This class contains a very large number of public methods (see the Java API description for `Component`) that provide access to and control of attributes of a visual component such as its position, size, colors, and so forth. Subclasses of `Component` can override some of these methods and also include new methods to deal with the specific features of the new type of component.

Another very important class, `Container`, which is itself a subclass of `Component`, provides for holding instances of other components. The `Window` subclass of `Container`, for example, provides for the *top-level* visible containers `Frame` and `Dialog` that usually hold multiple visible components.

Containers can also hold other containers. The `Panel` class, in particular, is used within a top-level container to hold and arrange its sub-components, which often are also panels. An elaborate GUI display typically employs several panels to arrange the visible *atomic* (non-container) components such as labels, buttons, text fields and text areas. As seen in the diagram above, the AWT includes the basic atomic components needed to create fairly complete, if not particularly attractive, GUI displays.

6.3   Swing: lightweight beats heavyweight

As mentioned above, the graphics prior to Java 1.2 included only the AWT in the core language packages and so did not provide for thrilling user interfaces. The AWT components came with a number of shortcomings. For example, simply creating a subclass of the `Button` class to allow for a custom button that displays an icon was and remains impractical. Other AWT visual components had similar limitations. Java programs based on the AWT became known for their bland appearance and minimal capabilities.

The basic problem is that the AWT components are closely tied to the so-called *peer* component classes written in native code for the local operating system's graphical interface. This means that Java portability required a lowest common denominator approach in which no visible component could provide more capability than what was available on all platforms. This resulted in very limited options in the appearance and performance of the components. These basic AWT components are called *heavyweight* because they must drag along all the peer component baggage.

A far more flexible approach is to open a heavyweight top-level class, such as a window frame, and then simply draw all the visible sub-components without involving any other local peer components. Such *lightweight* components are very flexible, especially when combined with the more powerful event handling structure that came with Java version 1.1 (*events*, such as mouse clicks, are discussed in Chapter 7).

The Swing set of classes (available in the `javax.swing` and related packages) consists primarily of lightweight components. Swing first became available as an independent class library that worked with Java version 1.1. Later, its packages were included in Java 1.2. It is now generally recommended that programmers use Swing for all serious graphics development on desktop platforms. When developing programs for platforms with limited memory and display capability, such as those targeted by J2ME, the pure AWT framework remains a viable choice. (The Web Course Chapters 6 and 7 provide an introduction to graphics programming with the AWT.)

Note that the Swing package names start with `javax` rather than `java`. The packages whose names start with "`java`" are often referred to as the *core* Java language packages. The `javax` package hierarchy was invented to include "standard extensions." There are now dozens of `javax` packages. So while Swing is not core, it *is* considered a *standard* part of the language and is included with the Standard (J2SE) and Enterprise (J2EE) editions for all platforms. (A number of `javax` packages, such as the `javax.comm` package discussed in Chapter 23, are considered as *optional* packages and are not shipped with any of the official editions.) Smaller platforms, such as cell phones and PDAs, often do not contain sufficient memory for Swing to run and therefore Swing does not appear in the J2ME edition of Java. (We briefly describe J2ME in Chapter 24.)

Swing brought a huge improvement in the GUI with new capabilities that ranged from buttons showing icons to checkboxes to tables with multiple columns and rows. Swing came as part of the *Java Foundation Classes* (JFC) set of packages that also include:

- Pluggable Look and Feel – the style of the components, such as the color scheme and design, can be customized and can look the same regardless of the platform
- Accessibility API – modifies the interface for easier use by the disabled
- Drag and Drop – provides for moving data between Java programs and other programs running on the native host
- Java 2D – an expanded set of drawing tools

6.4   Swing class hierarchy

The Swing classes build upon the lower level classes of the original AWT graphics packages. As the diagram in Figure 6.2 shows, the Swing user interface components extend from the `Container` and `Component` classes in `java.awt`. The diagram shows only a subset of the Swing components and how they extend the AWT components (the diagram omits some Swing components like `JTable`). Most visual component classes begin with the letter "J" while there are various supporting classes in the `javax.swing` packages that don't begin with "J."

The `JComponent` subclasses are lightweight, so they run inside a single heavyweight high-level component, such as `JFrame` and `JDialog`, and draw and re-draw themselves completely with Java; no other native code peer components are involved. Combined with the event handling process described in Chapter 7, Swing components provide very flexible and elaborate GUI tools. One can also develop custom components in a straightforward manner by extending either `JComponent` or one of its subclasses.

The number of Swing classes and their depth and complexity is far greater than with the AWT. We cover a number of aspects of Swing in this and later chapters. However, we can only touch on a small fraction of Swing's capabilities. To describe all of the classes, one popular book for Swing programming requires more than 1600 pages [1]! Another useful Swing reference is Sun's Java Swing tutorial [2].

As we indicated earlier, a drawback of this huge graphics resource is the large amount of memory that Swing GUI programs can absorb when many components are involved. Another problem comes from the fact that the browser on many desktop machines still does not include a Java 1.2 (or later) JVM. So applets with Swing components will not run in all browsers. In the Web Course Chapter 1 we

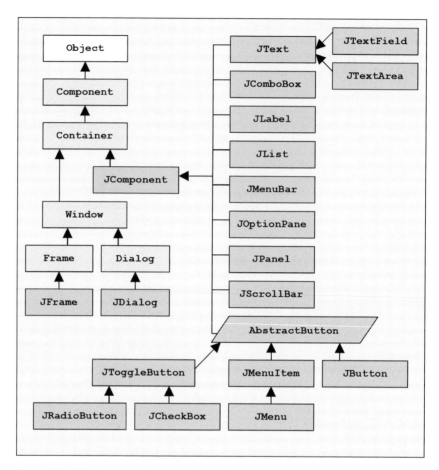

Figure 6.2 This diagram shows a subset of the components in the Swing class hierarchy. Light gray indicates classes from the `java.awt` package, dark gray classes from `javax.swing`.

provide instructions on how to set up the HTML for applets so that the browser will download an up-to-date Java Plug-in. The plug-in brings the Java support up to the current 1.4 or 5.0 level but requires a large download the first time it is used (then it is cached for later use). The long initial download can be a problem for users on slow connections. Another promising trend is that many desktop computer manufacturers (including Dell and HP) now include a recent version of the JRE and Java Plug-in with their Windows operating systems.

6.5 Containers

`Container` components can, as the name implies, hold other components. The sub-components can be *atomic* components that serve a single purpose, such as a button, or they can be containers themselves.

The top-level containers include `JApplet`, `JFrame`, and `JDialog`. These are never held inside other containers. `JApplet` is the Swing version of `Applet`. It extends the `Applet` class as shown by this diagram:

```
java.lang.Object
    |
    +--java.awt.Component
            |
            +--java.awt.Container
                    |
                    +--java.awt.Panel
                            |
                            +--java.applet.Applet
                                    |
                                    +--javax.swing.JApplet
```

The following example, which just displays a button, illustrates the essentials of creating an instance of `JApplet`:

```java
import javax.swing.*;
import java.awt.*;

/** Simple demo of adding a JComponent subclass, here a
  * JButton, to the content pane.
 **/
public class SwingButtonApplet extends JApplet
{
    public void init () {
        // Swing adds JComponents to the container's
        // "content pane" rather than directly to the panel
        // as with the AWT.
        Container content_pane = getContentPane ();

        // Create an instance of JButton
        JButton button = new JButton ("A Button");

        // Add the button to the content pane.
        content_pane.add (button);
    }
} // SwingButtonApplet
```

As shown here, the basic steps to creating a Swing interface are not very complicated. Instances of components are created and then added to a container. We add the components to a container referred to as the *content pane*. The top-level

Swing containers – `JFrame`, `JApplet`, `JDialog`, `JWindow` – are conceptually constructed of several such overlapping *panes*. The panes organize the display of the components, the interception of events, the z-ordering of components, and various other tasks. If you ever want to build a custom component, it is necessary to understand the details of the panes system. However, for most GUI building you only need to deal with the content pane.

We note that for J2SE 5.0, calling `getContentPane()` is no longer required for Swing components. This was accomplished by enhancing the `add()`, `remove()`, and `setLayout()` methods so that they forward all calls to the content pane automatically for the `JFrame`, `JDialog`, `JWindow`, `JApplet` and `JInternalFrame` components. In the above example, you would thus only need to invoke `add (button)` to place the button on the applet panel. We continue to use the content pane explicitly in this book for those readers who have not yet upgraded their Java environment to the 5.0 platform.

6.5.1 `JPanel` and `JButton`

We typically build an interface with one or more instances of `JPanel`. A `JPanel` is a container that holds other components. The following applet creates a subclass of `JPanel` called `ActionButtonsPanel` that holds two `JButton` objects. An instance of this `JPanel` is then added to the applet's content pane. Figure 6.3 shows the resulting display.

```java
import java.awt.*;
import javax.swing.*;

/** Demonstrate the Creation of a JPanel subclass. **/
public class ButtonsPanelApplet extends JApplet
{
   public void init () {
      Container content_pane = getContentPane ();
      // First create a panel of buttons
      ActionButtonsPanel buttons_panel =
        new ActionButtonsPanel ();

      content_pane.add (buttons_panel);
   }
} // class ButtonsPanelApplet

// JPanel subclass with two buttons
class ActionButtonsPanel extends JPanel
{
   ActionButtonsPanel () {
      // Create two buttons
```

Figure 6.3 The program `Buttons-PanelApplet` puts two `JButton` components on a `JPanel`.

```
        JButton add_but = new Jbutton ("Add");
        JButton mult_but = new Jbutton ("Mult");

        // Put a button in each grid cell
        add (add_but);
        add (mult_but);
    } // ctor
} // class ActionButtonsPanel
```

Note that the buttons here do nothing when clicked. We explain in Chapter 7 how to add actions to buttons and other active components.

6.5.2 More about components and laying them out

We introduced the JButton and JPanel user interface components in the previous section. Two other components commonly needed are:

- JLabel – for text and icon labels
- JTextField – to provide for input and output of text in a single line display

For both components you can pass a string via a constructor parameter to provide the initial text. The classes contain a number of methods but the two primary ones for basic operation are:

```
public String getText ()
public void setText (String str)
```

While Jlabel objects do not interact with the user, JTextField is an active component like JButton that can respond to user actions. We discuss adding behavior to text fields in Chapter 7. You can allow or disallow user modification of the text in a text field display by invoking setEnabled (boolean flag). An enabled state of true means that the text can be modified while false prevents modification.

For our next applet we want to create a subclass of JPanel with two text fields. We put labels beside each so that the user can identify the text field. To insure that the labels and text fields are arranged in a logical manner we need to use a *layout manager*.

Java interfaces require great flexibility in how they arrange the components. The interfaces must be portable to different platforms and graphical environments and to displays with different screen sizes and resolutions. The width and height of the container may change with the tag settings for an applet or by the user shrinking or expanding an application window.

For these reasons absolute positioning is not suitable for Java GUIs. Instead, a Java container uses one of the several layout manager classes, which arrange the components according to general design guidelines rather than fixed coordinate settings. We discuss layout managers in more detail in Chapter 7 but for the example programs here we introduce two of the simplest layout managers:

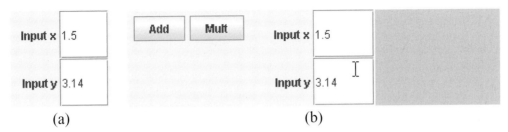

(a)   (b)

Figure 6.4 (a) The program `InputsPanelApplet` arranges two `JLabel` and two `JTextField` components with a `GridLayout`. (b) This program `MultiPanelApplet` combines the two button panel shown in Figure 6.3 (buttons are side by side due to the extra horizontal space available) with the panel from `InputsPanelApplet` and a `JTextArea` to display the results of the operations.

- `FlowLayout` – the default layout manager for `JPanel`. Components are placed in the order they are added, starting horizontally until all the space is filled and then shifting down to the next line, starting again at the left. The exact arrangement depends on the amount of area available and the minimum and maximum sizes allowed for each component.
- `GridLayout` – the components are distributed on a uniform grid or table. The size of the grid is set by the number of rows and columns passed in the constructor. As they are added, components are placed horizontally left to right until the row cells are filled and then continue with the next row down.

The following applet shows how to arrange four components using a `GridLayout` with two rows and two columns (see Figure 6.4(a)). It also shows how to initialize the text in a `JTextField` object and a `JLabel`.

```java
import java.awt.*;
import javax.swing.*;

/** Demonstrate the display of a JPanel subclass. **/
public class InputsPanelApplet extends JApplet {
  /** Create the interface in the init method **/
  public void init () {
    Container content_pane = getContentPane ();

    // Next create a panel of input fields and labels
    InputsPanel inputs_panel =
      new InputsPanel ("Input x", "1.5",
                       "Input y", "3.14");

    content_pane.add (inputs_panel);
  }
}// InputsPanelApplet
```

This applet creates an instance of the `InputsPanel` class, shown below, and adds it to the content pane. Using `JPanel` subclasses like `InputsPanel` allows for a flexible modular approach to graphical interface building.

```java
import java.awt.*;
import javax.swing.*;
/** Panel to hold input text fields. **/
public class InputsPanel extends JPanel
{
  JTextField fTextfieldTop;
  JTextField fTextfieldBot;

  /** Constructor builds panel with labels and text
   * fields.
   **/
  InputsPanel (String label_strtop, String init_top,
               String label_str_bot, String init_bot) {

    // Set the layout with 2 rows by 2 columns
    setLayout (new GridLayout (2, 2));

    // Create two text fields with the initial values
    fTextfieldTop = new JTextField (init_top);
    fTextfieldBot = new JTextField (init_bot);

    // Create the first label and right justify the text
    JLabel label_top =
     new JLabel (label_str_top, SwingConstants.RIGHT);

    // Insert the label and textfield into the top grid
    // row
    add (label_top);
    add (fTextfieldTop);

    // Create the second label and right justify the text
    JLabel label_bot =
      new JLabel (label_str_bot, SwingConstants.RIGHT);

    // Insert the second label and textfield into
    // the bottom grid row
    add (label_bot);
    add (fTextfieldBot);
  } // ctor
} // class InputsPanel
```

The text in the labels is right justified so that each label clearly refers to the values in the corresponding text field. The text field references are instance variables so that in a more ambitious program that uses InputsPanel other methods could access them and display text on the text fields or grab user input text from them.

The following program – MultiPanelApplet – shows how to create a more complex user interface by combining multiple panels and components. We set the layout manager for the content pane to a GridLayout of one row by three columns. We create instances of our ActionButtonsPanel and InputsPanel classes discussed above and also an instance of a JTextArea. The JTextArea component can display multiple lines of text input or output. Here we set the text area so that it only shows output and accepts no input, and we set its background color to light gray. These components are added in the order that we wish them to appear left to right.

```java
import java.awt.*;
import javax.swing.*;

/** Demonstrate the use of multiple JPanel subclasses. **/
public class MultiPanelApplet extends JApplet
{
  InputsPanel fInputsPanel;
  JTextArea fTextOutput;

  /** Build the interface with InputsPanel
    * and ActionButtonsPanel.
    **/
  public void init () {
    Container content_pane = getContentPane ();

    // Set the layout as before with 1 row of 3 columns
    // but now in one step.
    content_pane.setLayout (new GridLayout (1, 3));

    // First create a panel of buttons
    ActionButtonsPanel buttons_panel =
      new ActionButtonsPanel ();

    // Next create a panel of input fields and labels
    fInputsPanel =
      new InputsPanel ("Input x ", "1.5",
                       "Input y ", "3.14");

    // Use a JTextArea for the output of the calculations.
    fTextOutput = new JTextArea ();
```

```
        fTextOutput.setBackground (Color.LIGHT_GRAY);
        fTextOutput.setEditable (false);

        // The grid fills the 3 columns sequentially.
        content_pane.add (buttons_panel);
        content_pane.add (fInputsPanel);
        content_pane.add (fTextOutput);
    } // init
} // class MultiPanelApplet

/** JPanel subclass with two buttons. **/
class ActionButtonsPanel extends JPanel
{
    ActionButtonsPanel () {
        // Create two buttons
        JButton add_but = new JButton ("Add");
        JButton mult_but = new JButton ("Mult");

        // Put a button in each grid cell
        add (add_but);
        add (add_but);
    } // ctor
} // class ActionButtonsPanel
```

Figure 6.4(b) shows the resulting display. These components now comprise a user interface with text fields for user input, buttons to control the operation, and a text area to display the results of the operations. The program only lacks the event handling that we add in the next chapter.

6.5.3 Text display

In the previous section we introduced three text related components. The JLabel component is a static component for labeling items on the interface. JTextField and JTextArea provide for both the display and input of text. JTextField displays a single line of text while JTextArea can display multiple lines.

In the previous chapters our example programs sent their output to the Java console with the print methods available with the System.out object. We now show how to use a JTextArea to display text in a fashion similar to the Java console but on a graphical interface. We put the JTextArea component on a JPanel subclass, which we name TextOutputPanel. We also put it into a JScrollPane, which is a Swing component that provides scroll bars when text goes beyond the boundaries.

Our class implements a custom interface called `Outputable`. It holds two methods `print (String)` and `println (String)`:

```java
public interface Outputable {
    static final char CR = '\n';

    // A method to print a string
    public void print (String str);

    // A method to print a string with a carriage return
    public void println (String str);
}
```

These methods are intended to correspond to the print methods in `System.out` except that they display their strings on the graphics display rather than on the console. The code below shows how we implement the `Outputable` methods with the `JTextArea` component:

```java
import java.awt.*;
import javax.swing.*;

/**
 * This JPanel subclass holds a JTextArea object in a
 * JScrollPane area. It also implements our Outputable
 * interface to provide print (String) and println (String)
 * methods similar to those in System.out.
 **/
public class TextOutputPanel extends JPanel
                    implements Outputable
{
    // A Swing textarea for display of string info
    JTextArea fTextArea;

    public TextOutputPanel () {

        // A BorderLayout would be more appropriate here but
        // it isn't discussed until chapter 7.
        setLayout (new GridLayout (1,1));

        // Create an instance of JTextArea
        fTextArea = new JTextArea ();
        fTextArea.setEditable (false);

        // Add to a scroll pane so that a long list of
```

```
      // computations can be seen.
      JScrollPane area_scroll_pane = new JScrollPane (fTextArea);
      add (area_scroll_pane);
   } // ctor

   /** Display a string + carriage return on the JTextArea. **/
   public void println (String str) {
      fTextArea.append (str + CR);
   }

   /** Display a string on the JTextArea. **/
   public void print (String str) {
      fTextArea.append (str);
   }
} // class TextOutputPanel
```

The panel uses a one cell `GridLayout`, which causes the `JTextArea` component to fill the entire area of the panel. The default `FlowLayout` would result in unused space around the text area. (`BorderLayout` would actually be more appropriate here but we wait to discuss it in Chapter 7 along with the other layout managers.)

We can now add the `TextOutputPanel` component to an applet panel and send print messages there instead of to the console. The code below illustrates this technique (see Figure 6.5):

```
import javax.swing.*;
import java.awt.*;

/** Use a JPanel subclass TextOutputPanel to show text output
  * on the applet area.
  **/
public class TextOutputApplet extends JApplet
{
   public void init () {

      Container content_pane = getContentPane ();

      TextOutputPanel output_panel = new TextOutputPanel();

      // Add panel with JTextArea to show output
      content_pane.add (output_panel);

      output_panel.println (
         "TextOutputPanel implements the Outputable interface");
```

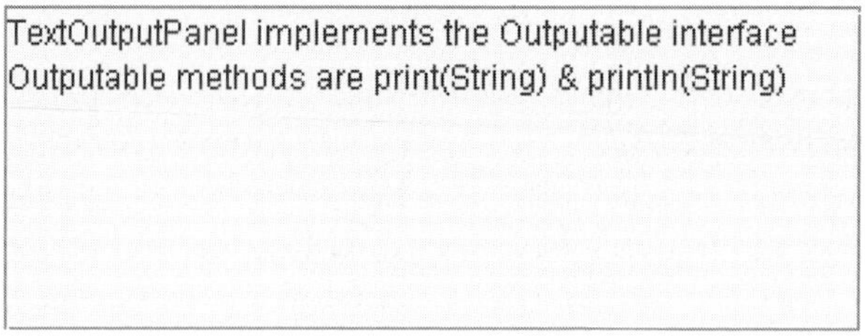

Figure 6.5 The program
`TextOutputApplet` uses
an `Outputable` text area
component to display
print messages.

```
TextOutputPanel implements the Outputable interface
Outputable methods are print(String) & println(String)
```

```
    output_panel.println (
      "Outputable methods are print(String) &
      println(String)");
  }
} // class TextOutputApplet
```

6.6 Drawing

A graphics interface often needs more than just buttons and text display. The
display of dynamic figures such as graphs and diagrams is a common requirement.
A simulation typically requires an animation created by a sequence of drawn
pictures. Furthermore, the interface needs the capability to draw images and text
on the displays. The Java graphics system provides a large assortment of drawing
tools to perform all of these tasks. In the following sections we discuss the basics
of drawing in a Swing environment.

In Swing the `JPanel` class usually serves as the drawing board. You override
the panel's `paintComponent()` method to perform the drawing operations.
Since all of the Swing components are lightweight, you can override their painting
methods and create custom features. For example, you could make your own
button component that displays a custom appearance in the *pressed* and *unpressed*
states.

6.6.1 Graphics contexts

The `paintComponent (Graphics g)` method receives an instance of the
`Graphics` class as a parameter. This class is referred to as the *graphics context*
since it provides the context under which the graphics commands operate for a
component. The `Graphics` object essentially represents a drawing surface or
tablet along with all of the settings such as the current foreground and background
colors and the font selection for the strings. In addition to representing the current

display of a component, a graphics context can represent an off-screen image as well.

Beginning with Java version 1.2, the object reference passed in the paint (Graphics g) and paintComponent (Graphics g) methods became a subclass of Graphics called Graphics2D. Since it is a subclass, you can still treat it as a Graphics object or, for more functionality, you can cast the Graphics object type to a Graphics2D type to use the Graphics2D methods. The Graphcs2D class provides the drawing context for the Java 2D API, which offers a greatly expanded set of drawing capabilities. The following packages comprise Java 2D: java.awt, java.awt.color, java.awt.font, java.awt.geom, java.awt.image, and java.awt.print. Some of these packages existed from Java 1.0 but they were expanded to provide additional capabilities for Java 2D.

We first briefly describe the basic drawing capabilities of the plain vanilla Graphics class and then look at some of the features of Graphics2D. Note that you can draw with both the old and new methods in the same context. In Chapter 11 we discuss image handling and processing with Java 2D tools.

6.6.2   Graphics coordinate system

The coordinate system (x = horizontal, y = vertical) for the drawing surface goes as follows:

- origin (0,0) at top-left corner
- x increases towards the right, beginning at 0
- maximum x = width − 1
- y increases towards the bottom, beginning at 0
- maximum y = height − 1

We see later that with the Java 2D tools, the origin can be moved or *translated*. For example, it can be mathematically convenient in some cases to work with the origin at the center of the component's drawing area rather than the top-left corner.

We can obtain the dimensions of a component in two ways. The getSize() method returns an instance of the Dimension class, which provides direct access to its height and width variables. As of Java 1.2 the component class provided the methods getHeight() and getWidth().

Java 2D makes a clear distinction between the user coordinate space where the drawing methods operate and the device coordinate space of computer monitors and printers. While the Graphics methods work only with integer pixel coordinates and dimensions, Graphics2D works with floating-point values in a user space that is transformed to the device space of a screen display or printer. The *rendering* system does the work of transforming the floating-point values in the drawing method arguments to integer pixel or dot numbers for drawing on a device.

For low-resolution devices, such as monitor screens, the conversion from user space units goes as the screen resolution, e.g. 96.0 units give 1 inch on a 96 pixels/inch screen, 2 inches for 48 pixels/inch, etc. On a high resolution device such as a printer, 72 user units always gives 1 inch regardless of the dots-per-inch setting [3]. You can modify the conversions with the scale() method in Graphics2D.

6.6.3 Color

The class java.awt.Color defines the properties of a color in Java. The default color space in Java is the sRGB (standard RGB), which offers "a simple and robust device independent color definition" [4]. In this space a color is defined by its RGB (Red-Green-Blue) color component values. It spans a subset of the standard CIEXYZ color space that includes all visible colors [5].

A color model, based on a particular color space, specifies exactly how a color is represented. In Java the default ARGB model uses eight bits for each of the RGB components plus another eight bits for an alpha transparency factor that specifies what happens when a color is drawn over another. Other color models, such as a gray scale model, can also be obtained.

For convenience, there are several Color constructors. Some constructors take int values between 0 and 255 for each color component. Some take float values between 0.0f and 1.0f. There are four-parameter constructors that let you specify the R, G, B, and alpha values and three-parameter versions that default to opaque alpha. In integer format, alpha is 0 for completely transparent and 255 for opaque and in floating-point these are 0.0f/1.0f, respectively.

An example of a three-parameter int constructor is shown here:

```
Color red = new Color (0xFF, 0, 0); // R, G, B,
                           // default alpha = 255
```

Note that it is common to use hexadecimal values to specify color components and here we used 0xFF instead of the decimal 255 for the red component. The three color components and alpha, each represented by a byte value, can be packed into an int. One constructor has a single int argument for packed RGB (alpha defaults to 0xFF) and another constructor has an int plus a boolean that is true if the integer value includes an alpha component, as shown here:

```
int intRed = 0xFF0000; // r=0xFF, g=0x00, b=0x00;
Color red = new Color(intRed); // default alpha = 0xFF
int intGreen = 0x8800FF00; // alpha=0x88, r=0x00, g=0xFF,
                   // b=0x00
Color green = new Color(intGreen, true);
```

Using hex numbers makes for a compact representation that allows you to quickly see the component values. In Section 10.14 we discuss bit handling and how to access and modify the bytes in a pixel integer value.

Here is an example using the `float` three-parameter constructor:

```
Color blue = new Color (0.0f, 0.0f, 1.0f);// R,G,B values
                   // between 0.0 and 1.0
```

Partially transparent red and blue colors are created in these two examples:

```
Color transparentRed = new Color (0xFF, 0, 0, 0x33);
                               // R,G,B, Alpha
Color transparentBlue = new Color (0.0f, 0.0f, 1.0f, 0.5f);
                               // R,G,B, Alpha
```

The transparency factor on the image color determines what percentage of the component's background color shows through.

For convenience the `Color` class definition provides several colors as class constants, such as

```
Color.BLUE, Color.WHITE, Color.RED
```

See the `Color` class in the Java API Specifications for a list of all the color constants. The original `Color` class in Java 1.0 used lower-case color constant names – `Color.blue`, `Color.white`, etc. This violated the convention of using all upper-case letters in names for constants, and so in Java 1.4 the upper-case names were added while the lower-case names remain for backward compatibility.

To set the color for the graphics context to use during drawing (i.e. the current "pen" color):

```
g.setColor (Color c); // where g = Graphics object
```

The background color of a component, such as a panel, can be set using

```
setBackground (Color c)
```

Similarly, there are methods to obtain the colors currently in use. For example,

```
Color c = g.getColor ();
```

gives the current color in use by the graphics context object. To obtain the background and foreground colors for a component, use

```
Color bg = getBackground ();
Color fg = getForeground ();
```

6.7 Drawing with the `Graphics` class

Here we illustrate drawing operations using the methods available in the `Graphics` class. The `Graphics2D` class offers more capabilities, which we discuss briefly in Section 6.8. Drawing typically begins with an invocation of the `setColor (Color c)` method in `Graphics` that tells the graphics context what color to use when it draws a line or primitive shape like a rectangle or when it fills a primitive shape with a solid color. The methods in the `Graphics` class for drawing and filling include the following (all parameters are `int` types):

- `drawLine (x1, y1, x2, y2)` – draws a line between points `(x1,y1)` and `(x2,y2)`.
- `drawRect (x, y, width, height)` and `fillRect (x, y, width, height)` – draws (fills) a rectangle, `(x,y)` are the coordinates of the top-left corner, the bottom-right corner will be at `(x+width,y+height)`.
- `drawOval (x, y, width, height)` and `fillOval (x, y, width, height)` – draws (fills) an oval *bounded* by the rectangle specified by these parameters.
- `draw3DRect (x, y, width, height)` and `fill3DRect (x, y, width, height)` – draws (fills) a rectangle with shaded sides that provide a 3-D appearance.
- `drawRoundRect (x, y, width, height)` and `fill3DRect (x, y, width, height)` – draws (fills) a rectangle with rounded corners.
- `drawPolyline (int[] x, int[] y, int n)` – draws lines connecting the n points given by the x and y arrays.
- `drawPolygon (int[] x, int[] y, int n)` – draws lines connecting the points given by the x and y arrays. Connects the last point to the first if they are not already the same point.

Note that `Graphics` does not provide a method to set the width of a line. The line is always one pixel wide and continuous (i.e. no dot-dash options). Nevertheless, these methods are simple and often convenient to use and can be used along with the `Graphics2D` methods.

Figure 6.6
Example
`DrawApplet`
uses drawing
methods in the
`Graphics` class.

6.7.1 Drawing demo

The applet `DrawApplet` shown in Figure 6.6 illustrates some of the drawing methods from the `Graphics` class. The applet first creates a `JPanel` subclass called `DrawingPanel` and then adds it to the `JApplet`'s content pane. The applet's `init()` method is shown here:

```
import javax.swing.*;
import java.awt.*;

/** Illustrate basic drawing in a Swing applet. **/
public class DrawApplet extends JApplet
{
   public void init () {
      Container content_pane = getContentPane ();

      // Create an instance of DrawingPanel
      DrawingPanel drawing_panel = new DrawingPanel ();

      // And add the DrawingPanel to the content pane.
      content_pane.add (drawing_panel);
   }
} // class DrawApplet
```

The `DrawingPanel` class overrides the `paintComponent()` method. It first invokes the superclass method so that the background is painted. (If the overriding method paints over the whole area then this is not necessary.) The center coordinates of the panel area are determined and then rectangles and circles are drawn relative to the center of the panel.

```
import javax.swing.*;
import java.awt.*;

/** Draw on this JPanel rather than on the JApplet. **/
public class DrawingPanel extends JPanel {

   DrawingPanel () {
      // Set background color for the applet's panel.
      setBackground (Color.WHITE);
   }

   public void paintComponent (Graphics g) {
      // Paint background
      super.paintComponent (g);

      // Get the drawing coordinates
      int dy = getSize ().height;
      int dx = getSize ().width;
```

```
     int mid_y = dy/2;
     int mid_x = dx/2;
     int rect_x = 3 * dx/4;
     int rect_y = 3 * dy/4;

     // Set current drawing color
     g.setColor (Color.BLACK);

     // Draw a rectangle centered at the mid-point
     g.drawRect (mid_x-rect_x/2, mid_y-rect_y/2,
                 rect_x, rect_y);

     // Set a new drawing color
     g.setColor (Color.LIGHT_GRAY);

     // Fill a rectangle centered at the mid-point. Put it
     // within the previous rectangle so that border shows.
     g.fillRect (mid_x-rect_x/2 + 10, mid_y-rect_y/2+10,
                 rect_x-20, rect_y-20);

     // Set a new drawing color
     g.setColor (Color.DARK_GRAY);

     // Draw a circle around the mid-point
     g.drawOval (mid_x-rect_x/6, mid_y-rect_y/6,
                 rect_x/3, rect_y/3);

     // Fill an oval inside the circle
     g.fillOval (mid_x-rect_x/6+10, mid_y-rect_y/6+10,
                 rect_x/3-20, rect_y/3-20);
  } // paintComponent
} // class DrawingPanel
```

6.7.2 Drawing text

Often in a drawing you want to include text such as placing labels on the axes
of a graph or adding a title. Here we discuss the basic text drawing capabilities
of the Graphics class. Java2D provides much more extensive text drawing
capabilities, such as drawing Arabic text right to left, but that is beyond the scope
of this book [5].

The first step in drawing text involves choosing the font. With Java 2D all
the fonts available on the platform can be used, but for nominal text you can
just specify one of the five standard fonts given by the following logical names:
Serif, SansSerif, Monospaced, Dialog, and DialogInput. These fonts

can be displayed in one of three styles: plain, bold, or italic. The JVM assigns an actual font to each of these logical names from the fonts available on the system.

You pass the font specification in the constructor of the Font class. For example, the following code draws the word "Test" with the font for Monospaced, plain, and 24-point size:

```
public void paintComponent (Graphics g) {
    g.setFont (new Font ("Monospaced", Font.PLAIN, 24);
    g.drawString ("Test", 10, 10);
}
```

Pay close attention to the parameter list of the Font constructor. The first parameter is a string. To use one of the logical font names, it must exactly match one of the five logical names given above. Unfortunately, the Font class does not define constants for those names, so you must be sure to type it exactly correctly or unexpected results occur. The second parameter is an int identifying the font style. The constants PLAIN, BOLD, and ITALIC are provided in the Font class to specify the style. The third parameter is an int giving the point size.

When a string str is drawn with g.drawString (str, x, y), the x and y parameters (both int type) give the position of the baseline of the first character in the string (see Figure 6.7(a)). It will be drawn with the graphic context's current color and font.

You might assume that the Font class includes methods to provide detailed information about the font, but instead the FontMetrics class provides that information. After the font is set, an instance of FontMetrics can be obtained from the graphics context using the getFontMetrics() method. The

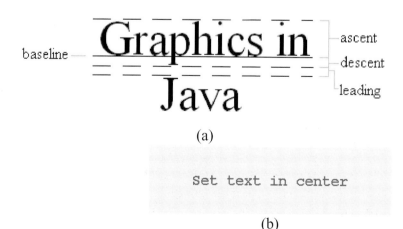

(a)

Set text in center

(b)

Figure 6.7 (a) Font metrics definitions. (b) This applet displays text with the DrawTextPanel class.

FontMetrics class provides various methods to obtain various measurements of characters and strings:

- getHeight () – total line height
- getMaxAscent () – height above the baseline
- getMaxDescent () – total
- stringWidth () – width of a string

Such information helps to position strings according to their size. The following class, DrawTextPanel, which we can display with an applet as usual, illustrates the basics of drawing a string according to its size specifications (see Figure 6.7(b)). It creates a Font object for the Monospaced type in the plain style and a 24-point size. After this Font is set as the current font for the graphics context, the FontMetrics object is obtained from the context. It provides various measurements of the characters, and these measurements are used to center the string.

```java
import javax.swing.*;
import java.awt.*;

/** JPanel subclass to demonstrate a drawing text. **/
public class DrawTextPanel extends JPanel
{
   public void paintComponent (Graphics g) {
      // First paint background
      super.paintComponent (g);

      // Add your drawing instructions here
      g.setColor (Color.red);
      String msg = "Set text in center";

      // Create the font and pass it to the Graphics context
      g.setFont (new Font ("Monospaced", Font.BOLD, 24));

      // Get measures needed to center the message
      FontMetrics fm = g.getFontMetrics ();

      // How many pixels wide is the string
      int msg_width = fm.stringWidth (msg);

      // How far above the baseline can the font go?
      int ascent = fm.getMaxAscent ();

      // How far below the baseline?
      int descent= fm.getMaxDescent ();
```

```
    // Use the string width to find the starting point
    int msg_x = getSize ().width/2 — msg_width/2;

    // Use the vertical height of this font to find
    // the vertical starting coordinate
    int msg_y = getSize ().height/2 — descent/2 + ascent/2;

    g.drawString (msg, msg_x, msg_y);
  } // paintComponent
} // class DrawTextPanel
```

Note that if the `paintComponent()` method does not draw over the entire surface, it should first invoke `super.paintComponent()`, which will cause the super class component to draw itself. For a `JPanel`, this would result in filling the area with its current background color.

6.8 Drawing in the Java 2D API

The Java 2D drawing capabilities are extensive so we only have space here for a brief introduction. We focus on the general techniques while the Web Course Chapter 6 provides additional discussion and numerous example programs.

6.8.1 Drawing setup

We saw with the `Graphics` class that, prior to a drawing operation, the color can be set for the "pen" that draws a line or fills a shape. Similarly the font can be selected before invoking the `drawString()` method. With `Graphics2D` this type of preparatory setup is taken to a much higher level. Several conditions can be defined that determine the features of the subsequent drawing operation.

6.8.1.1 Paint
Before a drawing operation you define the *paint* that it will use with the `set-Paint()` method. The paint could be a color but it could also be a gradient between two colors or a texture made by tiling an image:

```
public void paintComponent (Graphics g) {
    Graphics2D g2 = (Graphics2d) g;
    g2.setPaint (Color.RED);
    draw operation . . .

    Gradient gradient = new Gradient
      (x1, y1, Color.YELLOW, x2, y2, Color.GREEN, true);
    g2.setPaint (gradient)
    draw operation . . .
}
```

Here the paint was initially set to the color RED before the first drawing operation. Then the paint was changed to a color gradient beginning with YELLOW at point (x1,y1) and gradually changing to GREEN along a straight path that extends to point (x2,y2). The final boolean parameter to the Gradient constructor determines whether the gradient repeats (cycles) along the path away from each point or remains constant.

The drawing operations that are inherited from Graphics (see Section 6.7) ignore the paint setting and use the current color setting instead. Only the new methods in Graphics2D use the paint setting.

6.8.1.2 Stroke

Prior to drawing a graphic primitive shape such as a line or rectangle, you can define the *stroke* with the setStroke() method. A class that implements the Stroke interface describes the features of the outline. For most drawing requirements you can use the available BasicStroke class, which defines such features as the width of a line, whether it is solid or dashed, whether the ends are flat or rounded, and the appearance of the "joins" between two lines, as in the corner of a rectangle:

```
Rectangle2D rect = new Rectangle2D.Double (25., 50., 100.,
200.);
g2.setPaint (Color.RED);
Stroke stroke = new BasicStroke (5.0f, BasicStroke.CAP_ROUND,
                          BasicStroke.JOIN_ROUND);
g2.setStroke (stroke);
g2.draw (rect);
```

This code creates an instance of Rectangle2D (see Section 6.8.2) and sets the paint to a solid color. The BasicStroke constructor sets the width of the line to 5.0. The next two parameters specify that the end of a line is rounded and that the corners where two lines meet are rounded. The stroke is set into the Graphics2D and finally the rectangle is drawn.

See the BasicStroke class specifications for a listing of the many options for these and other stroke features. The class includes several constructors for creating different types of strokes.

6.8.1.3 Rendering hints

The Java 2D process begins with the set of drawing method invocations that involve various shapes. A rasterizer then converts these shapes to a 2D array of pixel values for rendering the complete drawing onto the screen or printer. You can influence the style of the rendering with the setRenderingHits() method. The RenderingHints class provides several different types of hints, but the most common hints deal with the edges. If only a single color is used for the pixels of a curved shape, the edge will be jagged or *aliased*. An anti-aliasing algorithm adds pixels along the edges with graded transparency that gives the edges a much smoother, continuous appearance. This process, however, takes longer to calculate and so can be turned off if not needed:

```
g2.setRenderingHints (
   new RenderingHints (RenderingHints.KEY_ANTIALISASING,
                       RenderingHints.VALUE_ANTIALIASING_OFF));
```

6.8.1.4 Transformation

Affine transformations are matrix operations on the 2D coordinates that keep parallel lines parallel. These transformations include translations, rotations, scaling, and shearing. Such transformations can be made to individual shapes or to the user space in general. A common technique is to move the origin to the center of the component area:

```
double cx = getWidth () / 2.0;
double cy = getHeight () / 2.0;
g2.translate (cx, cy);
```

or

```
AffineTransform trans =
AffineTransform.getTranslateInstance (cx, cy);
g2.transform (trans);
```

6.8.1.5 Compositing

When a shape (see Section 6.8.2) such as rectangle is drawn over another shape, then a compositing rule is used to decide how they are combined. The default is that the source shape (the one currently being drawn) simply draws over a destination shape (the one already drawn). However, with the `java.awt.AlphaComposite` class you can choose from eight compositing rules such as "destination-over" where the destination shape will be "on top" of the source shape. The compositing formula also takes into account the transparency factors of the source and destination shapes.

6.8.1.6 Clipping

Drawing can be restricted to just the area within a given shape's perimeter. This masking can reduce the number of operations and thus speed up the drawing. This can be particularly useful for animations where often only a part of the scene changes from frame to frame.

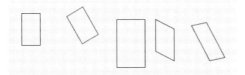

Figure 6.9 The `TransformsPanel` demonstrates the effects of the affine transforms by performing them on a rectangle shape like that shown on the far left. A translation operation shifts the drawing coordinates further to the right for each example. The shape second from the left is created by a *rotation* transform on the rectangle, the third a *scale* transform, the fourth a *shear*, and the shape on the far right is created with a compound transform of a shear and then a rotation.

is turned on (for good practice, but not crucial for the straight edged shapes displayed here). A `Rectangle2D.Double` object is created and drawn. Then a set of transforms are carried out, each time creating a new `Shape` object (therefore the original rectangle is not altered). Figure 6.9 shows the display of this panel when it is added to a `JApplet`.

```java
import javax.swing.*;
import java.awt.*;
import java.awt.geom.*;

/** Demonstrate different AffineTransforms of a rectangle. **/
class TransformPanel extends JPanel
{
   public void paintComponent (Graphics g) {
      // First paint background
      super.paintComponent (g);
      Graphics2D g2 = (Graphics2D) g;

      // Turn on anti-aliasing.
      g2.setRenderingHint (RenderingHints.KEY_ANTIALIASING,
                           RenderingHints.VALUE_ANTIALIAS_ON);

      // Create a rectangle.
      Shape shape1 = new Rectangle2D.Double (20.0, 20.0,
                           30.0, 50.0);
      // Now draw it.
      g2.draw (shape1);

      // Shift the drawing origin 72 units (1 inch) to the
      // right to obtain room for the next drawing
```

```
     AffineTransform at =
       AffineTransform.getTranslateInstance (72, 0);
     g2.transform (at);

     // Create a rotation transform of 30 degrees CCW around
     // the top left corner of the rectangle.
     AffineTransform atx =
       AffineTransform.getRotateInstance (—Math.PI/6, 30, 20);
     // Take the shape object and create a rotated version
     Shape atShape = atx.createTransformedShape (shape1);
     g2.draw (atShape);

     // Another 72 unit shift.
     g2.transform (at);

     // Create a scaling transform
     atx = AffineTransform.getScaleInstance (1.5, 1.5);
     // Take the shape object and create a scaled version
     atShape = atx.createTransformedShape (shape1);
     g2.draw (atShape);

     // Another 72 unit shift.
     g2.transform (at);

     // Create a shear transform
     atx = AffineTransform.getShearInstance (0.0, 0.5);
     // Take the shape object and create a sheared version
     atShape = atx.createTransformedShape (shape1);
     g2.draw (atShape);

     // Another 72 unit shift.
     g2.transform (at);

     // Illustrate compound transforms
     // First get a transform object
     atx = new AffineTransform ();
     // Then set to a shear transform
     atx.setToShear (0.0, 0.5);
     // and then rotate about the current origin
     atx.rotate (-Math.PI/5, 40, 50);
     // Now apply to the rectangle
     atShape = atx.createTransformedShape (shape1);
     g2.draw (atShape);

   } // paintComponent
} // class TransformPanel
```

In the first four transforms, we used static methods in the `AffineTransform` class to obtain instances of the class for the particular type of transform desired. For example, a translation transform is obtained with

```
AffineTransform at = AffineTransform.getTranslateInstance (72, 0);
```

Applying this to the graphics context moves its origin horizontally by 72 units:

```
g2.transform (at);
```

All subsequent coordinate values are relative to that new position on the drawing surface. Here the translation lets us draw the rectangle at a different position without changing its internal coordinate values. Then the rectangle is operated on with a rotation transform obtained with the `getRotateInstance()` static method:

```
AffineTransform atx =
    AffineTransform.getRotateInstance (−Math.PI/6, 40, 50);
```

Then a new rectangle in the rotated orientation is obtained with the `create-TransformedShape()` method as follows:

```
Shape atShape = atx.createTransformedShape (shape1);
```

Finally, the new rotated rectangle is drawn with:

```
g2.draw (atShape);
```

This same process is repeated for scaling and shear transforms. Finally, a combined transform is created by creating an instance of `AffineTransform` and then its `setToShear()` and `rotate()` methods are applied to the original rectangle to obtain a new transformed one.

6.9 Images

We introduce the basics of image handling here so you can begin to use them in your programs. We return to images again in Chapter 11 with a much more detailed discussion.

The base class for images is `java.awt.Image`. With Java 1.2 came the more capable subclass `java.awt.image.BufferedImage`. It works with the image processing tools of the Java 2D API and so we wait to discuss `Buffered-Image` in Chapter 11.

As of Java 1.4 you can load and draw image files encoded as JPEG, GIF, and PNG. With Java 5.0 you can also load bitmap formats BMP and WBMP. To load an image into an applet, you can use one of the overloaded `getImage()` methods in the `Applet` class for locating the file with a URL, as in

```
Image img = getImage ("http://www.someschool.edu/anImage.gif");
```

or

```
Image img = getImage (getCodeBase (), "anImage.gif");
```

In the latter case the `Applet` class method `getCodeBase()` provides the web address for the location of the applet's class file. The file name in the second parameter is then appended to the codebase and this combined address is used to locate the image file.

To load an image from within an application, as opposed to an applet, you can use

```
Image img = Toolkit.getDefaultToolkit ().getImage (URL or
filename);
```

Here the method parameter is either a `java.net.URL` or a `String` containing the filename of the image file. The `Toolkit` is a class in the `java.awt` package that provides various resources and tools for the graphics system. One of the `Toolkit` methods is `getImage()`, which functions much like the `Applet.getImage()` method. It is overloaded to take either a `String` filename parameter specifying the location of the image file or a URL parameter identifying the image file. (See Chapter 13 for information about the `java.net.URL` class.) Before calling `getImage()`, one must have a reference to the `Toolkit` instance in use. The static method `Toolkit.getDefaultToolkit()` returns a reference to that `Toolkit`.

You can obtain an image from a JAR file by using the static `getResource()` method from the `Class` class. This takes advantage of the *class loader* in the JVM that reads in a class and loads it for running. The class loader knows how to load files so it can also be used for loading image files and other resources. For example, if you were running an application named `YourApp`, you could obtain an image as follows:

```
URL url = YourApp.class.getResource ("myPhoto.gif");
Image img = Toolkit.getDefaultToolkit ().getImage (url);
```

Or you could just use `this.getClass()` to get the `Class` of the current object:

```
URL url = this.getClass ().getResource ("myPhoto.gif");
Image img = Toolkit.getDefaultToolkit ().getImage (url);
```

You can then draw the image with this method in the `Graphics` class:

```
void drawImage (Image img, int x, int y, ImageObserver io);
```

As we discuss further in Chapter 11, the `getImage()` method returns immediately. The actual loading of the image does not begin until the program attempts to draw the image or to obtain the dimensions of the image with the `getWidth()` and `getHeight()` methods. This approach was designed to avoid slowing a program while waiting for images to arrive over slow network links.

In Chapter 11 we discuss techniques for monitoring the image loading and to signal when the loading has finished. Here we just note the `ImageObserver` parameter. The image loading machinery uses this reference to *call back* via the `imageUpdate()` method of the `ImageObserver` interface. The callbacks occur periodically during the loading and provide information about the status of the loading. The `Component` class implements the `ImageObserver` interface and overrides the `imageUpdate()` method. So you commonly see the following type of invocation of `drawImage()` where "`this`" references the component on which the image is to be drawn:

```
drawImage (img, int x, int y, this);
```

6.10   Java and tech graphics

Programs for engineering and science applications frequently employ graphics for tasks such as:

- **Charting**, such as in a histogram, pie chart, or other arrangement that displays data in some informative manner
- **Plotting** functions, such as the trajectory of a ballistic projectile in altitude versus horizontal distance
- **Animating** a simulation of some device or physical process (see Chapter 8)
- **Image processing** to bring out features of interest (see Chapter 11)

A graphical display program would also typically include various control and data entry components to provide a user interface to allow the user to interact with the presentation. As we have seen in this chapter, Java provides lots of tools and components for building such graphical displays and user interfaces.

Of course, many other programs also provide for data display and manipulations. Why use Java for this purpose? With Java you can build custom graphical programs for tasks where you can integrate all of the Java features together. For example, an application program to control and monitor a remote experiment via the Web could use both the graphical capabilities of Java and the networking capabilities of Java (see Part II). We see in Chapter 8 that you can use the easy thread-processing capabilities of Java to create animations for demonstrations and simulations.

Note that you can now obtain libraries, both commercial and freeware, that provide Java classes and visible components for charting, image processing, and other graphics-related tasks. However, you will still find it useful to know how to write your own graphics classes for customized purposes.

6.11   Histogram graphics

In the Web Course Chapter 6: *Tech* section, we demonstrate Java graphics with the `PlotPanel` and `HistPanel` classes. `PlotPanel` is an abstract class that provides a framed area for plotting. It includes a title for the plot, scale values for the axes, and a horizontal axis label. `PlotPanel` extends `JPanel` so it can be added easily to the layouts of applets and applications. See the Web Course for a detailed description of the code and a complete listing. Here we give a brief overview and snippets of the code.

`PlotPanel` includes four methods that a subclass must override. These methods set the title and label, they calculate the scaling along the axes, and they draw a histogram or other type of plot inside the frame. For the axes scale values, we use a static utility method in the class `PlotFormat` that converts double values to strings in decimal or scientific format using the techniques discussed in the Chapter 5.

The `paintComponent()` method in `PlotPanel` draws the frame, title, label, and scale values and then invokes `paintContents()`, which the subclass must provide:

```java
public void paintComponent (Graphics g) {
   // First paint background
   setBackground (bgColor);
   super.paintComponent (g);

   // Get the positions for the titles, labels, etc.
   getPositions ();

   // Draw the top title
   g.setColor (titleColor);
   drawText (g, getTitle (), titleX, titleY,
            titleWidth, titleHeight, 0,CENTER);

   // Draw the bottom label
   drawText (g, getXLabel (), horzLabelX, horzLabelY,
             horzLabelWidth, horzLabelHeight,0,CENTER);

   // Draw the plot frame.
   paintFrame (g);

   // Draw the plot within in the frame.
   // This method must be overriden.
   paintContents (g);
} // paintComponent
```

The `HistPanel` class extends `PlotPanel` and displays the contents of the `Histogram` class (see below). It provides a `paintContents()` method that draws the values of the histogram bins as vertical bars. In the Web Course *Tech* and *Physics* tracks we use `HistPanel` in many of the demonstration programs.

Figure 6.10 A demonstration of histogram plotting with the `HistPanel` class.

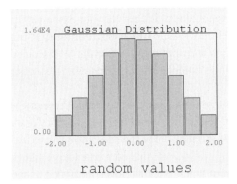

Figure 6.10 shows an example of a Gaussian random number distribution displayed with `HistPanel`.

`PlotPanel` and `HistPanel` help to illustrate both the graphics capabilities of Java and also the modularity of its objected-oriented design. A `HistPanel` object can be added whenever we need to look at the distribution of values of some quantity. Multiple `HistPanel` instances can be used when we want to look at multiple quantities. We also see in later chapters and in the Web Course that `PlotPanel` can be extended by other classes that draw different types of plots and graphs besides histograms.

To simplify our histogram class hierarchy, we created a new base class called `Histogram`. This class combines the attributes and methods of the earlier `BasicHist` (see Chapter 3) and `BetterHist` (see Chapter 4) classes, which were used to illustrate class and inheritance concepts. We added several new methods to our histogram class to get and set values describing the histogram. We provide the full code listing of `Histogram.java` here:

```java
/** This class provides the bare essentials for a
  * histogram. **/
public class Histogram
{
    protected String fTitle = "Histogram";
    protected String fXLabel = "Data";

    protected int[] fBins;
    protected int fNumBins;
    protected int fUnderflows;
    protected int fOverflows;

    protected double fLo;
    protected double fHi;
    protected double fRange;
```

```
/**
  * The constructor creates an array of a given
  * number of bins. The range of the histogram is given
  * by the upper and lower limit values.
  **/
public Histogram (int numBins, double lo, double hi) {
  // Check for bad range values.
  // Could throw an exception but will just
  // use default values;
  if (hi < lo) {
    lo = 0.0;
    hi = 1.0;
  }
  if (numBins <= 0) numBins = 1;

  fNumBins = numBins;
  fBins = new int[fNumBins];
  fLo = lo;
  fHi = hi;
  fRange = fHi — fLo;
} // ctor

/**
  * This constructor includes the title and horizontal
  * axis label.
  **/
public Histogram (String title, String xLabel,
                  int fNumBins, double lo, double hi) {
  this (fNumBins, lo, hi);
    // Invoke overloaded constructor
  fTitle = title;
  fXLabel = xLabel;
}

/** Get the title string. **/
public String getTitle ()
{ return fTitle; }

/** Set the title. **/
public void setTitle (String title)
{ fTitle = title; }

/** Get the horizontal axis label. **/
public String getXLabel ()
```

```java
{ return fXLabel; }

/** Set the horizontal axis label. **/
public void setXLabel (String xLabel)
{ fXLabel = xLabel; }

/** Get the low end of the range. **/
public double getLo ()
{ return fLo; }

/** Get the high end of the range.**/
public double getHi ()
{ return fHi; }

/** Get the number of entries in the largest bin. **/
public int getMax () {
  int max = 0;

  for (int i=0; i < fNumBins; i++)
    if (max < fBins[i]) max = fBins[i];

  return max;
}

/**
  * This method returns a reference to the fBins array.
  * Note that this means the values of the histogram
  * could be altered by the calling code.
  **/
public int[] getBins () {
  return fBins;
}

/** Get the number of entries in the smallest bin.**/
public int getMin () {
  int min = getMax ();

  for (int i=0; i < fNumBins; i++)

    if (min > fBins[i]) min = fBins[i];
  return min;
}

/** Get the total number of entries not counting
  * overflows and underflows.
  **/
```

```
public int getTotal () {
  int total = 0;
  for (int i=0; i < fNumBins; i++)
    total += fBins[i];

  return total;
}


/**
  * Add an entry to a bin.
  * @param x value added if it is in the range:<br>
  * lo <= x < hi
  **/
public void add (double x) {
  if (x >= fHi) fOverflows++;
  else if (x < fLo) fUnderflows++;
  else {
    double val = x - fLo;

    // Casting to int rounds off to lower
    // integer value.
    int bin = (int) (fNumBins * (val/fRange));

    // Increment the corresponding bin.
    fBins[bin]++;
  }
}


/** Clear the histogram bins and the over and under flow
  * counts. **/
public void clear () {
  for (int i=0; i < fNumBins; i++)
    fBins[i] = 0;
  fOverflows = 0;
  fUnderflows = 0;
}


/**
  * Provide access to the value in the bin element
  * specified by bin_num.<br>
  * Return the underflows if bin value negative,
  * Return the overflows if bin value more than
  * the number of bins.
  **/
```

```java
public int getValue (int bin_num) {
  if (bin_num < 0)
    return fUnderflows;
  else if (bin_num >= fNumBins)
    return fOverflows;
  else
    return fBins[bin_num];
}


/**
  * Get the average and standard deviation of the
  * distribution of entries.
  * @return double array
  */
public double [] getStats () {

  int total = 0;
  double wt_total = 0;
  double wt_total2 = 0;
  double [] stat = new double[2];
  double bin_width = fRange/fNumBins;

  for (int i=0; i < fNumBins; i++) {
     total += fBins[i];
  double bin_mid = (i - 0.5) * bin_width + fLo;
  wt_total += fBins[i] * bin_mid;
  wt_total2 += fBins[i] * bin_mid * bin_mid;
}

if (total > 0) {
  stat[0] = wt_total / total;
  double av2 = wt_total2 / total;
  stat[1] = Math.sqrt (av2 - stat[0] * stat[0]);
}
else {
  stat[0] = 0.0;
  stat[1] = -1.0;
}

  return stat;
} // getStats

/**
  * Create the histogram from a user provided array
  * along with the under and overflow values.
```

```
 *  The low and high range values that the histogram
 *  corresponds to must be passed in as well.<br>
 *
 *  @param user_bins    array of values for histogram.
 *  @param under    number of underflows.
 *  @param over    number of overflows.
 *  @param lo    value of the lower range limit.
 *  @param hi    value of the upper range limit.
 **/
  public void pack (int[] user_bins,
                       int under, int over,
                       double lo, double hi) {
    fNumBins = user_bins.length;
    fBins = new int[fNumBins];
    for (int i = 0; i < fNumBins; i++) {
      fBins[i] = user_bins[i];
    }

    fLo = lo;
    fHi = hi;
    fRange = fHi-fLo;
    fUnderflows = under;
    fOverflows = over;
  } // pack

} // class Histogram
```

We extend Histogram with several subclasses that are described in subsequent chapters in the Web Course *Tech* track but not in the book due to space limitations. In a demonstration program in Chapter 15 of the book we use one of these subclasses, HistogramAdapt, which widens its range dynamically so as to include all data entries with no under or overflows.

6.12 Web Course materials

The Web Course Chapter 6: *Supplements* section includes an introduction to basic AWT user interface design (as opposed to the Swing emphasis here). It also looks at drawing with the Java 2D methods. A more detailed description of the composting rules, for example, is provided as well.

 The graphical interfaces shown here are only intended to demonstrate the basic programming techniques and look rather crude. Swing, in fact, provides a number of optional features, such as a selection of borders for components, to create beautiful interfaces. The Web Course *Supplements* sections for Chapters 6 and 7 present ways to refine the appearance of your user interfaces.

The *Tech* and *Physics* tracks look at using the graphics techniques to plot data graphs and histograms using the `PlotPanel` and `HistPanel` classes discussed above.

References

[1] David M. Geary, *Graphic Java 2: Mastering the JFC*, Prentice Hall, 1999.

[2] *Trail: Creating a GUI with JFC/Swing – The Java Tutorial*, at Sun Microsystems, `http://java.sun.com/docs/books/tutorial/uiswing/`.

[3] Baldwin, *Java 2D Graphics, The Graphics2D Class*, Developer.Com, February 9, 2001, `www.developer.com/java/other/print.php/626071`.

[4] M. Stokes, M. Anderson, S. Chandrasekar and R. Motta, *A Standard Default Color Space for the Internet – sRGB*, 1996, World Wide Web Consortium (W3C), `www.w3.org/Graphics/Color/sRGB.html`.

[5] Jonathan Knudsen, *Java 2D Graphics*, O'Reilly, 1999.

Chapter 7
Graphical User Interfaces

7.1 Introduction

In the previous chapter we introduced Java graphics and focused on the display of components, drawings, and images. Here we discuss how to build a graphical display that interacts with the user. A Graphical User Interface (GUI) requires dynamic components like buttons and menus that cause something to happen when the user activates them with the mouse or keyboard. These components generate messages called *events* that signal what action occurred. We will first look in this chapter at the underlying structure for generating and processing events.

We then introduce several more Swing components such as checkboxes and sliders and present demonstration programs that illustrate how to receive and process the events they generate. We discuss in more detail how components are arranged on the interface with the use of layout managers. For event handling and other tasks, we examine the use of inner classes and adapter classes. After a discussion of frames and menu bars, we demonstrate GUI construction and event techniques with a couple of programs involving histograms.

7.2 Events

The graphical user interface (GUI) offers a profoundly different programming environment than the old step-by-step, linear world of a procedural program communicating via a console. In the latter approach, the user simply starts a program and waits for it to churn through its algorithm and eventually reach the end and stop. In the GUI environment, the program instead waits for the user to select some task and then carries out that selected operation. When it finishes the operation, the program returns to a wait state. Furthermore, with thread processes (see Chapter 8) the program can carry on multiple operations and interactions with the user, all appearing to happen simultaneously even though they typically run in a single processor.

The JVM communicates with the operating system to monitor the program's interface. When the user moves a mouse, clicks one of its buttons, hits a key,

7.2.3   Button events

Buttons provide the most common event generating components in user interfaces. (See the `JButton` and other button classes in the Swing hierarchy in Figure 6.2.) Clicking on a button should initiate some action and in fact the term *action* is used in the names for the button event handling machinery. A button maintains a list of objects that implement the `ActionListener` interface. When the button is pressed, it invokes the `actionPerformed()` method for all of the `ActionListener` instances in its list. To include an `ActionListener` in its list, you invoke the button's `addActionListener(ActionListener al)` method.

ActionListener classes must provide an implementation of the method

```
void actionPerformed (ActionEvent ae)
```

You use this method to code for the particular operation required when the user clicks the button. An `ActionEvent` object is passed as the parameter of `actionPerformed()` and from it you can extract information about the event such as which button initiated the event.

We mentioned above the `getSource()` method inherited from `EventObject` that can identify what object sent the event. With buttons you can also use the method `getActionCommand()` in `ActionEvent` to identify the button that sent the event. This method returns either the text string displayed on the button or a string set directly with the `setActionCommand()` method (inherited from `AbstractButton` class).

The `PlainButtonApplet` program below implements the `ActionListener` interface and so provides an `actionPerformed()` method. We create two instances of `JButton` and then add the applet to the buttons' `ActionListener` lists with the statements:

```
buttonA.addActionListener (this);
buttonB.addActionListener (this);
```

Here the "`this`" reference refers to the applet instance.

Each time the GUI button is clicked with the mouse button, the GUI button's `actionPerformed()` method is invoked. From the `ActionEvent` we obtain the identity of the button via the `getActionCommand()` method and decide

B pushed 2 times

Figure 7.1   The `PlainButton` applet running illustrates event handling for buttons. Each click on a button sends a message to the status bar.

what action to perform. Here we use the Applet showStatus() method to display a string on the browser status line. Figure 7.1 shows the program running in the applet viewer program.

```java
import javax.swing.*;
import java.awt.*;
import java.awt.event.*;

/** Demonstrate event handling with two buttons. **/
public class PlainButtonApplet extends JApplet
                          implements ActionListener
{
   int fPushACount = 0;
   int fPushBCount = 0;

   /** Build the interface with two buttons. **/
   public void init () {
      Container content_pane = getContentPane ();
      content_pane.setLayout (new FlowLayout ());

      // Create an instance of JButton
      JButton button_a = new JButton ("A");

      // Create an instance of JButton
      JButton button_b = new JButton ("B");

      // Add this applet to each button's ActionListener
      // list
      button_a.addActionListener (this);
      button_b.addActionListener (this);

      JPanel panel = new JPanel ();

      // Add the buttons to the content pane.
      content_pane.add (button_a);
      content_pane.add (button_b);

   } // init

   /** Count each button click and post total on status
    * line. **/
   public void actionPerformed (ActionEvent event) {

      String cmd = event.getActionCommand ();
```

```
     if (cmd.equals ("A")) {
       fPushACount++;
       showStatus (event.getActionCommand () +
                   "pushed" + fPushACount + "times");
     } else {
       fPushBCount++;
       showStatus (event.getActionCommand () +
                   "pushed" + fPushBCount + "times");
     }
   } // actionPerformed

} // class PlainButtonApplet
```

The `ActionEvent` class also includes the method

```
public int getModifiers ()
```

The integer returns with a bitwise OR of the values indicating what modifier keys were held down when the event occurred. For example, a value of 0x09 would indicate the OR of `ActionEvent.ALT_MASK` (0x08) and `ActionEvent.SHIFT_MASK` (0x01) for the Alt and Shift keys, respectively.

7.2.4   Simple GUI

In Chapter 6 we created a very basic display with two buttons, two text fields, corresponding labels, plus a text area. However, that applet could not perform any operation. Now that we know about button event handling, we can create a genuine user interface with these components.

Figure 7.2 shows the display from the `MultiPanelWithEvents` applet whose code is given below. It corresponds to the `MultiPanelApplet` example from Chapter 6 and uses the same `InputsPanel` class. However, now the applet implements the `ActionListener` interface and the `ActionButtonsPanel` is modified to pass an `ActionListener` reference in the constructor. Each button adds this reference to its list of listeners to send events.

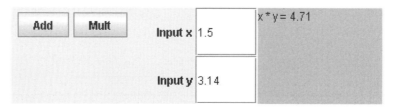

Figure 7.2   With a layout like that in Figure 6.4(b), the `MultiPanelWithEvents` applet can handle events generated by the buttons and execute the selected operation on the data values in the text fields. It then displays the result in the text area.

```
import java.awt.*;
import java.awt.event.*;
import javax.swing.*;

/** Demonstrate GUI building with multiple panels
  * and event handling.
 **/
public class MultiPanelWithEvents extends JApplet
              implements ActionListener
{
  InputsPanel fInputsPanel;
  JTextArea fTextOutput;

  /** Build the interface with InputsPanel
    * and ActionButtonsPanel.
   **/
  public void init () {
    Container content_pane = getContentPane ();
    content_pane.setLayout (new GridLayout (1, 3));

    // First create a panel of buttons
    ActionButtonsPanel buttons_panel =
        new ActionButtonsPanel (this);

    . . . Rest same as in init() in MultiPanelApplet in
        Chapter 6 . . .

  } // init

  // Handle the button events. **/
  public void actionPerformed (ActionEvent ae) {

    /** Get the values in the two text fields. **/
    String str1 = fInputsPanel.fTextfieldTop.getText ();
    String str2 = fInputsPanel.fTextfieldBot.getText ();

    double val1=0.0;
    double val2=0.0;

    try {
      val1 = Double.parseDouble (str1);
      val2 = Double.parseDouble (str2);
    } catch (NumberFormatException nfe) {
      System.out.println ("Improper input");
    }
```

```
      if (ae.getActionCommand ().equals ("Add")) {
         fTextOutput.setText ("x + y = " + (val1+val2));
      } else {
         fTextOutput.setText ("x * y = " + (val1*val2));
      }
   } // actionPerformed

} // class MultiPanelWithEvents
```

```
import java.awt.event.*;
import javax.swing.*;

/** JPanel subclass with two buttons. **/
public class ActionButtonsPanel extends JPanel
{
   /** Constructor adds 2 buttons to the panel and
    * adds the listener passed as an argument to the
    * action listener list in each button.
    **/
   ActionButtonsPanel (ActionListener listener) {

      // Create two buttons
      JButton add_but = new JButton ("Add");
      JButton mult_but = new JButton ("Mult");

      add_but.addActionListener (listener);
      mult_but.addActionListener (listener);

      // Put a button in each grid cell
      add (add_but);
      add (mult_but);
   } // ctor
} // class ActionButtonsPanel
```

The `actionPerformed()` method in the applet grabs the strings from the text fields and converts them to floating-point values. It then examines the action command string (by default this is the text on the button) to determine which button sent the event. It then carries out the chosen operation and displays the result in the text area.

7.2.5   Mouse events

As the following diagram shows, the `MouseEvent` class inherits from several classes:

```
java.lang.Object
    |
    +-java.util.EventObject
           |
           +-java.awt.AWTEvent
                  |
                  +-java.awt.event.ComponentEvent
                         |
                         +-java.awt.event.InputEvent
                                |
                                +-java.awt.event.MouseEvent
```

These include a number of methods that provide information about an event. (See the Java 2 API Specification for a detailed listing.) A sampling of the methods include:

- `getComponent ()` – identifies the `Component` object that generated the event
- `getX ()`, `getY ()`, `getPoint ()` – provide the coordinates of the mouse location
- `getClickCount ()` – gives number of times the mouse button was clicked
- `getModifiers ()` – indicates what modifier keys were held down during the event

Both the motion of the mouse and its buttons create events. There are two separate types of mouse event listeners:

- **MouseMotionListener**

  Each move of the mouse generates a motion event. To listen for mouse motion events, a class needs to implement the `MouseMotionListener` interface. The implementing class will need to provide two methods:
  - `mouseMoved (MouseEvent e)` – mouse motion
  - `mouseDragged (MouseEvent e)` – mouse motion when mouse button is held down
- **MouseListener**

  The `MouseListener` interface provides these methods for mouse button and cursor events:
  - `mousePressed (MouseEvent e)` – mouse button is pressed
  - `mouseReleased (MouseEvent e)` – mouse button is released
  - `mouseClicked (MouseEvent e)` – button clicked (press and release counts as one action)
  - `mouseEntered (MouseEvent e)` – the cursor enters the area of the component
  - `mouseExited (MouseEvent e)` – the cursor exits the area of the component

The following example program `CaptureEvtApplet` illustrates how to use the `MouseListener` interface to monitor when a mouse cursor enters or exits the area of the panel and when clicks are made over the panel. The program displays messages in a `JTextArea` to indicate the different ways that mouse events

Figure 7.3 The
`CaptureEvtApplet`
program illustrates mouse
event handling. Whenever
the mouse moves the
cursor into the top area or
clicks while the cursor is
over the top panel, a
message will be displayed
in the lower text area.

Mouse exited detected on javax.swing.JPanel.
Mouse clicked (# of clicks: 1) detected on javax.swing.J
Mouse released; # of clicks: 1 detected on javax.swing.
Mouse pressed; # of clicks: 1 detected on javax.swing.J
Mouse clicked (# of clicks: 1) detected on javax.swing.J
Mouse released; # of clicks: 1 detected on javax.swing.
Mouse pressed; # of clicks: 1 detected on javax.swing.J
Mouse clicked (# of clicks: 1) detected on javax.swing.J
Mouse released; # of clicks: 1 detected on javax.swing.

can be generated (see Figure 7.3). The `CaptureEventPanel` implements the
`MouseListener` interface and so must provide all five methods.

Note that in the `saySomething()` method, the invocation of `getClass()`.
`getName()` provides the name of any class as a string. Here we have used this
technique to find the identity of the component that generated the event. (See the
Web Course Chapter 5: *Supplements* section for more about the `Class` class and
the information it provides about a class definition.)

```java
import javax.swing.*;
import java.awt.*;
import java.awt.event.*;

/** Demonstrate a MouseListener component. **/
public class CaptureEvtApplet extends JApplet
{
  /** Create the interface with CaptureEventPanel. **/
  public void init () {
    Container content_pane = getContentPane ();

    // Create an instance of the JPanel subclass
    CaptureEventPanel cap_evt_panel =
      new CaptureEventPanel ();

    // And the panel to the JApplet panel.
    content_pane.add (cap_evt_panel);

  } // init
} // class CaptureEvtApplet
```

```
import javax.swing.*;
import java.awt.*;
import java.awt.event.*;

/** This panel uses MouseListener to capture mouse events **/
public class CaptureEventPanel extends JPanel
           implements MouseListener
{
  JTextArea fTextOutput;
  String newline;

  /**
    * Constructor adds this class to the MouseListener
    * list for a panel and sends messages to a text area
    * whenever an event occurs over the panel.
    **/
  CaptureEventPanel () {
    setLayout (new GridLayout (2,1));

    JPanel p = new JPanel ();
    p.setBackground (Color.LIGHT_GRAY);
    add (p);
    //Register to receive mouse events on the panel.
    p.addMouseListener (this);

    fTextOutput = new JTextArea ();
    fTextOutput.setEditable (false);
    add (fTextOutput);

  } // ctor

  // Implementation of Mouse Listener requires overriding
  // all five of its methods.

  public void mousePressed (MouseEvent e) {
   saySomething ("Mouse pressed; # of clicks: "
                 + e.getClickCount (), e);
  }

  public void mouseReleased (MouseEvent e) {
    saySomething ("Mouse released; # of clicks: "
                 + e.getClickCount (), e);
  }

  public void mouseEntered (MouseEvent e) {
    saySomething ("Mouse entered", e);
  }
```

```
public void mouseExited (MouseEvent e) {
   saySomething ("Mouse exited", e);
}
public void mouseClicked (MouseEvent e) {
   saySomething ("Mouse clicked (# of clicks: "
                 + e.getClickCount () + ")", e);
}

/** On JTextArea print messages describing mouse events. **/
void saySomething (String event_description, MouseEvent e){
   fTextOutput.insert (event_description + "detected on"
                       + e.getComponent ().getClass ()
                       .getName () + "." + "\n",0);
}

} // class CaptureEventPanel
```

7.3   More user interface components

Now that we know how to handle events, we introduce four more event-generating components. In each of the cases discussed here, two panels are displayed on an applet. The OutputPanel simply displays a solid color. The other panel contains the UI components that set the color of OutputPanel. The components that we discuss are:

- JCheckBox
- JRadioButton
- JList
- JSlider

In the four sections that follow, we create JPanel subclasses that hold the above components. First we show the code for the program UiTestApplet, which selects one of these panels according to a parameter passed from the applet tag in the web page. It then displays both the selected panel and OutputPanel, whose code is also shown below.

```
import javax.swing.*;
import java.awt.*;
import java.awt.event.*;

/**
   * Display one of four types of component panels
   * to control the background color on a separate panel.
   * Applet parameter selects the type of panel. **/
```

```
public class UiTestApplet extends JApplet
{
   /** Create the GUI with OutputPanel and  ControlPanel. **/
   public void init () {
      Container content_pane = getContentPane ();

      // Create an instance of OutputPanel
      OutputPanel output_panel = new OutputPanel ();

      // Find out from the applet tag which of the four
      // panels to use. Each panel demonstrates a different
      // type of control component.
      String panel_choice = getParameter ("Choose Panel");
      JPanel control_panel;

      // Create an instance of the control panel
      if (panel_choice.equals ("Checkboxes"))
         control_panel = new CheckBoxesPanel (output_panel);
      else if (panel_choice.equals ("Radiobuttons"))
         control_panel = new RadioButtonsPanel (output_panel);
      else if (panel_choice.equals ("List"))
         control_panel = new ListPanel (output_panel);
      else
         control_panel = new SliderPanel (output_panel);

      // Add the panels to applet's pane.
      content_pane.add (output_panel, BorderLayout.CENTER);
      content_pane.add (control_panel, BorderLayout.SOUTH);

   } // init
} // class UiTestApplet
```

Here is the code listing for OutputPanel, whose job is just to display the color currently selected by the user:

```
import javax.swing.*;
import java.awt.*;

/** Create a JPanel subclass called OutputPanel.
  * It provides a method to set its color plus
  * a paintComponent method. **/
public class OutputPanel extends JPanel
```

```
{
   Color fBgColor = Color.RED;

   /** Set the color by passing the 3 RGB component
    * values (in range 0-255).
    **/
   void setColor (int cr, int cg, int cb) {
      fBgColor = new Color (cr,cg,cb);
      repaint ();
   }

   /** For Swing components you must override
    * paintComponent () rather than paint ().
    **/
   public void paintComponent (Graphics g) {
      super.paintComponent (g);
      // Now we fill a rectangle the size of the
      // panel with the chosen color.
      g.setColor (fBgColor);
      g.fillRect (0, 0, getWidth (), getHeight ());
   }
} // class OutputPanel
```

7.3.1 JCheckBox

Checkboxes allow the user to select from several non-exclusive options. Check-boxes were available in Java 1.0 in the original AWT graphics package, but we use the improved Swing version known as JCheckBox. (Notice the spelling: the original AWT checkbox component is Checkbox – lower case "b" – while the new Swing component is JCheckBox, with uppercase "B".)

Figure 7.4 shows the display of the UITestApplet program for the case where the applet parameter instructed it to use an instance of

Figure 7.4 The JCheckBox component allows for the selection of more than one item in a set of checkboxes. With CheckBoxesPaneldisplayed by UITestApplet, the user selects one or more of three color components to create the background color for the top panel.

CheckBoxesPanel to control the color of the top OutputPanel. The CheckBoxesPanel holds three checkboxes, and implements the ItemListener interface. A reference to the CheckBoxesPanel instance is added to the ItemListener list for the checkboxes.

JCheckBox inherits from the Swing AbstractButton class and one ActionEvent and one ItemEvent are fired each time a checkbox is clicked with the mouse button. You could listen for both event types but it only makes sense to listen for one type or the other. In most applications, it makes more sense to use an item listener because ItemEvents have the advantage of providing an easy way to tell whether the event was a selection or de-selection event.

The ItemEvent objects are handled by the itemStateChanged() method in the ItemListener interface. Our implementation of this method first looks to see if the event is due to a selection or de-selection. Then it examines each checkbox using the isSelected() method to determine if it is in a selected state or not.

If a checkbox is selected, then the corresponding color component (red, green, or blue) is added to the combined color for the background of the OutputPanel using the setColor (int red, int green, int blue) method. (See the Web Course demos for the non-gray scale versions!)

```java
import javax.swing.*;
import java.awt.*;
import java.awt.event.*;


/** The CheckBoxesPanel holds the checkboxes that set
  * the color of the output panel.
 **/
public class CheckBoxesPanel extends JPanel
                  implements ItemListener
{
    JCheckBox fRed, fGreen, fBlue;
    OutputPanel fOutputPanel;


    /** Constructor adds 3 checkboxes to the panel. **/
    CheckBoxesPanel (OutputPanel output_panel) {

        fOutputPanel = output_panel;

        // Initial color is red so select this button.
        fRed = new JCheckBox ("Red", true);
        fRed.addItemListener (this);
        add (fRed);
```

```
    fGreen = new JCheckBox ("Green");
    fGreen.addItemListener (this);
    add (fGreen);

    fBlue = new JCheckBox ("Blue");
    fBlue.addItemListener (this);
    add (fBlue);

  } // ctor

  /**
    * An item event from a checkbox comes here when
    * the user clicks on one.
    **/
  public void itemStateChanged (ItemEvent evt) {
    if (evt.getStateChange () == ItemEvent.SELECTED)
      System.out.println ("SELECTED");
    else
      System.out.println ("DESELECTED");
    int cr = (fRed.isSelected ()  ? 0xFF: 0);
    int cg = (fGreen.isSelected ()? 0xFF: 0);
    int cb = (fBlue.isSelected () ? 0xFF: 0);

    fOutputPanel.setColor (cr,cg,cb);
  } // itemStateChanged

} // class CheckBoxesPanel
```

7.3.2 JRadioButton

Radio buttons allow for situations where only one of several choices can be accepted. You must add the particular set of radio buttons into a `ButtonGroup`, which enforces the exclusionary rule. Figure 7.5 shows `UiTestApplet` again, this time for the case where the applet parameter specified the use of the

Figure 7.5   The `JRadioButton` component allows for the selection of only one button in a set of buttons. With `RadioButtonsPanel` displayed by `UITestApplet`, the user selects one of three color components to create the background color for the top panel.

RadioButtonsPanel class. This class creates three radio buttons in the constructor and adds them to a ButtonGroup instance and to the panel. The class implements the ActionListener interface, and the buttons send their events to actionPerformed(). That method uses getSource() to determine which of the radio buttons generated the event and then sets the color accordingly. The code listing for RadioButtonsPanel is given below:

```java
import javax.swing.*;
import java.awt.*;
import java.awt.event.*;
import javax.swing.event.*;

/**
  * The RadioButtonsPanel holds radio buttons that set
  * the color of the output panel.
 **/
   public class RadioButtonsPanel extends JPanel
                    implements ActionListener
{
   JRadioButton fRed, fGreen, fBlue;
   OutputPanel fOutputPanel;

   RadioButtonsPanel (OutputPanel output_panel) {

     fOutputPanel = output_panel;

     // RadioButtons need to be organized with a
     // ButtonGroup object.
     ButtonGroup group = new ButtonGroup ();

     fRed = new JRadioButton ("Red", true);
     fRed.addActionListener (this);

     // Add the JRadioButton instance to both the
     // ButtonGroup and the panel.
     group.add (fRed);
     add (fRed);

     fGreen = new JRadioButton ("Green", false);
     fGreen.addActionListener (this);
     group.add (fGreen);
     add (fGreen);

     fBlue = new JRadioButton ("Blue", false);
     fBlue.addActionListener (this);
```

```
      group.add (fBlue);
      add (fBlue);

   } // ctor

   /** Action events come here when the user clicks
    * on a button.
    **/
   public void actionPerformed (ActionEvent evt) {

      // Default color component values.
      int cr=0;
      int cg=0;
      int cb=0;
      // Only one button can be selected at a time
      // so find the button and set the corresponding
      // output panel color.
      Object source = evt.getSource ();
      if (source == fRed)
         cr = 0xFF;
      else if (source == fGreen)
         cg = 0xFF;
      else if (source == fBlue)
         cb = 0xFF;
      fOutputPanel.setColor (cr,cg,cb);

   } // actionPerformed

} // class RadioButtonsPanel
```

7.3.3 JList

The JList component provides a list of items from which the user can select.
Depending on how the JList is configured, either a single item or multiple items
can be selected. This time we tell UiTestApplet to use ListApplet, which
creates the interface shown in Figure 7.6. The user can select one of five colors
in a JList. The ListPanel code is given below.

One constructor of JList takes an Object array as a parameter. The elements of the array become the list items. We'll use a String array. The
number of items that will be visible depends on the value passed to the
setVisibleRowCount() method. The setSelectionModel() is used to
configure the list to be either single- or multiple-selection mode, and the initial
item selected is set with setSelectedIndex().

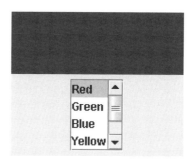

Figure 7.6 The `JList` component provides a list of items for the user to choose. The component can allow for single or multiple item selections. Here, for our `ListPanel` class in `UITestApplet`, only one item can be selected.

Instead of an `ActionListener`, the `JList` sends its events to a `ListSelectionListener` (which appears in the `javax.swing.event` package). This version of our `ControlPanel` implements the `ListSelec-tionListener` interface and provides a `valueChanged()` method. The selected item is determined from the list and the `OutputPanel`'s background is set to that color.

```
import javax.swing.*;
import java.awt.*;
import java.awt.event.*;
import javax.swing.event.*;

/**
  * The ListPanel holds the JList component with items
  * that determine the color of the output panel.
  **/
public class ListPanel extends JPanel
                implements ListSelectionListener
{
    OutputPanel fOutputPanel;

    /** Constructor adds a JList to the panel. **/
    List1Panel (OutputPanel output_panel) {

        fOutputPanel = output_panel;

        String [] colors = {
            "Red","Green","Blue","Yellow","White","Black"};

        JList color_list = new JList (colors);

        // Show only 4 items in the list at a time.
        color_list.setVisibleRowCount (4);
```

```
    // Allow only one of the items to be selected.
    color_list.setSelectionMode (
      ListSelectionModel.SINGLE_SELECTION);

    // Select initially the top item.
    color_list.setSelectedIndex (0);

    color_list.addListSelectionListener (this);

    // Add to a JScrollPane so that we can have
    // a scroller to view other items.
      JScrollPane scroll_pane = new JScrollPane (color_list);
      add (scroll_pane);

} // ctor

// This class implements the ListSelectionListener
// so events come to this valueChanged method.
public void valueChanged (ListSelectionEvent evt) {

    // Default color component values.
    int cr=0;
    int cg=0;
    int cb=0;
    // Get the reference to the JList object
    JList source = (JList)evt.getSource ();
    // and get an array of the selected items.
    Object [] values = source.getSelectedValues ();

    // In this case only one value can be selected
    // so just look at first item in array.
    String color_selected = (String)values[0];
    if (color_selected.equals ("Red"))
      cr = 0xFF;
    else if (color_selected.equals ("Green"))
      cg = 0xFF;
    else if (color_selected.equals ("Blue"))
      cb = 0xFF;
    else if (color_selected.equals ("Yellow")) {
      cr = 0xFF;
      cg = 0xFF;
    }
    else if (color_selected.equals ("White")) {
      cr = 0xFF;
```

```
        cg = 0xFF;
        cb = 0xFF;
    }
    fOutputPanel.setColor (cr,cg,cb);
  } // valueChanged

} // class ListPanel
```

Note the use of the JScrollPane class, which provides scroll bars (vertical and horizontal) for those components, such as JList and JTextArea, that implement the Scrollable interface. The bars appear whenever the content extends beyond the visible area of the component.

7.3.4 JSlider

Sliders allow the user to choose from a continuous range of values rather than from discrete values as with a set of buttons or a list. In Figure 7.7, UiTestApplet has selected SliderPanel as its control panel. The panel holds three instances of JSlider. The user can move the sliders to select the values for the Red–Green–Blue component values, each on a 0 to 255 scale. The combination of these three color values becomes the background color of the OutputPanel. Because of the continuous color changes possible, this is an applet that really should be experienced live. To see the sliders in action, we encourage you either to see the program on the Web Course site or to create the program from this source code and run it on your own.

The constructor parameters set the orientation of the slider (horizontal or vertical) and the upper and lower limit values. The last parameter sets the initial value of the slider.

A slider sends instances of ChangeEvent to each of the entries added to its ChangeListener list. The panel therefore implements the ChangeListener interface and provides a stateChanged() method. The current value setting on the slider is obtained easily with its getValue() method.

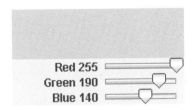

Figure 7.7 The JSlider component allows the user to select from a continuous range of values. With the SliderPanel class displayed in UITestApplet, the three sliders set the Red-Green-Blue component values from 0 to 255 each. The combined RGB values determine the color of the background of the top area.

```java
import java.awt.*;
import java.awt.event.*;
import javax.swing.*;
import javax.swing.event.*;

/**
  * The SliderPanel holds three JSlider widgets that set
  * the color of the output panel.
  **/
public class SliderPanel extends JPanel
                              implements ChangeListener
{
    OutputPanel fOutputPanel;
    JLabel fRedLabel,fGreenLabel,fBlueLabel;
    JSlider fRed,fGreen,fBlue;

    SliderPanel (OutputPanel output_panel) {
        fOutputPanel = output_panel;
        setLayout (new GridLayout (3, 2));
        add (fRedLabel = new JLabel (
            "Red 0 ",SwingConstants.RIGHT));

        // The JSlider constructor parameters:
        //
        // orientation, minimum, maximum, inital value
        //
        // The sliders are set horizontally.
        // The values range from 0 to 255.
        // Set the red slider to max initially to match the
        // initial Red color for the output panel.
        //
        add (fRed = new JSlider (
            JSlider.HORIZONTAL, 0, 255, 255));
        fRed.addChangeListener (this);

        add (fGreenLabel = new JLabel (
            "Green 0 ",SwingConstants.RIGHT));
        add (fGreen = new JSlider (
            Adjustable.HORIZONTAL, 0, 255, 0));
        fGreen.addChangeListener (this);

        add (fBlueLabel = new JLabel (
            "Blue 0 ",SwingConstants.RIGHT));
        add (fBlue = new JSlider (
            Adjustable.HORIZONTAL, 0, 255, 0));
```

```
        fBlue.addChangeListener (this);

    } // ctor

    /** This class is the AdjustmentListener for the
      * slider. So the events come here when the
      * slider is moved.
      **/
    public void stateChanged (ChangeEvent evt) {

        // Use the labels to show the numerical values of the
        // scroll bar settings.
        fRedLabel.setText ("Red " + fRed.getValue ());
        fGreenLabel.setText ("Green " + fGreen.getValue ());
        fBlueLabel.setText ("Blue " + fBlue.getValue ());

        // Get the values from each scroll bar and pass
        // them as the color component values.
        fOutputPanel.setColor (fRed.getValue (),
                               fGreen.getValue (),
                               fBlue.getValue ());
        fOutputPanel.repaint ();

    } // stateChanged

} // class SliderPanel
```

7.4 Layout managers

You are now familiar with several Swing components and the basics of event handling. You can build quite elaborate interfaces. However, before proceeding you need to know how to control the arrangement, or *layout*, of the components.

With Java you do not normally give components fixed numerical dimensions and coordinate locations. Java is intended to be portable to different platforms with different graphical operating systems and with different types of display devices. A flexible approach to component arrangement is thus required. This flexibility is achieved with *layout managers*, which we briefly mentioned in Chapter 6. A layout manager follows a general set of design rules as to how it should arrange components. You give it instructions and the layout manager tries its best to follow them within its own framework for how components are arranged.

Note that the layout manager can try to expand or shrink the size of a component to fit the layout. However, it cannot arbitrarily modify the size of a component

Figure 7.8 Arrangement of components with `FlowLayout`.

since that component may have a fixed size or a limited range of sizes. The user can override the default size settings with the methods `setMinimumSize()`, `setMaximumSize()`, and `setPreferredSize(Dimension)` methods. The layout manager calls the corresponding *getter* method to find out the minimum, maximum, and preferred dimensions for a component. A label or button, for example, wants enough room to display its text or icon. If there is too little room and the manager cannot satisfy the program's instructions, then it will do the best it can, according to its own rules. In some cases this will mean that one or more components will be partially seen or left out of the interface altogether.

There are several layout manager classes to choose from and we will give here a general description of how they work. (We discussed `GridLayout` in Chapter 6.) For more details, see the specifications in the Java 2 API, the Web Course, and the Sun tutorials [2].

Note that you typically build complex interfaces with multiple panels, each with its own layout manager that best suits the particular subgroup of components. The `FlowLayout` is the default for `JPanel` but you can specify one of the other layout managers in its constructor or with the `setLayout()` method.

Some layout managers have been available since Java 1.0 and appear in the `java.awt` package. The newer layout managers appear in the `javax.swing` package.

7.4.1 FlowLayout

The following applet code demonstrates the coding for the `FlowLayout`. The program produces the arrangement shown in Figure 7.8.

```
import javax.swing.*;
import java.awt.*;

/** Demo of the FlowLayout manager. **/
public class FlowApplet extends JApplet
{
    // Add an instance of FlowPanel to the applet.
    public void init () {
        Container content_pane = getContentPane ();

        // Create an instance of our FlowPanel class.
        FlowPanel flow_panel = new FlowPanel ();

        // And add it to the applet's panel.
        content_pane.add (flow_panel);
    } // init
} // class FlowApplet
```

```
/** A simple example of a Panel with five buttons. **/
class FlowPanel extends JPanel
{

  FlowPanel () {

    // Default for JPanel is FlowLayout
    add (new JButton ("One"));
    add (new JButton ("Two"));
    add (new JButton ("Three"));
    add (new JButton ("Four"));
    add (new JButton ("Five"));
  } // ctor

} // class FlowPanel
```

The `FlowPanel` is a subclass of `JPanel`, which uses `FlowLayout` as its default layout. Each component added to the panel will be inserted from left to right until there is no room and then the components will go to the next row down and continue left to right. By default, `FlowLayout` attempts to center the group of components. Left- or right-alignment can be specified with one of the overloaded constructors or with the `setAlignment()` method and the constants `FlowLayout.LEFT` or `FlowLayout.RIGHT`.

7.4.2 `BoxLayout` and `Box`

`FlowLayout` arranges components horizontally until it runs out of space and then shifts down vertically to the next row. Before Swing, there was no good way to arrange components in a vertical manner. The `javax.swing.BoxLayout` solved that omission. `BoxLayout` arranges components sequentially like the `FlowLayout` manager but it will set them either horizontally or vertically as instructed. Unlike `FlowLayout`, however, it will not continue the components on the next line or column when there is insufficient room. Components out of range will not be shown. The following code produces the arrangement shown in Figure 7.9(a):

```
public class BoxLayoutApplet extends JApplet
{
    . . . init() builds the interface . . .
} // class BoxLayoutApplet

/** Arrange components with a BoxLayout manager. **/
class BoxPanel extends JPanel
```

Figure 7.9 Arranging components with (a) the BoxLayout, (b) the Box container, and (c) the Box plus glue and struts.

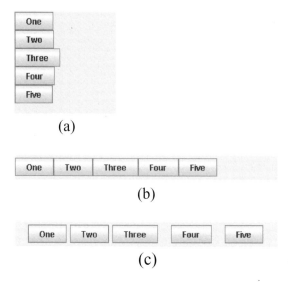

(a)

(b)

(c)

```
{
  BoxPanel () {
    setLayout (new BoxLayout (this,BoxLayout.Y_AXIS));
    // Default for JPanel is BoxLayout
    add (new JButton ("One"));
    add (new JButton ("Two"));
    add (new JButton ("Three"));
    add (new JButton ("Four"));
    add (new JButton ("Five"));
  } // ctor
} // class BoxPanel
```

The Box is a container like JPanel except that it uses BoxLayout as a default instead of the FlowLayout used by JPanel. In addition, Box provides for three special *invisible* elements that insert spacing between components. You can create a Box object with its constructor or you can use two static methods to produce Box instances with either horizontal or vertical alignment:

```
Box horizontalBox = Box.createHorizontalBox();
Box verticalBox = Box.createVerticalBox();
```

Figure 7.9(b) shows the buttons arranged according to the following code:

```
public class BoxApplet extends JApplet
{
  public void init () {
    Container content_pane = getContentPane ();
```

```
        // Create a Box with horizontal alignment
        Box box = Box.createHorizontalBox ();

        // Add the components to the Box
        box.add (new JButton ("One"));
        box.add (new JButton ("Two"));
        box.add (new JButton ("Three"));
        box.add (new JButton ("Four"));
        box.add (new JButton ("Five"));

        // And add the Box to the JApplet panel.
        Content_pane.add (box);
    } // init
} //class BoxApplet
```

In Figure 7.9(a) the components bunch toward the top and leave a gap at the bottom and in Figure 7.9(b) the components bunch to the left. Whenever the individual components have a maximum size in vertical or horizontal dimensions, this unattractive bunching will occur. To allow for arranging the components with spacing between them, the Box provides methods to create three types of invisible elements:

- Glue – surplus space in between components or between a component and container side.
- Strut – a fixed width or height spacing for horizontal or vertical alignments.
- RigidArea – both width and height dimensions are fixed values.

Figure 7.9(c) shows the buttons arranged in a Box according to the following code where we use some horizontal glue and struts to specify the spacing between the components:

```
import javax.swing.*;
import java.awt.*;

/** Demo of Box with glue and struts. **/
public class BoxSpacingApplet extends JApplet
{
    public void init () {
        Container content_pane = getContentPane ();

        // Create a Box with horizontal alignment.
        Box box = Box.createHorizontalBox ();

        // Add the buttons plus spacing components.
        box.add (Box.createHorizontalGlue ());
```

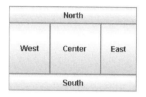

Figure 7.10 Arranging components with `BorderLayout`.

```
    box.add (new JButton ("One"));
    box.add (Box.createHorizontalStrut (5));
    box.add (new JButton ("Two"));
    box.add (Box.createHorizontalStrut (5));
    box.add (new JButton ("Three"));
    box.add (Box.createHorizontalGlue ());
    box.add (new JButton ("Four"));
    box.add (Box.createHorizontalGlue ());
    box.add (new JButton ("Five"));
    box.add (Box.createHorizontalGlue ());

    // And add box to the Applet's panel.
    content_pane.add (box);
  } // init
} // class BoxSpacingApplet
```

We see that this technique produces a more attractive arrangement of the components. Similar code can be used for vertical layouts.

7.4.3   `BorderLayout`

If you have two to five components to group together then a `BorderLayout` is often a convenient layout mananger. Figure 7.10 shows the result of the following code that uses a `BorderLayout`:

```
public class BorderApplet extends JApplet
{
. . . init() builds the interface . . .
} // class BorderApplet

/** Arrange five buttons using a BorderLayout. **/
class BorderPanel extends JPanel
{
  BorderPanel () {

    setLayout (new BorderLayout ());
    add (BorderLayout.EAST, new JButton ("East"));
    add (BorderLayout.WEST, new JButton ("West"));
    add (BorderLayout.NORTH, new JButton ("North"));
    add (BorderLayout.SOUTH, new JButton ("South"));
    add (BorderLayout.CENTER, new JButton ("Center"));
  } //ctor
} //class BorderPanel
```

(a)   (b)

Figure 7.11 (a) A `CardLayout` only shows one component at a time. (b) A `JTabbedPane` component, like this one with three tabbed pages, makes clear how many panes are overlapping.

The components go into the five locations labeled according to the four compass directions plus the center. The components that go into the "NORTH" and "SOUTH" locations will fill the horizontal space while maintaining their preferred vertical dimensions. The components that go into the "WEST" and "EAST" locations will fill the available vertical space while maintaining their preferred horizontal dimensions. The component that goes into the center will fill up the rest of the space in both horizontal and vertical dimensions.

If you want to maintain both the preferred horizontal and preferred vertical dimensions of the components, you can put each component into its own `JPanel` and then in turn add these panels to a panel that uses the `BorderLayout`.

7.4.4 `CardLayout` and `JTabbedPane`

The `CardLayout` arranges components into a "vertical" stack where only the top component is visible at a given time. Figure 7.11(a) shows the display for the following code where we stack three buttons on top of each other:

```
public class CardApplet extends JApplet
{
. . . init() builds the interface . . .
} // class CardApplet


/** Stack three buttons using CardLayout. **/
class CardPanel extends JPanel
                 implements ActionListener
{
  CardLayout fCards;

  /** Constructor adds three buttons to the panel
    * and uses CardLayout.
    **/
  CardPanel () {
    fCards = new CardLayout ();
    setLayout (fCards);
```

```
    add ("one", makeButton ("one"));
    add ("two", makeButton ("two"));
    add ("three", makeButton ("three"));
  } // ctor
  /** Create a JButton here and add this object
    * to its action listeners list.
    **/
  JButton makeButton (String name) {
    JButton b = new JButton (name);
    b.addActionListener (this);
    return b;
  }

  /** Flip to the next card when a button pushed. **/
  public void actionPerformed (ActionEvent e) {
    fCards.next (this);
  }
} // class CardPanel
```

Clicking on one of the buttons leads to the invocation of the next () method in
the CardLayout. This will display the next card in the stack, which circles back
to the beginning when the last card is reached. The cards are stacked according
to the order they are added.

The JTabbedPane is a component rather than a layout manager but it provides an alternative to CardLayout for overlaying a set of components. It provides a set of tabbed pages in which each page can hold a component. Selecting
the tab for a particular page will bring that page to the top. Using a container such
as JPanel can, of course, hold many sub-components for a page. Figure 7.11(b)
shows the interface created by the following program that uses a JTabbedPane
subclass:

```
public class TabbedApplet extends JApplet
{
  public void init() {
    Container contentPane = getContentPane ();
    // Create an instance of the JTabbedPane subclass
    Tabs tabs = new Tabs (this);

    // Add the Tabs object to the applet's panel.
    contentPane.add (tabs);
  } // init
} // class TabbedApplet
```

```
/** This JTabbedPane subclass holds three panes,
  * each with one button.
 **/
class Tabs extends JTabbedPane
                   implements ActionListener
{
  JApplet fApplet;
  /** Put a button on each of three pages. **/
  Tabs (JApplet applet) {
    fApplet = applet;
    add (makeButton ("1"), "One");
    add (makeButton ("2"), "Two");
    add (makeButton ("3"), "Three");
  } // ctor

  /** Make a button here and add this object
    * to its action listeners list.
    **/
  JButton makeButton (String name) {
    JButton b = new JButton (name);
    b.addActionListener (this);
    return b;
  }

  /** When button pushed, show message on the
    * browser status bar.
    **/
  public void actionPerformed (ActionEvent e) {
    JButton but = (JButton) (e.getSource ());
    String str = but.getText ();
    fApplet.showStatus ("Pushed button "+ str);
  }

} // class Tabs
```

The JTabbedPane class provides the add() method for adding a component along with a label that goes on the page tab. You can also add icons and mnemonics for the tabbed labels as well. (*Mnemonics* allow you to select a page with key combination such as "Alt-a".)

7.4.5 SpringLayout

The layout manager javax.swing.SpringLayout came with Java 1.4 and works by placing constraints on the distances between edges of components. The

Figure 7.12 Arranging
components with the
SpringLayout.

constraint can be placed on the distance between any two edges. This includes
the distance between the edge of one component and another, between the side
of a container and a component edge, and also between two edges of the same
component. That is, the width of a component can be constrained by setting the
distance between the left and right edges, while its height can be constrained by
setting the distance between its top and bottom edges.

When a container uses a SpringLayout you must put constraints explicitly
on each component. Otherwise, each component will put its top left corner at
the (0, 0) point of the container and the arrangement will become an overlapping
mess.

The layout specifies the edges of a component with a compass nomenclature
similar to that for BorderLayout. That is, the left edge is SpringLayout.
WEST, the right edge is SpringLayout.EAST, the bottom edge is Spring-
Layout.SOUTH, and the top is SpringLayout.NORTH. The Spring-
Layout relies on two helper classes called Spring and SpringLayout.
Constraints. The spring tries to keep a component at its preferred dimen-
sion while it resists stretching to its maximum and compressing to its minimum.

You can just use the putConstraint() method to set the distance in pixels
between the edges. The following code produces the layout shown in Figure 7.12:

```
public class SpringApplet extends JApplet
{
    . . . init () builds the interface . . .
} // class SpringApplet


/** Arrange five buttons using a SpringLayout. **/
class SpringPanel extends JPanel
{
    /**
     * Constructor creates 5 button interface with
     * SpringLayout and constrains each to a particular
     * position relative to the panel frame and to its
     * neighbors.
     **/
    SpringPanel () {
```

```
SpringLayout layout = new SpringLayout ();
setLayout (layout);

JButton one = new JButton ("One");
JButton two = new JButton ("Two");
JButton three = new JButton ("Three");
JButton four = new JButton ("Four");
JButton five = new JButton ("Five");

add (one);
add (two);
add (three);
add (four);
add (five);

// Now set the distances between the edges

// Put the first button at pixel coords (5,5)
// relative to the panel's frame.
layout.putConstraint (SpringLayout.WEST, one,
                      5, SpringLayout.WEST, this);
layout.putConstraint (SpringLayout.NORTH, one,
                      5, SpringLayout.NORTH, this);

// Put the second button 5 pixels to the right of the
// first button and 40 pixels below the top panel edge
layout.putConstraint (SpringLayout.WEST, two,
                      5, SpringLayout.EAST, one);
layout.putConstraint (SpringLayout.NORTH, two,
                      40,SpringLayout.NORTH, this);

// Put the third button 100 pixels to the left of the
// panel edge and 5 pixels above the second button
layout.putConstraint (SpringLayout.WEST, three,
                      100, SpringLayout.WEST, this);
layout.putConstraint (SpringLayout.NORTH, three,
                      5, SpringLayout.SOUTH, two);

// Put the fourth button 15 pixels to the right of the
// first button and 5 pixels below the top panel edge
layout.putConstraint (SpringLayout.WEST, four,
                      15, SpringLayout.EAST, one);
layout.putConstraint (SpringLayout.NORTH, four,
                      5, SpringLayout.NORTH, this);

// Put the fifth button 25 pixels to the right of the
```

```
        // third button and 5 pixels below it
        layout.putConstraint (SpringLayout.WEST, five,
                              25, SpringLayout.EAST, three);
        layout.putConstraint (SpringLayout.NORTH, five,
                              5, SpringLayout.SOUTH, three);


    } // ctor


} // class SpringPanel
```

The program uses the `putConstraint()` method to arrange the locations of five buttons. The following code, for example, puts five pixels between the left side of the first button and the left side of the container:

```
layout.putConstraint (SpringLayout.WEST, one, 5,
SpringLayout.WEST, this);
```

For a few components this layout can be quite convenient, but the coding becomes tedious to deal with a large number of components. The `SpringLayout` is intended primarily for use with so-called *GUI builder* programs with which a user interactively builds an interface and the program takes care of all the code details. The Sun tutorial *Creating a GUI with JFC/Swing* [3] provides a `SpringUtilities` class that assists with building interfaces with the `SpringLayout`.

7.4.6 `GridBagLayout`

The most powerful layout manager is the `GridBagLayout`. You can arrange the components in virtually any possible manner with it. However, it can be rather tricky and often requires several iterations of adjusting the settings to get the display just as you want it. In fact, if your interface has grown to the point you need a `GridBagLayout`, you may find it a lot easier to use a GUI builder. We will have room here only to give a brief introduction to this layout along with an example to illustrate its capabilities. See the Web Course and the references for more details.

   As its name indicates, the `GridBagLayout` manager uses a grid like `GridLayout` does but it works more like the hypertext table in the HTML language. That is, instead of a fixed number of rows and cells with fixed cell dimensions, this layout lets components span more than one cell vertically and horizontally. Also, the relative widths of columns and the heights of rows can vary.

   The `GridBagLayout` class does not provide an extensive set of methods to specify the layout. Instead, each component is paired with an instance of the helper class `GridBagConstraints` that specifies several parameters for the layout to set the location and size of the component. For example, unlike `GridLayout`, you do not set explicit dimensions for the grid in the constructor. For each component

Table 7.1 *GridBagConstraints Parameters.*

Parameter	Description	Values
`int gridx, gridy`	Cell column, row number	Values start at 0,0
`int gridheight, gridwidth`	Number of rows and columns taken up by the component	≥ 1
`int weightx, weighty`	Allotment weight for setting width & height of a cell	Arbitrary scale. Settings go according to relative values. See text for more details
`int fill`	Determines whether component fills cell in x, y, both, or none	HORIZONTAL, VERTICAL, BOTH, NONE
`int anchor`	Position of component within its cell when fill is not set to BOTH	CENTER, EAST, SOUTHEAST, SOUTH, SOUTHWEST, WEST, NORTHWEST, NORTH, NORTHEAST
`int ipadx, ipady`	Padding added between a component and cell borders	≥ 0
`Insets insets`	Defines padding between components	≥ 0

you set the `gridx` and `gridy` integer variables in the `GridBagConstraints` object for the cell position of the component. The layout examines these values for all the components to find the number of cells needed in the horizontal and vertical dimensions.

Table 7.1 lists the parameters in `GridBagConstraints` along with descriptions and allowed values. The purpose of the parameters is straightforward except for `weightx` and `weighty`. These two parameters determine the allocation of the available space for a column's width and a row's height when the container is resized. If the weights are all the same then the column and row widths will be the same (except for the variations due to the different minimum widths and heights of components in the cells).

You can vary the weight values to force the columns and rows to span different widths and heights. For example, the column with the cell with the largest `weightx` gets the biggest allocation of horizontal space. The column with the next biggest `weightx` gets the next biggest amount of horizontal space. And so on. The `weighty` values work in a similar manner for vertical space allocations to the rows.

The following code leads to the layout shown in Figure 7.13. The program demonstrates the use of several of the `GridBagConstraints` parameters. Note that the same `GridBagConstraints` object is used throughout. Only the cell coordinates and one or more of its other values are changed when it is used with the addition of the next component.

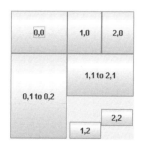

Figure 7.13 Arranging components with the `GridBagLayout`.

Note how the multi-cell spanning works. The component placed in the cell at (column, row) = (0, 1) is assigned `gridheight=2`. This means it will span 2 rows (row 1 to row 2). Similarly, the component placed at (1, 1) has `gridwidth=2` so it spans two columns.

```
public class GridBagApplet extends JApplet
{
 . . . init () sets up interface . . .
} // class GridBagApplet

/** Creates a panel with 5 buttons using GridBagLayout. **/
class GridBagPanel extends JPanel
{
  GridBagConstraints constraints = new GridBagConstraints ();

  GridBagPanel () {
    setLayout (new GridBagLayout ());

    // Create a 3 row grid

    // Fill the grid squares with the component
    // in both x and y directions
    constraints.fill = GridBagConstraints.BOTH;

    // Keep same weight in vertical dimension
    constraints.weighty = 1.0;

    // Top row will include three components, each
    // weighted differently in x
    // 0,0
    constraints.weightx = 1.0;
    constraints.gridx = 0; constraints.gridy = 0;
    add (new JButton ("0,0"), constraints);

    // 0,1
    constraints.weightx = 0.5;
    constraints.gridx = 1; constraints.gridy = 0;
    add (new JButton ("1,0"), constraints);

    // 0,2
    constraints.weightx = 0.1;
    constraints.gridx = 2; constraints.gridy = 0;
    add (new JButton ("2,0"), constraints);

    // Middle row has two components. First takes up two
```

```
        // rows, second takes up two columns
        // The first component on second row will span
        // vertically to third row
        // 0,1 to 0,2
        constraints.weightx = 1.0;
        constraints.gridx = 0; constraints.gridy = 1;
        constraints.gridheight = 2;
        add (new JButton ("0,1 to 0,2"), constraints);

        // The second component on second row will span
        // horizontally to third column
        // 1,1 to 2,1
        constraints.weightx = 1.0;
        constraints.gridx = 1; constraints.gridy = 1;
        constraints.gridheight = 1;
        constraints.gridwidth = 2;
        add (new JButton ("1,1 to 2,1"), constraints);

        // Bottom row has 2 components with fill set to NONE
        // Use anchor.
        constraints.fill = GridBagConstraints.NONE;

        // 1,2
        constraints.anchor = GridBagConstraints.SOUTHEAST;
        constraints.weightx = 0.5;
        constraints.gridx = 1; constraints.gridy = 2;
        constraints.gridheight = 1;
        constraints.gridwidth = 1;
        add (new JButton ("1,2"), constraints);

        // 2,2
        constraints.anchor = GridBagConstraints.WEST;
        constraints.weightx = 0.1;
        constraints.gridx = 2; constraints.gridy = 2;
        constraints.gridheight = 1;
        constraints.gridwidth = 1;
        add (new JButton ("2,2"), constraints);
    } // ctor
} // class GridBagPanel
```

7.5 Convenience classes

With Java 1.1 came several types of classes that facilitate the writing of Java code. Some programmers feel that these convenience classes violate the spirit of the object-oriented approach in which classes are strictly modular and separate. However, if these classes are kept small and only used where best needed, they can make Java code easier both to write and to read.

7.5.1 Inner classes

A technique that came with Java 1.1 allowed the nesting of classes. That is, *inner classes* can be inserted inside other classes. For example, in the following code we show how to put the definition of the class AnInnerClass inside the constructor:

```
public class AnOuterClass
{
   int fVar;
   public AnOuterClass () {
      . . . other code . . .

      class AnInnerClass {
         int b;
         AnInnerClass () {
            b = 4. * fVar;
         }
      } // class AnInnerClass

      AnInnerClass inner = new AnInnerClass ();
      . . . other code . . .
   }

   void someMethod () {
      . . . code . . .
   }
} // class AnOuterClass
```

The *scope*, which refers to the variables and methods accessible to a given section of code, for the inner classes includes the instance variables and methods of the parent class.

The most common use of inner classes is with event handling. The event model allows us to make any class into a listener for our events. However, this can lead to listener classes spread throughout the code, which makes the programs less readable. With inner classes we can put the listener class definition immediately adjacent to the code for the component that uses the listener. For example, in the following code segment, an ActionListener class is placed next to where an instance of it is added to a button:

```
public class AnOuterClass extends JApplet
{
   int fVar = 0;
   JButton fBut = new JButton ("OK");
   public AnOuterClass () {

      class AnInnerClass implements ActionListener
      {
         public void actionPerformed (ActionEvent e) {
            fVar++;
            System.out.println ("Pressed " + fVar + "times");
         }
      } // class AnInnerClass

      fBut.addActionListener (new AnInnerClass ());
      add (fBut);
   } // ctor
} // class AnOuterClass
```

The compiler will separate out these inner classes and create separate `class` files for them. The names will be preceded with the outer class name, as in

```
AnOuterClass$AnInnerClass.class
```

for the above example.

7.5.2 Anonymous inner classes

It is common in Java programming to find situations where you need to create an object but don't need to bother with giving it an explicit name. For example, to specify the size of a panel you could use code like this:

```
Dimension d = getSize ();
int width = d.width;
int height = d.height;
```

where a `Dimension` object is returned from a method and then used to specify the width and height values. But why bother to create a variable name for this `Dimension` object since it will never be used again? Instead, you could replace the code with

```
int width = getSize ().width;
int height = getSize ().height;
```

where the compiler bytecode will keep track of the `Dimension` object that provides the width and height.

We can in fact cascade several such calls, using the object returned in the left method to call the next method on the right:

```
a = getA ().methodB ().aMethodC ().variableX;
```

This *anonymity* eliminates from the code a lot of unnecessary variables and makes it more readable.

With the inner classes we can take this concept to another level by creating and instantiating a class without bothering to give the class a name. In the code below an instance of the `ActionListener` class is created in the parameter of the `addActionListener` method:

```java
public class AnOuterClass extends JApplet
{
   int fVar = 0;
   JButton fBut = new JButton ("OK");

   public AnOuterClass () {
     fBut.addActionListener
     (// Left parenthesis of method
       new ActionListener () {// no name given for this
                              // ActionListener object
       // Override actionPerformed as usual
       public void actionPerformed (ActionEvent e) {
         fVar++;
         System.out.println ("Pressed " + fVar + "times");
         }
       }
   ); // Right parenthesis of method
     add (fBut);
   } // ctor
} // class AnOuterClass
```

Here in one step we created a class that implemented the `ActionListener` interface and created an instance of it for use by the button. The compiler will create a class file name `AnOuterClass$1.class` where a number, in this case "1", is used to identify the class file name for an anonymous inner class.

7.5.3   Adapter classes

The Java class library also includes *adapter classes*, which make writing listener classes more convenient. While the listener architecture greatly improves the efficiency and capabilities of event handling, there are some complications and annoyances that come along with it. In particular, we find that the listener interfaces hold up to six methods that must be implemented. (Remember that all methods in an interface are abstract and so concrete implementations for each must be provided.) If you need only one or a few of these methods, you must nevertheless implement all of the remaining methods with empty code bodies.

In the `AnonListenterApplet` example given here, we use an anonymous `MouseListener` class but only one of the five methods declared in the `MouseListener` interface is actually needed:

```
. . . In the AnonListenerApplet class . . .
  void init () {
    Container content_pane = getContentPane ();
    // Create an instance of JPanel
    JPanel panel = new JPanel ();

    // User an anonymous MouseListener object with the
    // panel
    panel.addMouseListener (
      new MouseListener () {
        public void mouseEntered (MouseEvent e) {
          System.out.println (e.toString ());
        }
         // Give empty code bodies to these unneeded
         // methods.
        public void mouseClicked (MouseEvent e) {}
        public void mousePressed (MouseEvent e) {}
        public void mouseReleased (MouseEvent e) {}
        public void mouseExited (MouseEvent e) {}
      }
    );

    // Add the panel to the content pane.
    content_pane.add (panel);

  } // init

. . . rest of code in AnonListenerApplet . . .
```

To avoid writing empty versions of all those unneeded methods, the `java.awt.event` package provides adapter classes for seven of the listener interfaces. These adapters have implemented the interface methods with empty methods. Since the adapters are concrete (non-abstract) classes, we only need to override the methods of interest.

The `AdapterApplet` program here is essentially the same as the `Anon-ListenerApplet` above but with a `MouseAdapter`:

```
. . . In the AdapterJApplet class . . .
  Container content_pane = getContentPane ();
  // Create an instance of JPanel.
  JPanel panel = new JPanel ();
  // Use an anonymous MouseAdapter object for the panel
  panel.addMouseListener (
    new MouseAdapter () {
      public void mouseEntered (MouseEvent e) {
        System.out.println (e.toString ());
      }
    }
  );
  // Add the panel to the content pane.
  content_pane.add (panel);
  . . .
```

This saves writing a lot of code but more importantly makes the code easier to read and understand. However, be careful that you properly spell the method that you override in the adapter. Otherwise, you are just adding another method to the subclass. It can be difficult to track down this bug in your program.

7.6   Frames and menus

The application programs that we have demonstrated so far in this course send their output to the console. We can instead create a window frame and build a graphical user interface in the window using the components and GUI techniques discussed thus far with regard to applets. As with other programs, we can also include a menu bar with drop-down menus to provide various options for the user such as the common *File* menu on the left-most position holding items such as *Open file*, *Save file*, and *Exit*.

An application and an applet can also open other window frames. For example, a program may need for the user to select from a bigger set of options and settings than a menu can provide. So it brings up a new frame with a graphical interface for the user to make the desired settings.

In the follow sections we first discuss how to add a menu bar to an applet. Then we look at how to open a frame from within an application or an applet. We then discuss how to create a GUI program that can run either as an applet or in a standalone application window.

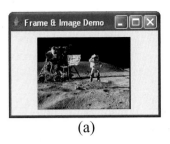
(a)

(b)

Figure 7.14 (a) The `JFrame` class provides a framed window. (b) `JFrame` with a menu bar and a single drop down menu.

7.6.1 Frame for an application

The following program illustrates the basics of creating a window frame as shown in Figure 7.14(a). The `main()` method creates an instance of `JFrame` (the string parameter passed to the `JFrame` constructor becomes the title in the frame's title bar). When the frame is closed, the program is told to exit:

```
f.setDefaultCloseOperation (JFrame.EXIT_ON_CLOSE);
```

(If you create a daughter frame from within an applet or application, use `DISPOSE_ON_CLOSE` to close the frame without exiting the program.) A panel that displays an image in the center is added to the panel's content pane. The `setSize()` method sets the dimensions of the frame. Then the frame is made visible.

```java
import javax.swing.*;
import java.awt.*;

/** This app displays an image on a JPanel subclass.**/
public class FrameApp
{
  /** Create a frame and display image with DrawingPanel. **/
  public static void main (String[] args) {

    // Use the AWT toolkit to obtain the image
    Image img = Toolkit.getDefaultToolkit ().getImage (
                               "Apollo16Lander.jpg");

    // Pass the image to an instance of DrawingPanel
    DrawingPanel drawing_panel = new DrawingPanel (img);

    // Create an instance of JFrame with a title string.
    JFrame f = new JFrame ("Frame & Image Demo");

    // Set mode for closing frame with the window exit button.
```

```
      f.setDefaultCloseOperation (JFrame.EXIT_ON_CLOSE);

      // Add the DrawingPanel object, set dimensions.
      f.getContentPane ().add (drawing_panel);
      f.setSize (new Dimension (240,160));
      f.setVisible (true);
  } // main
} // class FrameApp

/** This JPanel subclass display an image.**/
class DrawingPanel extends JPanel
{
  Image fImg;

  DrawingPanel (Image img) {
    fImg = img;
  }

  public void paintComponent (Graphics g) {
    super.paintComponent (g);

    // Put the image at the center of the panel
    int img_x = getSize ().width/2 - fImg.getWidth (this)/2;
    int img_y = getSize ().height/2 - fImg.getHeight (this)/2;

    //Draw image at centered in the middle of the panel
    g.drawImage (fImg, img_x, img_y, this);

  } // paintComponent
} // class DrawingPanel
```

7.6.2   A menu bar for a frame

For most applications with frames, you will want to add a menu bar. Creating a useful menu bar involves at least three classes. The JMenuBar class represents the menu bar itself while JMenu holds instances of JMenuItem. The following code shows how to create the frame and menu bar shown in Figure 7.14(b):

```
import javax.swing.*;
import java.awt.*;
import java.awt.event.*;

/** This app displays an image and includes a menu bar.**/
public class FrameMenuApp extends JFrame
                          implements ActionListener
```

```
{

  /** Constructor passes title to super class. **/
  public FrameMenuApp (String title) {
    super (title);
  }


  /** Create a frame and display image with DrawingPanel. **/
  public static void main (String[] args) {

    // Get an image and pass it to an instance of DrawingPanel.
    Image img = Toolkit.getDefaultToolkit ().getImage (
                                   "Apollo16Lander.jpg");
    DrawingPanel drawingPanel = new DrawingPanel (img);

    // Create an instance of this JFrame subclass.
    FrameMenuApp f =
        new FrameMenuApp ("Frame with Menu Bar Demo");
    f.setDefaultCloseOperation (JFrame.EXIT_ON_CLOSE);

    // Build a menu bar.
    JMenuBar mb = new JMenuBar ();

    // Create a standard "File" drop-down menu.
    JMenu menu = new JMenu ("File");

    // Include for key selection of the menu item.
    menu.setMnemonic (KeyEvent.VK_F);

    // Add the menu to the menu bar
    mb.add (menu);

    // Create an Open item for the menu.
    JMenuItem menuOpen = new JMenuItem ("Open", KeyEvent.VK_O);
    menu.add (menuOpen);
    // The FrameMenuApp implements ActionListener
    // to respond to the menu item selections.
    menuOpen.addActionListener (f);

    // Include a separate line on the menu.
    menu.addSeparator ();

    // Add the exit item
    JMenuItem menuExit = new JMenuItem ("Exit", KeyEvent.VK_X);
    // Includes the accelerator key combo to select it.
    menuExit.setAccelerator (KeyStroke.getKeyStroke (
```

```
      KeyEvent.VK_Q, ActionEvent.CTRL_MASK));
   menu.add (menuExit);
   menuExit.addActionListener (f);


   // Add the menu bar to the frame.
   f.setJMenuBar (mb);


   f.getContentPane ().add (drawingPanel);
   f.setSize (new Dimension (240,180));
   f.setVisible (true);


} // main


/** Respond to the menu items. **/
public void actionPerformed (ActionEvent e){
   String cmd = e.getActionCommand ();
   // Exit program
   if (cmd.equals ("Exit"))
     System.exit (0);
   else if (cmd.equals ("Open"))
     // Put code here to open a file
     System.out.println ("Open");
 }// actionPerformed


}// class FrameMenuApp


  . . . DrawingPanel class same as for FrameApp example . . .
```

This program goes as the previous one except that it creates a menu bar with a single menu labeled "File". The File menu holds two items – Open and Exit – and a separator.

The menu bar has a simple no argument constructor and is added to the frame with the `setMenuBar()` method. The `JMenu` constructor passes the title as a parameter. A mnemonic allows for the selection of a menu or menu item via the keyboard. The following line assigns the `Alt-f` key pair to bring the focus on the File menu:

```
menu.setMnemonic (KeyEvent.VK_F);
```

The "F" in the File label will be underlined to indicate that it is the mnemonic key. A constructor for `JMenuItem` provides for setting the mnemonic with the second parameter, as in

```
JMenuItem menuOpen = new JMenuItem ("Open", KeyEvent.VK_O);
```

This line defines the "Open" item in the File drop-down menu with the mnemonic key pair set to `Alt-O`.

The mnemonic brings the focus on a menu or menu item but it doesn't select it. Pressing the Return or Enter key on the keyboard then selects that menu item. To create a shortcut keystroke to fire a menu item, use an *accelerator key*, which causes the item to be selected immediately. The following line will cause the `Ctrl-q` pair to select the Exit item:

```
menuExit.setAccelerator (KeyStroke.getKeyStroke (
    KeyEvent.VK_Q, ActionEvent.CTRL_MASK));
```

The `FrameMenuApp` class implements the `ActionListener` class, and the frame object is added to both menu items. A separator line is inserted into the menu with the following:

```
menu.addSeparator ();
```

Note that a nice feature of the Swing `JApplet` is that it allows a menu bar on the panel on the browser page. Previously, the heavyweight menu bar component could not be put onto the applet within the browser. The lightweight Swing menus are simply drawn on the panel like any other Swing component.

7.7   User interface with histogram display

In this section we add interactive graphics components to the histogram display discussed in Chapter 6 so that the user can control with the program. Figure 7.15 shows the histogram user interface display obtained from the program `UIDrawHistApplet` listed below. We use the same `Histogram`,

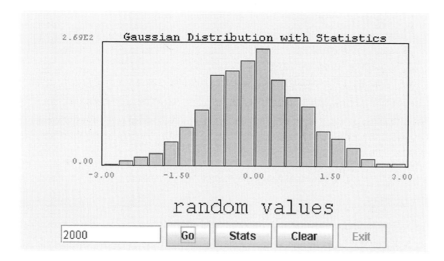

Figure 7.15   This graphical user interface combines a histogram with a set of control components. The user enters into a `JTextField` the number of Gaussian random numbers to generate and the buttons initiate the generation, display statistics about the distribution, clear the histogram, and end the program.

HistFormat, and HistPanel classes, illustrating the modularity of working with classes. Here the program adds the histogram panel HistPanel to its
content pane and also a panel with three buttons and a text field. The text field
displays the number of values to be generated for the histogram when the "Go"
button is pushed. The user can change this value. The "Clear" button empties the
histogram.

```java
import javax.swing.*;
import java.awt.*;
import java.awt.event.*;

/** A demonstration of the Histogram class. **/
public class UIDrawHistApplet extends JApplet
        implements ActionListener
{
    // Use the HistPanel JPanel subclass here
    HistPanel fOutputPanel;

    Histogram fHistogram;
    int fNumDataPoints = 100;

    // A text field for input strings
    JTextField fTextField;

    // Flag for whether the applet is in a browser
    // or running via the main () below.
    boolean fInBrowser = true;

    //Buttons
    JButton fGoButton;
    JButton fClearButton;
    JButton fExitButton;

    /**
      * Create a User Interface with a text area with
      * scroll bars and a Go button to initiate processing
      * and a Clear button to clear the textarea.
      **/
    public void init () {
        Container content_pane = getContentPane ();
        JPanel panel = new JPanel (new BorderLayout ());

        // Create a histogram with Gaussian distribution.
        makeHist ();
```

```
      // JPanel subclass here.
      fOutputPanel = new HistPanel (fHistogram);

      panel.add (fOutputPanel,"Center");

      // Use a textfield for an input parameter.
      fTextField =
        new JTextField (Integer.toString (fNumDataPoints), 10);

      // If return hit after entering text, the
      // actionPerformed will be invoked.
      fTextField.addActionListener (this);

      fGoButton = new JButton ("Go");
      fGoButton.addActionListener (this);

      fClearButton = new JButton ("Clear");
      fClearButton.addActionListener (this);

      fExitButton = new JButton ("Exit");
      fExitButton.addActionListener (this);

      JPanel control_panel = new JPanel ();
      control_panel.add (fTextField);
      control_panel.add (fGoButton);
      control_panel.add (fClearButton);
      control_panel.add (fExitButton);

      if (fInBrowser) fExitButton.setEnabled (false);
      panel.add (control_panel,"South");

      // Add text area with scrolling to the content pane.
      content_pane.add (panel);
  } // init

/** Respond to buttons. **/
public void actionPerformed (ActionEvent e){
    Object source = e.getSource ();
    if (source == fGoButton || source == fTextField) {
       String strNumDataPoints = fTextField.getText ();
       try {
          fNumDataPoints = Integer.parseInt (strNumDataPoints);
        } catch (NumberFormatException ex) {
          // Could open an error dialog here but just
          // display a message on the browser status line.
```

```
              showStatus ("Bad input value");
              return;
          }

              makeHist ();
              repaint ();

        } else if (source == fClearButton) {
              fHistogram.clear ();
              repaint ();

        } else if (!fInBrowser)
              System.exit (0);

  } // actionPerformed

    /** Create a histogram if it doesn't yet exit. Fill it
     * with Gaussian random distribution.
     **/
  void makeHist () {
      // Create an instance of the Random class for
      // producing our random values.
      java.util.Random r = new java.util.Random ();

      // The method nextGaussian in the class Random
      // produces a value centered at 0.0 and a standard
      // deviation of 1.0.

      // Create an instance of our histogram class. Set the
      // range so that it includes most of the distribution.
      if (fHistogram == null)
         fHistogram = new Histogram ("Gaussian Distribution",
                          "random values", 20,-3.0,3.0);

      // Fill histogram with Gaussian distribution
      for (int i = 0; i < fNumDataPoints; i++) {
         double val = r.nextGaussian ();
         fHistogram.add (val);
      }
  } // makeHist

  /** Create a frame and add the applet to it. **/
  public static void main (String[] args) {
      // Dimensions for our frame
      int frame_width = 450;
      int frame_height = 300;
```

```
          // Create an instance of the applet to add to the frame.
          UIDrawHistApplet applet = new UIDrawHistApplet ();
          applet.fInBrowser = false;
          applet.init ();

          // Create the frame with the title
          JFrame f = new JFrame ("Histogram with Gaussian");
          f.setDefaultCloseOperation (JFrame.EXIT_ON_CLOSE);

          // Add applet to the frame and display the frame.
          f.getContentPane ().add (applet);
          f.setSize (new Dimension (frame_width,frame_height));
          f.setVisible (true);

     } // main

} // class UIDrawHistApplet
```

The conversion of the string obtained from the text field to an integer can throw an exception. Here we indicate a bad value by showing an error message on the browser status bar. A more sophisticated approach would open a message dialog to warn the user of the bad input (see the Web Course Chapter 7: *Supplements* section for a discussion and several demonstrations of message dialogs).

When building a GUI you should arrange for the controls to behave in way that will be natural and intuitive for the user. For example, if the user enters a value into the text field and does not hit Enter, the value in the text field should still be used when the "Go" button is selected. This ensures that the value showing in the text file is always the parameter used in the calculation, as most users would expect. Furthermore, we added the `ActionListener` (the applet in this case) to the list for the `JTextfield`. Thus, when the user enters text and hits Enter, an action event will be generated and the `actionPerformed()` method invoked just as if the "Go" button had been selected.

The above example also shows how to create a GUI program that runs either as an applet or as a standalone frame. When run as an applet the `main()` method is ignored. When run as an application, the `main()` method creates an instance of `UIDrawHistApplet` and adds it to the frame content pane. The `Applet` class `getImage()` method will not work properly in application mode so we use a flag to indicate which of the two image loading techniques to use.

7.8   Web Course materials

All of the programs discussed here are available as applets or applications in the Web Course Chapter 7: *Java* section along with additional discussions on event handling, layout managers, and Swing components. The Chapter 7: *Supplements* section provides a tutorial on creating dialogs in the Swing framework. In addition, it discusses GUI development with just the basic AWT components.

The Chapter 7: *Tech* section discusses the above `Histogram` applet with user interface controls. It also discusses the creation of non-uniform random number distributions with techniques such as the transformation and the rejection methods. A `Histogram` subclass called `HistogramStat` that provides more extensive statistical measures of the distribution is developed.

The Chapter 7: *Physics* section looks further at the generation of non-uniform distributions and presents the histogram technique for creating custom distributions. It also gives a simple example of a Monte Carlo integration.

References

[1] *Lesson: Writing Event Listeners – The Java Tutorial*, Sun Microsystems, `http://java.sun.com/docs/books/tutorial/uiswing/events/`.

[2] *Lesson: Laying Out Components within a Container – The Java Tutorial*, Sun Microsystems, `http://java.sun.com/docs/books/tutorial/uiswing/layout/`.

[3] *Trail: Creating a GUI with JFC/Swing – The Java Tutorial*, Sun Microsystems, `http://java.sun.com/docs/books/tutorial/uiswing/`.

Chapter 8
Threads

8.1 Introduction

Threads in Java are processes that run in parallel within the Java Virtual Machine. When the JVM runs on a single real processor the parallelism is, of course, only apparent because of the high speed switching of threads in and out of the processor. Yet even in that case, threading can provide significant advantages. For example, while one thread deals with a relatively slow I/O operation such as downloading a file over the network, other threads can do useful work. Threads can assist in modularizing program design. An animation can assign different threads to rendering the graphics, to sound effects, and to user interactions so that all of these operations appear to take place simultaneously. Furthermore, a JVM can assign Java threads to native OS threads (but isn't required to) and on a multiprocessor system it could thus provide true parallel performance.

Java makes the creation and running of threads quite easy. We will concentrate on the basics of threading and only briefly touch on the subtle complications that arise when multiple threads interact with each other and need to access and modify common resources. Such situations can result in *data race*, *deadlock*, and other interference problems that result in distorted data or hung programs.

8.2 Introduction to threads

In Java you can create one or more threads within your program just as you can run one or more programs in an operating system [1–4]. Most JVMs, in fact, take great advantage of threads for such tasks as input/output operations and user-interface event handling. Since the Java garbage collector always runs in a separate thread, even the simplest Java program is actually multithreaded.

In the previous chapters we saw that Java applications begin when the JVM invokes the `main()` method. (The application itself runs as a thread in the JVM.) Instead of a `main()`, the thread processes begin and end with a method named `run()`. You place code in `run()` to control the operations that you wish to accomplish with the thread. The thread lives only as long as the process remains within `run()`. When the thread process returns from the `run()`, the thread is dead and cannot be resurrected.

You create a thread class in one of two ways:

1. Create a subclass of the Thread class and override the run() method.
2. Create a class that implements the Runnable interface, which has only one method: run(). Pass a reference to this class in the constructor of Thread. The thread then calls back to this run() method when the thread starts.

In the following sections we examine these two thread creation techniques further.

8.2.1 Thread creation: subclass

Creating a subclass of Thread offers the most conceptually straightforward approach to threading. In this approach the subclass overrides the run() method with the code you wish to process. The following code segments illustrate this approach.

The class MyThread extends the Thread class and overrides the method run() with one that contains a loop that prints out a message until a counter hits 20.

```
public class MyThread extends Thread
{
   public void run () {
      int count = 0;
      while (true) {

         System.out.println ("Thread alive");

         // Print every 0.10sec for 2 seconds
         try {
            Thread.sleep (100);
         }
         catch (InterruptedException e) {}
         count++;
         if (count >= 20) break;
      }
      System.out.println ("Thread stopping");
   } // run
} // class MyThread
```

In MyApplet shown below, the start() method creates an instance of the MyThread class and invokes the thread's start() method. This will in turn invoke the run() method. The thread goes into a loop and prints a message every 100 ms using the Thread class static method sleep(long time), where time is in milliseconds. The thread then dies – i.e. it cannot be restarted – once the process exits from `run()`.

```
/** Demo threading with Runnable implementation.**/
public class MyApplet extends java.applet.Applet
{
   /** Applet's start method creates and starts a thread.
    **/
   public void start () {

      // Create an instance of MyThread.
      MyThread myThread = new MyThread ();

      // Start the thread
      myThread.start ();
   } // start

   public void paint(java.awt.Graphics g) {
      g.drawString("Thread Demo 1",20,20);
   }
} // class MyApplet
```

The diagram in Figure 8.1 shows schematically how the main thread and `MyThread` thread run in parallel.

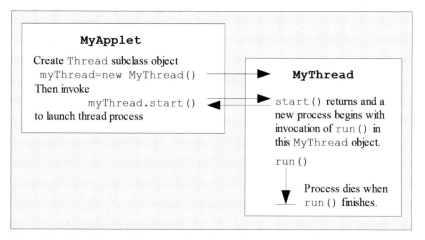

Figure 8.1   This diagram illustrates threading with a `Thread` subclass. `MyApplet` creates an instance of `MyThread`, which extends `Thread` and overrides `run()`. The applet invokes the `start()` method for the `MyThread` object. The `start()` method returns but the thread process continues independently with the invocation of the `run()` method in `MyThread`. When the thread process returns from `run()` the thread dies (i.e. cannot be started again).

8.2.2   Thread creation: Runnable

In the second threading technique a class implements the Runnable interface and overrides its run() method. This approach is often convenient, especially for cases where you want to create a single instance of a thread, as in an animation for an applet. You pass a reference to the Runnable object via the constructor of Thread and when it starts, the thread *calls back* to the run() method. As before, the thread process dies after exiting run().

The following code segment illustrates this approach. Here MyRunnableApplet implements the Runnable interface. The start() method creates an instance of the Thread class and passes a reference to itself (with the "this" reference) in the thread's constructor. When it invokes the start() method for the thread, the thread will invoke the run() method in MyRunnableApplet.

```
/** Demo threading with Runnable implementation. **/
public class MyRunnableApplet extends java.applet.Applet
                            implements Runnable
{
  /** Applet's start method creates a thread. **/
  public void start () {

    // Create an instance of Thread with a
    // reference to this Runnable instance.
    Thread thread = new Thread (this);

    // Start the thread
    thread.start ();
  } // start

  /** Override the Runnable run() method. **/
  public void run () {
    int count = 0;
    while (true) {
      System.out.println ("Thread alive");
      // Print every 0.10sec for 5 seconds
      try{
        Thread.sleep(100);
      } catch (InterruptedException e) {}
      count++;
      if (count >= 50) break;
    }
    System.out.println ("Thread stopping");
  } // run

  public void paint (java.awt.Graphics g) {
```

```
     g.drawString ("Thread demo 2",20,20);
   }
} // class MyRunnableApplet
```

The diagram in Figure 8.2 shows schematically how this approach works.

8.2.3 `Thread` subclass vs. `Runnable`

The choice between these two thread creation techniques depends on the particular application and what seems most appropriate and convenient for it. Since Java does not allow multiple inheritance, an applet class that is already a subclass of `Applet` or `JApplet` can become multithreaded by implementing `Runnable`. The `run()` method will have access to the variables and methods of the class. For example, an applet animation may need parameters for initialization and may also need to invoke methods from the applet.

Extending `Thread` applies well to the situation where you want to create a specialized thread class that does not need to extend any other class. A common case is where many *worker* threads are needed such as in a server program that

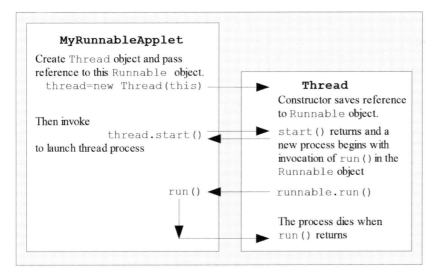

Figure 8.2 This diagram illustrates threading with a `Runnable` class. `MyRunnableApplet` implements the `Runnable` interface and it creates an instance of `Thread` and passes in the constructor a reference to itself as a `Runnable` object. (We use the name "`runnable`" for the reference variable to the `Runnable` object.) The applet invokes the `start()` method for the thread and it returns while the thread process continues independently with the invocation of the `run()` method *in the* applet object. When the thread process returns from the applet's `run()` the thread dies.

assigns a worker to service each client that connects to it. When the client signs off, the threaded worker process assigned to it dies.

8.3   Stopping threads

A thread dies in three ways:

- it returns from `run()`
- the `stop()` method is invoked (this method is now deprecated)
- it is interrupted by a runtime exception

The first approach is always the preferred way for a thread to die. In the examples shown above in Section 8.2, we used a flag variable in a loop to tell the run method to finish. We recommend this approach to killing a thread.

Do *not* use the `Thread` method `stop()` to kill a thread. The `stop()` method has been *deprecated*. That means that it still exists in the class definition but is officially marked as obsolete and should be avoided. The `stop()` method causes the thread to cease whatever it is doing and to throw a `ThreadDeath` exception. This can leave data in an unknown state. For example, the thread might be midway through setting the values of a group of variables when the thread was stopped. This will leave some variables with new values and some with old values. Other processes using those variables might then obtain invalid results. Furthermore, an instruction involving a `long` or `double` type value can require two operations in the JVM, which moves data in 32-bit chunks. So a thread stop might occur after the first operation and leave the value in an indeterminate state. These kinds of errors will be difficult to track down since the effect may not be seen until the processing reaches another part of the program.

As mentioned earlier, the best way to stop a thread is to signal that the processing should return from `run()`. Setting a flag can usually accomplish this. A loop can check the flag after each pass and escape from the loop with the flag switches. This allows for the process to finish what it is doing in a controlled manner. In previous examples we set a `boolean` flag. In the applet below we use the thread reference instead of a separate flag variable. Setting the reference to `null` signals for the end of a loop in `run()` and also allows the garbage collector to reclaim the memory used by the thread.

```
public class MyApplet extends Applet implements Runnable
{
  Thread fMyThread;
  public void init () {
    . . .
```

```
   }
   . . .
   public void start () {
      if (fMyThread!= null) {
         fMyThread.start ();
      }
      else
         fMyThread = new Thread (this);
   }

   public void stop () {
      fMyThread = null;
   }

   void run () {
      while (fMyThread!= null) {

         . . .

      }
   }
} // MyApplet
```

Remember that the start() and stop() methods in the Applet class are unrelated to methods with the same names in the Thread class. Like the init() method in the Applet class, these are just methods that allow the browser to control the applet. The browser invokes start() each time the applet page is loaded (note that init() is only invoked the first time the applet web page is loaded). The applet's stop() is a good place to do housecleaning such as killing any live threads. Always explicitly stop your threads in applets when the applet stop() is called. Otherwise, they may continue running even when the browser loads a new web page.

Furthermore, do *not* use the deprecated suspend() and resume() methods in the Thread class for the same reasons given for not using the stop() method. You can obtain effective suspend/resume operations by killing the thread (that is, signaling for it to return safely from the processing in the run() method) and creating a new one with the same values of the variables as when the previous thread died. The new thread will then simply continue from where the last one finished.

8.4 Multiprocessing issues

An operating system executes multiple processes in a manner similar to that for multithreading except that each process stack refers to a different program in memory rather than code within a single program. The Java Virtual Machine (JVM) controls the threads within a Java program much as the machine operating system controls multiple processes.

In some JVM implementations, threads are directly assigned to native processes in the operating system. Furthermore, in operating systems that support multiple physical processors, the threads can actually run on different processors and thus achieve true parallel processing.

Multiprocessing in Java with threads is relatively straightforward and provides for great flexibility in program design. The JVM handles most of the details of running threads but your program can influence the sharing of resources and setting priorities for multiple threads.

8.4.1   Sharing resources

Just as in an operating system, when multiple threads need to share a processor or other resources, the JVM must provide a mechanism for a thread to pause and allow other threads the opportunity to run. The two basic designs for *context switching* of threads are:

- *preemptive* or *time-slicing* – give each thread fixed periods of time to run
- *non-preemptive* or *cooperative* – a thread decides for itself when to surrender control

Generally, the preemptive approach is the most flexible and robust. A misbehaving thread cannot hog all the resources and possibly freeze the whole program. Unfortunately, the context switching design is not specified currently for Java and so different JVMs do it differently. Thus you should design your multithreaded code for either possibility if you expect to distribute your program for general use.

For example, you can explicitly add pauses to your threads to ensure they share the processing. The static method `yield()` in the `Thread` class tells the currently executing thread to pause momentarily and let other threads run. The static method `sleep(long millis)`, where `millis` is in milliseconds, tells the currently executing thread to pause for a specific length of time. There is also the overloaded version method `sleep(long millis, int nanos)`, where the sleep time equals `millis` in milliseconds plus `nanos` in nanoseconds. (Most platforms, however, cannot specify time that accurately.) With these two methods, you can ensure that when your program runs in a non-preemptive environment, the threads will release control at suitable points in the thread code.

The resources needed for each thread is another aspect of multiprocessing to consider when creating a high number of threads. The maximum number of threads depends on the stack space needed per thread and the amount of memory available. The stack size default is 400 KB. For a 1 gigabyte address space this should allow up to 2500 tiny threads, but in practice, because the thread code itself plus any memory allocated for objects a thread uses takes up memory too, an `OutOfMemoryError` will usually occur far sooner.

8.4.2   Setting priorities

Every thread has an integer priority value between 1 and 10 that can be controlled using methods in the `Thread` class. Generally, higher priority threads can be expected to be given preference by the thread scheduler over lower priority threads. However, the implementation of thread scheduling is left up to the JVM implementation. This lack of specificity provides maximum flexibility to JVM designers since Java can be implemented on platforms with limited speed and resources and also on platforms with multiple processors and extensive resources.

The JVM implementation must work within the native platform's multithreading capabilities, which might or might not include native multithreading features. Even among host operating systems that natively support multiple threads, the details of that support are sure to be different among different operating systems and perhaps among different hardware platforms. About all that can be said for certain is that higher priority threads *should* receive preferential treatment by the thread scheduler compared to threads with lower priority. However, if two or more threads are waiting for processor resources, the thread scheduler may also take into account how long the threads have been waiting. The highest priority thread is perhaps likely to be the first to be scheduled, though not necessarily. Over a long enough sampling time, higher priority threads will, on average, be scheduled more often than lower priority threads, but that does not mean that at any given time a lower priority thread might have control of the CPU while a higher priority thread is waiting. In general, changing Java thread priorities is not a reliable way to attempt to force one thread to *always* have preference over another. (See Section 24.4 for a discussion of the real-time specification for Java, which expands the number of priority levels to 28 and requires strict enforcement of thread execution according to priority settings.)

With those caveats, you can get and set a thread's priority with the `getPriority()` and `setPriority()` methods in the `Thread` class. The `Thread` class defines three constants:

- `MIN_PRIORITY`
- `NORM_PRIORITY`
- `MAX_PRIORITY`

The default priority is `Thread.NORM_PRIORITY`, which is 5, although new threads always inherit the priority value of the creating thread. The following code increments a thread's priority to one unit higher than the normal priority:

```
 .   .   .
Thread threadX = new Thread (this);
threadX.setPriority (Thread.NORM_PRIORITY + 1);
threadX.start ();
 .   .   .
```

Attempting to set a thread's priority below MIN_PRIORITY or above MAX_PRIORITY results in an IllegalArgumentException.

For multiple threads in a non-preemptive system, once one of them starts running it will continue until one of the following happens:

- sleeps via an invocation of sleep()
- yields control with yield()
- waits for a lock in a synchronized method (synchronization is discussed in the next section)
- blocks on I/O such as a read() method waiting for data to appear
- terminates with a return from run()

We will discuss synchronization and the wait() method in the following section.

8.5   Using multiple threads

Programs for some tasks become much easier to design with threads, sometimes with lots of threads. We've already mentioned animations, and in Part II we will see that client/server systems lend themselves naturally to multithreaded design – the server can spin off a new thread to service each client. Some mathematical algorithms, such as sorting and prime searching, also work well with multiple threads working on different segments of the problem. On multiprocessor systems, JVMs can take advantage of true parallel processing and provide significant speedups in performance for multithreaded applications.

There are basically four situations in which multiple threads operate:

1. **Non-interacting threads** – the actions of the threads are completely independent and do not affect each other.
2. **Task splitting** – each thread works on a separate part of the same problem, such as sorting different sections of a data set, but do not overlap with each other.
3. **Exclusive thread operations** – the threads work on the same data and must avoid interfering with each other. This requires *synchronization* techniques.
4. **Communicating threads** – the threads must pass data to each other and do it in the correct order.

The latter two cases can entail complex and often subtle interference problems among the threads. We look in more detail at these four cases in the following sections.

8.5.1 Non-interacting threads

The simplest situation with multiple threads is when each thread runs independently without interacting with any other thread. Below is a simple example of such a case. We have one `Thread` subclass called `IntCounter` that prints out the values of an integer counter. We could do a more interesting calculation but for demonstration purposes this will suffice.

```
/** Demo Thread class to show how threads could do
  * calculations in parallel. **/
class IntCounter extends Thread
{
    int fId=0;
    int fCounter = 0;
    int fMaxIter = 0;
    Outputable fOutput;

/** Constructor to initialize parameters. **/
IntCounter (int id, Outputable out) {
    fId = id;
    fMaxIter = 100000
    fOutput = out;
} // ctor

/** Simulate a calculation with an integer sum.**/
public void run () {
    while (fCounter < maxIter) fCounter++;
    fOutput.println ("Thread" + fId + ": sum = " + fCounter);
    }
} // class IntCounter
```

The program `NonInteractApplet`, which implements the `Outputable` interface discussed in Chapter 6, provides a text area and "Go" and "Clear" buttons (see Figure 8.3). Clicking on "Go" invokes the applet's `start()` method, which creates three instances of this `Thread` subclass and starts them:

```
import javax.swing.*;
import java.awt.*;
import java.awt.event.*;

public class NonInteractApplet extends JApplet
                    implements Outputable, ActionListener
{
    . . . Build interface . . .
```

```
// Pushing Go button leads to the invocation of this button
public void start () {
   // Create 3 instances of each of the Thread subclass
   IntCounter ic1 = new IntCounter (1, this);
   IntCounter ic2 = new IntCounter (2, this);
   IntCounter ic3 = new IntCounter (3, this);

   // Start the threads
   ic1.start ();
   ic2.start ();
   ic3.start ();
   } // start
} // class NonInteractApplet
```

In the output shown in Figure 8.3 you see that the threads can finish in a different order for each press of "Go". The order depends on the time allocated to each thread and on what kind of thread scheduling the JVM uses. You can experiment with different thread priorities by adding code to set the priorities of the three thread instances differently. For instance, set ic1 to a high priority and ic3 to a low priority before starting the threads

```
ic1.setPriority (Thread.MAX_PRIORITY);
ic2.setPriority (Thread.NORM_PRIORITY);
ic3.setPriority (Thread.MIN_PRIORITY);
```

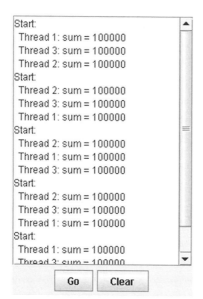

Figure 8.3 Display of the NonInteractApplet program. The Go button has been pushed several times to illustrate the different order in which the threads finish.

8.5.2 Task splitting

The next level in complexity involves multiple threads working on the same problem but on separate, non-interfering parts. For example, given a particular integer value, a program could find the number of primes up to that value by using different threads to work on different sections of the range between 1 and the specified value.

In the example here, we use the task-splitting technique to scan a matrix. The snippet below shows a class that searches a matrix and counts the number of positive non-zero elements:

```
/**
 * Thread class to count the number of non-zero elements
 * in a section of a matrix.
 **/
class MatHunter extends Thread
{
    int [][] fMatrix;
    int fIlo, fIhi, fJlo, fJhi;
    int fOnes=0;
    Outputable fOutput;

/** Constructor gets the matrix and the indices specifying
 * what section to examine.
 **/
MatHunter (
    int [][] imat,
    int i1, int i2, int j1, int j2,
    Outputable out
) {
    fIlo=i1; fIhi=i2;
    fJlo=j1; fJhi=j2;
    fMatrix = imat;
    fOutput = out;
    } // ctor

    /** Examine a section of a 2D matrix and
     * count the number of non-zero elements.
     **/
    public void run () {
        for (int i=fIlo; i <= fIhi; i++) {
            for (int j=fJlo; j <= fJhi; j++) {
                if (fMatrix[i][j] > 0) fOnes++;
            }
            yield ();
```

```
    }
    fOutput.println ("# ones =" + fOnes + "for i =" +
       fIlo + "to" + fIhi + "& j =" + fJlo + "to" + fJhi);
  } // run
} // class MatHunter
```

The program `TaskSplitApplet` creates a matrix with a random distribution of zero and non-zero elements. It then creates four instances of `MatHunter`, one for each quadrant of the matrix. Each instance works on the same problem but in a separate, independent section of the matrix.

```
public class TaskSplitApplet extends JApplet
            implements Outputable, ActionListener
{
   . . . Build the interface . . .

  public void start () {
    int[][] imat = new int[2000][2000];

    for (int i=0; i < 2000; i++) {
      for (int j=0; j < 2000; j++) {
        if (Math.random() > 0.5) imat[i][j] = 1;
      }
    }
    MatHunter mh1 = new MatHunter (imat,0,999,0,999,this);
    MatHunter mh2 =
      new MatHunter (imat,0,999,1000,1999,this);
    MatHunter mh3 =
      new MatHunter (imat,1000,1999,0,999,this);
    MatHunter mh4 =
      new MatHunter (imat,1000,1999,1000,1999,this);

    Println ("Start:");
    mh1.start ();
    mh2.start ();
    mh3.start ();
    mh4.start ();
  } // start
} // class TaskSplitApplet
```

Figure 8.4 shows the results of different threads finishing in a different order each time the "Go" button is pressed.

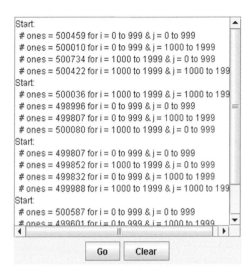

Figure 8.4 Display of TaskSplitApplet program. Pressing the "Go" button can result in a different sequence in the completion times of the thread each time.

8.5.3 Exclusive thread operations

Threading becomes trickier when threads perform operations that can conflict with each other. For example, Figure 8.5 depicts a situation where two thread processes both want to access an object but for different purposes (this is derived from an example in the Sun Java Tutorial). The Filler thread wants to put a number into the bin variable in the Box. It can only do so when the cavity is empty. The Getter, on the other hand, wants to retrieve the number from Cavity and leave the Cavity empty. Ideally, Filler and Getter would alternate their calls to the methods put() and get(). However, if no special steps are taken, it is quite easy for Getter to invoke get() when the Cavity is empty and for Filler to invoke put() when the Cavity is still full.

This type of situation is called a *data race* because each thread is racing to do its task without waiting for the other thread to finish its activity. A *synchronization* scheme prevents this problem. Synchronization forces threads to wait in single file at the method or code block of an object where the conflict can occur.

In this case, this means that the Box object only allows one thread at a time to invoke either its put() or get(). It is as if only one thread object owns the *lock* on the door to a Box object. That thread must give up the lock before any other thread can access *any* synchronized method *on the object*. (Note that the *lock* terminology is by convention. Giving up the *key* might be more illuminating. The term *monitor* is also used.)

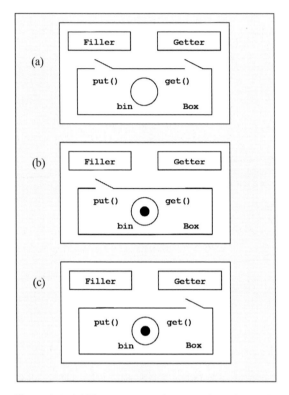

Figure 8.5   (a) The `Filler` and `Getter` threads need to access the bin in the `Box`. The `Getter` needs, however, to wait till the bin is filled. (b) While the `Filler` places a value in the bin via the synchronized `put()` method, the `Getter` cannot invoke the synchronized `get()` method. (c) Similarly, the `Filler` must wait till the `Getter` finishes invoking the `get()` method before it can invoke `put()`.

In the following code for the `Box` class, we see that the `get()` and `put()` methods are prefaced by the modifier `synchronized`. This indicates that only one thread can invoke either of these methods at the same time for the same object. That is, during the time that a thread executes, say, the `get()` method, no other thread can execute either the `get()` or `put()` method for the same `Box` object.

```
public class Box
{
   private int fBin;
   private boolean fFilled = false;
   Outputable fOutput;
   /** Constructor obtains reference to Outputable object.
    **/
   Box (Outputable out) {
      fOutput = out;
```

```
} // ctor

/** If bin is not filled, wait for it to be. **/
public synchronized int get () {
  while (!fFilled){
    try {
      wait ();
    }
    catch (InterruptedException e) {}
  }
  fFilled = false;
  fOutput.println ("Get value:" + fBin);
  notifyAll ();
  return fBin;
} // get

/** If bin is filled, wait for it to be emptied. **/
public synchronized void put (int value){
  while (fFilled) {
    try {
      wait ();
    }
    catch (InterruptedException e) {}
  }
  fBin = value;
  fFilled = true;
  fOutput.println ("Put value: " + fBin);
  notifyAll ();
} // put
} // class Box
```

We want to emphasize that each instance of Box has its own lock. There is no interference problem among different Box objects. If a thread owns the lock on one Box object, this does not prevent another thread from owning the lock on a different Box object.

This code also illustrates the wait() and notifyAll() methods. When a thread invokes put() or get(), it will wait until it is granted the lock for that object because of the presence of the synchronized keyword. Once the thread is granted the lock, it continues on through the method. Inside the method, a check is made on the fFilled flag. When attempting a put(), if fFilled is already true (i.e. if the bin is already full), then an explicit wait() is invoked. Similarly, during a get(), if fFilled is false (i.e. if the bin is empty), then wait() is invoked. Invoking wait() means that the thread gives up the lock and remains at that point in the method until notified to continue.

Let's suppose that the `Filler` thread finds that fFilled = true during the put() method; that thread will go into a wait state. Since fFilled is true, the `Getter` thread passes the fFilled test in the get() method, obtains the fBin value, sets the fFilled flag to false, and invokes notifyAll() before it returns. The notifyAll() method causes all threads in a wait state to attempt to acquire the lock. When the lock is released by the `Getter` in the synchronized get() method, the `Filler` thread can acquire the lock and continue on through the put() method and fill the bin again.

The following code shows the `Filler` class. In the run() method, a loop puts a value into the box and then pauses for a random period of time before doing it again. For each pass of the loop, the put() invocation results in the printing of a message via the `Outputable` reference.

```
public class Filler extends Thread
{
    private Box fBox;

    public Filler (Box b) {
        fBox = b;
    }

    public void run () {
        for (int i=0; i < 10; i++) {
            fBox.put (i);
            try {
                sleep ((int)(Math.random () * 100));
            }
            catch (InterruptedException e) {}
        }
    } // run
} // class Filler
```

The following code shows the `Getter` class. The loop in its run() method will continue until it gets ten values from the box. Note, however, that the process will experience occasional wait states in the get() method in Box to give time for the `Filler` to do its job.

```
public class Getter extends Thread
{
    private Box fBox;
    private int fNumber;

    public Getter (Box b) {
        fBox = b;
```

```
Put value: 0
Get value: 0
Put value: 1
Get value: 1
Put value: 2
Get value: 2
Put value: 3
Get value: 3
Put value: 4
Get value: 4
Put value: 5
Get value: 5
Put value: 6
Get value: 6
Put value: 7
Get value: 7
Put value: 8
Get value: 8
Put value: 9
Get value: 9

        Go       Clear
```

Figure 8.6 The output of the `Filler` and `Getter` threads for the `ExclusiveApplet` as they fill and retrieve a bin value in a `Box` object.

```
    }
    public void run () {
        int value = 0;
        for (int i=0; i < 10; i++) {
            fNumber = fBox.get ();
        }
    } // run
} // class Getter
```

The snippet from `ExclusiveApplet` shown below creates a `Box`, a `Filler`, and a `Getter` object and then starts the two threads. Figure 8.6 shows a typical output. We see that the synchronization prevents a data race situation and the two threads each complete their respective tasks.

```
public class ExclusiveApplet extends JApplet
            implements Outputable, ActionListener
{
    . . . Build the interface . . .
```

```
/** Create Filler and Getter thread instances and start
 * them filling and getting from a Box instance. **/
public void start () {

   Box b = new Box (this);
   Filler f1 = new Filler (b);
   Getter b1 = new Getter (b);
   f1.start ();
   b1.start ();
} // start
   . . .
} // class ExclusiveApplet
```

8.5.4 Communications among threads

In the previous section, we discussed the case where multiple threads try to access an object and can step on each other if not properly synchronized. Here we look at the even trickier situation where a thread needs to access data in another thread and must also avoid a data race situation.

The standard example for communicating threads is the *producer/consumer* paradigm. The producer object invokes its own synchronized method to create the data of interest. The consumer cannot invoke the producer's get() method, which is also synchronized, until the producer has finished with its creation method. The producer, in effect, locks its own door to the consumer until it finishes making the data. (Imagine a physical store that locks its doors and does not allow shoppers in while restocking the shelves.) Similarly, while the consumer gets the data from the producer, it obtains the *lock* and prevents the producer from generating more data until the consumer is finished.

Below we illustrate this paradigm with a program in which the Sensor class represents the producer thread and DataGetter represents the consumer thread. An instance of Sensor obtains its data (here just clock readings) in a loop in run() via calls to the synchronized sense() method. The data goes into a buffer array. A thread can invoke get() in Sensor to obtain the oldest data in the buffer. The indices are set up to emulate a FIFO (First-In-First-Out) buffer. When the buffer is full, the Sensor thread waits for data to be read out (that is, it gives up the lock by calling the wait() method).

To obtain the data, a DataGetter instance invokes the synchronized get() method in the Sensor instance. If no new data is available, it will give up the lock and wait for new data to appear (that is, when notifyAll() is invoked in the sense() method).

This snippet from DataSyncApplet creates the sensor and starts it. Then a DataGetter is created and started.

```
public class DataSyncApplet extends JApplet
              implements Outputable, ActionListener
{
    . . . Build the interface . . .

  /** Create Sensor and DataGetter thread instances and
    * start them filling and getting from a Box instance.
    **/
  public void start() {
    // Create the Sensor and start it
    Sensor s = new Sensor (this);
    s.start ();

    // Create DataGetter and tell it to obtain
    // 100 sensor readings.
    DataGetter dg = new DataGetter (s, 100, this);
    dg.start ();
  } // start
    . . .
} // class DataSyncApplet
```

The Sensor (see code below) produces one data value (just a string containing the number of milliseconds since the program began) and stores it in an element of a buffer array. The fBufIndex keeps track of where the next value should go. When it reaches the end of the array, it will circle back to the start. The fGetIndex marks the value in the buffer that will be sent next to the DataGetter. The fGetIndex should never fall farther behind fBufIndex than the MAXGAP value (set here to 8). If the lag reaches the value of fMaxGap then the sensor goes into a loop with an invocation of wait() for each pass. When the DataGetter invokes the get() method, the notifyAll() will wake the Sensor thread from its wait state and it will check the lag again. If it is no longer at the maximum, the process leaves the wait loop and produces more data. Otherwise, it loops back around and invokes wait() again.

```
import java.util.*;
/**
  * This class represents a sensor producing data
  * that the DataGetter objects want to read.
  */
public class Sensor extends Thread
{
  // Size of the data buffer.
  private static final int BUFFER_SIZE = 10;
```

```java
// Don't let data production get more than
// 8 values ahead of the DataGetter
private static final int MAXGAP = 8;
private String [] fBuffer;
private int fBufIndex = 0; // sensor data buffer index
private int fGetIndex = 0; // data reading index
private final long fStart = System.currentTimeMillis ();

boolean fFlag = true;
Outputable fOutput;

/** Constructor creates buffer. Gets Outputable ref. **/
Sensor (Outputable out) {
  fOutput = out;
  fBuffer = new String [BUFFER_SIZE];
}
/** Turn off sensor readings. **/
public void stopData () {
  fFlag = false;
}
/** Take sensor readings in a loop until flag set false.
 **/
public void run () {
  // Measure the parameter of interest
  while (fFlag) sense ();
}

/** Use clock readings to simulate data. **/
private final String simulateData () {
    return "" + (int) (System.currentTimeMillis () —
    start);
}

/** Use indices fBufIndex, fGetIndex, and the lag()
  * method to implement a first-in-first-out (FIFO)
  * buffer. **/
synchronized void sense () {
  // Don't add more to the data buffer until the getIndex
  // has reached within the allow range of bufIndex.
  while (lag () > MAXGAP) {
    try {wait ();}
    catch (Exception e) {}
  }
  fBuffer[fBufIndex] = simulateData ();
  fOutput.println("Sensor["+ (fBufIndex) + "] = "
                  + fBuffer[fBufIndex]);
```

```
      // Increment index to next slot for new data
      fBufIndex++;
      // Circle back to bottom of array if reaches top
      if (fBufIndex == BUFFER_SIZE) fBufIndex = 0;
      notifyAll ();
   } // sense

   /** Calculate distance the DataGetter is running behind
     * the production of data. **/
   int lag () {
      int dif = fBufIndex - fGetIndex;
      if (dif < 0) dif += BUFFER_SIZE;
      return dif;
   }
   /** Get a data reading from the buffer. **/
   synchronized String get () {
      // When indices are equal, wait for new data.
      while (fBufIndex == fGetIndex) {
         try{ wait(); }
         catch (Exception e) {}
      }
      notifyAll ();

      // Get data at current index
      String data = fBuffer[fGetIndex];

      // Increment pointer of next datum to get.
      fGetIndex++;
      // Circle back to bottom of array if reaches top
      if (fGetIndex == BUFFER_SIZE) fGetIndex = 0;
      return data;
   } // get
} // class Sensor
```

The DateGetter grabs a data value from the sensor after random delay until it gets its maximum number of data values. Figure 8.7 shows typical output from DataSync.

```
import java.util.*;

/** This class obtains sensor data via the get () method.
  * To simulate random accesses to the sensor, it will
  * sleep for brief periods of different lengths after
  * every access. After the data is obtained, this thread
```

```
   * will stop the sensor thread. **/
public class DataGetter extends Thread
{
   Sensor fSensor;
   Outputable fOutput;
   int fMaxData = 1000;
   int fDataCount = 0;

   DataGetter (Sensor sensor, int maxNum, Outputable out) {
      fSensor = sensor;
      fMaxData = maxNum;
      fOutput = out;
   }
   /** Loop over sensor readings until data buff filled. **/
   public void run () {
      Random r = new Random ();
      while (true) {
         String data = fSensor.get();
         fOutput.println(fDataCount++ + ". Got: " + data);

         // Stop both threads if data taking finished.
         if (fDataCount >= fMaxData) {
            fSensor.stopData ();
            break;
         }
         // Pause briefly before access the
         // data again.
         try {
            sleep (r.nextInt () % 300);
         }
         catch (Exception e) {}
      }
   } // run
} // class DataGetter
```

8.6 Animations

A popular task for a thread in Java is to control an animation. A thread process can direct the drawing of each frame while other aspects of the interface, such as responding to user input, can continue in parallel.

The `Drop2DApplet` program below illustrates a simple simulation of a bouncing ball using Java 2D drawing tools. The applet creates a thread to direct the drawing of the frames of the animation as the ball falls and bounces on the

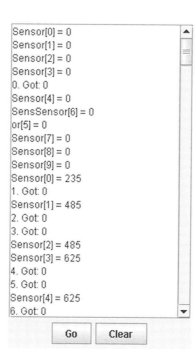

Figure 8.7 Output of the `DataSyncApplet` program with `Sensor` and `DataGetter` classes.

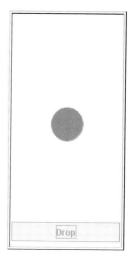

floor and gradually comes to a rest (see Figure 8.8). The interface consists of a subclass of `JPanel` called `Drop2DPanel` and a button to initiate a new drop of the ball. `Drop2DPanel` displays the ball and calculates its position.

The applet implements `Runnable` and in the `start()` method it creates a thread to which it passes a reference to itself. The applet's `run()` method first does some initialization and then enters a loop that draws each frame of the animation. The loop begins with a 25 millisecond pause using the static `sleep()` method in the `Thread` class. Then the `Drop2DPanel` is told to paint the next frame. If the drop is done, the process jumps from the loop and exits the `run()` method, thus killing this thread.

```java
import javax.swing.*;
import java.awt.*;
import java.awt.event.*;


/** This applet implements Runnable and uses a thread
  * to create a simple dropped ball demonstration.**/
public class Drop2DApplet extends JApplet
                    implements Runnable, ActionListener
{
   // Will use thread reference as a flag
```

Figure 8.8 The `Drop2DApplet` program demonstrates how to create a simple animation by simulating a dropped ball that bounces when it hits the floor and gradually comes to rest.

```java
Thread fThread = null;
Drop2DPanel fDropPanel;
JButton fDropButton;

/** Build the interface. **/
public void init () {
  Container content_pane = getContentPane ();
  Content_pane.setLayout (new BorderLayout ());

  // Create an instance of DropPanel
  fDropPanel = new Drop2DPanel ();
  // Add the Drop2DPanel to the content pane.
  Content_pane.add (BorderLayout.CENTER, fDropPanel);
  // Create a button and add it
  fDropButton = new JButton ("Drop");
  fDropButton.addActionListener (this);
  Content_pane.add (BorderLayout.SOUTH, fDropButton);
} // init

/** Start when browser is loaded or button pushed. **/
public void start () {
  // If the thread reference not null then a
  // thread is already running. Otherwise, create
  // a thread and start it.
  if (fTthread == null) {
    fThread = new Thread (this);
    fThread.start();
  }
} // start

/** Applet's stop method used to stop thread. **/
public void stop () {
  // Setting thread to null will cause loop in
  // run() to finish and kill the thread.
  fThread = null;
} // stop

/** Button command, **/
public void actionPerformed (ActionEvent ae){
  if (fDropPanel.isDone ()) start ();
}

/** The thread loops to draw each frame of drop. **/
public void run () {
  // Disable button during drop
  fDropButton.setEnabled (false);
```

```
      // Initialize the ball for the drop.
      fDropPanel.reset ();

      // Loop through animation frames
      while (fThread!= null) {
          // Sleep 25msecs between frames
        try{Thread.sleep (25);
        }
        catch (InterruptedException e) {}
        // Repaint drop panel for each new frame
        fDropPanel.repaint ();
        if (fDropPanel.isDone ()) fThread = null;
      }
      // Enable button for another drop
      fDropButton.setEnabled (true);
   } // actionPerformed
} // class DropApplet
```

The `Drop2DPanel` class is shown below. The panel calculates the position of the ball for each increment of time between the frames and redraws the ball. It reverses the ball when it hits the floor and also subtracts some speed to simulate friction. Eventually, the ball comes to a rest and sets a flag that the drop simulation is done.

```
import javax.swing.*;
import java.awt.*;
import java.text.*;
import java.util.*;

/** This JPanel subclass displays a falling ball. **/
public class Drop2DPanel extends JPanel
{
  // Parameters for the drop
  double fY = 0.0, fVy = 0.0;
  // Conversion factor from cm to drawing units
  double fYConvert = 0.0;
  double fXPixel= 0.0, fYPixel = 0.0,
  double fRadius = 0.0, fDiam = 0.0;
  // starting point for ball in cm
  double fY0 = 1000.0;
  // Frame dimensions.
  double fFrameHt, fFrameWd;
  // Flag for drop status
  boolean fDropDone = false;
```

```
Ellipse2D fBall;

/** Reset parameters for a new drop. **/
void reset () {
  fFrameHt = getHeight ();
  fFrameWd = getWidth ();

  fXPixel = getWidth ()/2;
  fY = fY0; fVy = 0.0;

  // Conversion factor from cm to pixels
  // Start the ball about 20% from the top.
  fYConvert = fFrameHt / (1.2 * fY0);

  // Choose a size for the ball relative
  // to height of drawing area.
  fRadius = (int) ((0.1 * fY0) * fYConvert);
  fDiam = 2 * fRadius;

  // Make the ball
  fBall = new Ellipse2D.Double(fXPixel-fRadius,
                               fYPixel-fRadius,
                               fDiam, fDiam);
  setBackground (Color.WHITE);
  fDropDone = false;
} // reset

/** Draw the ball at its current position. **/
public void paintComponent (Graphics g) {
  super.paintComponent (g);
  Graphics2D g2 = (Graphics2D)g;
  // Antialiasing for smooth surfaces.
  g2.setRenderingHint(RenderingHints.KEY_ANTIALIASING,
        RenderingHints.VALUE_ANTIALIAS_ON);

  // Determine position after this time increement
  calcPosition ();
  // Move the ball.
  fBall.setFrame(fXPixel-fRadius, fYPixel-fRadius,
                 fDiam,fDiam);
  // Want a solid red ball.
  g.setColor (Color.RED);
  g2.fill(fBall);
  // Now draw the ball
  g2.draw (fBall);
```

```
}// paintComponent

/** Calculate the ball position in the next frame. **/
void calcPosition () {
  // Increment by 25 millseconds per frame
  double dt = 0.025;

  // Calculate position and velocity at each step
  fY = fY + fVy * dt - 490.* dt * dt;
  fVy = fVy - 980.0 * dt;

  // Convert to the pixel coordinates
  fYPixel = fFrameHt - (int)(fY * fYConvert);

  // Reverse direction when ball hits bottom.
  if ((fYPixel + fRadius) >= (fFrameHt-1)) {
      fVy = Math.abs (fVy);
      // Subtract friction loss
      fVy - = 0.1 * fVy;
      // Stop when speed at bottom drops
      // below an arbitrary limit
      if (fVy < 15.0) {
          fDropDone = true;
      }
  }
} // calcPosition

/** Provide a flag on drop status. **/
public boolean isDone () {
  return fDropDone;
}
} // class Drop2DPanel
```

8.7 Timers

A timer provides for periodic updates and scheduling of tasks. For example, a timer can:

- signal the redrawing of frames for an animation
- issue periodic reminders as with a calendar application
- trigger a single task, e.g. an alarm, to occur at a particular time in the future

As we saw in the previous section, with the Thread class you could create your own simple timer using the Thread.sleep (long millis) method to delay action for a given amount of time. This approach, however, has some drawbacks.

For periodic events, if the duration of processing in between the sleep periods varies significantly, then the overall timing will vary with respect to a clock. Also, if you need several timer events, the program will require several threads and this will use up system resources.

Java provides two timer classes [5–7]:

- `javax.swing.Timer` came with the Swing packages and is useful for such tasks as prompting the updating of a progress bar
- `java.util.Timer` and its helper class `java.util.TimerTask` provide for general purpose timers with more features than the Swing timer

These timers can provide multiple timed events from a single thread and thus conserve resources. They also have useful methods such as `scheduleAt-FixedRate(TimerTask task, long delay, long period)` in `java.util.Timer`. This method will set events to occur periodically at a fixed rate and ties them to the system clock. This is obviously useful for many applications such as a countdown timer and an alarm clock where you don't want the timing to drift relative to absolute time.

8.7.1 `java.util.Timer` and `TimerTask`

The `Timer` and `TimerTask` combo in `java.util` offers the most general purpose timing capabilities and includes a number of options. A `Timer` object holds a single thread and can control many `TimerTask` objects. The `TimerTask` abstract class implements the `Runnable` interface but it does not provide a concrete `run()` method. Instead you create a `TimerTask` subclass to provide the concrete `run()` method with the code to carry out the task of interest.

In the example below, we create a digital clock using a timer to redraw a time display every second. The clock display uses `DateFormatPanel`, which we describe in Chapter 10 when discussing the date classes. Whenever this panel is drawn it displays the current time. The applet adds an instance of this panel to its content pane and in the `start()` method creates an instance of `java.util.Timer`.

A subclass of `TimerTask` called `UpdateTask` overrides the `run()` method and simply tells the panel to redraw itself. `UpdateTask` is defined as an inner class here and has access to the clock panel reference. The timer schedules calls to the `UpdateTask` every 1000 milliseconds. Figure 8.9 shows the clock display.

7:16:10 PM

Figure 8.9 The `ClockTimer1` and `ClockTimer2` programs, which both provide a current time display like that shown here, illustrate the use of `java.util.Timer` and `javax.swing.Timer`, respectively.

```
import javax.swing.*;
import java.awt.*;
import java.util.*;

/** This applet implements Runnable and uses a thread
```

```
    * to create a digital clock. **/
  public class ClockTimer1 extends Japplet
  {
    java.util.Timer fTimer;
    // Need panel reference in run().
    DateFormatPanel fClockPanel;

    public void init () {
      Container content_pane = getContentPane ();

      // Create an instance of DrawingPanel
      fClockPanel = new DateFormatPanel ();

      // Add the DrawingPanel to the contentPane.
      content_pane.add (fClockPanel);
    }

    public void start () {
      // Create a timer.
      fTimer = new java.util.Timer ();

      // Start the timer immediately and then repeat calls
      // to run in UpdateTask object every second.
      fTimer.schedule (new UpdateTask (), 0, 1000);
    }

    /** Stop clock when web page unloaded. **/
    public void stop () {
      // Stop the clock updates.
      fTimer.cancel ();
    }

    /** Use the inner class technique to define the
      * TimerTask subclass to update the clock.**/
    class UpdateTask extends java.util.TimerTask {
      public void run () {
        fClockPanel.repaint ();
      }
    }
  } // class ClockTimer1
```

(Note that since we import both `javax.swing.*` and `java.util.*` it is necessary to use the fully qualified type `java.util.Timer` when declaring the `fTimer` variable. Without the full qualification, the compiler would not know whether we wanted `javax.swing.Timer` or `java.util.Timer`.)

8.7.2 `javax.swing.Timer`

Although it has fewer options, the `javax.swing.Timer` can do some of the same basic timing tasks as `java.util.Timer`. Below we show another version of the digital clock except that it uses `javax.swing.Timer`. This timer contacts an `ActionListener` after every time period rather than a `TimerTask` object. Here the applet implements the `ActionListener` interface. The constructor for the timer takes as arguments the update period value and the reference to the applet. The timer is then started and after every second the `actionPerformed()` method will be invoked and the clock panel repainted. The applet's `stop()` method stops the timer.

```java
import javax.swing.*;
import java.awt.*;
import java.awt.event.*;
import java.util.*;

/** This applet implements Runnable and uses a thread
  * to create a digital clock. **/
public class ClockTimer2 extends JApplet
        implements ActionListener
{
    javax.swing.Timer fTimer;

    // Need panel reference in run().
    DateFormatPanel fClockPanel;

    public void init () {
        Container content_pane = getContentPane ();

        // Create an instance of DrawingPanel
        fClockPanel = new DateFormatPanel ();

        // Add the DrawingPanel to the contentPane.
        content_pane.add (fClockPanel);
    }

    public void start () {
        // Send events very 1000ms.
        fTimer = new javax.swing.Timer (1000, this);

        // Then start the timer.
        fTimer.start ();
    }
```

```
    // Timer creates an action event.
    public void actionPerformed (ActionEvent e) {
       Object source = e.getSource ();
       if (source == fTimer)
          fClockPanel.repaint ();
    }

    // Stop clock when web page unloaded
    public void stop () {
       // Stop the clock updates.
       fTimer.stop ();
    }
} // class ClockTimer2
```

8.8 Concurrency utilities in J2SE 5.0

Java Release 5.0 adds numerous enhancements to the threading control and concurrency features of Java. Some of the enhancements are advanced features beyond the scope of this book, and others require an understanding of the new generics feature of 5.0. So we defer discussion of these until after we have explained generics in Chapter 10.

8.9 Web Course materials

The Web Course Chapter 8: *Supplements* section provides additional information and examples dealing with threading. This includes additional discussion of the new `java.util.concurrent` tools available with Java 5.0.

In the Chapter 8: *Tech* section we expand the number of histogram classes and subclasses as we add new capabilities. For example, we create an adaptive histogram class that can expand its range limits as new data arrives. We use timers to simulate the reading of data to plot in a histogram. We also discuss sorting tools in Java and use them to sort the bins in a histogram according to the number of entries in the bins. We use a thread to animate the sorting of a histogram.

This increase in histogram classes illustrates a common challenge in object-oriented programming: when to modify existing classes, when to create subclasses, and when to create whole new classes. Subclasses would seem the logical answer for an OOP environment but many small revisions for every new option that comes along can quickly lead to an unmanageable plethora of subclasses. Eventually, your entire class design may need to be re-worked (also called *refactoring*, with the implication that common parts are *factored out* into a common superclass). We discuss class design and refactoring further in the *Tech* section. The *Physics* section looks at issues involved in animating simulations.

References

[1] Scott Oaks, Henry Wong, *Java Threads,* 2nd edn, O'Reilly, 1999.

[2] *Lesson: Threads: Doing Two or More Tasks At Once – The Java Tutorial*, Sun Microsystems, `http://java.sun.com/docs/books/tutorial/essential/threads/`.

[3] *How to Use Threads in Creating a GUI with JFC/Swing – The Java Tutorial*, Sun Microsystems, `http://java.sun.com/docs/books/tutorial/uiswing/misc/threads.html`.

[4] *AWT Threading Issues – Java 2 Platform, Standard Edition, API Specification*, `http://java.sun.com/j2se/1.5/docs/api/`.

[5] *Using the Timer and TimerTask Classes – The Java Tutorial*, Microsystems, `http://java.sun.com/docs/books/tutorial/uiswing/misc/timer.html`.

[6] *Using Timers to Run Recurring or Future Tasks on a Background Thread*, JDC Tech Tips, May 30, 2000, `http://java.sun.com/developer/TechTips/2000/tt0530.html#tip2`.

[7] John Zukowski, *Using Swing Timers*, JDC Tech Tips, May 21, 2002, `http://java.sun.com/developer/JDCTechTips/2002/tt0521.html`.

Chapter 9
Java input/output

9.1 Introduction

Java provides a consistent framework for all input/output. That framework centers on the concept of a *stream*. A stream in Java is a sequential flow of bytes in one direction. There is an output stream that carries text to the console and a corresponding input stream that brings text from the keyboard. Another type of output stream carries data to a file, and a corresponding input stream brings data into a program from a file. There is another output stream that sends data through a network port to another computer on the network while an input stream brings in data through a network port from such a source.

The bulk of Java I/O classes belong to the `java.io` package. (See diagram in Figure 9.1.) The class hierarchy builds on the base classes `InputStream`, `OutputStream`, `Reader`, and `Writer` to provide a wide range of input and output tasks. In addition, this package holds stream classes that *wrap* a stream to add more capabilities to it. Some I/O classes provide a destination or source, such as a file or an array. Others process the stream in some way such as buffering or filtering the data.

Packages involving I/O include:

- `java.io` – the primary Java I/O classes
- `java.nio`, `java.nio.*` – a set of five packages new with Java 1.4 based on the concept of *channels* that represent an open connection to a hardware device, file, or other entity. Channels don't supplant streams but rather work with them to add additional capabilities and enhanced scaling when working with large numbers of connections
- `java.net` – I/O over the network
- `java.util.zip` – methods for reading from and writing to ZIP and GZIP files
- `java.util.jar` – methods for reading from and writing to JAR (Java Archive) files
- `javax.imageio`, `javax.imageio.*` – this set of six packages deals with image I/O, including the encoding/decoding of images in particular formats (discussed in Chapter 11)

Java I/O is an enormous topic that involves many packages, classes, and methods [1,2]. Here we give an overview of some of the basic aspects of Java I/O with

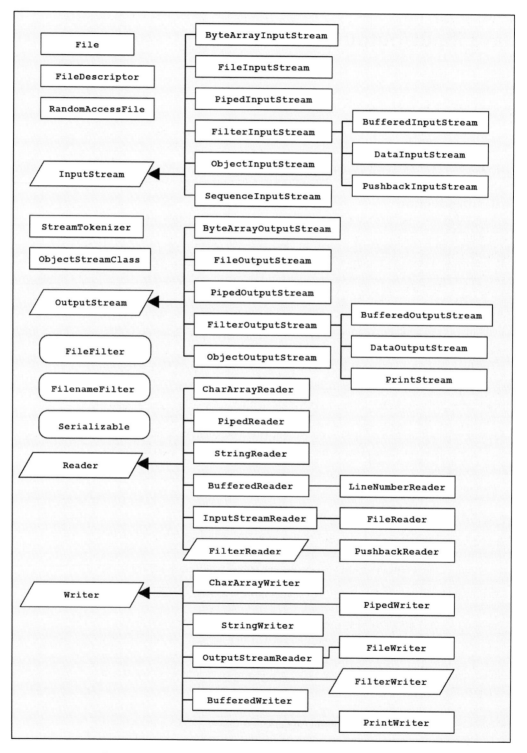

Figure 9.1 The diagram shows most of the members of the `java.io` package. Rectangles are classes, parallelograms are abstract classes, and ovals are interfaces.

a focus on text and numerical data going to and from the console and disk files. Part II deals with I/O over networks, and Chapter 23 deals with I/O via serial and parallel ports. The Web Course Chapter 9: *Supplements* section looks at the `java.nio` capabilities.

9.1.1 Java I/O challenges

Unfortunately, when one first encounters Java I/O it seems to violate a primary goal of Java – *provide simpler and more transparent code than the C and C++ languages*! The elegant abstraction that is so powerful when dealing with complex I/O tasks can seem clumsy and overly complicated for basic tasks such as writing and reading text with the console.

Until now we avoided the complexities by sending text output to the console with the `System.out.println()` method. The `System.out` object is defined in the `java.lang.System` class. It is an instance of `PrintStream`, which provides higher level methods such as `println()` than the basic stream classes. The JVM automatically opens this standard output stream so no user setup is required. The `System.in` object, also defined in `java.lang.System` is an instance of `InputStream`, which only provides low-level methods and requires extra steps to convert input to a string. We discuss console I/O in Section 9.4 for text and numbers. These and other features of Java I/O can seem rather confusing at first. We will provide lots of examples here and in the Web Course so that one can find guides for many different kinds of I/O tasks.

As we discussed in Chapter 5, Java separated formatting of text and its output until J2SE 5.0, which introduced the `printf()` method into the `PrintStream` class. This method provides C/C++ style formatting with similar format specifiers. In addition, there is a new `Scanner` class with methods to read formatted text from the console and from files.

9.2 Streams

The primary conceptual component of the Java I/O framework is the stream:

stream = a one way sequential flow of bytes

Input operations begin by opening a stream from the source and then invoking a `read()` method, or an overloaded version of it, to obtain the data. Similarly, output operations begin by opening a stream to the destination and then using a `write()` method to send the data.

The base stream classes are the abstract classes:

- `InputStream` – byte input stream base class
- `OutputStream` – byte output stream base class
- `Reader` – 16-bit character input stream base class
- `Writer` – 16-bit character output stream base class

The Reader/Writer classes were added in Java 1.1 to deal with 16-bit character encoding. Java uses the 16-bit *Unicode* format to represent all text characters. (We discuss Unicode and other encoding schemes in Section 9.7.)

Classes that inherit from the above abstract classes provide specialized streams such as for keyboard input and file I/O. The Java I/O classes either extend or wrap lower level classes to provide additional capabilities. See the java.io package specifications in the Java 2 API Specifications for a list of its many stream and ancillary classes.

A frequent criticism of Java I/O is that it involves too many classes. Often an entire class, such as PushbackInputStream, which puts data back into the stream, is required to do a task that might well have been done by a method within another class.

9.3   Stream wrappers

The Java I/O framework builds specialized I/O streams by combining classes together. The goal is to provide flexibility and modularity by *wrapping* an instance of a lower level stream class with an instance of a class that provides additional features. Here wrapping refers to passing a stream object as a parameter in a constructor of another stream class. A wrapper provides its own set of methods for input or output operations and uses the wrapped stream internally to carry them out.

For example, the following code segment shows an instance of the Input-StreamReader class wrapping the System.in stream:

```
InputStreamReader in = new InputStreamReader (System.in);
```

This wraps an 8-bit character stream, System.in, with a 16-bit character Reader class. This provides for more efficient handling of non-ASCII characters. (See Section 9.7.)

Similarly, we can wrap an instance of InputStreamReader with a *buffered* class wrapper:

```
BufferedReader buf_in = new BufferedReader (in);
```

Buffered classes improve the performance of I/O by providing intermediate data storage buffers. The data fills the buffer to a certain level before it is sent in bulk to the next stage, thus performing fewer time-consuming operations. (This requires *flushing* at the end of the transmission to ensure that no data remains in the buffer.)

In addition to buffering the data input, a BufferedReader provides several useful methods. For example, the read() methods in InputStreamReader return only one character at a time so we would need a loop to read a whole line of characters. Instead, we can wrap this class with a BufferedReader and use the BufferedReader.readLine() method:

```
String str_line = buf_in.readLine ();
```

This method reads the entire input line (up to an end-of-line character) and returns it as a string.

9.4 Console I/O

The term *console* typically refers to a command line interface on the host platform such as a *cmd* or *command* window when using Windows or Linux shell. For a browser it refers to a window where standard output (`System.out`) and error messages (`System.err`) from an applet are displayed (no console input allowed). In this day of graphical interface domination, many consider console I/O anachronistic. However, Java is intended for many different platforms and for some systems a console is the most convenient, or perhaps the only way, for a user to interact with a program. (See, for example, some of the small embedded processor systems discussed in Chapter 24.) Console I/O also provides a nice introduction to a number of techniques that apply to all Java I/O.

When the JVM starts, it automatically creates three stream objects accessible with these static references:

- `System.in` – `InputStream` object for keyboard input
- `System.out` – `PrintStream` object for console output
- `System.err` – `PrintStream` object for console output of error messages

These are all 8-bit streams. The `PrintStream` constructors have actually been deprecated but not the class's methods since they are so convenient and common. As you have seen with our example codes, you can still use `System.out.println()` without a deprecation warning message during compilation.

Java 1.0 provided only for 8-bit encoded character I/O. Java 1.1 upgraded to 16-bit character I/O, but to remain compatible with previous code, the 8-bit streams remain and are still available for the console and keyboard. The wrapper classes `InputStreamReader` and `OutputStreamWriter` were provided to convert the 8-bit streams to 16-bit and vice versa.

For binary input, the data is read as bytes and then combined into appropriate values according to the data types involved. We first discuss the `System` streams and then the `Reader/Writer` streams.

9.4.1 Text output

The following simple program, `PrintWriterApp`, demonstrates both text output and the wrapping of streams to gain greater capabilities. We first wrap the `System.out` object, which is an 8-bit stream, with an `OutputStreamWriter` to obtain 16-bit streaming. However, `OutputStreamWriter` provides only a few methods. So we wrap our `OutputStreamWriter` object in turn with `PrintWriter`, which offers a full set of `print()` and `println()` methods for printing strings and primitive types.

Destinations include OutputStream, an instance of File (see Section 9.5), and any class that implements the new Appendable interface.

The program FormatWriteApp shows how we can use Formatter to send formatted numerical values to the console rather than using the printf() method:

```java
import java.io.*;
import java.util.Formatter;

/**
  * Demonstrate the java.util.Formatter capabilities for
  * formatting primitive types.
  **/
public class FormatWriteApp
{
   public static void main (String arg[]) {

      // Send formatted output to the System.out stream.
      Formatter formatter =
         new Formatter ((OutputStream)System.out);

      formatter.format ("Text output with Formatter.%n");
      formatter.format ("Primitives converted to strings:%n");

      boolean a_boolean = false;
      byte    a_byte   = 114;
      short   a_short  = 1211;
      int     an_int   = 1234567;
      long    a_long   = 987654321;
      float   a_float  = 983.6f;
      double  a_double = -4.297e-15;

      formatter.format ("boolean = %9b %n",   a_boolean);
      formatter.format ("byte    = %9d %n",   a_byte);
      formatter.format ("short   = %9d %n",   a_short);
      formatter.format ("int     = %9d %n",   an_int);
      formatter.format ("long    = %9d %n",   a_long);
      formatter.format ("float   = %9.3f %n", a_float);
      formatter.format ("double  = %9.2e %n", a_double);

      // Need to flush the data out of the buffer.
      formatter.flush ();
      formatter.close ();

   } // main
} // class FormatWriteApp
```

The output to the console then looks like:

```
Text output with Formatter.
Primitives converted to strings:
boolean =      false
byte    =        114
short   =       1211
int     =    1234567
long    = 987654321
float   =    983.600
double  = -4.30e-15
```

When we call the `Formatter` constructor, `System.out`, which references an instance of `PrintStream`, must be cast to `OutputStream` because otherwise there is an ambiguity over which constructor to use (since `PrintStream` implements `Appendable`).

Note that you can directly obtain the formatted string created by the `Formatter` by invoking the `toString()` method. Also, if you simply want a formatted string, such as for a graphical text component, and don't want to send it to an output destination, you can create a `Formatter` with the no-argument constructor. The `Formatter` uses a `StringBuilder` (see Chapter 10) internally to create the string, and you can access it via the `toString()` method.

9.4.3 Text input

While the print methods of `System.out` are simple and convenient to use, a common complaint of new Java users has been the lack of a comparably simple set of text input methods. Fortunately, J2SE 5.0 comes with the new `Scanner` class, which makes input much simpler. We discuss it in the next section. First we look at the text input techniques available with Java 1.4 and earlier.

`System.in` references an object with only the very limited capabilities of an `InputStream`. You can, for example, read in a single byte, which returns as an `int` value, or a byte array. You must cast each byte obtained from the keyboard to a `char` type. For example:

```
try {
   int tmp = System.in.read ();
   char c = (char) tmp;
}
catch (IOException e) {}
```

Here a byte value is read and returned as the first byte in an `int` value. This value is then converted to a `char` value. This operation assumes that the byte corresponds to an 8-bit encoded character, such as an ASCII character. As for

```
try {
   String s = buf_reader.readLine (); // read the number as a
                             // string
   // Trim the whitespace before parsing.
   tmp = Integer.parseInt (s.trim ()); // Convert string to int
   System.out.println (" echo = " + tmp);
}
catch (IOException ioe) {
   System.out.println ("IO exception = " + ioe);
}
```

Similar code is needed for parsing other numerical input according to the type of values. The following text discusses an easier technique for numerical input using the `Scanner` class of J2SE 5.0.

9.4.4 Text input with `Scanner`

The `java.util.Scanner` class is a very useful new tool that can parse text for primitive types and substrings using regular expressions [4,5]. It can get the text from various sources such as a `String` object, an `InputStream`, a `File` (Section 9.5), or a class that implements the `Readable` interface.

The `Scanner` splits its input into substrings, or tokens, separated by delimiters, which by default are any white space. The tokens can then be obtained as strings or as primitive types if that is what they represent. For example, the following code snippet shows how to read an integer from the keyboard:

```
Scanner scanner = new Scanner (System.in);
int i = scanner.nextInt ();
```

For each of the primitive types there is a corresponding `nextXxx()` method that returns a value of that type. If the string cannot be interpreted as that type, then an `InputMismatchException` is thrown. There is also a set of `hasNextXxx()` methods, such as `hasNextInt()`, that return true or false according to whether the next token matches the specified type.

```
import java.io.*;
import java.util.*;

/** Demonstrate the Scanner class for input of numbers.**/
public class ScanConsoleApp
{
   public static void main (String arg[]) {

      // Create a scanner to read from keyboard
      Scanner scanner = new Scanner (System.in);
```

```
    try {
        System.out.printf("Input int (e.g. %4d): ",3501);
        int int_val = scanner.nextInt ();
        System.out.println ("You entered" + int_val +"\n");

        System.out.printf ("Input float (e.g. %5.2f): ", 2.43);
        float float_val = scanner.nextFloat ();
        System.out.println ("You entered" + float_val +"\n");

        System.out.printf ("Input double (e.g. %6.3e): ",
                           4.943e15);
        double double_val = scanner.nextDouble ();
        System.out.println ("You entered" + double_val +"\n");
    }
    catch (InputMismatchException e) {
        System.out.println ("Mismatch exception:" + e);
    }
  } // main
} // class ScanConsoleApp
```

A session with the program `ScanConsoleApp` is shown here:

```
Input int (e.g. 3501): 23431
   You entered 23431

Input float (e.g. 2.43): 1.2343
   You entered 1.2343

Input double (e.g. 4.943e+15): -2.34e4
   You entered -23400.0
```

There are a number of other useful methods in the `Scanner` class such as `skip()` to skip over some input, `useDelimiter()` to set the delimiter, and `findInLine()` to search for substrings. The `Scanner` class uses tools from the `java.util.regex` package for pattern matching with regular expressions. We don't have space here to describe these very powerful text-matching tools but you can find more info in the Java 2 API Specifications.

9.5 The `File` class

Files and directories are accessed and manipulated via the `java.io.File` class. The `File` class does not actually provide for input and output to files. It simply provides an *identifier* of files and directories. Remember that just because a `File` object is created, it does not mean there actually exists on the disk a file with the identifier held by that `File` object.

The `File` class includes several overloaded constructors. For example, the following instance of `File` refers to a file named `myfile.txt` in the current directory where the JVM is running:

```
File file = new File ("myfile.txt");
```

Again, the file `myfile.txt` may or may not exist in the file system. An attempt to use a `File` object that refers to a file that does not exist will cause a `FileNot-FoundException` to be thrown.

The next example creates a `File` object from a file name that includes the full directory path:

```
File file = new File ("/tmp/myfile.txt");
```

Another overloaded constructor allows the separate specification of path and file name:

```
File file = new File ("/tmp", "myfile.txt");
```

`File` objects can also represent directories. For example,

```
File tmp_directory = new File ("/tmp");
```

There are a number of useful methods in `File` such as the following:

- `Boolean exist ()` – indicates if the file referred to actually exists
- `Boolean canWrite ()` – can the file be written to
- `Boolean canRead ()` – can the file be read
- `Boolean isFile ()` – does the `File` object represent a file?
- `Boolean isDirectory ()` – or a directory?

Given an instance of `File`, the class provides methods to obtain the file name and path components, to make a directory, to get a listing of the files in a directory, and so forth:

- `String getName ()` – get the name of the file (no path included)
- `String getPath ()` – get the abstract file path
- `String getCanonicalPath ()` – get the name of the file with path
- `String getAbsolutePath ()` – get the absolute file path

Java must run on different types of platforms, and file and directory systems differ among platforms. For example, path names use different separator characters on different hosts. Windows uses the backslash ("\") and Unix uses the forward slash ("/"). To obtain platform independence, instead of explicit separator characters, you can use these static constants defined in the `File` class:

- `File.separator` – `String` with file separator
- `File.separatorChar` – `char` with file separator
- `File.pathSeparator` – `String` with path separator
- `File.pathSeparatorChar` – `char` with path separator

For example, here we can build a path for a file:

```
String dir_name = "dataDir";
String file_name= "data.dat";
File file_data  = new File (dir_name + File.separator
                                + file_name);
```

Other talents of the `File` class include the method

```
boolean mkdir ()
```

This method will create a directory with the abstract path name represented by the `File` object if that `File` object represents a directory. The method returns `true` if the directory creation succeeds and `false` if not. A case in which the directory cannot be created is when the `File` object refers to an actual file that already exists in the file system.

9.6   File I/O

The `java.io` package includes these base file stream classes:

- `FileInputStream` – base class for binary file input
- `FileOutputStream` – base class for binary file output
- `FileReader` – read text files
- `FileWriter` – write to text files

These classes treat a file as the source or destination. They each have several overloaded constructors, including constructors that take a string file name as the parameter and constructors that take an instance of `File` as a parameter. For example, an input stream from a file can be created as follows:

```
File file_in = new File ("data.dat");
FileInputStream in = new FileInputStream (file_in);
```

If the file does not exist, the `FileInputStream(File)` constructor will throw an instance of `FileNotFoundException`.

Usually, the `FileInputStream` object is wrapped with another stream to obtain greater functionality. For example, `BufferedInputStream` is used to smooth out the data flow as shown previously with the other "buffer" wrapper classes.

Output streams to files are opened in a similar manner:

```
File file_out = new File ("tmp.dat");
FileOutputStream out = new FileOutputStream (file_out);
```

In this case, if the file doesn't exist, it will be created rather than throwing an exception. As with input file streams, output streams are also usually wrapped with one or more other streams to obtain greater functionality such as that provided by `BufferedOutputStream`.

The following snippet shows how to append data to an existing file:

```
File file_out = new File ("old.dat");
FileOutputStream out = new FileOutputStream (file_out, true);
```

The second parameter in the constructor indicates that the file should be opened for appending. `RandomAccessFile` can also be used for appending. It is discussed in the Web Course Chapter 9: *Supplements* section.

9.6.1 Text output to a file

We can use the `FileWriter` stream class to write character data to a file. The following `PrintWriterFileApp` example gets the name for an output file either from a parameter on the command line or from a default. The program follows closely to `PrintWriterApp` in Section 9.4.1 except that instead of starting with the `PrintStream` object referenced by `System.out`, we create a `FileWriter` for a `File` object and then wrap this with a `BufferedWriter` and a `PrintWriter`. These two wrappers give us the efficiency of stream buffering and a handy set of print methods.

```java
import java.io.*;

/**
  * Demonstrate wrapping streams to send text and primitive
  * type values to a file.
  **/
public class PrintWriterFileApp
{
   public static void main (String[] args) {

      File file = null;

      // Get the file from the argument line.
      if (args.length > 0) file = new File (args[0]);
      if (file == null) {
         System.out.println ("Default: textOutput.txt");
         file = new File ("textOutput.txt");
      }

      // Create a FileWriter stream to a file and then wrap
      // a PrintWriter object around the FileWriter to print
      // primitive values as text to the file.
      try {
         // Create a FileWriter stream to the file
         FileWriter file_writer = new FileWriter (file);
```

```
        // Put a buffered wrappter around it
        BufferedWriter buf_writer =
            new BufferedWriter (file_writer);

        // In turn, wrap the PrintWriter stream around this
        // output stream and turn on the autoflush.
        PrintWriter print_writer =
            new PrintWriter (buf_writer, true);

        // Use the PrintWriter println methods to send
        // strings and primitives to the file.
        print_writer.println ("Text output with PrintWriter.");
        print_writer.println (
            "Primitives converted to strings:");

        boolean a_boolean = false;
        byte    a_byte    = 114;
        short   a_short   = 1211;
        int     an_int    = 1234567;
        long    a_long    = 987654321;
        float   a_float   = 983.6f;
        double  a_double  = -4.297e-15;

        print_writer.println (a_boolean);
        print_writer.println (a_byte);
        print_writer.println (a_short);
        print_writer.println (an_int);
        print_writer.println (a_long);
        print_writer.println (a_float);
        print_writer.println (a_double);

        // PrintWriter doesn't throw IOExceptions but instead
        // offers the catchError () method
        if (print_writer.checkError ()) {
            System.out.println ("An output error occurred!");
        }
    }
    catch (IOException ioe){
        System.out.println ("IO Exception");
    }
    } // main
} // class PrintWriterFileApp
```

Running this code produces a file that holds text exactly like that shown as the output from PrintWriterApp in Section 9.4.1.

9.6.2   `Formatter` output to a file

The following program, `FormatWriteFileApp`, follows closely to that of `FormatWriteApp` discussed in Section 9.4.2 in which we used the `java.util.Formatter` class to send formatted output to the console. However, this program sends the output to a file. The `Formatter` class does much of the work. We don't even have to create a `FileWriter` stream as we did in the previous section. We simply pass the `File` object to the `Formatter` constructor and it takes care of all the output mechanics. If the file doesn't exist yet, then the `Formatter` will create it, though we need to catch an exception that is thrown in such a case.

```java
import java.io.*;
import java.util.Formatter;
/**
  * Demonstrate the java.util.Formatter capabilities for
  * formatting primitive types and sending them to a file.
 **/
public class FormatWriteFileApp
{
   public static void main (String[] args) {
      Formatter formatter = null;
      File file = null;

      // Get the file from the argument line.
      if (args.length > 0) file = new File (args[0]);
      // Else use a default file name.
      if (file == null) {
         System.out.println ("Default: textOutput.txt");
         file = new File ("textOutput.txt");
      }

      // Send formatted output to the file.
      try {
         formatter = new Formatter (file);
      }
      catch (FileNotFoundException e) {
         // File not found exception thrown since this is a
         // new file name. However, Formatter will create
         // the new file.
      }

      formatter.format ("Text output with Formatter. %n");
      formatter.format ("Primitives converted to
                         strings: %n");
```

```
        boolean a_boolean = false;
        byte      a_byte    = 114;
        short     a_short   = 1211;
        int       an_int    = 1234567;
        long      a_long    = 987654321;
        float     a_float   = 983.6f;
        double    a_double  = -4.297e-15;

        formatter.format ("boolean = %9b %n",    a_boolean);
        formatter.format ("byte     = %9d %n",    a_byte);
        formatter.format ("short    = %9d %n",    a_short);
        formatter.format ("int      = %9d %n",    an_int);
        formatter.format ("long     = %9d %n",    a_long);
        formatter.format ("float    = %9.3f %n", a_float);
        formatter.format ("double   = %9.2e %n", a_double);

        // Need to flush the data out of the buffer.
        formatter.flush ();
        formatter.close ();

    } // main
} // class FormatWriteFileApp
```

This program creates a file that contains the same output as `FormatWriteApp` in Section 9.4.2.

9.6.3 Text input from a file

The following example, `TextFileReadApp`, illustrates how to use the `FileReader` stream to read strings from a text file. The goal is to read a file and count the number of lines in which a particular string occurs at least once. We wrap the `FileReader` stream with a `BufferedReader` class and take advantage of its `readLine()` method to read a whole line at a time. We use the `indexOf()` method in the `String` class to search for the string of interest.

As usual, we enclose the stream reading within a `try-catch` statement to catch the `IOException` or one of its subclass exceptions that can be thrown by the stream constructors and the read methods.

```
import java.io.*;
import java.util.*;

/** Demonstrate reading text from a file. **/
```

```java
public class TextFileReadApp
{
  public static void main (String[] args) {
    // Count the number of lines in which this string occurs
    String string_to_find = "new";

    File file = null;
    // Get the file from the argument line.
    if (args.length > 0) file = new File (args[0]);
    if (file == null ||!file.exists ()) {
      System.out.println ("Default: TextFileReadApp.java");
      file = new File ("TextFileReadApp.java");
    }

    // Count the number of lines with the string of interest.
    int num_lines = 0;
    try {
      // Create a FileReader and wrap it with BufferedReader.
      FileReader file_reader = new FileReader (file);
      BufferedReader buf_reader =
        new BufferedReader (file_reader);
      // Read each line and look for the string of interest.
      do {
        String line = buf_reader.readLine ();
        if (line == null) break;
        if (line.indexOf (string_to_find)!= -1) num_lines++;
      } while (true);

      buf_reader.close ();
    }
    catch (IOException e) {
      System.out.println ("IO exception = " + e);
    }

    System.out.printf (
      "Number of lines containing \"%s\"= % 3d %n",
      string_to_find, num_lines);
  } // main
} // class TextFileReadApp
```

Output from the program goes as:

```
Default: TextFileReadApp.java
Number of lines containing "new" = 5
```

Note that we close the stream explicitly. For this short program that ends very soon after finishing the read, closing the input stream is not so important, but in general, it is good practice to always close input and output streams for files when I/O transfers are completed.

9.6.4 Input from a file with `Scanner`

The `Scanner` class, discussed in Section 9.4.4, can read from a file just as easily as it can from the console. The example program `ScanFileApp` expects an input file of the type produced by `FormatWriteApp` (see Section 9.6.2) containing the output as shown for `FormatWriteAppFile` (see Section 9.4.2). It uses `Scanner` to scan past the text at the beginning of the file and then it reads each of the primitive type values.

There are many options with the pattern-matching capabilities of `Scanner` to jump past the initial text. We choose a simple technique of looking for the first primitive type value, which is a `boolean` type. So we loop over calls to `hasNextBoolean()` until we find a `boolean`. This method and the similar ones for other primitive types look ahead at the next token and return true or false depending on whether the token is of the type indicated. It does not jump past the token. So, if the next token is not a `boolean`, we use the `next()` method to skip this token and examine the next one. When we find the `boolean` we break out of the loop.

```java
import java.io.*;
import java.util.*;

/** Demonstrate using Scanner to read a file. **/
public class ScanFileApp
{
    public static void main (String[] args) {

        File file = null;

        // Get the file from the argument line.
        if (args.length > 0) file = new File (args[0]);
        // or use the default
        if (file == null) {
            file = new File ("textOutput.txt");
        }

        Scanner scanner = null;
        try {
            // Create a scanner to read the file
```

```
            scanner = new Scanner (file);
        }
        catch (FileNotFoundException e) {
            System.out.println ("File not found!");
            // Stop program if no file found
            System.exit (0);
        }

        try {
            // Skip tokens until the boolean is found.
            while (true) {
                if (scanner.hasNextBoolean ()) break;
                scanner.next ();
            }

            System.out.printf ("Skip strings at start of%s%n",
                        file.toString ());
            System.out.printf (
              "and then read the primitive type values:%n%n");

            // Read and print the boolean
            System.out.printf("boolean = %9b %n",
                              scanner.nextBoolean ());

            // and then the set of numbers
            System.out.printf ("int = %9d %n",
                              scanner.nextInt ());
            System.out.printf ("int = %9d %n",
                              scanner.nextInt ());
            System.out.printf ("int = %9d %n",
                              scanner.nextInt ());
            System.out.printf ("long = %9d %n",
                              scanner.nextLong ());
            System.out.printf ("float = %9.1f %n",
                              scanner.nextFloat ());
            System.out.printf ("double = %9.1e %n",
                              scanner.nextFloat ());

        }
        catch (InputMismatchException e) {
            System.out.println ("Mismatch exception:" + e);
        }
    } // main
} // class ScanFileApp
```

The output of reading the file then looks like:

```
Skip strings at start of textOutput.txt
and then read the primitive type values:
boolean     =           false
int         =             114
int         =            1211
int         =         1234567
long        =       987654321
float       =           983.6
double      =        -4.30e-15
```

9.6.5 Binary output to a file

Numerical data transfers faster and more compactly in a raw binary format than as text characters. Here we look at examples of writing numerical data to a binary file and reading numerical data from a binary file.

In the example program below called `BinOutputFileApp`, we first create some data arrays with some arbitrary values. We then open a stream to a file with the binary `FileOutputStream` class. We wrap this stream object with an instance of the `DataOutputStream` class, which contains many useful methods for writing primitive types of the `writeX()` form, where X is the name of a primitive type. We use the `writeInt (int i)` and the `writeDouble (double d)` methods, to write the data to the file as pairs of `int`/`double` type values. In the next section we will show next how to read the binary data from this file.

```java
import java.io.*;
import java.util.*;

/** Write a primitive type data array to a binary file. **/
public class BinOutputFileApp
{
  public static void main (String[] args) {
    Random ran = new Random ();
    // Create an integer array and a double array.
    int [] i_data = new int[15];
    double [] d_data = new double[15];
    // and fill them
    for (int i=0; i < i_data.length; i++) {
      i_data[i] = i;
      d_data[i] = ran.nextDouble() * 10.0;
    }
```

```
File file = null;
// Get the output file name from the argument line.
if (args.length > 0) file = new File (args[0]);
// or use a default file name
if (file == null) {
    System.out.println ("Default: numerical.dat");
    file = new File ("numerical.dat");
}

// Now write the data array to the file.
try {
    // Create an output stream to the file.
    FileOutputStream fileOutput =
        new FileOutputStream (file);
    // Wrap the FileOutputStream with a DataOutputStream
    DataOutputStream dataOut =
        new DataOutputStream (fileOutput);

    // Write the data to the file in an integer/
    // double pair
    for (int i=0; i < i_data.length; i++) {
        dataOut.writeInt (i_data[i]);
        dataOut.writeDouble (d_data[i]);
    }
    // Close file when finished with it. .
    fileOutput.close ();
}
catch (IOException e) {
    System.out.println ("IO exception = " + e);
}
} // main
} // class BinOutputFileApp
```

9.6.6 Binary input from a file

In the example program below called `BinInputFileApp`, we read a binary file created by `BinOutputFileApp` discussed in the previous section. We begin by first opening a stream to the file with a `FileInputStream` object. Then we wrap this with a `DataInputStream` class to obtain the many `readX()` methods, where X represents the name of a primitive data type as in `readInt()` and `readDouble()`. Here we read pairs of integer and double values.

Rather than test for the return of a `-1` value as we did in the text input streams, we simply continue to loop until the read method throws the `EOFException`. In the `catch` statement for this exception you can carry out the final housekeeping chores before closing the file stream.

```java
import java.io.*;
import java.util.*;

/** Demonstrate reading primitive type values from a
 * binary file. **/
public class BinInputFileApp
{
  public static void main (String[] args) {
    File file = null;
    int i_data = 0;
    double d_data = 0.0;

    // Get the file from the argument line.
    if (args.length > 0) file = new File (args[0]);
    if (file == null) {
      System.out.println ("Default: numerical.dat");
      file = new File ("numerical.dat");
    }

    try {
      // Wrap the FileInputStream with a DataInputStream
      FileInputStream file_input =
        new FileInputStream (file);
      DataInputStream data_in =
        new DataInputStream (file_input);

      while (true) {
        try {
          i_data = data_in.readInt ();
          d_data = data_in.readDouble ();
        }
        catch (EOFException eof) {
          System.out.println ("End of File");
          break;
        }
        // Print out the integer, double data pairs.
        System.out.printf (" %3d. Data = %8.3e %n",
                           i_data, d_data);
      }
      data_in.close ();
    }
    catch (IOException e) {
      System.out.println ("IO exception = " + e);
    }
  } // main
} // class BinInputApp
```

We illustrate the output and input of binary data by first running `BinOutput-FileApp` to produce the data file `numerical.dat`. We then run `BinInput-FileApp`, which reads the file `numerical.dat` and produces the following output on the console. Your output will vary since `BinOutputFileApp` uses `Random` to generate random values.

```
Default: numerical.dat
    0. Data = 2.633e+00
    1. Data = 7.455e+00
    2. Data = 2.447e+00
    3. Data = 7.046e+00
    4. Data = 2.652e+00
    5. Data = 5.120e+00
    6. Data = 1.754e+00
    7. Data = 7.489e+00
    8. Data = 7.386e-01
    9. Data = 6.036e+00
   10. Data = 7.002e-01
   11. Data = 9.625e+00
   12. Data = 5.966e+00
   13. Data = 8.535e+00
   14. Data = 2.744e+00
End of File
```

9.7 Character encoding

For text I/O, each character is specified by an encoded value. The particular type of encoding, the number of bits and bytes required for the encoding, transformations between encodings, and other issues thus become important, especially for a language like Java that is aimed towards worldwide use. So we give a brief overview of character encodings here.

The 7-bit ASCII code set is the most famous, but there are many extended 8-bit sets in which the first 128 codes are ASCII and the extra 128 codes provide symbols and characters needed for other languages besides English. For example, the ISO-Latin-1 set (ISO Standard 8859-1) provides characters for most West European languages and for a few other languages such as Indonesian.

Java itself is based on the 2-byte Unicode representation of characters. The 16 bits provide for a character set of 65 535 entries and so allows for broad international use.

The first 256 entries in 2-byte Unicode are identical to the ISO-Latin-1 set. That makes the 2-byte Unicode inefficient for programs in English since the second byte is seldom needed. Therefore, a scheme called UTF-8 is used to encode text characters (e.g. string literals) for the Java class files. The UTF code varies from

1 byte to 3 bytes. If a byte begins with a 0 bit, then the lower 7 bits represent one of the 128 ASCII characters. If the byte begins with the bits 110, then it is the first of a 2-byte pair that represent the Unicode values for 128 to 2047. If any byte begins with 1110, then it is the first of a 3-byte set that can hold any of the other Unicode values.

Java typically runs on platforms that use 1-byte extended-ASCII-, encoded characters. Therefore, text I/O with the local platform, or with other platforms over the network, must convert between the encodings. As we mentioned in the previous section, the original 1-byte streams were not convenient for this so the `Reader/Writer` classes for 2-byte I/O were introduced.

The default encoding is typically ISO-Latin-1, but your program can find the local encoding with a static method in `System`:

```
String local_encoding = System.getProperty ("file.encoding");
```

The encoding can be explicitly specified in some cases via the constructor, such as in the following file output:

```
FileOutputStream out_file = new FileOutputStream
("Turkish.txt");
OutputStreamWriter file_writer = new OutputStreamWriter
(out_file, "8859_3");
```

A similar overloaded constructor is available for `InputStreamReader`. See the book by Harold [2] for more information about character encoding in Java.

If a character is not available on your keyboard, it can be specified in a Java program by its Unicode value. This value is represented with four hexadecimal numbers preceded by the ("u") escape sequence. For example, the "ö" character is given by \u00F6 and "è" by \u00E8.

We note finally that even the 65 535 entries of the version of Unicode used by Java are not enough to encompass all of the language characters and symbol sets in the world. Therefore, Java will gradually make the transition to Unicode 4.0, which uses 32 bits. This is a challenge for many reasons, including the fact that the `char` primitive is only 16-bit. Java 5.0 has some tools for dealing with 32-bit supplementary characters but we don't have space here to discuss them. We refer the reader to the article by Lindenberg and Okutsu for further information on 32-bit character support in Java [6].

9.8 Object I/O

So far we have seen that we can do I/O with primitive data types and with text, which involves `String` objects. By means of the `ObjectInputStream` and `ObjectOutputStream` you can also do I/O with other types of objects. The `writeObject (Object)` method in `ObjectOutputStream` grabs the data from the class fields of an existing object and sends that data through the stream.

The readObject() method from ObjectInputStream can read this data from the stream and fill the data fields in a new instance of the class.

The data is sent sequentially, or *serially*, a byte at a time on the stream, so this process is also referred to as *serialization*. For a class to work with these methods, it must implement the Serializable interface. This interface has no methods but is intended to tag the class as suitable for serializing. There are, for example, security concerns with regard to this process since internal data of an object may become vulnerable as it travels through the I/O system. So, not all classes are suitable for serializing. Many core language classes implement Serializable but you should check the Java 2 API specifications to make sure for a particular class of interest.

Within a class there may be data that is only temporary. These data fields can be labeled transient and will not be included in the serialization. When an object to be serialized holds references to other objects, those objects will also be serialized if they implement Serializable. Otherwise, they should be labeled transient.

So a class for serializing could look as follows:

```
public class MyClass implements Serializable {
    int fI,fJ;
    double fValue
    transient int fTmpValue;
    String fTitle;
    OtherClass fOtherClass;
    . . . constructors & methods . . .
}
```

Instances of this class could be saved to a file using a method like the following:

```
static public void saveMyClass (MyClass my_object, File file) {
    FileOutputStream file_out = new FileOutputStream (file);
    ObjectOutputStream obj_out_stream = new ObjectOutputStream
      (file_out);
    obj_out_stream.writeObject (my_object);
    obj_out_stream.close ();
}
```

Similarly, to read the object back in from a file we could use a method as follows:

```
static public MyClass getMyClass (File file) {
    FileInputStream file_in = new FileInputStream (file);
    ObjectInputStream obj_in_stream = new ObjectInputStream
      (file_in);
    MyClass my_obj = (MyClass)(obj_in_stream.readObject ());
    obj_in_stream.close ();
    return my_obj;
}
```

Table 9.1 *Input stream examples.*

Source	Stream wrappings
Console	`BufferedReader (new InputStreamReader (System.in))`
Text in a byte stream	`BufferedReader (new InputStreamReader (InputStream))`
Text file	`BufferedReader (new FileReader (new File ("f.txt"))`
Binary Data File	`DataInputStream (new FileInputStream (new File ("f.bin"))`

Table 9.2 *Output stream examples.*

Destination	Stream wrappings
Console	`PrintWriter (new OutputStreamWriter (System.out))`
Text in a byte stream	`PrintWriter (new BufferedWriter (new OutputStreamWriter (OutputStream)))`
Text file	`PrintWriter (new BufferedWriter (new FileWriter (new File ("f.txt")))`
Binary Data File	`DataOutputStream (new FileOutputStream (new File ("f.bin"))`

There are a number of other issues regarding serialization that are beyond the scope of this chapter and book. More discussion and example programs are available in the Web Course: Chapter 9 and serialization also is a technique important for the distributed computing discussions in Part II.

9.9 Choosing a stream class

All of these stream classes for text and binary data, for console and files, can get rather confusing. To help decide what stream classes and wrappers to use for a particular kind of data, we provide two tables. Table 9.1 suggests input stream combinations and Table 9.2 suggests output streams according to the type of data and its source or destination. You can check these tables to find the starting class for your data I/O and possible wrapper classes that add buffering, filtering, and other useful methods.

9.10 Primitive types to bytes and back

We have seen that Java is very strict with data types. Every data element must be assigned a specific type. Casting can convert one type into another but a data value can never simultaneously act as more than one type. This differs from C/C++

where, say, a 32-bit integer type value can easily be accessed as two short values or four separate bytes

For binary I/O with files as discussed above, the underlying flow of data occurs as a sequence of bytes. We saw that by wrapping `FileOutputStream` with a `DataOutputStream` we obtained methods such as `writeInt (int)` and `writeDouble (double)` that put the bytes of wider types like `int` or `double` onto the outgoing stream. Similarly, by wrapping a `FileInputStream` with a `DataInputStream`, the `readInt()` and `readDouble()` methods convert 4 bytes in the stream into an `int` value and 8 bytes into a `double` value, respectively.

Note that on all platforms, Java uses the so-called *big-endian* representation. This means that for a primitive longer than 1 byte, the most significant byte is at the lowest address (i.e. big end first.) So, for example, in an output stream of `int` values, the high-order byte goes out first and in an input stream the high-order byte arrives first. This order is preserved by the JVM even on underlying platforms that use the reverse representation known as *little-endian*.

What if you obtain an array of bytes and need to extract data values of different types from it? For example, a byte array of 19 bytes might contain one `char` type, two `int` values, one `double` value, and one `byte` type. In C/C++ you can use memory pointers to reference different parts of the array and cast a group of bytes there to a particular type. In Java that type of direct memory referencing is not allowed. Instead, you can make the array the source of a stream with the `ByteArrayInputStream` class. You then wrap this stream with `DataInputStream` and use its selection of methods for reading different data types. For example, to extract our data from the 19 elements of a byte array:

```java
public void getData (byte[] data) {
  ByteArrayInputStream byte_in =
    new ByteArrayInputStream (data);
  DataInputStream data_in = new DataInputStream (byte_in);
  char c   = data_in.readChar ();
  int i    = data_in.readInt ();
  int j    = data_in.readInt ();
  double d = data_in.readDouble ();
  byte b   = data_in.readByte ();
    . . .
}
```

Conversely, you can send a stream of data of different types to a `ByteArray-OutputStream` and then extract its internal byte array. It is convenient to wrap this stream with `DataOutputStream`, which provides a selection of methods for different types of such as `writeInt (int)`, `writeDouble (double)`, and `writeChars()`.

```
ByteArrayOutputStream byte_out = new ByteArrayOutputStream ();
DataOutputStream data_out = new DataOutputStream (byte_out);
```

The byte array from the stream can be obtained with

```
byte_out.toByteArray ();
```

In Section 9.13.1 we see that these techniques are useful for I/O with our histogram classes where we want to save a mix of data types into one big byte array. Also, in Chapter 23 we discuss serial I/O with a Java hardware processor that requires low-level I/O with bytes. In an example, we use a byte array stream to convert a set of data bytes to wider types.

9.11 Sources, destinations, and filters

We have seen that I/O stream classes can represent sources of input data and destinations for output data. For example, disk files become sources via `FileInputStream` and a byte array can become the destination of a stream via the `ByteArrayOutputStream` class.

We emphasized that wrapper classes add greater functionality to the streams they contain. However, you can also look at many of the wrapper classes as filters. A stream filter monitors, transforms, or is some way processes the data as the stream flows through it. The `BufferedInputStream` and `BufferedOutputStream` classes, for example, hold data in buffers until they are full before letting the data out. They extend the classes named `FilterInputStream` and `FilterOutputStream`, respectively (see Figure 9.1).

The `FilterIntputStream` class wraps an `InputStream` object passed via its constructor:

```
protected FilterInputStream (InputStream in_stream)
```

The `FilterInputStream` class overrides all of the same methods in `InputStream` but they simply invoke the corresponding methods in the `in_stream` object. The `FilterInputStream` does nothing itself and is meant to be extended. A subclass overrides some or all of the methods to carry out the desired action on the data. For example, `BufferedInputStream` overrides all but one of the `read()` methods in `FilterInputStream`. Similarly, `FilterOutputStream` is intended to be subclassed by a class such as `BufferedOutputStream` that overrides some or all of its methods to carry out operations on the outgoing data.

Java I/O can be somewhat overwhelming at first but it allows for a great deal of modularity and high-level abstraction that can actually bring clarity to program design. For example, we will discuss in Section 9.13 how to make histograms

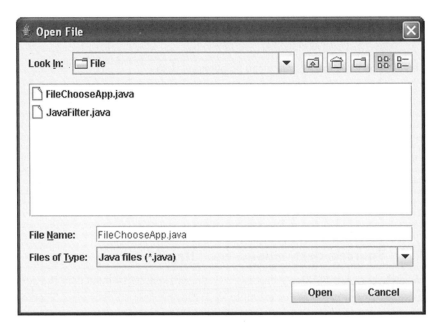

Figure 9.2   An instance of `FileChooser` in action.

into stream sources and stream destinations and how to filter data going to a histogram.

9.12   The `JFileChooser` dialog

To find files and directories on local media for I/O with a graphical user interface, Swing provides the `JFileChooser` class. This file browser component (see Figure 9.2) allows the user to search through directories and select one or more files and directories, which are then available via instances of `File`.

The code snippet below illustrates how to use a `JFileChooser` to select files for reading. The title, selection mode, and current directory are set with methods from the class. The chooser allows for a filter that selects particular file types for the list of files. This requires a subclass of the `FileFilter` class from the `javax.swing.filechooser` package. As shown with the class `JavaFilter`, by overriding the `accept()` method you can examine the current file or directory and return `true` or `false` to accept or reject the current selection. You can override `getDescription()` to return a string that describes the particular file types that the filter selects.

```
. . . This method uses a file chooser to locate a file to
      read . . .
   boolean openFile () {
```

```
      JFileChooser fc = new JFileChooser ();
      fc.setDialogTitle ("Open File");

      // Choose only files, not directories
      fc.setFileSelectionMode (JFileChooser.FILES_ONLY);

      // Start in current directory
      fc.setCurrentDirectory (new File ("."));

      // Set filter for Java source files.
      fc.setFileFilter (fJavaFilter);

      // Now open chooser
      int result = fc.showOpenDialog (this);

      if (result == JFileChooser.CANCEL_OPTION) {
         return true;
      }
      else if (result == JFileChooser.APPROVE_OPTION) {
         fFile = fc.getSelectedFile ();
         // Invoke the readFile method in this class
         String file_string = readFile (fFile);

         if (file_string!= null)
            fTextArea.setText (file_string);
         else
            return false;
      }
      else {
         return false;
      }
      return true;
   } // openFile
```

```
import javax.swing.*;
import java.io.*;
/** Class to filter file types for JFileChooser. **/
public class JavaFilter extends
   javax.swing.filechooser.FileFilter {

   public boolean accept (File f) {
      return f.getName ().toLowerCase ().endsWith (".java")
```

```
        || f.isDirectory ();
    }

  public String getDescription () {
    return "Java files (*.java)";
  }
}
```

Web Course Chapter 9 discusses a similar approach for setting up a JFileChooser to save a file.

9.13   Histogram I/O

Here we use our histogram classes to illustrate some of the I/O topics discussed in this chapter. We add the capability to save the histograms to disk files and to read them back from the files. We also create a histogram stream class in which a histogram becomes the destination of a stream of data. Similarly, we create a stream filter than calibrates the data heading for a histogram.

9.13.1   Saving histograms

An obviously useful feature for our histogram classes would be the ability to save a histogram to a disk file and to read it back later. The data in the class Histogram includes the number of bins, the bin array, the upper and lower range values, the underflow and overflow values, plus the title and axis label. (Note that this data involves double, int, and String types.)

One approach to saving the histogram is to obtain the data from the histogram and use the file output techniques discussed above to write the various data values to a file. The data could be made directly accessible or provided with a set of "get" (or "getter") methods, such as getTitle() to obtain the title string and getBins() to return the integer array holding the bin contents. The histogram could be rebuilt by reading in the data with the file input techniques, creating a new histogram object, and then filling its values with "set" (or "setter") methods, such as setTitle (String title).

The class HistIOTools (see the listing in the Web Course Chapter 9: *Tech* section) provides a group of static methods to do this except that it first writes the data into a byte array using the techniques discussed in Section 9.10 and then writes this array to a file. Conversely, it provides methods to read this array back from a file, extract the data, and load the data into a new histogram object. Packaging the data into a single-byte array reduces the I/O operations with the file.

The most elegant approach, however, for adding I/O to our histograms is to use the object serialization techniques and simply stream a Histogram object to a file. For example, if Histogram were made Serializable, we could create a method like the following to save a histogram object to a file:

```
static public void saveHistogram (Histogram histogram, File
file) {
   FileOutputStream fos = new FileOutputStream (file);
   ObjectOutputStream oos = new ObjectOutputStream (fos);
   oos.writeObject (histogram);
   oos.close ();
}
```

Similarly, to read a histogram back in from a file we could use a method as follows:

```
static public Histogram getHistogram (File file) {
   FileInputStream fis = new FileInputStream (file);
   ObjectInputStream ois = new ObjectInputStream (fis);
   Histogram histogram = (Histogram)(ois.readObject ());
   ois.close ();
   return histogram;
}
```

An complete example program using this technique is given in the Web Course Chapter 9: *Tech* section.

9.13.2 Histograms as stream destinations

We saw above how a byte array can become the destination of a stream with the `ByteArrayOutputStream` class. We can apply a similar technique to histograms: we make a histogram into the destination of a stream. Perhaps in a data monitoring program there is extensive handling of dozens or more data channels, each assigned its own histogram. In such a case, `Histogram` streams might offer a neat approach to organizing the filling of the histograms.

In Chapter 7 we discussed the `HistPanel` component that displays an instance of our `Histogram` class. In the code snippet shown below, the `StreamedHistPanel` extends `HistPanel` and includes an inner class called `HistogramOutputStream` that extends `OutputStream`. `HistogramOutputStream` overrides the `write (int b)` method of `OutputStream` with a method that adds data to the histogram in the `StreamedHistPanel` object (it can access the histogram since it is an inner class of `StreamedHistPanel`).

```
import java.io.*;
/** This class provides provides a HistPanel destination
  * for a output stream. **/
public class StreamedHistPanel extends HistPanel
{
   OutputStream fHistStream;
   boolean fNewDataFlag = false;
```

```
/** Create the panel with the histogram. **/
public StreamedHistPanel (Histogram histogram) {
  super (histogram);
  fHistStream = new HistogramOutputStream ();
}


public OutputStream getOutputStream () {
  return fHistStream;
}


/** This inner class provides the stream to write
  * data.**/
class HistogramOutputStream extends OutputStream {

  /** In OutputStream write (int b), only the lower byte
    * of the int value is used but here we use the full
    * value. **/
  public synchronized void write (int data) {
    // Convert back to double and shift down two
    // decimal places.
    histogram.add (((double)data)/100.0);
    fNewDataFlag = true;
  }
} // class HistogramOutputStream
} // class StreamedHistPanel
```

Such panels for each histogram can be added to the program display. The Web
Course Chapter 9: *Tech* section includes the demonstration program Hist-
Stream.java, which creates three instances of the StreamedHistPanel
and obtains the output stream for each, as shown in this snippet:

```
. . . GUI building in HistStreamApplet . . .
  for (int i=0; i < NUM_HISTS; i++) {
    histogram[i] = new Histogram ("Sensor " + i,
                                  "Data", 25,0.0,10.0);
    histPanel[i] = new StreamedHistPanel (histogram[i]);

    // Add the histogram panels to the container panel
    histsPanel.add (histPanel[i]);

    // Get the output streams for each panel.
    dataOut[i] = histPanel[i].getOutputStream ();
  }
  . . .
```

The applet creates simulated sensor data and streams the data to the histograms simply with

```
. . . loop to generate data for each sensor histogram . . .
try {
  // Send data to stream destination
  dataOut[j].write (ival);
}
catch (IOException ioe) {}
. . .
```

(`StreamedHistPanel` follows an example in Harold [2] in which a `TextArea` class becomes the destination for a stream.)

9.13.3   Filtering histogram data streams

We discussed in Section 9.11 how a wrapper class can act as a filter on the data streaming through it. We can take advantage of this technique to process data as it streams to a histogram. The class `HistFilterOutputStream`, shown below, extends `FilterOutputStream`. The job of this filter is to "calibrate" the data as it streams through.

The filter stream wraps an instance of `OutputStream` and overrides the `write (int b)` method in `FilterOutputStream`. Before it writes the datum to the `OutputStream` it does a pedestal (i.e. a constant offset) correction and a slope correction to the value.

```
import java.io.*;
public class HistFilterOutputStream extends
  FilterOutputStream
{
  double fSlope = 0.0, fPedestal = 0.0;

  HistFilterOutputStream (OutputStream out,
                          double slope,
                          double pedestal) {
    super (out);
    fSlope = slope;
    fPedestal = pedestal;
  } // ctor

  /** Override the write (int b) method but use the full
    * integer value rather than only the lowest byte as
    * usual with this method. Carry out the calibration and
    * then write the resulting value as an integer to the
    * output stream. **/
```

```
public void write (int b) {
  double val = ((double)b)/100.0;
  int ival = (int) (100.0 * ((val - fPedestal)/fSlope));
  try {
    out.write (ival);
  }
  catch (IOException ioe) {}
} // write

// Not overrriding other methods in FilterOutputStream.

} // class HistFilterOutputStream
```

The applet `HistFilterStream` in the Web Course Chapter 9: *Tech* section displays data from three sensors (artificially generated data) in instances of the `StreamHistPanel` class discussed in the previous section. The data for each sensor is generated with a different slope and pedestal to simulate real world variations in measurement data. By wrapping the streams going into the histogram with `HistFilterOutputStream`, we can calibrate the data before it reaches the histogram.

The applet displays a row of `StreamHistPanel` histograms with raw data and a row below this one showing histograms with the calibrated data. In the interface building section of the program, the code snippet here shows the section that creates the top row of histogram panels and then the bottom row (`NUM_HISTS = 3`):

```
. . . GUI building in HistFilterStreamApplet . . .

for (int i=0; i < NUM_HISTS; i++) {
  fHistogram[i] = new Histogram ("Sensor " + i,
                                 "Data", 25,0.0,10.0);
  fHistPanel[i] = new StreamedHistPanel (fHistogram[i]);

  // Add the histogram panels to the container panel
  fHistsPanel.add (fHistPanel[i]);

  // Get the output streams for each panel.
  fDataOut[i] = fHistPanel[i].getOutputStream ();
}

// Create HistPanels for the calibrated sensor
// histograms.
for (int i=0; i < NUM_HISTS; i++) {
```

```
      fHistogram[i+NUM_HISTS] =
         new Histogram ("Calibrated Sensor" + i,
                        "Data", 25,0.0,10.0);
      fHistPanel[i+NUM_HISTS] =
         new StreamedHistPanel (fHistogram[i+NUM_HISTS]);

      // Add the histogram panels to the container panel
      fHistsPanel.add (fHistPanel[i+NUM_HISTS]);

      // Get the output streams for each panel and wrap them
      // in a filter that will calibrate the stream data.
      fDataOut[i+NUM_HISTS] =
        new HistFilterOutputStream (
           fHistPanel[i+NUM_HISTS].getOutputStream (),
           fSlope[i],fPedestal[i]);
   }
   . . .
```

We see that the panel setup is similar for both rows except the output stream for the bottom row is wrapped with `HistFilterOutputStream`. In the sensor-generating section, the loop is the same as before (except over six rather than three histograms):

```
. . . loop to generate data for each sensor histogram . . .
try {
  // Send data to stream destination
  dataOut[j].write (ival);
}
catch (IOException ioe) {}
. . .
```

The polymorphic feature of our objects means that the `write()` method for the particular subclass of `OutputStream` is invoked.

9.14 More Java I/O

In the following chapters we will frequently discuss Java I/O and give a number of examples. In Chapter 11 we discuss transmission of images over the network and reading and writing images in disk files. The chapters of Part II involve many aspects of I/O with other computers over network connections. Part III also involves I/O with Chapter 23 discussing communications with external devices via serial ports and Chapter 24 examining communications with embedded Java devices.

9.15   Web Course materials

There are several other I/O classes that we don't have space here to discuss. In the Web Course Chapter 9: *Supplements* section we examine some of the other I/O techniques and classes including:

- more about serialization
- `RandomAccessFile`
- `PipedInputStream`, `PipedOutputStream`
- Zip and Gzip files

Also, with Java 1.4 came a new set of I/O classes in the `java.nio.*` packages. These involve the concept of I/O *channels*. We introduce channels and other features of `java.nio` in the *Supplements* section.

As discussed in Section 9.13, the Web Course Chapter 9: *Tech* section demonstrates a number of I/O techniques using histograms. The *Physics* section provides an extensive demonstration of an experimental simulation with Java.

References

[1] *Lesson: I/O: Reading and Writing – The Java Tutorial* Sun Microsystems, `http://java.sun.com/docs/books/tutorial/essential/io`.

[2] Elliotte R. Harold, *Java I/O*, O'Reilly, 1999.

[3] `java.util.Formatter` class specification, `http://java.sun.com/j2se/1.5.0/docs/api/java/util/Formatter.html`.

[4] `java.util.Scanner` class specification, `http://java.sun.com/j2se/1.5.0/docs/api/java/util/Scanner.html`.

[5] *Regular Expression – The Java Tutorial*, Sun Microsystems, `http://java.sun.com/docs/books/tutorial/extra/regex/`.

[6] Norbert Lindenberg and Masayoshi Okutsu, *Supplementary Characters in the Java Platform*, Sun Microsystems, Inc., May 2004, `http://java.sun.com/eveloper/technicalArticles/Intl/Supplementary/`.

Chapter 10
Java utilities

10.1 Introduction

Any language should come with a well-stocked toolbox of utilities to ensure programming efficiency and convenience. The Java core language, in fact, includes a large package named, quite sensibly, `java.util` that holds classes to handle arrays, hash tables, time keeping, and other common tasks. In this chapter, we focus mostly on this package but also discuss some handy tools in other packages such as `String` and `StringBuffer` in `java.lang` and `StringBuilder` and enum, that were added to `java.lang` for J2SE 5.0 [1,2]. We also discuss the Collections Framework and the new Generics feature of J2SE 5.0.

We also look at classes and techniques for handling numbers at two extremes. For bits we have the `java.util.BitSet` class. For arbitrarily large numbers and for those that with very long decimal fractions we have the `java.math.BigInteger` and `java.math.BigDecimal` classes.

We can only briefly describe here the general workings of these classes, many of which contain a great number of methods. See the class descriptions in the Java 2 API Specifications for the full details.

10.2 The `java.util` Package

The package `java.util` has been part of the core language since Java 1.0. However, several classes have been added in subsequent versions. As the name suggests, the classes in this package serve a number of useful utility purposes, and we discuss several of them in this chapter. These include:

- `Vector` – unlike an array, `Vector` objects contain a list of objects that can grow or shrink
- `Enumeration`, `Iterator` – convenience classes for cycling one item at a time through a list
- `Hashtable`, `Properties` and `HashMap` – associative arrays that hold key/value pairs
- `Preferences` – a set of classes to maintain user preference settings in a program
- `StringTokenizer` – search and parse strings
- `Date`, `Calendar` – tools for handling dates and time values

- `Arrays` – includes tools for handling arrays such as fast filling, searching and sorting
- `BitSet` – manipulation of bits

Note that in Chapter 8 we already examined the `Timer` and `TimerTask` classes, also from `java.util`. In Chapter 4 we discussed the `java.util.Random` class for generating random numbers.

10.3 `Vector` and `Enumeration`

An instance of the `Vector` class provides a list of objects. The elements of a `Vector` are references to `Object` types, which, of course, include all objects in Java. A `Vector` differs considerably from an `Object` array in that it can grow and shrink whereas the number of elements in an array cannot be changed from the array size initially created.

For example, we can create a `Vector` and add different objects to it and then remove an element from the middle:

```
Vector list = new Vector ();
list.addElement ("A string");
list.addElement ("Another string");
list.addElement ("Yet another string ");
list.addElement (new Date ());
list.addElement (new Date ());
list.removeElementAt (3);
  . . .
```

There are several methods that provide access to the entries in a `Vector`. These include:

```
Object get (int index)
Object firstElement ()
Object lastElement ()
Object [] toArray ()
```

The `get()` method returns the element at the specified index. The next two methods listed return the first and last elements in the `Vector`. The last method shown dynamically builds and returns an array of references to the objects in the `Vector`.

Note that these methods return as `Object` type references. To treat the reference as any type other than `Object`, you must cast it to its proper class or superclass. For example, if you know that you've put a `String` object into the `Vector`, then you first retrieve the `Object` and then cast it to a `String`:

```
String str = (String) list.firstElement ();
```

If you cast the returned object to a class that it does not belong to, then a runtime `ClassCastException` occurs:

```
// Error: the object returned is of the Date type, not a
// String object.
String date = (String) list.lastElement ();
```

You can use the operator `instanceof` to query the object for what type of object it is:

```
Object o = list.lastElement ();
if (o instanceof String)
   String date = (String) o;
else if (o instanceof Date)
   Date aDate = (Date) o;
```

The `Vector` class has a number of other methods, such as a search for the index number of a particular object in the vector, as in

```
int i = list.indexOf ("Yet another string");
```

A `Vector` can also return an `Enumeration`. An `Enumeration` provides for a one-time scan through a list of `Objects`. (We note that an `Enumeration` has no relationship at all with the enumerated type of J2SE 5.0, which is explained in Section 10.9.)

```
Enumeration e = list.elements ();
while (e.hasMoreElements ()) {
   System.out.println (e.nextElement ().toString ());
}
```

After `hasMoreElements()` returns `false`, the `Enumeration` cannot be used again.

With the Java 1.2 Collections Framework (see Section 10.6) came a preferred alternative to `Enumeration` called `Iterator`. The `Iterator` class differs from `Enumeration` in that the iterator permits removing one or more elements from the `Vector` without damaging the `Iterator`. In addition, the iterator methods `hasNext()` and `next()` have more concise names than the corresponding methods `hasMoreElements()` and `nextElement()` in `Enumeration`. We explain more about the `Iterator` interface in Section 10.6.

10.4 Hashtable, Properties, and HashMap

The `Hashtable` class provides associative arrays with key-value pairings. Presentation of the key retrieves the value. The keys and values must be objects, not primitive type values. If you want to include numerical values, then use the corresponding wrapper classes such as `Integer` for an `int` value (or rely on the autoboxing support in J2SE 5.0 as explained in Chapter 3).

The methods `put (Object key, Object value)`, `get (Object key)`, and `remove (Object key)`, respectively, enter a new key/value pair into the table, retrieve a value for a given key, and remove a key/value pair.

For example,

```
Hashtable mass_table = new Hashtable ();
mass_table.put ("photon", new Double (0.0);
mass_table.put ("electron", new Double ("9.10938188E-28");
mass_table.put ("proton", new Double (1.67262158E—24)");
Float mp = mass_table.get ("proton");
mass_table.remove ("photon");
```

The `Hashtable` can return enumerations for both the keys and the values. For example,

```
for (Enumeration e = dates.keys (); e.hasMoreElements ();) {
   System.out.println (e.nextElement ().toString ());
}
```

You can test for the presence of a particular key as follows:

```
if (dates.containsKey ("Thanksgiving")) {. . .}
```

The `Hashtable` class relies on the `hashcode()` method derived from `Object`. This hash code is a unique numerical value that is the same for "equal" objects and different for "unequal" objects. Just what "equal" and "unequal" mean is determined by the `Object.equals()` method. Unless the `equals()` method is overridden by subclasses, Java requires that equality of object references means that the references refer, in fact, to the very same object – i.e. that the references are equal to one another.

Whenever you create a custom class, the `hashcode()` and `equals()` methods can and usually should be overridden. Otherwise, your custom object uses the superclass `hashcode()` and `equals()` methods from `Object`. If you want to allow objects with equivalent "contents" to be considered as equivalent, then you must override `equals()` to do a "deep" comparison since the `Object.equals()` superclass method only compares the references. When you override `equals()` you should almost always override `hashCode()` too in order to ensure that equal objects (as defined by the `equals()` method returning `true`) have equal hash codes, as is expected by hash tables.

The `Properties` class extends `Hashtable`. Instances of `Properties` typically represent persistent values such as system settings, and the keys, which can be any object type in `Hashtable`, are generally expected to be `String` objects. We discuss in Chapter 23 the method `System.getProperties()` that returns the properties settings for the host platform on which a program is running.

You can create your own properties tables to maintain, for example, the configuration settings for a large complex program with many parameters and options. The tables can be saved to or loaded from a file. However, with J2SE 1.4 came the more elaborate Preferences API for this task, which we discuss in the following section.

The `HashMap` class came with the Collections Framework in Java 1.2. It provides essentially the same capabilities as `Hashtable` except that it can store `null` references and its methods are not synchronized. When synchronization is not needed, then the `HashMap` is to be preferred over `Hashtable` since the former performs faster. See Section 10.6 for more information about Collections.

10.5 Preferences

The Preferences API is one of those small features in the Java 2 class library that is easy to ignore as unimportant but yet proves to be enormously helpful when its services are needed. Preferences are unlikely to be useful for custom scientific programs that you write purely for your own purposes, since you will probably just hard-code all your preferences into the code itself. However, if an application has more than one user, you can be sure that some of the users will wish to modify the default choices for such things as font style, size, and color or window sizes. An application might provide a "Font" menu in which the font style and size is controlled and so go a long way toward providing some of the control that users desire. But your users will be disappointed to find that the application reverts to your hard-coded choices each time the application is restarted.

One solution to this problem is to use a configuration file that a user edits to configure an application. To use such a configuration file, however, introduces several difficulties. First, the user must manually edit the file and this means that he must learn the format of the file, get all the spelling exactly correct, know the proper place to store the file, etc. Second, the application must find and read the configuration file. Devising a platform-independent scheme to specify where a configuration file is located in a file system is not a problem with an obvious or easy solution, though Java's built-in system properties `user.home` and `user.name` offer some help. It is also not trivial to invent a format for a configuration file, or to write the code to read and parse that format, or to document the chosen format so that users will know how to create and edit the file. And you must also deal with configuration preferences of multiple users of the same application on the same machine. So storing a single configuration file is not sufficient since multiple users will overwrite each other's preferences. A valuable technique used in the past to handle the formatting, reading, and writing of configuration information is the `java.util.Properties` class, mentioned in the previous section, and its `load()` and `store()` methods. The `Properties` class has been a part of Java since version 1.0. Still, the problem of knowing where to store the configuration file remains.

The Preferences API is in the `java.util.prefs` package and has been a part of Java since version 1.4. Amazingly, it provides the tools to easily store and retrieve user preferences that persist across application invocations (i.e. the preferences are "remembered" from one run of an application to the next). It automatically maintains separate preference lists for multiple users, and transparently handles storing the preferences information (the "configuration file," if you wish) in a way that the programmer need never worry about. The actual place that the preferences data is stored is platform *dependent*, but the Preferences API transparently handles access to the preferences information in a platform-*independent* manner. Put another way, source code that uses the Preferences API behaves the same way on any platform and hides the fact that the preferences data might be stored differently on different platforms. The API even includes a platform-independent import/export facility so that preferences can be backed up or moved from one machine to another.

10.5.1 Easy to use

The Preferences API is very easy to use, despite what you might think after a first brush with the online API documentation [3]. The most important class in `java.util.prefs` is the `Preferences` class. Alas, the online documentation is not written in a tutorial manner. That documentation says that the `Preferences` class is "a node in a hierarchical collection of preference data" and continues later with "there are two separate trees of preference nodes, one for user preferences and one for system preferences." While nodes and trees may not sound simple or easy to use, they really are, and we explain how to use the Preferences API through a simple example.

One first obtains a `Preferences` object by using the static method `userNodeForPackage()`. This method requires a `Class` object as its only parameter. The system then determines the package in which the specified class resides and returns the `Preferences` object to be used to access user preferences information for that package. Since all applications generally have their own unique package names, preferences based on package names do not conflict with other packages. The `userNodeForPackage()` method returns a different `Preferences` object for each user. In that way, multiple users of the same application on the same machine do not conflict with each other. That is, different users, as distinguished by different `user.name` system property values, have separate storage locations for their preferences data.

The description so far may not sound simple and easy to use yet, but the API really is simple, as the following code snippets illustrate. So far, we have described one line of code to obtain the `Preferences` object:

```
Preferences prefs = Preferences.userNodeForPackage
  (getClass ());
```

To store a preferences item takes one more line of code – the `prefs.put()` method. To retrieve a stored item at a later time, say the next time the application is run, takes just one line of code – the `prefs.get()` method.

Preferences information is stored as key/value pairs by the `Preferences` object, similar to the way system properties are stored, so the call to the `put()` method must provide both the name of the key under which the item is to be stored and its value, both as `String` objects. An example is

```
prefs.put ("color", "red");
```

Here the key "`color`" is stored with the value "`red`". You choose the key names to be whatever makes sense. Then, the next time the same Java application is run by the same user, the previously stored value can be retrieved with

```
String preferred_color = prefs.get ("color",
"some-default-value");
```

All of the preferences methods that retrieve values require a second parameter that provides a default value in case nothing is found under the named key. In that way, the application can continue running with default values, although perhaps with slightly reduced functionality.

In addition to the general `put()` and `get()` methods described above, there are convenience methods to store and retrieve `int` values:

```
prefs.putInt ("width", 500);
int preferred_width = prefs.get ("width", 700);
```

Here, the integer value 500 is stored under the key name `width`. Later the value stored is retrieved and loaded into the `int` variable `preferred_width`, using a default value of 700. There are similar convenience methods for `boolean` (`putBoolean()` and `getBoolean()`), `long` (`putLong()` and `getLong()`), `float` (`putFloat()` and `getFloat()`), and `double` (`putDouble()` and `getDouble()`) values. There are even convenience methods to put and get a byte array. Another useful method is `clear()`, which removes all key/value pairs in the preferences node.

10.5.2 Working examples

The Web Course Chapter 10 includes an example of the Preferences API usage in a simple command-line Java application. The `PrefsDemo` application stores values under the key name "`PrefsValue`". The simple application is set up to accept either "put", "get", or "clear" on the command line. These command line parameters may put the next value supplied on the command line into the preferences node, or they may get and echo a previously stored value or clear all key/value pairs. Because we always provide a default value when attempting to retrieve stored preferences, as is required, any attempt to retrieve a value before one is stored results in a suitable default value being returned. You may download and compile the `PrefsDemo` application and play with the "put", "get", and "clear" parameters to get a feel for how the Preferences API works. In addition to the "`PrefsValue`" key name, the demo also uses the `putInt()` and `getInt()` convenience methods to keep track of the total number of "put" operations that have been performed. This counter is incremented by one each time a "put" is performed and, more importantly, the value is maintained across application invocations, even across recompilations and reboots! If the first command line parameter is "clear", then all key/value pairs are removed from the preferences node, effectively resetting the counter back to 0.

10.5.3   Exporting and importing preferences

Okay, we've shown that it's easy to store and retrieve preferences. Doing so, and having it work with so little effort, is almost magical – as long as your application always runs on the same machine. But what about moving to another machine? How can we not lose the preferences already stored on the first machine? The answer is to use the export and import facilities of the Preferences API. The `exportNode()` method creates an XML document (see a discussion of XML in Chapter 21) on the specified `OutputStream` that can then be transferred and imported to same application running on another machine. Because of the cross-platform nature of XML, the preferences XML file can even be moved to the same Java application running on a completely different platform.

If the last parameter to the `javatech.prefs.PrefsDemo` application is "export", then the app uses `exportNode()` to output the preferences XML file onto `System.out`. In this way, you can examine the tree and node structure of the preferences tree.

Use of the `importPreferences()` method to read in an XML preferences file is not shown in the demo application, but its use is straightforward after referring to the online documentation.

10.5.4   System preferences

So far we've talked about user-level preferences – i.e. preferences stored for a particular user. The Preferences API also supports system-level preferences that apply to all users. To reach the system-wide portion of the preferences storage system, use

```
Preferences prefs = Preferences.systemNodeForPackage
   (getClass ());
```

The typical situation that system-level preferences would be used is to store system-wide defaults when an application is first installed.

10.5.5   Other services

Our brief discussion above describes most of the features of the Preferences API, including the most typical usage. There are a few other services provided by the API, such as *listeners* that listen for preference value changes. These additional features are less likely to be used, especially in a scientific application. You can find good documentation of these features in the online Java 2 API specifications for the `java.util.prefs` package.

10.5.6   Where is the preferences data really stored?

The actual storage of preferences information is implementation dependent. We programmers don't need to know where the data is stored, as long as the Preferences API always works the same on all platforms, which it does. The important thing to know is that the data really is stored persistently *somewhere*.

Nevertheless, the curious might want to know where the data is *really* stored. In practice, the Sun J2SE implementation on Windows platforms utilizes the Windows Registry as can be verified by examining the registry. On Solaris and Linux, the user node is normally stored in a hidden file in the user's home directory. Other implementations might use directory servers or SQL databases. It really is implementation dependent. And it really is unimportant.

10.6   The Collections Framework

The Collections Framework was first added in Java 1.2. It includes the original `Vector`, `Hashtable`, and `Properties` classes from Java 1.0 and also adds several important classes designed to make handling of "collections" of objects much easier. We referred to these collections as "object containers" in Chapter 1 (not to be confused with graphics containers in Chapters 6 and 7). We do not have space to devote to a full discussion of all the interfaces and classes in the Collections Framework, but we do describe some of the basic functionality. To quote from the online documentation [4],

> The collections framework is a unified architecture for representing and manipulating collections, allowing them to be manipulated independently of the details of their representation. It reduces programming effort while increasing performance. It allows for interoperability among unrelated APIs, reduces effort in designing and learning new APIs, and fosters software reuse.

How does it do all that? Some of the increased performance of the Collections Framework comes about because of the use of non-synchronized versions of some of the original object container classes. For example, we've already seen the non-synchronized `HashMap` class that generally replaces the original `Hashtable` class when thread safety is not an issue. There is also a non-synchronized replacement for `Vector` called `ArrayList`.

In addition, the Collections Framework includes optimized high-performance implementations of several useful and important algorithms and data structures – sorting algorithms for instance. Since these are provided by the framework, you don't have to write them yourself.

10.6.1　Collections Framework organization

The Collections Framework adds interfaces for several new container types. The root interface is `Collection`, which represents a group of objects. There are three main sub-interfaces – `Set`, `List`, and `SortedSet`. A `Set` is a collection that has no duplicate elements. The order of appearance of elements in the `Set` is unspecified and may change. A `List` is an ordered collection such that each element in the list has a distinct place in the list. An element in a list can be retrieved by its integer index, somewhat like an array. Most `Lists` permit duplicate elements. A `SortedSet` is a `Set` stored in such a way that when you iterate through the `SortedSet` with an `Iterator` (see next section), the elements are returned in ascending "order."

These are interfaces, not concrete classes. The Collections Framework has several concrete implementations. For example, `Stack`, `LinkedList`, and `ArrayList` all implement `List`. A concrete `SortedSet` is `TreeSet`, and a concrete `Set` is `HashSet`. There are several other concrete implementations as well for specialized purposes. We illustrate only a few of these in this book.

There are two other base interfaces in the Collections Framework – `Map` and `SortedMap`. All maps use key/value pairs, as in the `Hashtable` seen above. Other concrete implementations of `Map` are the `Properties` and `HashMap` classes seen earlier. A concrete version of `SortedMap` is the `TreeMap`.

J2SE 5.0 adds the `Queue` interface to the Collections Framework, and the `LinkedList` class has been retrofitted to implement `Queue`. That means you can use a `LinkedList` as a plain `List` or as a `Queue`. Queues are useful for when you need first-in, first-out behavior, though some queues provide different orderings. All `Collections` support `add()` and `remove()` methods to insert and remove elements. The `Queue` interface adds the preferred `offer()` and `poll()` methods. They are functionally equivalent to `add()` and `remove()` except for their behavior in exceptional situations. While `add()` fails by throwing an unchecked exception if the queue is full, `offer()` returns `false`. Similarly, remove() fails with an exception if the queue is empty while poll() returns null.

10.6.2 Using `java.util.Iterator`

We mentioned above that the `Iterator` interface is preferred over the old `Enumeration` interface. The original object containers from Java 1.0 days used the `Enumeration` interface to provide a way to loop over the elements in the container. All the old container classes were retro-fitted with the Java 1.2 release to support the `Iterator` interface as well. The new container classes added in JDK 1.2 along with the Collections Framework support only the `Iterator` interface. As mentioned, `Iterator` is preferable to `Enumeration` because `Iterator` allows the caller to remove elements from the underlying collection during an iteration without corrupting the `Iterator`.

The syntax for iterating over the elements in a `Vector` is as follows:

```
Vector vec = new Vector;
// Populate it . . . Then later, iterate over its elements
Iterator it = vec.iterator ();
while (it.hasNext ()) {
   Object o = it.next ();
}
```

Here, since we haven't been specific about what kinds of objects were put into the `Vector`, we show retrieving the elements as plain `java.lang.Object` types.

Next we illustrate another popular iteration style, this time using an `ArrayList`, which is preferred over `Vector` in a thread-safe situation:

```
ArrayList a_list = new ArrayList ();
. . .
for (Iterator it = a_list.iterator (); it.hasNext ();) {
   Object o = it.next ();
}
```

Again, we retrieve plain `java.lang.Object` types from the `Iterator`. All these container objects, `Vector`, `ArrayList`, and all the others, accept input of any kind of object. They can do this because the `add()` method receives `java.lang.Object` as the parameter type, and since `java.lang.Object` is the superclass of all other object types, any kind of object can be added. But the containers don't know what kinds of objects are being stored in them (see, however, the generics feature of J2SE in the next section). Therefore, when retrieving an object from one of the containers, it can only be returned as a `java.lang.Object` type. In most cases, you need to cast the `Object` received into the specific object type you desire. You, as the programmer, should know what kind of objects you store into a container, so you can do the cast correctly.

With 5.0, the collection object "knows" what type it holds, so it returns that type instead of the `Object` type.

In the pre-5.0 days, iterating through a list of strings and printing each out required code like this:

```
List list_of_strings = new ArrayList ();
. . .
for (Iterator it = list_of_strings.iterator (); it.hasNext
();) {
  String s = (String) it.next ();
  Sytem.out.println (s);
}
```

But when `list_of_strings` is a `List<String>` type, this simplifies to:

```
for (Iterator it = list_of_strings.iterator (); it.hasNext
();) {
  String s = it.next ();
  Sytem.out.println (s);
}
```

No more explicit cast! With autoboxing and unboxing, we can insert and retrieve primitive types without explicitly using the wrapper objects:

```
List<Integer> list_of_ints = new ArrayList<Integer> ();
. . .
for (Iterator it = list_of_ints.iterator (); it.hasNext
();) {
  int i = it.next ();
    . . .
}
```

We should warn you of a nuisance with the J2SE 5.0 compiler. The use of these special type-safe containers removes a significant source of errors. For backward compatibility, all the old containers are still available. So any pre-5.0 code continues to work exactly the same in 5.0 and beyond. However, because the old container types are *potentially* unsafe, the 5.0 *javac* compiler now issues warnings about possible unsafe usage whenever it encounters one of the old container types. These are just warnings and can be ignored if you are sure that the code is safe. The best way to get rid of the warnings is to switch to the use of the generic types where appropriate.

Of course, sometimes the old non-type-safe containers are needed, such as when you really *want* to insert a mix of object types. In this case, you can just ignore the warnings. Alternatively, the metadata system discussed in Chapters 1 and 4 provides an `@SuppressWarnings` annotation to explicitly suppress such warnings.

10.8 Concurrency utilities in J2SE 5.0

As mentioned in Chapter 8 on threads in Java, release 5.0 adds numerous enhancements to the threading control and concurrency features of Java. Some of the enhancements are advanced features beyond the scope of this book, but we explain some of the simpler new features here.

10.8.1 The Executor class

The most important new feature for the casual developer of multithreaded applications is the new `Executor` framework. A `java.util.concurrent.Executor` is an object that executes submitted `Runnable` tasks. In that regard, it is similar to calling

```
new Thread (aRunnable).start ();
```

For a single new thread, there is perhaps not much reason to use an `Executor`. However, most multithreaded applications involve several threads. Threads need stack and heap space, and, depending on the platform, thread creation can be expensive. In addition, cancellation and shutdown of threads can be difficult, as seen in Chapter 8, and a well-designed and implemented thread pooling scheme is not simple. The new `Executor` framework solves all those problems in a way that decouples task submission from the mechanics of how each task will be run, including details of thread use, scheduling, etc.

An `Executor` can and should be used instead of explicitly creating threads. For example, rather than creating a new thread and starting it as above, you can use:

```
Executor executor = some Executor factory method;
exector.execute (aRunnable);
```

Notice that our `Executor` object was returned by an `Executor` factory. (As we discuss in Chapter 16, a "factory" is a standard name for a method that is used to obtain an instance of a class or subclass rather than making it with a constructor.) There are several static `Executor` factory methods available in the `java.util.concurrent.Executors` class. If you have several `Runnable` tasks to carry out, but do not have specific requirements about the order in which they occur, as is commonly the case, then a simple thread pool arrangement is needed:

```
Executor executor = Executors.newFixedThreadPool (5);
executor.execute (new RunnableTask1 ());
executor.execute (new RunnableTask2 ());
executor.execute (new RunnableTask3 ());
  . . .
```

Here the tasks run and complete under the control of the Executor, which reuses threads from the thead pool as needed without incurring the overhead of always creating new threads. There are several more Executor factories in the Executors class for other needs beyond the scope of this book. Refer to the J2SE 5.0 API docs for complete information [1–3].

10.8.2   The Callable interface

The new java.util.concurrent.Callable interface is much like Runnable but overcomes two drawbacks with Runnable. The run() method in Runnable cannot return a result (it must be declared to return void) and cannot throw a checked exception. If you try to throw an exception in a run() method, the javac compiler insists that you use a throws clause in the method signature. But if you then declare the run() method to throw an exception, javac tells you that the overridden run() method does not throw any exceptions. That is, the superclass run() method, or in this case the method defined in the Runnable interface, does not throw any exceptions, so any overriding methods cannot throw exceptions either.

If you need a result from a Runnable task, you have to provide some external means of getting that result. A common technique is to set an instance variable in the Runnable object and provide a method to retrieve that value. For example,

```java
public MyRunnable implements Runnable
{
  private int fResult = 0;
  public void run () {
    . . .
    fResult = 1;
  }
  public int getResult () {return fResult;}
} // class MyRunnable
```

The Callable interface solves these problems. Instead of a run() method the Callable interface defines a single call() method that takes no parameters but is allowed to throw an exception. A simple example is

```java
import java.util.concurrent.*;
public class MyCallable implements Callable
{
  public Integer call () throws java.io.IOException {
    return 1;
  }
} // MyCallable
```

This `call()` method returns an `Integer`. (Note that we have conveniently used the autoboxing support in J2SE 5.0 to have the `int` literal 1 automatically boxed into an `Integer` return value.)

Getting the return value from a `Callable` depends upon the new generics feature:

```
FutureTask<Integer> task = new FutureTask<Integer> (new
MyCallable ());
ExecutorService es = Executors.newSingleThreadExecutor ();
es.submit (task);
try {
   int result = task.get ();
   System.out.println ("result from task.get () = " + result);
}
catch (Exception e) {
   System.err.println (e);
}
es.shutdown ();
```

Here, we use the `FutureTask` class that supports an `Integer` return value. Then the task is submitted using the `ExecutorService submit()` method, and the result is obtained from the `FutureTask get()` method, again using auto-unboxing to convert the `Integer` to an `int`. See the API documentation for more information on `ExecutorService`, and `FutureTask`.

10.8.3 Other concurrency enhancements

Other enhancements in the `java.util.concurrent` package not discussed here include advanced synchronization techniques, atomic types, and new high-performance thread-safe collections `ConcurrentHashMap`, `CopyOn-WriteArrayList`, and `CopyOnWriteArraySet`. See the API documentation for more information on these new collections and other features.

10.9 Enumerated types in J2SE 5.0

Until Release 5.0, Java did not have a facility like the `enum` of C and C++. In those languages, `enum` is used to create a set of named integer values to use as constants. This helps prevent accidentally using an illegal value where a group of predefined constant values, and nothing else, is expected. We've seen a related feature in `BorderLayout`'s constants NORTH, SOUTH, CENTER, etc (see Chapter 6). Those constants are just `final static Strings`, and it is perfectly permissible, though not advisable, to pass in literal `Strings` when using a `BorderLayout`. Doing so is not advisable because a misspelled string literal ("north" instead of "North", for instance) is not interpreted as you hoped.

What if we could insist at compile time that only a certain set of constant values be accepted? The new enumerated type feature of J2SE 5.0 does just that and much more. Before explaining enumerated types, let's see how we might use constants in J2SE 1.4 and below. Consider a class that accepts one of the four seasons and returns the average temperature of that season. We might implement such as class as follows:

```java
public class FourSeasons
{
   public static final int SPRING = 1;
   public static final int FALL = 2;
   public static final int SUMMER = 3;
   public static final int WINTER = 4;

   public float getAverageTemp (int season) {
      switch (season) {
        case SPRING:
           return calculateSpringAverageTemp ();
        case FALL:
           return calculateFallAverageTemp ();
      } . . .
   }
} // class FourSeasons
```

A user of this class could call it like this:

```java
FourSeasons s4 = new FourSeasons ();
float average_temp = s4.getAverageTemp (FourSeasons.SPRING);
```

But there is no way to prevent a user from calling

```java
s4.getAverageTemp (5);
```

resulting in unpredicatable behavior.

Using enums we can catch such errors at compile time. For this example, we can define

```java
public enum Season {SPRING, SUMMER, FALL, WINTER}
```

and then implement a method

```java
public float getAverageTemp (Season s) {
   . . .
}
```

Here the method parameter is the enum type Season, and the compiler catches any attempt to call getAverageTemp() with any parameter other than one of the constant names defined in the enum declaration. Note that the enum is like a class definition. There is no need for a semicolon at the end of the closing brace, though including one is not an error.

Enumerated types can do a lot more than replace constants. One nice feature is the ability to obtain an array of all the enumerated type values within an enum:

```
Season[] seasons = Season.values ();
```

For more about the enumerated type, we refer you to the online API documentation [3] and the books referenced in Chapter 1 that cover the new features in J2SE 5.0 [1,2].

10.10 The Arrays class

The class `java.util.Arrays` provides a number of static methods for filling, searching, sorting, and comparing arrays. For each method name, there are several overloaded versions that differ according to the type of the array or array arguments. We examine several of these in more detail here.

10.10.1 Arrays.equals()

Let's consider the `equals()` method (actually *methods*, since there are several overloaded versions). There is one for each primitive type plus one more for `Object` arrays:

```
boolean equals (type[] array1, type[] array2)
```

These compare two input arrays of the same type for equality. The methods return true if each array holds the same number of elements and each element in `array1` equals the value in the corresponding element in `array2`. For the case of `Object` arrays, the references in the two corresponding elements must either point to the same object or both equal `null`.

10.10.2 Arrays.fill()

There are two forms of the `fill()` methods:

```
void fill (type[] array, type value)
void fill (type[] array, int fromIndex, int toIndex, type value)
```

The first version of the overloaded `fill` methods, one for each primitive type and for `Object`, fill all elements in `array` of the same type with the `value` argument. The second version fills those elements between the given indices.

10.10.3 `Arrays.sort()` and `Arrays.binarySearch()`

The `sort` methods for the primitive types look like:

```
void sort (type[] array)
void sort (type[] array, int fromIndex, int toIndex)
```

There is a `sort` method for each primitive type except `boolean`. The elements in the array are sorted in ascending order using a variant of the QuickSort algorithm. (See the Java 2 API Specifications for the `java.util.Arrays` class for details.) The second version sorts those elements between the given indices.

For sorting `Object` types, there are two approaches. For the first approach there are two methods available:

```
void sort (Object[] array)
void sort (Object[] array, int fromIndex, int toIndex)
```

Here the objects are sorted into ascending order using the "natural ordering" of the elements. The array elements must implement the `java.lang.Comparable` interface and provide the method

```
int compareTo (Object)
```

This method defines the "natural ordering" such that comparing object x to y results in a negative number if x is lower than y (on a scale that makes sense for the class), equal to 0 if the objects are identical, and a positive number if x is greater than y. Your `compareTo()` method must be written to return a negative, zero, or positive integer according to the ordering rules that make sense for the nature of the objects that the class describes.

For example,

```
MyClass my_class1 = someMethodThatBuildsAMyClassInstance ();
MyClass my_class2 =
someMethodThatBuildsADifferentMyClassInstance ();

// Now compare the two instances of MyClass
if (my_class1.compareTo (my_class2) < 0)
  System.out.println ("my_class1 is 'less than' my_class2");
else if (my_class1.compareTo (my_class2) == 0)
  System.out.println ("my_class1 is 'equal to' my_class2");
else
  System.out.println (
    "my_class1 is 'greater than' my_class2");
```

For the alternative approach to comparing `Object` arrays, you create an auxiliary class that implements the `java.util.Comparator` interface and that knows how to compare and order two objects of interest. This technique is particularly useful if you need to sort classes that you cannot re-write to implement the comparable interface. The `Comparator` class has two methods:

```
int compare (Object obj1, Object obj2)
boolean equals (Object obj1, Object obj2)
```

The `compare()` method should return a negative, zero, or positive integer, according to the ordering rules that you implement, if `obj1` is less than, equal to, or greater than `obj2`. Similarly, the `equals()` method compares the objects for equality according to the rules you implement. When using the `Comparator` technique, the two object array sorting methods are:

```
void sort (Object[] array, Comparator comp)
void sort (Object[] array, int fromIndex, int toIndex,
Comparator comp)
```

Another useful set of overloaded methods in the `Arrays` class is the set of binary searching methods:

```
int binarySearch (type[] array, type key)
int binarySearch (Object[] array, Object key,
Comparator comp)
```

There is an overloaded `binarySearch()` method for each primitive type except `boolean` and one more for `Object`. These methods search an array for the given key value. The array to be searched must be sorted first in ascending order, perhaps by using one of the `sort()` methods just discussed. The methods return the index to the position where the key is found, or if it is not found, to the insertion point. The insertion point is the place in the array where all the elements are less than the key and all the elements above it are greater than or equal to it. The methods return `array.length()` if all elements are less than the key.

For the case of `Object` arrays and the first type of `binarySearch()` method above, the elements must implement the `Comparable` interface. If you cannot change the classes to implement this interface, then you can provide a `Comparator` and use the second type of `binarySearch()` method shown above.

10.10.4 `Arrays.toString()`

One small, but extremely useful, addition to J2SE 5.0 is the static `Arrays.toString()` method. It conveniently returns a string representation of the contents of an array. Consider the now familiar `public static void main (String[] args)` method. Almost anyone who has used Java for any time has written a loop like the following to echo the input parameters:

```
for (int i=0; i < args.length; i++)
   System.out.println (args[i] + ", ");
```

Wouldn't it be nice to avoid writing that loop ever again? `Arrays.`
`toString()` provides just what you need. It accepts an array as a parameter and
returns a formatted string representation of the contents of the array. There are
overloaded versions for arrays of all the primitive types and a version for object
types. For the object types, each element's own `toString()` is used. Therefore
for `String[]` types, you get the contents of the string. Here is a simple example:

```
public class Demo {
  public static void main (String[] args) {
    System.out.println (Arrays.toString (args));
  }
} // class Demo
```

If you run this code with the following command-line parameters:

```
java Demo Arrays.toString is really cool
```

you are greeted with

```
[Arrays.toString, is, really, cool]
```

10.10.5 `Arrays.deepToString()` and `deepEquals()`

There is also a recursive `Arrays.deepToString()` that works as might be
expected for multidimensional arrays. The following code snippet:

```
int[][] a = new int[3][4];
a[0][0] = 1;
a[0][1] = 2;
a[0][2] = 3;
a[0][3] = 4;

a[1][0] = 5;
a[1][1] = 6;
a[1][2] = 7;
a[1][3] = 8;

a[2][0] = 9;
a[2][1] = 10;
a[2][2] = 11;
a[2][3] = 12;

System.out.println (Arrays.deepToString (a));
```

produces

```
[[1, 2, 3, 4], [5, 6, 7, 8], [9, 10, 11, 12]]
```

Finally, for comparing multidimensional arrays, there is the `deepEquals()` method. It returns a `boolean` and works recursively on nested arrays of any depth. Two array references are considered deeply equal if they contain the same number of elements and all corresponding pairs of elements in the two arrays are deeply equal. (If both array references are `null` they are also considered deeply equal.)

10.11   Tools for strings

We briefly discussed the `String` class in Chapter 2 and in Chapter 3. We also used strings frequently in the various demonstration programs. Here we look further at tools for dealing with strings including the many useful methods in the `String` class itself and in `StringBuffer`, `StringBuilder`, and `StringTokenizer`.

10.11.1   `String` class methods

In Chapter 3 we discussed the `valueOf()` methods in the `String` class for converting primitive type values to strings. The `String` class contains a large number of other useful methods. Here we briefly examine a sample of these methods.

10.11.1.1 `int length ()`
This method returns the number of characters in the string, as in

```
String str = "A string";
int x = str.length ();
```

This results in variable x holding the value 8.

10.11.1.2 `String trim ()`
Removes white space from the leading and trailing edges of the string:

```
String string = " 14 units ";
String str = string.trim ();
```

This results in the variable `str` referencing the string "14 units". As always, `String` objects are immutable. The `trim()` method always returns a new `String` object containing the trimmed version of the original `String`.

10.11.1.3 `int indexOf (int ch),int lastIndexOf (int ch)`
The `indexOf()` method returns the index, starting from 0, for the location of the given character in the string. (The `char` value is widened to `int`.) For example,

```
String string = "One fine day";
int x = string.indexOf ('f');
```

This results in a value of 4 in the variable x. If the string holds no such character, the method returns -1. To continue searching for more instances of the character, you can use the method

```
indexOf (int ch, int fromIndex)
```

This starts the search at the fromIndex location in the string and searches to the end of the string. The additional overloaded methods

```
indexOf (String str)
indexOf (String str, int fromIndex)
```

provide similar functions but search for a substring rather than just for a single character. Similarly, the methods

```
lastIndexOf (int ch)
lastIndexOf (int ch, int fromIndex)

lastIndexOf (String str)
lastIndexOf (String str, int fromIndex)
```

search "backwards" for characters and strings starting from the right end and moving from right to left. (The fromIndex second parameter still counts from the left, with the search continuing from that index position toward the beginning of the string.)

10.11.1.4 *boolean startsWith (String prefix),boolean endsWith (String str)*
These two methods test whether a string begins or ends with a particular substring. For example,

```
String [] str = {"Abe", "Arthur", "Bob"};
for (int i=0; i < str.length (); i++) {
   if (str[i].startsWith ("Ar")) doSomething ();
}
```

10.11.1.5 *String toLowerCase (),String toUpperCase ()*
The first method returns a new string with all the characters set to lower case while the second returns a string with the characters set to upper case:

```
String[] str = {"Abe", "Arthur", "Bob"};
for (int i=0; i < str.length (); i++) {
   if (str[i].toLowerCase ().startsWith ("ar")) doSomething ();
}
```

See the Java 2 Platform API Specifications to examine the other methods available with the String class [3].

10.11.2 `java.lang.StringBuffer`

String objects cannot be altered. Once they are created, they are *immutable* objects. Concatenating two strings together creates a whole new string object in memory:

```
String str = "This is ";
str = str + "a new string object";
```

The `str` variable now references a new `String` object that holds "This is a new string object". The `String` class maintains a pool of strings in memory. String literals are saved there and new strings are added as they are created. Extensive string manipulation with lots of new strings created with the `String` append operation can therefore result in lots of memory taken up by unneeded strings. Note, however, that if two string literals are the same, the second string reference will point to the string already in the pool rather than create a duplicate.

The class `java.lang.StringBuffer` provides a much more memory efficient tool for building strings. The class works as follows:

```
StringBuffer strb = new StringBuffer ("This is ");
strb.append ("a new string object");
System.out.println (strb.toString ());
```

You first create an instance of the `StringBuffer` class and then append new strings to it using the `append()` method. Internally, the class holds a large character array. Appending strings just involves filling places in the array with new `char` values. It does not need to create any new objects. If the array is already filled when you try to do an `append()`, then `StringBuffer` creates a larger array and copies the old characters into the new one. All this happens internally, completely transparently to you.

10.11.3 `java.lang.StringBuilder`

Java Version 5.0 added the `StringBuilder` class, which is a drop-in replacement for `StringBuffer` in cases where thread safety is not an issue. Because `StringBuilder` is not synchronized, it has better performance than `StringBuffer`. In general, you should use `StringBuilder` in preference over `StringBuffer`. In fact, the J2SE 5.0 `javac` compiler normally uses `StringBuilder` instead of `StringBuffer` whenever you perform string concatenation, as in

```
System.out.println ("The result is " + result);
```

All the methods available on `StringBuffer` are also available on `StringBuilder`, so it really is a drop-in replacement.

10.11.4 `StringTokenizer`

The `String` class provides methods for scanning a string for a particular character or substring. The class `java.util.StringTokenizer` allows you to break a string into substrings, or *tokens*, in one operation. (For J2SE 1.4, much of the value of `StringTokenizer` has been largely replaced with the `String.split()` method, which we discuss in the next section.) The tokens are separated by *delimiters*, which are the characters defined as separators of the tokens. The default delimiters are the white space characters such as spaces and line returns. Since `StringTokenizer` implements the `Enumeration` interface, the standard `Enumeration` methods `hasMoreElements()` and `nextElement()` step through the tokens. Since `nextElement()` returns an `Object` type that requires casting to a `String` type, `StringTokenizer` also includes the `nextToken()` method that returns a `String` type automatically. For naming consistency, there is also a `hasMoreTokens()` method.

For example, if a string contains a sentence, you can use `StringTokenizer` to provide a list of the words. For example,

```
String str = "This is a string object";
StringTokenizer st = new StringTokenizer (str);
while (st.hasMoreTokens ()) {
   System.out.println (st.nextToken ());
}
```

This results in a console output as follows:

```
This
is
a
string
object
```

An overloaded constructor allows you to specify the delimiters. For example,

```
String str = "A*bunch*of*stars";
StringTokenizer st = new StringTokenizer (str,"*");
```

This breaks the string into the tokens separated by the "*" character.

10.11.5 `String.split()`

J2SE 1.4 added the `split()` method to the `String` class to simplify the task of breaking a string into substrings. This method uses the concept of a "regular expression" to specify the delimiters. A regular expression is a remnant from the Unix *grep* tool ("grep" meaning "general regular expression parser"). A full discussion of regular expressions is beyond the scope of this book; see almost any introductory Unix text or the Java API documentation for the `java.util.regex.Pattern` class for complete documentation [5,6]. In its simplest form, searching for a regular expression consisting of a single character finds a match of that character. For example, the character 'x' is a match for the regular expression "x".

The `split()` method takes a parameter giving the regular expression to use as a delimiter and returns a `String` array containing the tokens so delimited. Using `split()`, the first example above becomes

```
String str = "This is a string object";

String[] words = str.split (" ");

for (int i=0; i < words.length; i++)

    System.out.println (words[i]);
```

To use "`*`" as a delimiter, simply specify "`*`" as the regular expression:

```
String str = "A*bunch*of*stars";

String[] starwords = str.split ("*");
```

For most string-splitting tasks, the `String.split()` method is much easier and more natural to use than the `StringTokenizer` class. However, `StringTokenizer` is still useful for some tasks. For example, an overloaded `StringTokenizer` constructor allows you to specify that the tokens to be returned include the delimiter characters themselves.

10.12   `Calendar`, `Date`, **and** `Time`

Java offers several classes for dealing with dates and times. Some are in the `java.util` package:

- `GregorianCalendar` (a subclass of the abstract `Calendar` class)
- `Date`

These two are in `java.text`:

- `DateFormat` and its subclass `SimpleDateFormat`

The `GregorianCalendar` class can be used to represent a specific date according to the Gregorian calendar (and the Julian calendar before that). Methods are provided to compare calendar objects such as whether one date came before or after another. The `Calendar` base class, which is abstract, holds static methods that give information such as the current date and time:

```
Calendar this_moment = Calendar.getInstance ();
```

The hierarchy implies that there could be other calendar subclasses, e.g. Chinese, but none has appeared in the core language as of version 5.0.

If you look at the Java 2 API Specifications for the `Date` class you will see that most of its methods are deprecated. As of version 1.1, `java.text.DateFormat` took over most of duties of the `Date` class, which is now essentially relegated to simply holding a date/time value. The class `DateFormat` can generate strings with many variations in the formats for dates and times. There are also a number of methods for dealing with time zone settings and internationalization of the strings.

The class `DateFormatPanel`, shown below, displays the current time. Each time the panel is repainted, it displays the current time. In Section 8.7 we discussed how to use timers to redraw this panel every second to create a digital clock.

```java
import javax.swing.*;
import java.awt.*;
import java.text.*;
import java.util.*;

/** This JPanel subclass uses the DateFormat class
  * to display the current time. **/
class DateFormatPanel extends JPanel {
  DateFormat fDateFormat;
  boolean fFirstPass = true;
  int fMsgX = 0, fMsgY = 0;
  Font fFont = new Font ("Serif", Font.BOLD, 24);

  /** Get the DateFormat object with the default time
    * style. **/
  DateFormatPanel () {
    fDateFormat =
      DateFormat.getTimeInstance (DateFormat.DEFAULT);
  } // ctor

  /** Draw the time string on the panel center. **/
  public void paintComponent (Graphics g) {
    super.paintComponent (g);
    // Get current date object
    Date now = new Date ();

    // Format the time string.
    String date_out = fDateFormat.format (now);

    // Use our choice for the font.
    g.setFont (fFont);

    // Do the size and placement calculations only for
```

```
      // the first pass (assumes the applet window never
      // resized).
      if (fFirstPass) {
        // Get measures needed to center the message
        FontMetrics fm = g.getFontMetrics ();

        // How many pixels wide is the string
        int msg_width = fm.stringWidth (date_out);

        // Use the string width to find the starting point
        fMsgX = getSize ().width/2 - msg_width/2;
        // How far above the baseline can the font go?
        int ascent = fm.getMaxAscent ();

        // How far below the baseline?
        int descent= fm.getMaxDescent ();

        // Use the vertical height of this font to find
        // the vertical starting coordinate
        fMsgY = getSize ().height/2 - descent/2 + ascent/2;
      }
      g.drawString (date_out, fMsgX, fMsgY);
      fFirstPass = false;
    } // paintComponent
  } // class DateFormatPanel
```

The `DateFormatPanel` class uses an instance of `DateFormat` obtained with the factory method `getTimeInstance (int style)`. The style of the date and time output is obtained with the `DateFormat` constants. The Java 2 API Specifications for this class list the style constants. Here we chose for the time format the `DateFormat.DEFAULT` style. What this actually means varies somewhat with the locale setting (see the Sun tutorial on internationalization of Java programs [7]). In the USA the default style results in a time string such as "5:30:33 PM".

The `DateFormat` subclass `SimpleDateFormat` offers a somewhat more explicit technique for setting the format. Date and time formats are chosen with a string pattern. The Java 2 API Specification for the `SimpleDateFormat` provides a table of symbols to use in the format patterns. For example, in the example below, the pattern

```
"EEE, d MMM yyyy HH:mm:ss Z"
```

results in a time format that goes as

```
Wed, 26 Mar 2003 15:34:35 −0500
```

Here the last number, −0500, indicates a time zone 5 hours earlier than GMT. The `JPanel` subclass `SimpleDateFormatPanel` shown below displays the date and time whenever the panel is repainted:

```java
import java.swing.*;
import java.awt.*;
import java.text.*;
import java.util.*;

public class SimpleDateFormatPanel extends JPanel
{
  SimpleDateFormat fSDateFormat;
  boolean fFirstPass = true;
  int fMsgX = 0, fMsgY = 0;
  Font fFont = new Font ("Serif", Font.BOLD, 24);

  /** Use SimpleDateFormat with the given date & time
    * style. **/
  SimpleDateFormatPanel () {
    fSDateFormat = new SimpleDateFormat (
                        "EEE, d MMM yyyy HH:mm:ss Z");
  }

  /** Paint the date and time at center of panel. **/
  public void paintComponent (Graphics g) {
    super.paintComponent (g);
    // Get current date
    Date now = new Date ();

    // And create a date/time string in the desired format
    String date_out = fSDateFormat.format (now);

    . . . rest of the code is the same as in
    DateFormatPanel . . .
```

Another time related tool is the static method `System.currentTime-Millis()`. It provides the current time as a `long` value containing the number of milliseconds since midnight, January 1, 1970 UTC. This is not very informative on its own but usually you look at the difference between two values. It is particularly useful for program timing and performance studies.

10.13 Arbitrary precision numbers

The 64 bits in the `long` type correspond to about 19 decimal digits. There are applications, however, that deal with numbers containing many more digits than provided by these types. For example, cryptography requires very large prime numbers. As many as 2000 bits are used for the public/private key encryption algorithms.

Floating-point values can represent large values but with limited precision. The `double` type contains a 53-bit signed significand, which translates to 15 to 17 digits decimal precision. As we discussed in Chapter 2, it is often the case that a finite number of digits in the binary base cannot represent a finite fraction in a decimal value (1/10 is a typical example). So there can be a loss of precision for a decimal value in floating-point format. Techniques to minimize and quantify the accumulation of such errors over repeated calculations are available from error analysis methods. A brute force approach, however, is simply to use extremely high precision number representations that avoid the error build up altogether. Financial calculations might need to use this approach, for example. Some mathematical exercises, such as calculating π to higher and higher precision, require indefinitely wide fractional values.

For applications like these that require very large integer values and extremely precise decimal values, the package `java.math` provides the `BigInteger` and `BigDecimal` classes. Instances of these classes hold values of arbitrary precision. The values internally are *immutable*. Like `String` objects, once created, they cannot change.

Since the object values are immutable, any operation on a `BigInteger` or `BigDecimal` value can only return a new instance of the class. This is something to consider when implementing algorithms with these classes that involve many operations. Unneeded values should be de-referenced so that the garbage collector can retrieve the memory space. Of course, operations with such classes are much slower than those with primitive types.

In the following two sections we briefly describe these two classes. See the Java 2 API Specifications for details about their methods [1]. See also the book by Mak for a more extensive discussion of these classes and for code that provides additional mathematical functions with them [8].

10.13.1 `BigInteger`

The `BigInteger` class contains methods that provide all the basic arithmetic operations such as add, subtract, multiply, and divide for indefinitely long integer values. For example,

```
BigInteger bi1 = new BigInteger
("11432095234905439534412323238479");
BigInteger bi2 = new BigInteger
("34548738754398454599999997876786578479");
BigInteger bi3 = bi2.add (bi1);    // = bi2 + bi1
BigInteger bi4 = bi2.divide (bi1); // = bi2 / bi1
```

Note that bi3 and bi4 are new BigInteger objects created by the add() and
divide() methods, respectively. There are four BigInteger instances in the
four lines of code above. If bi1 and bi2 are no longer needed, they could be
de-referenced as follows:

```
bi1 = bi2 = null;
```

With no more references to the BigInteger objects formerly known as bi1
and bi2, the garbage collector is free to reclaim that memory when needed.

Several other arithmetic methods are included in BigInteger as well, such
as abs(), negate(), and pow(). Plus there are methods to return values as
int, long, float, or double primitives. These are, of course, narrowing
conversions that lose information about the value if the BigInteger value is
too large to represent as the primitive type. See the API documentation for full
details.

Since the BigInteger instances are mostly used for encryption tasks, there
are several methods related to prime number generation and modulo arithmetic.
This includes

```
static BigInteger probablePrime (int numberBits,
    Random ran)
```

This method generates a number that has a high likelihood of being prime. The
parameters include the bit length of the prime number and a reference to an
instance of the Random class that the method uses to select candidates for pri-
mality. The method

```
boolean isProbablePrime (int certainity)
```

gives a confidence test for the value. The certainty parameter indicates the desired
level of confidence that the number really is prime. If the probability exceeds
$(1 - 1/2^{\text{certainty}})$, the method returns true.

The BigInteger class also contains methods to carry out various bitwise
operations, as in

```
BigInteger shiftLeft (int n)
BigInteger shiftRight (int n)
```

These return new BigInteger values shifted left or right by the number of
bits indicated. Other bitwise methods include: and, or, XOR, not, andNot,
testBit, setBit, clearBit, and flipBit.

We note that BigInteger implements Comparable, which makes arrays
of BigIntegers sortable with the java.util.Arrays methods described
in Section 10.10.3. The compareTo() method, required by Comparable, is
useful for comparing two BigInteger values outside of the sort() methods.

10.13.2 `BigDecimal`

`BigDecimal` contains internally an arbitrary precision integer – `unscaled-Value` – and a 32-bit scale value – `scale`. The value represented by a `BigDecimal` object then corresponds to

```
unscaledValue / 10^scale
```

So for the following instance of `BigDecimal`:

```
BigDecimal one = new BigDecimal (new BigInteger("1"),
Integer.MAX_VALUE);
```

the decimal point is $2^{31} - 1$ (i.e. 2 147 483 648) places to the left of the 1. You can obtain a new `BigDecimal` value with the scale increased by n places with the method

```
BigDecimal movePointLeft (int n)
```

Similarly, you can obtain a new value with the scale decreased by n places using

```
BigDecimal movePointRight (int n)
```

Note that in both cases the precision of the unscaled value remains unchanged. Only the decimal point has moved.

Other methods provide the scale value (as an `int`) and the unscaled value (as an instance of `BigInteger`). There are also methods that return a `float` and a `double` type value, with the precision truncated as necessary to fit the narrower fractional range of these floating-point types.

As with `BigInteger`, the `BigDecimal` class contains methods that provide all the basic arithmetic operations such as add, subtract, multiply, and divide. However, for the division, one of eight rounding options must be chosen. For example,

```
BigDecimal bd1 = new BigDecimal (
  new BigInteger ("11432095234905439534123238479", 12345);
BigDecimal bd2 = new BigDecimal
                ("3.45487387543984545999999997876786578479");
BigDecimal bd4
  = bd2.divide (bd1, BigDecimal.ROUND_ DOWN);
  // = bd2/bd1
```

Here the division rounds towards zero. Other rounding style constants defined in the `BigDecimal` class include `ROUND_UP`, `ROUND_FLOOR`, `ROUND_CEILING`, etc. See the API documentation for rounding details.

10.14   Bit handling

You may occasionally need to access and manipulate data at the bit level. For example, each bit in a set of data might represent the on/off state of a sensor or a pixel in a detector array. Representing the values as bits is much more memory efficient than assigning a full byte to each detector element if each element can be only in one of two states.

A more common application of bitwise operations deals with colors. We saw in Chapter 6 that in Java the RGB & alpha components (red, green, blue, and the transparency factor) for a color pixel are each assigned a byte value and the four bytes are packed in an `int` field. For image processing (see Chapter 11) and other graphics applications, you can use the bitwise operators to obtain these bytes from a color value, modify the component values, and then pack them back into an `int` value.

10.14.1   Bitwise operations

In Appendix 2 we display a table of the bitwise operators that act on the individual bits in integer type values. Four of the operators carry out Boolean operations on the bits. The ~x compliment operation flips each bit in x. The x & y, x | y, and x ^ y operations perform AND, OR, and XOR, respectively, between the corresponding bits in the x and y values. For the case where a bitwise operator involves two operands of different integer types, the wider type results.

The other three bit operators involve shifting of bits. The shift left operation, x << y, shifts x to the left by y bits. The high-order bits are lost while zeros fill the right bits. The signed shift right operation, x >> y, shifts x to the right by y bits. The low-order bits are lost while the sign bit value (0 for positive numbers, 1 for negative) fills in the left bits. The unsigned shift right, operation, x >>> y, shifts x to the right by y bits. The low-order bits are lost while zeros fill in the left bits regardless of the sign.

The following code shows how to place the four color component values into an `int` variable. We use the hexadecimal format for the literal constants as a compact way to specify byte values.

```
int[] aRGB = {0x56, 0x78, 0x9A, 0xBC};
int color_val = aRGB[3];
color_val = color_val | (aRGB[2] << 8);
color_val = color_val | (aRGB[1] << 16);
color_val = color_val | (aRBG[0] << 24);
```

This results in `color_val` holding the value: 56 78 9A BC (separating the byte values for clarity). Similarly, to obtain a particular byte from an `int` value, the code goes like:

```
int alpha val = (color_val >>> 24) & 0xFF;
int red_val   = (color_val >>> 16) & 0xFF;
int green_val = (color_val >>> 8)  & 0xFF;
int blue_val  = color_Val & 0xFF;
```

10.14.2 `java.util.BitSet`

A `BitSet` object represents a vector of bits whose size can grow as needed. The bits can be used, for example, to represent a set of Boolean values. This is more memory efficient than using a whole byte for each value as would be the case with an array of `boolean` primitive types. (A JVM implementation can in fact use bits to represent a `boolean` array but not all do.)

The `BitSet` methods might also be helpful if one is handling data in which individual bits represent information of interest. For example, each bit might represent the state of a relay in a large group of relays. Although internally the JVM might represent the `BitSet` with an array of `long` values, there is unfortunately no method in the `BitSet` class that converts an array of `long` values into a `BitSet` or vice versa.

The `BitSet` class provides methods to access a given bit in the array with an index value:

- `get (int index)`
- `set (int index)`
- `clear (int index)`
- `flip (int index)`

Two `BitSet` arrays can undergo Boolean operations:

- `and (BitSet bitset)`
- `or (BitSet set)`
- `xor (BitSet set)`
- `andNot (BitSet set)` – clears those bits in the `BitSet` object for which the corresponding bits are set in the parameter `bitset` object

The class includes a number of other methods such as `clone()` for making copies of a `BitSet`, `cardinality()` for finding the number of bits set to one, and `nextSetBit (int fromIndex)`, for finding the index of the next bit set to one at position `fromIndex` or higher.

10.14.3 More bit handling

What if data from another computer platform or an external device arrives as `int` or `long` type values but the bits actually represent `float` or `double` values? How would we change an integer type value to the corresponding floating-point type? Or conversely, what if we want to map the bits of a floating-point value into an integer type?

In Chapter 9 we used methods in the `DataInputStream` wrapper to read values from a byte array source as a particular primitive type. However, for a data set that is a mix of many types, it might be convenient to read the data into a single array of an integer type. You can then convert those elements of the array that represent `float` values with the `intBitsToFloat()` method in the `Float` class. There are also methods to map from `float` to `int`.

The `Float` class includes:

- `static int floatToIntBits (float x)` – Returns an `int` type whose bits match those of a float according to the IEEE 754 floating-point "single format" bit layout as described in Appendix 3.
- `static int floatToRawIntBits (float value)` – This is the same as `floatToInBits (float x)` except the NaN value can be other than the single IEEE 754 official value (`0x7fc00000`). As explained in Appendix 3, a value is NaN if all the bits in the exponent equal 1 and any of the bits in the significand equal 1.
- `static float intBitsToFloat (int x)` – Treats the bits in the x value as the bits in a `float` value and returns it as a `float` type. That is, this method converts the output of `floatToIntBits (float)` back to a `float` value.

The `Double` class has a corresponding set of methods for converting back and forth from `long` and `double` types.

The `Integer` wrapper also offers some bit handling methods:

- `static String toBinaryString (int i)` – Converts the value i to a string with 0 and 1 characters but no leading zeros.
- `static int parseInt (String s, int radix)` – Conversely, if s is a string representing a binary value, such as "110103," and radix is set to 2, then the method returns an `int` value equal to the binary value. (We previously used `parseInt (String s)`, which assumes a decimal value is represented by string.)

The `Long` wrapper has two methods with similar names for converting long values to binary strings and converting strings that represent binary values to long values.

As discussed in Section 10.13, the `BigInteger` class contains bitwise methods to test bits, shift bits, clear a particular bit, and so forth. However, remember that any change of a bit results in a new `BigInteger` object since instances of this class are immutable.

10.15   Other utilities

There are a number of other utility classes in the `java.util` package. The class `Stack`, for example, provides a last-in-first-out type vector with pop and push methods. We discussed the `Formatter` class in Chapter 5 that deals with formatting of numerical values into strings. In Chapter 9 we discussed `Scanner` for parsing input strings. These latter two classes were added to `java.util` in J2SE 5.0.

Internationalization tools include:

- `Locale` – specifies a particular geographical, political, or cultural region.
- `ResourceBundle` – holds locale specific information such as alternative language strings for buttons, menus, and other GUI text.
- `Currency` – provides symbols for strings representing currency values for the particular locale.

The sub-packages to `java.util` include:

- `java.util.jar` – package or tools for reading and writing the JAR (Java ARchive) file format, which is based on the Zip compression system. (see Section 5.6).
- `java.util.zip` – package provides various classes for packing and unpacking files with either the Zip or Gzip compression systems.
- `java.util.regex` – package with classes for matching character sequences against patterns specified by regular expressions.
- `java.util.logging` – provides the Java platform's core logging facilities that were added in version 1.4. The logging system is a large and complete system designed to systematically produce logging information that can be used by developers to debug an application under development as well as end users and field service engineers to support applications delivered to others. For single developers, the logging system is considerably more powerful than using `System.out.println()` repeatedly, but we do not have space to cover it in this book.

10.16   Web Course materials

The Chapter 10 Web Course: *Supplements* examines techniques for measuring and optimizing Java performance. It does this in the context of the JVM design and with regard to different JVM implementations.

The Web Course: *Tech* section provides demos for the arbitrary precision number classes. The *Physics* section continues with development of data analysis programs.

References

[1] Calvin Austin, *J2SE 1.5 in a Nutshell*, May 2005, `http://java.sun.com/developer/technicalArticles/releases/j2se15/`.
[2] Brett McLaughlin and David Flanagan, *Java 1.5 Tiger, A Developer's Notebook*, O'Reilly, 2004.
[3] Java 2 Platform, Standard Edition, API Specification, `http://java.sun.com/j2se/1.5/docs/api/`.

[4] *Collections Framework*, documentation at Sun Microsystems,
http://java.sun.com/j2se/1.5.0/docs/guide/collections/

[5] *Regular Expressions – The Java Tutorial*, Sun Microsystems,
http://java.sun.com/docs/books/tutorial/extra/regex/.

[6] java.util.regex.Pattern specification, http://java.sun.com/
j2se/1.5.0/docs/api/java/util/regex/Pattern.html.

[7] *Java Tutorial: Internationalizations*, Sun Microsystems,
http://java.sun.com/books/tutorial/i18n/index.html.

[8] Ronald Mak, *The Java Programmer's Guide to Numerical Computing*, Prentice Hall,
2003.

Chapter 11
Image handling and processing

11.1 Introduction

In Chapter 6 we presented the basic techniques for loading image files into applets and applications. We used instances of the `Image` class to hold images and invoked the `drawImage()` method in the `Graphics` class to display them. In this chapter we explore further the image handling and processing capabilities of Java.

In the first few sections we look in greater detail at the `Image` class and introduce its `BufferedImage` subclass, which offers many useful features. We discuss how to monitor the loading of images, how to scale image displays, how to create images, and how to save images to files.

We then switch to topics related to image processing. We show how to gain access to the pixels of an image, how to modify them, how to make an image from a pixel array, and how to use these techniques to create animations. We then discuss the standard filters provided with Java 2D and also give an example of a custom filter.

We give only a brief overview of the wide range of image tools available with Java. The classes mentioned here, for example, hold many overloaded constructors and methods that provide many options. See the Java 2 API Specifications for thorough descriptions of the classes. Also, see the book by Knudsen [1] and the other resources for in-depth discussions of images in Java [2,3].

11.2 The `Image` and `BufferedImage` classes

The `Image` class is abstract so you normally deal with instances of platform-specific subclasses obtained via methods such as `getImage()` in the `Applet` class. The `Image` class provides limited information about an image. The Java 2D API, which came with Java 1.2, introduced `java.awt.image.BufferedImage`, which is a non-abstract subclass of `Image` that provides much greater access to and control of image data.

In Chapter 6 we discussed colors in Java and the `java.awt.Color` class. You might guess that an image in Java consists internally of a two-dimensional array of `Color` objects each representing a pixel. This, however, is not practical

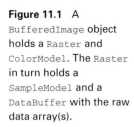

Figure 11.1 A `BufferedImage` object holds a `Raster` and `ColorModel`. The `Raster` in turn holds a `SampleModel` and a `DataBuffer` with the raw data array(s).

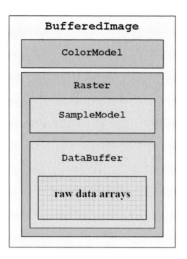

for reasons of space and speed. Instead raw image data is packed in a primitive type array or set of arrays and the colors are constructed from the array data as needed. The internal structure of `BufferedImage` is shown in Figure 11.1. It includes a `ColorModel` object and a `Raster` object. The `Raster` in turn contains `SampleModel` and `Databuffer` objects.

The `Raster` specifies the colors of the individual pixels of an image. It does this with the help of a `DataBuffer` that holds one or more raw data arrays (each array is referred to as a *bank*) and a `SampleModel` subclass that knows how the pixel data is packed in the `DataBuffer`. The abstract `SampleModel` has several subclasses such as `PixelInterleavedSampleModel` and `BandedSampleModel` that describe particular packing arrangements. The term *sample* refers to one of the values that describe a pixel. In RGB color space, for example, there are three samples – the red sample, the green sample, and the blue sample – that make up a pixel. The collection of values of a particular sample for all the pixels is referred to as a *band*. That is, if you filtered an image to show only the red values, this would display the red band.

The bands can be arranged in many different ways in the data buffers. For example, the band arrays could be assembled sequentially with the red array first, green array second, and blue array last. Or they could be interleaved within a single array in triplets of samples for each pixel. That is, the first three array elements hold the red, green, and blue samples for the first pixel and then repeat for the next pixel until the full image is complete. A binary black and white image might use just a zero or one for each pixel and be packed as bits in a byte array. The job of the sample model is to gather the samples for a given pixel and supply them to the color model.

The `ColorModel`, which must be compatible with the sample model, interprets the pixel samples as color components. The ARGB (alpha, red, green, blue) model is the default while other models include the gray-scale model with only one sample per pixel and index models where a data value holds an index into a palette of a limited number of colors. For example, a sample model for a binary black and white image packed as bits would send its zeros and ones to an index color model that would in turn assign zero to black (RGB 0x000000) and one to white (RGB 0xFFFFFF).

The `BufferedImage` is intended to handle the many different ways that images can be encoded and stored. Except for specialized cases, you don't need to deal directly with the details of the internal structure of the class. In Section 11.8.4 we show how to create an ARGB image and a gray-scale image from pixel data. Other examples are given in the Web Course Chapter 11.

11.3   Image loading

The Java AWT core language classes originally provided for loading and display of images in GIF and JPEG encoding format. With Java 1.4 came access to PNG encoding and with Java 5.0 came the bitmap formats BMP and WBMP.

We mentioned in Chapter 6 that an image does not actually begin loading when an applet invokes the `getImage()` method. The loading begins only when the program attempts to access the image such as by invoking `drawImage()` in the graphics context or by invoking the `getWidth()` or `getHeight()` methods of the `Image` class. Image loading typically requires a substantial amount of bandwidth on the network so the Java designers sought to postpone the loading of an image in case it was never needed.

Once the loading starts, it may require a substantial amount of time, especially over a slow link. Other threads in the program can continue to run in parallel while an internal thread takes care of the image loading. Java provides two ways to monitor the loading to know when it has finished:

1. Implement the `ImageObserver` interface and provide the `imageUpdate()` method.
2. Use an instance of `MediaTracker` to load one or more images simultaneously.

You can also use `ImageIcon` to load an image. This class holds an internal `MediaTracker` to monitor the loading. Another image loading option is to use the `read()` methods in the `javax.imageio.ImageIO` class. We discuss these techniques below.

11.3.1   `ImageObserver`

When a program invokes the `drawImage()` method of the `Graphics` class, the method returns before the image display has completed the loading and drawing of the image. A separate thread takes over this job while your program can go on to do other things. However, you can arrange for the internal image machinery to periodically call back to your program to let you know how it is doing. The last parameter in the method drawImage() is a reference to an object that implements the ImageObserver interface:

```
drawImage (Image img, int x, int y, ImageObserver io)
```

The image-handling process uses the ImageObserver object to allow the programmer to monitor the loading. The image-drawing process periodically invokes the imageUpdate() method of the ImageObserver object:

```
public boolean imageUpdate (Image img, int infoflags,
                            int x, int y, int w, int h)
```

Conveniently, the Component class already implements ImageObserver so we usually just put a this reference for the ImageObserver parameter in drawImage() in our applets and applications. The image-loading process thus calls back to the default imageUpdate() method in Component during the image loading and drawing operations. Since the Swing JComponent base class extends Container, which extends Component, this technique works for Swing applications too.

You can override the imageUpdate() method with one that provides you with status reports on the loading. The bits of the infoflags parameter indicate the status of the loading. For the situation of loading multiple images, you can identify which image is being updated from the first parameter. The method imageUpdate() returns true as long as the loading is not yet finished.

The image loading techniques in the following sections are much simpler to use so we do not say more about this approach. See the ImageObserver class description in the Java 2 API Specifications and the constant field values for more details about the infoflags values provided in the method's parameters. The Web Course Chapter 11: *Supplements* section includes additional information and an example program.

11.3.2 MediaTracker

The MediaTracker class provides a simple and elegant way to monitor the image loading. The class includes several methods to track one or more images, to check whether the loading has finished, and to check for errors during the loading.

The media tracker's constructor goes as

```
MediaTracker (Component comp)
```

The constructor needs the reference to the component on which the image will appear. So the loading of an image might go as follows:

```
int image_id = 1;
img = getImage (getCodeBase (), "m20.gif");
tracker = new MediaTracker (this);

// Pass the image reference and image ID nummber.
tracker.addImage (img, image_id);
```

Here we pass the `MediaTracker` constructor a reference to the component object and then add the image to the tracker with an ID number. We can use the image ID to find whether a particular image (or group of images if we use the same number for more than one image) has finished loading.

For checking on the status of an image, `MediaTracker` provides the method

```
int status (int image_id, boolean load)
```

If the last parameter is `true`, the tracker will start loading any images that are not yet being loaded. This method returns an integer that is an OR of four flags:

1. `MediaTracker.ABORTED`
2. `MediaTracker.COMPLETE`
3. `MediaTracker.ERRORED`
4. `MediaTracker.LOADING`

The `MediaTracker` is especially convenient when you need to load many images. The `statusAll()` method returns a similar value as above except that it is an OR of the status of all images currently loading.

In the following snippet the `run()` method uses the `waitForID()` method to wait for the tracker to signal that the image has loaded. (We could wait for several images to finish loading using the `waitForAll()` method.) If there is no error in the loading, the method allows the image to be painted.

```
public void run () {
  // Wait for the image to finish loading
  try {
    tracker.waitForID (fImageNum);
  }
  catch (InterruptedException e) {}

  // Check if there was a loading error
  if (tracker.isErrorID (fImageNum))
    fMessage= "Error";
  else
    fShow = true;
  // Draw the image if it loaded OK
  if (fShow) repaint ();
} // run
```

11.3.3 `ImageIcon`

A convenient shortcut is to use `ImageIcon` to load an image. This class, which came in version 1.2 with the `javax.swing` package, is nominally for obtaining a small image to use as an icon on a button and other component. However, it can be used to load any image file of any size. The class contains a `MediaTracker` internally, so it provides a compact way to load a single image with just a couple of lines of code:

```
.  .  .
ImageIcon img_icon = new ImageIcon (url);
Image image = img_icon.getImage ();
.  .  .
```

Here the constructor blocks until either the image is loaded or the address is deemed invalid. In the latter case, an image icon with zero dimensions is created. The method

```
int getImageLoadStatus ()
```

returns the `MediaTracker` status value to indicate whether the loading was successful. Note that the internal `MediaTracker` object is a static property used by all the `ImageIcon` instances.

11.3.4 Reading a `BufferedImage`

The `javax.imageio` package, which appeared in Java 1.4, includes the `ImageIO` class that offers several methods for reading and writing images in encoded formats. For example, the methods

```
public static BufferedImage read (File input)
public static BufferedImage read (URL input)
```

read an image from a given file or URL address and return it as a `BufferedImage` object. If loading from a particular URL might cause a substantial delay then it would be wise to use the image-loading techniques discussed above and then convert the `Image` to a `BufferedImage` (see Section 11.5).

As of Java 5.0 the image encodings JPEG, GIF, PNG, BMP, and WBMP (Wireless bitmap) can be read with these methods. Implementations on some platforms may offer additional types. The `ImageIO` class includes several methods to determine what file types are available. For example,

```
String[] format_names = ImageIO.getReaderFormatNames ();
```

provides a string array that would include "gif", "jpeg", "png", "bmp", and "wbmp" types. This method

```
format_names = ImageIO.getReaderMIMETypes ();
```

returns strings in the MIME format such as "image/jpeg" and "image/png".

11.4 Image display

To display an image on a component, we have used the `Graphics` class method

```
boolean drawImage (Image img, int x, int y, ImageObserver io)
```

Here the position of the top-left corner of the image is specified by the x and y values. The width and height of the drawn image go according to the image's dimensions. If this is less than the drawing area, the background color of the component is visible in the uncovered regions. If the image is larger than the component area, then the portion of the image outside the component won't be seen. The x and y values can actually be negative, in effect moving the top left corner of the image up and/or to the left of the drawing area. In such a case, only the region of the image within the component area will be visible.

The class contains several other overloaded versions of `drawImage()`. For example, the method

```
boolean drawImage (Image img, int x, int y, int width, int
                   height, ImageObserver io)
```

draws the image with the top left corner at the specified origin but scales the image such that its width and height fit the given values.

The method

```
boolean drawImage (Image img,
                   int dx1, int dy1, int dx2, int dy2,
                   int sx1, int sy1, int sx2, int sy2,
                   Color bg_color, ImageObserver io)
```

specifies that the rectangle between the corners `sx1,sy1` and `sx2,sy2` of the source image is drawn into the rectangle between the corners `dx1,dy1` and `dx2,dy2` of the destination image. The areas not covered in the destination component are painted with the color specified by the `bg_color` parameter.

While not specific to images, a section of a drawing area can be copied into other parts of the area using

```
void copyArea (int x, int y,
               int width, int height,
               int dx, int dy)
```

This can be used, for example, to tile an image.

You can also obtain a new `Image` object that represents a scaled version of an image with the `getScaledInstance()` method as in

```
. . .
Image img = getImage (url);
Image scaled_img = getScaledInstance (width, height, hints);
```

Here the new image referenced by `scaled_img` will have the width and height as given in the first two parameters. The type of scaling algorithm used depends on the third `hints` parameter. For example, the constant `Image.SCALE_AREA_AVERAGING` indicates that averaging around each point in the image should be used for the scaling. The `Image` class offers five options for scaling (see the class specification for details).

11.5 Creating images

Images can be created in Java in several ways. We can, for example, use the `createImage()` method from the `Component` class as in the following snippet:

```
. . .
Image image = createImage (width, height);
Graphics g = image.getGraphics ();
paint (g);
. . .
```

Here we created an `Image` object and obtained its graphics context. We then passed the context to our `paint()` method, which draws on the image rather than on a component. Such an off-screen image can be used for *double buffering* to speed up the graphics. Rather than sending lots of individual draw operations to the display device, a frame is first drawn on the off-screen image and then the image is sent to the display in a single operation. This saves a significant amount of time and is very useful for eliminating flickering in animations on AWT components. Swing components, however, provide double buffering by default.

The `BufferedImage` class provides a constructor in which you pass the desired dimensions and a parameter indicating the type of object:

```
BufferedImage buffered_image =
  new BufferedImage (width, height,
                       BufferedImage.TYPE_INT_RGB);
g = buffered_image.getGraphics ();
paint (g);
```

The overridden `getGraphics()` method for `BufferedImage` returns a `Graphics2D` object typed as its superclass `Graphics` type and so must be cast to `Graphics2D` to use its methods. This method is present for backward compatibility. The preferred alternative is to use `createGraphics()`, which explicitly returns a `Graphics2D` type.

The Java 2D API offers many image tools that work primarily with the `BufferedImage` class. You might encounter a situation where you get an `Image` object but you want to apply Java 2D operations to the image. You can obtain a `BufferedImage` object from an `Image` object by drawing the image onto the graphics context of the `BufferedImage`:

```
Graphics2D buf_image_graphics = buffered_image.createGraphics ();
buf_image_graphics.drawImage (image, 0, 0, null);
```

Another approach to image creation is to build them directly from pixel arrays. We discuss this approach in the image-processing sections later in this chapter.

11.6   Saving images

Although the core Java packages always provided for reading JPEG files, it wasn't until Java 1.4 that the core language allowed for saving images to disk files in JPEG format. Sun previously provided the package `com.sun.image.codec.jpeg` for writing to JPEG files but it belonged to the category of *optional* packages, which meant that it was only available for a limited number of platforms instead of all J2SE-compatible systems.

With Java 1.4 came the `javax.imageio` package and sub-packages that allow for both reading and writing of JPEG and PNG images. With Java 5.0 came BMP/WBMP image reading and writing. GIF files can be read but not saved. The GIF encoder involves the patented LZW compression algorithm so GIF encoding was not included in the standard Java packages. GIF encoders can be obtained from third party sources. See the Web Course Chapter 11 for links to several independent image handling packages that provide for saving images in these and other formats.

Java image I/O includes the `javax.imageio.*` packages. The `javax.imageio.ImageIO` class provides a number of static methods for image reading and writing with several encoding formats. For example,

```
    . . .
  BufferedImage bi = myImageAnalysis ();
  File file = new File ("c:\images\anImage.jpg");
  ImageIO.write (bi, "jpeg", file);
```

The second parameter of `write()` is a string that corresponds to a supported format. Some platforms may allow for other formats besides the standard ones. A list of supported writer formats can be obtained with

```
  String[] format_names = ImageIO.getWriterFormatNames ();
```

The `javax.imageio` classes also provide for more sophisticated image handling such as creating plug-ins for reading and writing custom image formats. See the Image I/O API Guide and the Java API Specifications for more about these classes [4,5].

11.7   Image processing

By the term *image processing* we refer both to manipulating images and to extracting information from them. With the tools available in Java we can manipulate images with several types of high level filters such as affine transforms and convolutions. We can also work at the pixel level of an image and directly examine and modify the colors that define each point in an image.

We might want to manipulate an image just to present it in some new and interesting manner. For example, we could sharpen an image so that it is more pleasing in appearance. In technical applications, a more common goal of image processing is to extract some information. For example, we might use a color filter on a photograph to look for a particular color in the field of view that indicates the presence of a material or object of interest. An edge detection filter can greatly simplify a complex scene so that searching for a particular shape in the image is much easier for pattern recognition tools.

The Java 2D API brought a much expanded array of tools and techniques for imaging processing. We can only touch upon a handful of these capabilities here. In the remaining sections we focus primarily on the Java 2D techniques available with `BufferedImage` but also look at pixel handling with the basic `Image` class.

11.8 Pixel handling

An image in Java consists essentially of an array of data in which each element describes the color of a point, or *pixel*, in the image. As discussed above, the interpretation of a value in the array element is determined by a color model. Both with `Image` and `BufferedImage` we can create images from pixel arrays and also access and modify the pixels of existing images.

11.8.1 ARGB Pixels

Java provides for different color models but the default model is the ARGB color model where a 32-bit integer value is packed with 8 bits for each of the three colors (Red, Green, and Blue) and the alpha transparency factor (see Section 6.6.3). The bits are packed as in

```
Bits  0−7 − blue
Bits  8−15 − green
Bits  16−23 − red
Bits  24−31 − alpha
```

You can use bit-handling operators (see Chapter 10) to obtain the individual component values as in the following code where the variable `pixel` is an `int` that contains an ARGB color value:

```
int alpha = (pixel >> 24) & 0xff;
int red   = (pixel >> 16) & 0xff;
int green = (pixel >> 8)  & 0xff;
int blue  = (pixel)       & 0xff;
```

Similarly, you can pack separate component values into an integer pixel variable:

```
int color = (alpha << 24) | (red << 16) | (green << 8) | blue;
```

The alpha, or transparency, factor is used when an image is overlaid on a background color or another image. This can be useful for various situations such as when placing an icon on a button. For example, you may want to allow the background color of the button to show through the blank areas of the icon.

11.8.2 PixelGrabber

The Image class does not offer direct access to its pixel array. Instead, you use the class java.awt.image.PixelGrabber, which, as its name implies, grabs the pixel data for you. The following code shows how to put the pixel values of an image into an array:

```
. . .
int[] pixels = new int[width * height];
boolean got_pixels = false;
PixelGrabber grabber =
  new PixelGrabber (img, x0, y0, width, height, pixels, 0,
scan_Width);

try {
  grabber.grabPixels ();
}
catch (InterruptedException e) {
  got_pixels = false;
  return;
}
if ((grabber.getStatus () & ImageObserver.ALLBITS)!= 0) {
  got_pixels = false;
  return;
}
```

First an array big enough to hold the pixels is created. The first parameter of the PixelGrabber constructor is the image reference. The pixels to be grabbed come from a rectangular section of the image as specified by the top left corner at (x0, y0) and by the width and height parameters. The pixel array reference is passed in the next parameter, and the last two parameters include an offset into the array to indicate where to begin putting the pixels (here set to 0) and the scan_Width value, which specifies the number of pixels per row. To capture a whole image in the pixels array, set the x0, y0 values to zero, use the width, height for the image and set scan_Width equal to the image width.

Invoking the `grabPixels()` method then initiates the filling of the pixel array. This method can return before the pixel array has finished being filled. In the snippet above, the status of the pixel filling is obtained and checked against the `ImageObserver` flags (see Section 11.3).

11.8.3 The `MemoryImageSource` class

You can also create images from pixel arrays. The class `MemoryImageSource` offers a convenient way to create images from pixel arrays and also for creating simple animations. Below we show the `ImagePanel` class, which gives an example of creating an image from a pixel array whose element values are created from a spectrum of RGB and alpha values:

```java
import javax.swing.*;
import java.awt.*;
import java.awt.image.*;

/** Create an image from a pixel array. **/
class ImagePanel extends JPanel
{
    Image fImage;
    int fWidth = 0, fHeight = 0;
    int[] fPixels;

    /** Create image with a pixel array and
     * MemoryImageSource. **/
    void init () {
        fWidth = getSize ().width;
        fHeight = getSize ().height;
        fPixels = new int [fWidth * fHeight];

        int i=0;
        int half_width = fWidth/2;

        // Build the array of pixels with the color components.
        for (int y = 0; y < fHeight; y++) {
            for (int x = 0; x < fWidth; x++) {
                // Start red on left and decrease to zero at center
                int red = 255 - (512 * x)/fWidth;
                if (red < 0) red = 0;

                // Green peaks in center
                int green;
                if (x < half_width)
                    green = (512 * x)/fWidth;
```

```
            else
               green = 255 − (255 * (x − half_width))/half_width;

            // Blue starts from center and peaks at right side.
            int blue = 0;
            if (x > half_width)
               blue = (255 * (x − half_width))/half_width;

            int alpha = 255; // non−transparent
            fPixels[i++] = (alpha << 24) | (red << 16)
                              | (green << 8) | blue;
         }
      }
      // Now create the image from the pixel array.
      fImage = createImage (
         new MemoryImageSource (fWidth, fHeight, fPixels,
                              0, fWidth));
   } // init

   /** Paint the image on the panel. **/
   public void paintComponent (Graphics g) {
      super.paintComponent (g);
      g.drawImage (fImage, 0, 0, this);
   }
} // class ImagePanel
```

The `MemoryImageSource` can be set so that if the pixel array is modified the image changes as well. This involves a subtle aspect of the way the `Image` class works. The `MemoryImageSource` implements the `ImageProducer` interface. The `Image` class obtains its pixel data from an `ImageProducer` rather than simply reading an array.

To animate the display, use `setAnimate()` as follows:

```
source = new MemoryImageSource (width, height, pixels, 0,
width);
source.setAnimate (true);
image = createImage (source);
```

To create frames for the animation, we could use a loop in a `run()` method of a threaded class as in the following code:

```
. . . in the run() method . . .
while (true) {
   // Invoke a method in the class that
   // modifies the pixels in some way
   modifyPixels ();
```

```
// Now inform the MemoryImageSource that the image has
// changed
source.newPixels ();

// Sleep between frames
try {
    Thread.sleep (dt);
}
catch (InterruptedException e) {}
```

Here `modifyPixels()` would be a method in the program that changes the pixel values in a desired manner. Then by invoking `newPixels()` the image is repainted in each pass of the loop, thus creating an animated effect.

11.8.4 Pixel handling with the `BufferedImage` class

As discussed in Section 11.3, the `BufferedImage` class offers much more access to and control of the image data than the `Image` class. Internally the `BufferedImage` consists of the `ColorModel`, `Raster`, `DataBuffer`, and `SampleModel` objects, which allow for a wide variety of image types and packing arrangements for the pixel data. However, for routine pixel handling tasks, you don't need to delve into the details of these classes to work with images at the pixel level.

For example, you can easily create an image from a pixel array packed with ARGB data (Section 11.8.1) as follows:

```
buffered_image =
  new BufferedImage (width, height,
                         bufferedImage.TYPE_INT_ARGB);
buffered_image.setRGB (0, 0, width, height, pixels,
                         0, width);
```

The constructor parameters specify the dimensions of the image and the type of image. In `setRGB()` the first four parameters specify the top left corner position and the width and height of the area of the image to be filled by the `pixels` array. The last two parameters give the offset into the pixel array where the data begins and the scan size for a row of pixels (usually just set to the image width).

Conversely, the ARGB pixel data for an image can be obtained via

```
int[] pixels = buffered_image.getRGB(0, 0, width, height,
                         array, 0, width);
```

This method returns an array with the ARGB data for the area of the image specified by the first four parameters. If not null, the fifth parameter should be an

Figure 11.2 The
`GrayBufImagePanel` class
creates a gray scale
`BufferedImage` from a
pixel array.

`int` array large enough to hold the pixel data. The next to last parameter specifies
the offset into the pixel array where the filling should begin and the last parameter
is again the scan size.

To deal with other types of images, a little more work must be done. The
following `GrayBufImagePanel` class creates a byte array with values from 0
to 255 that represent gray levels. A `BufferedImage` of the `TYPE_BYTE_GRAY`
is created. To fill it with our byte array we need to get the raster for the image and
it must be a `WritableRaster` type that allows us to modify the pixels. This is
obtained with

```
WritableRaster wr = fBufferedImage.getRaster ();
```

Then the pixel data is set with

```
wr.setDataElements (0, 0, fWidth, fHeight, fPixels);
```

Figure 11.2 shows the resulting image.

```
import javax.swing.*;
import java.awt.*;
import java.awt.image.*;

/** Create a gray scale BufferedImage from a pixel array. **/
public class GrayBufImagePanel extends JPanel
{
  BufferedImage fBufferedImage;
  int fWidth = 0, fHeight = 0;
  byte[] fPixels;

  /** Build a BufferedImage from a pixel array. **/
  void makeImage () {
    fWidth = getSize ().width;
    fHeight = getSize ().height;
    fPixels = new byte [fWidth * fHeight];

    // Create an array of pixels with varying gray values
    int i = 0;
    int half_width = fWidth/2;
```

```
    for (int y = 0; y < fHeight; y++) {
      for (int x = 0; x < fWidth; x++) {
        // Peaks white in middle
        int gray = (255 * x)/half_width;
        if (x > half_width) gray = 510 − gray;
        fPixels[i++] = (byte) gray;
      }
    }
    // Create a BufferedIamge with the gray values in
    // bytes.
    fBufferedImage =
      new BufferedImage (fWidth, fHeight,
                          BufferedImage.TYPE_BYTE_GRAY);

    // Get the writable raster so that data can be changed.
    WritableRaster wr = fBufferedImage.getRaster ();

    // Now write the byte data to the raster
    wr.setDataElements (0, 0, fWidth, fHeight, fPixels);
  } // makeImage

  /** Draw the image on the panel. **/
  public void paintComponent (Graphics g) {
    super.paintComponent (g);
    if (fBufferedImage!= null)
        g.drawImage (fBufferedImage, 0, 0, this);
  }
} // class GrayBufImagePanel
```

With the `BufferedImage` we can also create an animation by altering the pixel array for each frame, invoking the `setRGB()` method to reload the pixel data, and then invoking `repaint()`. A more sophisticated approach, which more closely resembles the `MemoryImageSource` animation technique, is to build the `BufferedImage` with a `DataBufferInt` object that holds the pixels plus a RGB `ColorModel` and a `WritableRaster` that allows direct modification of the raster data. For each frame of the animation, you modify the pixel array and then invoke `repaint()`. See the Web Course Chapter 11 for demonstrations of these two techniques.

11.9   Filtering

The Java 2D API provides a framework for filtering, or processing, images. That is, a source image enters a filter, the filter modifies the image data in some way, and a new image emerges out of the filter (the original is unaffected). Several filter classes are included in the `java.awt.image` package and you can also create your own in a straightforward manner. The filter classes implement the `java.awt.image.BufferedImageOp` interface. This interface holds five methods but the most important is

```
public BufferedImage filter (BufferedImage source_image,
                             BufferedImage dest_image)
```

This method acts upon but does not change the `source_image` and creates the processed image. If the destination image reference (`dest_image`) is not `null`, then the filter uses this image object to hold the processed image. If it is `null`, then the filter creates a new image object and returns it as the method return value. In some filters, but not all, the source and destination images can be the same. (This is referred to as *in-place* filtering.)

The five filtering classes provided with the `java.awt.image` package include:

- `ConvolveOP` – convolution filter that applies a given *kernel* operator to the image data for edge detection, sharpening, and other effects.
- `AffineTransformOp` – affine transforms include translation, scaling, flipping, rotation, and shearing. These map 2D structures in one space to another space while maintaining straight lines and the parallelism of the original image.
- `LookupOp` – instances of `LookupTable` are used to map source pixels to destination pixels according to the pixel component values (cannot be used with indexed color model images). Provides color transformation effects such as the inversion of gray scales.
- `RescaleOp` – apply a scaling factor to the color components so as to brighten or dim an image.
- `ColorConvertOp` – change to a different color space such as converting a color image to a grey scale image.

In the following sections we discuss these filters in more detail. See Chapter 11 in the Web Course for demonstration programs for each filter type.

11.9.1 Convolution

The convolution filter applies a *kernel* operator to the 2D image matrix. The kernel consists of a small square matrix (typically 3×3) that scans across the image matrix. The kernel is centered on a pixel and each kernel element multiplies the image pixel that it overlaps. The sum of these products for each color component then determines the new value of the pixel at the center of the kernel. (Lower limit on the sum is 0 and upper limit is 255 for RGB type pixels.)

For example, an edge detection kernel could consist of this 3×3 matrix:

0.0	−1.0	0.0
−1.0	4.0	−1.0
0.0	−1.0	0.0

A section of the data in a source image might appear as in this matrix:

1	1	1	1	1
1	1	1	1	0
1	1	1	0	0
1	1	0	0	0
1	0	0	0	0

For the sake of simplicity, we just give the pixels values of 0 and 1, thus creating a binary (or black and white) image. If we apply the kernel to the shaded region in the image section as shown below, the sum of the products results in a value of 0 in the corresponding center pixel in the destination matrix:

1	1	1	1	1
1	1	1	1	0
1	1	1	0	0
1	1	0	0	0
1	0	0	0	0

1	1	1	1	1
1	0	1	1	0
1	1	1	0	0
1	1	0	0	0
1	0	0	0	0

If we moved the kernel to the shaded area shown next, then the operation results in a non-zero value at the center pixel:

1	1	1	1	1
1	1	1	1	0
1	1	1	0	0
1	1	0	0	0
1	0	0	0	0

1	1	1	1	1
1	0	1	1	0
1	1	2	0	0
1	1	0	0	0
1	0	0	0	0

We want to apply the kernel to the entire image matrix. However, a problem occurs at the borders of the image because part of the kernel "hangs over" the edge and does not provide valid product values. The convolution filter allows for two choices: the image border values are set to 0 (EDGE_ZERO_FILL) or are left unchanged (EDGE_NO_OP).

If we choose the zero edge fill for our edge-finding convolution, the resulting image matrix becomes:

0	0	0	0	0
0	0	0	2	0
0	0	2	2	0
0	2	2	0	0
0	0	0	0	0

You can see that when this kernel is applied throughout a large complex image, the uniform areas are set to zero while borders between two areas of different intensities become enhanced.

Other kernels offer different effects. For example, a kernel such as this:

0.0	−1.0	0.0
−1.0	6.0	−1.0
0.0	−1.0	0.0

Figure 11.3 Edge detection convolution applied on the left image produces the image on the right.

enhances the edge regions but does not zero out the uniform regions, thus giving a sharpening effect.

To create an edge detection instance of the `ConvoleOp` class, we can use code like the following:

```
float edge_mat = {0.0, -1.0, 0.0,
                 -1.0, 4.0, -1.0,
                  0.0, -1.0, 0.0};
ConvoleOp edge_finder_op =
   new ConvoleOp (new Kernel(3,3,edge_mat),
   ConvoleOp.EDGE_NO_OP, null);
```

(The last parameter is for an optional `RenderingHints` object that you can use to adjust the color conversion.) We can then apply this convolution tool to an image with

```
BufferedImage edge_img = edge_finder_op.filter (an_image,
null);
```

The `ConvoleOp` class requires that the source and destination objects be different `BufferedImage` objects. Figure 11.3 shows an example of applying this convolution to an image.

11.9.2 Affine transforms

The affine transform filters map 2D structures in one space to another space while maintaining the straight lines and parallelism in the original image. (We discussed these transforms in Section 6.8 with regard to Java 2D drawing.) The operations include translation, scaling, flipping, rotation, and shearing. An `AffineTransformOp` object uses an instance of `AffineTransform` to apply the transform to a source image to create a new destination image (which must be different `BufferedImage` objects).

For example, the following code shows how to apply a shearing operation to an image:

```
AffineTransform shearer = AffineTransform.getShearInstance
(0.4, 0.0);
AffineTransFormOp shear_op = new AffineTransformOp
(shearer, interpolation);
BufferedImage dest_img = shear_op.filter (source_img, null);
```

Each point in the source image at (x, y) moves to (x + 0.4y, y) in the destination image.

After an affine transform, a single pixel in a destination image usually does not correspond directly to a single pixel in the source (i.e. it is split among two or more destination pixels). So the transforms require an algorithm to determine the colors of the destination pixels. The interpolation setting allows you to choose between a *nearest neighbor* algorithm and a *bilinear interpolation* algorithm. The nearest neighbor technique applies the color of the nearest transformed source pixel to the destination pixel. The bilinear interpolation instead uses a combination of colors from a set of transformed source pixels around the position of the destination pixel. (A *bicubic* option was added with J2SE 5.0.)

Note that these transforms can result in cutting off some parts of the source image that extend past the borders of the destination image. Also, some operations, such as a rotation, can leave some areas with zero color values (resulting in black for RGB images and transparent black for ARGB) where no image data remains.

11.9.3   Lookup tables

Lookup tables provide a very flexible approach to transforming the colors of an image. One can use a lookup table filter, for example, to create a negative of the source image. The LookupOp filter uses an instance of LookupTable (or, actually, one of its two subclasses) to map source pixels to destination pixels according to the source pixel component values. Note that this filter cannot be used with indexed color model images (defined in Section 11.2).

For example, the eight bits of the red component for an RGB pixel would need a table of up to 256 elements, each holding a value for the corresponding red component in the destination pixel. You can provide one table used by all three components or separate tables for each (four tables for ARGB pixels).

There are two subclasses of the abstract LookupTable. The ByteLookup-Table and the ShortLookupTable essentially offer the same features except for the type of arrays. Each provides a constructor for creating a single table for all color components and a constructor for creating multiple tables, one for each color component.

The following snippet shows how to create a table that selects only colors above a given threshold:

```
short[] threshold = new short[256];
for (int i = threshold_level; i < 256; i++)
   threshold[i] = (short) i;
LookupTable threshold_table = new ShortLookupTable (0,
threshold);
LookupOp threshold_op = new LookupOp (threshold_table,
null);
BufferedImage dest_image = threshold_op.filter
(source_image, null);
```

To invert the colors you could create an array of size 256 and fill the first element with the value 256 and then decrease each subsequent element by one until the value reaches 0 in the last element.

If we want to apply the above threshold filter to only the red component and leave the other components unchanged, we create a two-dimensional array to hold the threshold array plus an identity array that leaves the other components unchanged:

```
short[] identity = new short[256];
for (int i = 0; i < 256; i++) identity[i] = (short) i;
short[][] red_threshold = {threshold, identity, identity};
LookupTable red_threshold_table = new ShortLookupTable (0,
red_threshold);
LookupOp red_threshold_op = new LookupOp
(red_threshold_table, null);
BufferedImage dest_image = red_threshold_op.filter
(source_image, null);
```

There are obviously many such lookup table transforms you can create. Note that in-place filtering can be done with the `LookupOp` filter.

11.9.4 Rescaling

The `RescaleOp` filter applies a scaling factor and an offset to each color component so as to brighten or dim an image:

```
dest_color = source_color * scale_factor + offset;
```

You could do this also with a lookup table but you need less code with this filter. For example,

```
RescaleOp brighten_op = new RescaleOp (2.0f, 32f, null);
BufferedImage dest_image = brighten_op.filter
(source_image, null);
```

Here each color component of each pixel is multiplied by 2.0 and then added to 32. If the value exceeds the maximum for that component, the maximum is used.

11.9.5 Color conversion

The `ColorConvertOp` filter changes an image from one color space to another. A common requirement in image processing is to change to gray scale:

```
ColorSpace gray_space =
   ColorSpace.getInstance (ColorSpace.CS_GRAY);
ColorConvertOp convert_to_gray_op = new ColorConvertOp
(gray_space, null);
BufferedImage gray_img = convert_to_gray_op.filter
(source_image, null);
```

The `java.awt.color.ColorSpace` class offers a number of options including `TYPE_CMYK` and `TYPE_HSV`.

11.9.6 Custom filters

You can implement the `BufferedImageOp` interface to create your own custom filters. In addition to the `filter()` method, there are four other methods in the interface that must be implemented. The following `RotateOp` example illustrates the basics of creating a filter. (See the book by Knudsen [1] and the other references [2,3] for more details about creating filters.)

The `filter()` method offers the option of using an existing `BufferedImage` object passed via the second parameter to receive the output of the filter. If this reference is null, a `BufferedImage` must be created and it must possess the same dimensions as the source image and also have a similar raster and color model. The `createCompatible DestImage()` method does this job of making a suitable destination image. For the `BufferedImage` constructor it uses a color model either passed as a parameter or from the source. It also gets a raster suitable for this color model and it checks to see if the alpha transparency factor pre-multiplies the color components.

The `getBounds2D()` method returns the bounds object obtained from the source image. The `getPoint2D()` method, which asks for the point in the destination image that corresponds to the given point in the source image, just returns the same point as in the source image since in this filter the dimensions are unchanged. There are no `RenderingHints` provided for displaying the filter output image.

```
import javax.swing.*;
import java.awt.*;
import java.awt.image.*;
import java.awt.geom.*;

/** Shift the color components with the filter. **/
public class RotateOp implements BufferedImageOp {
```

```java
public final BufferedImage filter (BufferedImage source_img,
                                   BufferedImage dest_img) {

  // If no destination image provided, make one of same
  // form as source
  if (dest_img == null)
    dest_img = createCompatibleDestImage
      (source_img, null);

  int width = source_img.getWidth ();
  int height= source_img.getHeight ();

    for (int y=0; y < height; y++) {
      for (int x=0; x < width; x++) {
        int pixel = source_img.getRGB (x,y);

        // Get the component colors
        int red   = (pixel >> 16) & 0xff;
        int green = (pixel >> 8)  & 0xff;
        int blue  = pixel         & 0xff;

        // Rotate the values
        int tmp = blue;
        blue = green;
        green = red;
        red = tmp;

        // Put new value into corresponding pixel of
        // destination image;
        pixel = (255 << 24) | (red << 16) | (green << 8) |
          blue;
        dest_img.setRGB (x,y,pixel);
      }
    }
    return dest_img;
} // filter

/**
  * Create a destination image if needed. Must be same
  * width as source and will by default use the same
  * color model. Otherwise, it will use the one passed
  * to it.
  */
public BufferedImage createCompatibleDestImage (
  BufferedImage source_img,
  ColorModel dest_color_model
){
```

```
    // If no color model passed, use the same as in source
    if (dest_color_model == null)
      dest_color_model = source_img.getColorModel ();

    int width = source_img.getWidth ();
    int height= source_img.getHeight ();

    // Create a new image with this color model & raster.
    // Check if the color components are already
    // multiplied by the alpha factor.
    return new BufferedImage (
      dest_color_model,
      dest_color_model.createCompatibleWritableRaster
      (width,height),
      dest_color_model.isAlphaPremultiplied (),
      Null
      );
  } // createCompatibleDestImage

   /** Use the source image for the destination bounds
     * size. **/
   public final Rectangle2D getBounds2D (BufferedImage
                                         source_img) {
     return source_img.getRaster ().getBounds ();
   }

   /** Point in source corresponds to same point in
     * destination. **/
   public final Point2D getPoint2D (Point2D source_point,
                                    Point2D dest_point) {

     if (dest_point == null) dest_point =
        new Point2D.Float ();
     dest_point.setLocation (source_point.getX (),
                             source_point.getY ());
     return dest_point;
   }

   /** This filter doesn't provide any rendering hints. **/
   public final RenderingHints getRenderingHints () {
       return null;
   }

} // class RotateOp
```

11.10 Web Course materials

The Web Course Chapter 11: *Java* section provides demonstration applets and applications of the image classes and techniques discussed here. The *Supplements* section gives an overview of the *Java Advanced Imaging* (JAI) API [6]. JAI offers many additional tools for imaging processing and I/O beyond those discussed in this chapter. JAI is a Sun Microsystems product and is not part of the core Java language distribution. However, it is available for Windows, Linux, and Solaris platforms.

The *Tech* section looks further at image-processing techniques and explores some image-making examples such as fractal animations. The *Physics* section continues with development of experimental simulations and data analysis tools with Java.

References

[1] Jonathan Knudsen, *Java 2D Graphics*, O'Reilly, 1999.

[2] Patrick Niemeyer, Jonathan Knudsen, *Learning Java*, 2nd edn, O'Reilly, 2002.

[3] *Programmer's Guide to the Java 2D API Enhanced Graphics and Imaging for Java*, Sun Microsystems, 2001, `http://java.sun.com/j2se/1.4.0/pdf/j2d-book.pdf`.

[4] Java Image I/O API Guide, `http://java.sun.com/j2se/1.5.0/docs/guide/imageio/index.html`.

[5] The `javax.imageio` package specifications, `http://java.sun.com/j2se/1.5.0/docs/api/javax/imageio/package-summary.html`.

[6] Java Advanced Imaging (JAI) API, `http://java.sun.com/products/java-media/jai/`.

Resource

How to Use Icons – The Java Tutorial, Sun Microsystems, `http://java.sun.com/docs/books/tutorial/uiswing/misc/icon.html`.

Chapter 12
More techniques and tips

12.1 Introduction

In Part I of this book and Web Course we tried to provide an introduction to the essential elements of the Java language that allow you to begin creating useful programs in short order. In this chapter we discuss several practical techniques that will expand the capabilities of your programs. We begin with a discussion of how to print your graphics displays and then discuss several user interface features such as cursor icons and popup menus, handling keystrokes, and audio. We also review various ways to improve the speed of Java programs.

12.2 Printing

Java 1.1 provided the capability to print what is displayed on a Java component [1]. Java 1.2 added Java 2D, which expanded the print capabilities to support greater control over multiple page printing and other features. (We should point out that printing only works with applications since the `SecurityManager` in browser JVMs blocks printing from applets.)

In Java graphics the usual job of the `paint()` method in AWT and the `paintComponent()` method in Swing is to send drawing commands to the monitor screen. Java printing simply entails sending drawing commands to the printer instead of the monitor screen. In rendering Java components, we have seen that a `Graphics` context object is passed to the `paint()` and `paintComponent()` methods. To render to the printer, you obtain an instance of `PrintGraphics`, which is a subclass of `Graphics`, and pass it to the `paint()` or `paintComponent()` method. The graphics context drawing commands then work as usual except that the drawing will be on the printer paper rather than on the screen.

Figure 12.1 shows the user interface for the program `PrintTestApp`, which displays an image in the frame and holds a menu bar with a dropdown menu to select whether to print or quit the program. The following code snippet from the program shows the steps needed to obtain the print dialog from the host system

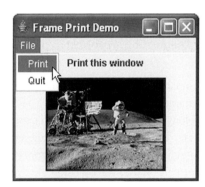

Figure 12.1 Display for the `PrintTestApp` program.

and then perform the printing:

```
. . .
public class PrintTestApp extends JFrame
        implements ActionListener {
. . .
  /** Execute the menu events here. **/
    public void actionPerformed (ActionEvent e) {
    String command = e.getActionCommand ();
    if (command.equals ("Quit")) {
        dispose ();
        System.exit (0);
    } else if (command.equals ("Print")) {
      print ();
    }
  }

  /** Do the print setup up here. **/
  public void print () {
    PrintJob pjb =
      getToolkit ().getPrintJob (this, "Print Test", null);

    if (pjb!= null) {
        Graphics pg = pjob.getGraphics ();
      if (pg!= null) {
        paint (pg); // Paint all components on this frame.
        // flush page when finished
        pg.dispose ();
      }
      pjob.end ();
    }
  } // print
. . .
```

First the program invokes the method

```
PrintJob getPrintJob (Frame frame, String title, Properties
                      props)
```

from the AWT toolkit. This displays the system print dialog on the screen. The first parameter in getPrintJob() references the frame to which the print dialog belongs. The title string, which appears on the top bar of the print dialog, comes next. The last parameter is a reference to a Properties object, which we set to null in this example. The properties were never standardized so they are not portable. With Java 1.3 came the overloaded method

```
PrintJob getPrintJob (Frame f, String t, JobAttributes
                      jA, PageAttributes pA)
```

Here the parameters include instances of the JobAttributes and PageAttributes classes. These classes provide methods to set a wide range of printer control parameters such as the number of copies, the page ranges, paper size, orientation, and so forth.

The standard printer dialog for the local platform appears after the invocation of the getPrintJob() method. The returned PrintJob object provides an instance of PrintGraphics via the getGraphics() method. This object, which is a subclass of Graphics, is then passed to the paint() or paint-Component() method, which proceeds to paint the frame display for the printer output.

As discussed in Section 6.6.2, a ratio of 72 user units to 1 inch is maintained for drawing commands regardless of the printer's resolution setting. If there are sub-components on the frame, such as buttons and labels, these also paint themselves to the printer.

12.3 Cursor icons

It can be useful when designing a user interface to alter the appearance of the cursor according to the current position and/or processing going on. The AWT has the Cursor class that comes with 14 different cursor icons. The Component class includes the setCursor() method that can change the cursor icon when it lies above a particular component.

In the following code snippet we show the class CursorPanel, which extends JPanel. On the panel we add a 7×2 grid of buttons and for each button we use the setCursor() method to set to an instance of a Cursor class. The particular type of cursor depends on the value of the constant in that class such as Cursor.WAIT_CURSOR passed in the Cursor constructor. See the applet CursorTestApplet in the Web Course Chapter 12 for a demonstration of this panel.

```java
/** Demonstrate the different cursor styles. **/
class CursorPanel extends JPanel
{
  CursorPanel() {
    setLayout (new GridLayout (7,2));

    JButton bt = new JButton ("Default");
    bt.setCursor (new Cursor (Cursor.DEFAULT_CURSOR));
    add (bt);

    bt = new JButton ("Busy");
    bt.setCursor (new Cursor (Cursor.WAIT_CURSOR));
    add (bt);

    bt = new JButton ("Hand");
    bt.setCursor (new Cursor (Cursor.HAND_CURSOR));
    add (bt);

    bt = new JButton ("Text");
    bt.setCursor (new Cursor (Cursor.TEXT_CURSOR));
    add (bt);

    bt = new JButton ("CrossHair");
    bt.setCursor (new Cursor (Cursor.CROSSHAIR_CURSOR));
    add (bt);

    bt = new JButton ("Move");
    bt.setCursor (new Cursor (Cursor.MOVE_CURSOR));
    add (bt);

    bt = new JButton ("East Resize");
    bt.setCursor (new Cursor (Cursor.E_RESIZE_CURSOR));
    add (bt);

    bt = new JButton ("North Resize");
    bt.setCursor (new Cursor (Cursor.N_RESIZE_CURSOR));
    add (bt);

    bt = new JButton ("West Resize");
    bt.setCursor (new Cursor (Cursor.W_RESIZE_CURSOR));
    add (bt);

    bt = new JButton ("South Resize");
    bt.setCursor (new Cursor (Cursor.S_RESIZE_CURSOR));
    add (bt);
```

```
        bt = new JButton ("NorthEast Resize");
        bt.setCursor (new Cursor (Cursor.NE_RESIZE_CURSOR));
        add (bt);

        bt = new JButton ("NorthWest Resize");
        bt.setCursor (new Cursor (Cursor.NW_RESIZE_CURSOR));
        add (bt);

        bt = new JButton ("SouthWest Resize");
        bt.setCursor (new Cursor (Cursor.SW_RESIZE_CURSOR));
        add (bt);

        bt = new JButton ("SouthEast Resize");
        bt.setCursor (new Cursor (Cursor.SE_RESIZE_CURSOR));
        add (bt);
    } // ctor
} // class CursorPanel
```

With Java 1.2 came the capability to create your own custom cursor icons. The method

```
Cursor createCustomCursor (Image cursor, Point hotSpot,
                                String name)
```

in the `java.awt.Toolkit` class includes an image for the cursor in the parameter list. The second parameter specifies the so-called hotspot that sets the pixel's (x, y) coordinates relative to the top-left corner of the image where the click occurs. (For example, the hotspot would specify the tip of an arrow cursor.) The last parameter provides the name of the cursor for the Java *Accessibility* system to use. (The Accessibility framework, not discussed here, provides enhancements to the GUI to assist handicapped users.)

12.4   Mouse buttons

In Chapter 7 we discussed mouse events that are produced by actions such as clicking on the primary mouse button. This is usually the left button for a two- or three-button mouse. You can detect clicks on the right mouse button, or its equivalent, on either a one-, two- or three-button mouse with the `getModifiers()` method of the `MouseEvent` class (inherited from the `Event` class). This code snippet illustrates the technique:

```
public void mouseClicked (MouseEvent e) {
   if (g.getModifiers () & InputEvent.BUTTON3_MASK)!= 0)
     doSomething ();
   . . .
```

The constant BUTTON3_MASK from the InputEvent class provides a bit mask with which to identify whether the third mouse button generated the event. Similarly, other buttons and button key combinations can be identified with these masks:

```
BUTTON1_MASK
BUTTON2_MASK
ALT_MASK
META_MASK
```

The latter two constants are used to determine if the keyboard ALT or META keys were held down during the mouse click event. (Not all keyboards have META keys.) The following code snippet shows a JPanel subclass from the applet named MouseButtonsApplet. An instance of a MouseAdapter subclass created via the inner class technique is added to the panel's MouseListener list. The adapter's mouseClicked() method uses getModifiers() to obtain the identity of the buttons that initiated the click and sends a message to a text area accordingly.

```
import javax.swing.*;
import java.awt.*;
import java.awt.event.*;
public class MouseButtonsApplet extends JApplet
{
    . . . Code to add a MouseButtonPanel instance to the applet . . .
}


class MouseButtonPanel extends JPanel
{
   JTextArea fTextArea;

   /** Build the panel interface and a mouse listener. **/
   MouseButtonPanel () {
     setLayout (new GridLayout (2,1));

     JPanel canvas = new JPanel ();
     add (canvas);
     canvas.setBackground (Color.red);

     fTextArea = new JTextArea ();
     fTextArea.setEditable (false);
     // Add to a scroll pane so that a long list of keyinputs can be seen.
     JScrollPane area_scroll_pane = new JScrollPane (fTextArea);

     add (area_scroll_pane);
```

```
      canvas.addMouseListener (
        new MouseAdapter () {
          public void mouseClicked (MouseEvent e) {
            if ((e.getModifiers () & InputEvent.BUTTON1_MASK)!= 0)
                    saySomething ("Left button pressed", e);

            if ((e.getModifiers () & InputEvent.BUTTON2_MASK)!= 0)
                    saySomething ("Middle button pressed",e);

            if ((e.getModifiers () & InputEvent.BUTTON3_MASK)!= 0)
                    saySomething ("Right button3 pressed",e);

            if ((e.getModifiers () & InputEvent.ALT_MASK)!= 0)
                    saySomething ("alt pressed",e);

            if ((e.getModifiers () & InputEvent.META_MASK)!= 0)
                    saySomething ("meta pressed",e);
          } // mouseClicked
          } // end anonymous class
        ); // end method call
      } // ctor

      /** Indicate what mouse event occurred. **/
      void saySomething (String eventDescription, MouseEvent e) {
        fTextArea.append (eventDescription + " on " +
              e.getComponent ().getClass ().getName () + "\n");
  }
} // class MouseButtonPanel
```

12.5 Popup menu

On all common graphical user interfaces a particular mouse button or combination of a mouse button and key strokes (such as CTRL or ALT keys) brings up a *popup* menu. A popup menu is a dialog window that appears close to the cursor and that typically lists a set of operations related to what you clicked on (such as *cut*, *copy*, *insert*, *paste* in an editor).

To provide for platform portability, the AWT provides the isPopup-Trigger () method in the MouseEvent class. This method returns a boolean value to indicate if the standard button/key combination for a popup menu on a particular platform was present during a mouse click event. You therefore don't need to test for the buttons or button/keystroke combinations yourself and, more importantly, you don't need to know and build in test code for all possible platforms to detect a popup menu request.

The demonstration applet `PopupApplet` shown below illustrates the creation of a popup menu (see Figure 12.2). Here the menu provides color options for the background of the component on which the menu was requested. The program also illustrates the use of an inner class for this kind of task.

```java
import javax.swing.*;
import java.awt.*;
import java.awt.event.*;

/** Demonstration of popup menus. **/
public class PopupApplet extends JApplet
{
  /** Create a simple interface with two panels. **/
  public void init () {
    Container content_pane = getContentPane ();
    PopupPanel popup_panel = new PopupPanel ();

    // Add the panel that will display the popup menu
    content_pane.add (popup_panel);
  }
} // class PopupApplet

/** Popup menu offers choices for colors of 2 subpanels. **/
class PopupPanel extends JPanel
                implements ActionListener
{
  MouseAdapter fAdapter;
  JPopupMenu fColorMenu;
  Component fSelectedComponent;
  Component fParent;

/** Constructor creates an interface with two panels.
 * A PopupMenu will offer a choice of colors for the
 * panels. **/
  PopupPanel () {
    setLayout (new GridLayout (2,1));

    JPanel canvas1 = new JPanel ();
    canvas1.setBackground (Color.RED);
    add (canvas1);

    JPanel canvas2 = new JPanel ();
    canvas2.setBackground (Color.GREEN);
    add (canvas2);
```

Figure 12.2 The `PopupApplet` program demonstrates the `JPopupMenu` component. The user moves the cursor over the top or bottom panel and executes the popup menu procedure appropriate for the platform, such as clicking on the right button of a two button mouse on a MS Windows system. The color of the panel will be set according to the menu item selected.

Figure 12.3 For the
`KeyTestApplet`
demonstration, when the
top panel has the focus,
keystrokes are captured
and reported on the text
area.

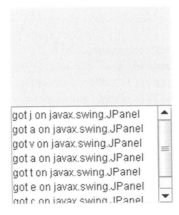

```
got j on javax.swing.JPanel
got a on javax.swing.JPanel
got v on javax.swing.JPanel
got a on javax.swing.JPanel
got t on javax.swing.JPanel
got e on javax.swing.JPanel
got c on javax.swing.JPanel
```

`keyTyped (KeyEvent e)` method. An instance of this `KeyListener` can
then be added to the list of such listeners maintained by a `Component` subclass.
When a key press occurs, the component invokes the `keyTyped()` method of
all the listeners in its list. From the `KeyEvent` object you can then obtain the
identity of the key pressed by the user:

```
public void keyTyped (KeyEvent e) {
   saySomething ("got " + e.getKeyChar (), e);
}
```

The `KeyEvent` object provides the key character via the `getKeyChar()`
method.

The example `KeyTestApplet` holds an instance of a `JPanel` subclass called
`KeyTestPanel`, which in turn holds a subpanel and a `JTextArea`. We add an
instance of a `KeyAdapter` to the subpanel's list of `KeyListener` objects. The
subpanel is made *focusable* so that it can receive the key events. (When the applet
runs, you should click on it and hit "tab" to put the focus on the subpanel.) The
adapter sends each `KeyEvent` object to the `keyTyped()` method whenever the
user presses a key (see Figure 12.3).

```
. . . Code in KeyTestApplet to display a KeyTestPanel
  object . . .

/** A JPanel class that detects key strokes on one subpanel
  * and displays messages about them in a text area. **/
class KeyTestPanel extends JPanel
{
   JTextArea fTextArea;

/** Create an interface with a text area and a blank panel.
  * Key strokes while the panel has focus will be detected
```

```
     * and a message printed in the text area.
    **/
  KeyTestPanel () {

      setLayout (new GridLayout (2,1));
      JPanel canvas = new JPanel ();
      add (canvas, BorderLayout.NORTH);
      canvas.setBackground (Color.YELLOW);

      fTextArea = new JTextArea ();
      fTextArea.setEditable (false);

      // Add to a scroll pane so that a long list of
      // keyinputs can be seen.
      JScrollPane area_scroll_pane = new JScrollPane
        (fTextArea);

      add (area_scroll_pane, BorderLayout.CENTER);

      // Add to the panel an anonymous KeyAdapter that will
      // respond to key strokes.
      canvas.addKeyListener (
        new KeyAdapter () {
          public void keyTyped (KeyEvent e) {
            saySomething ("got " + e.getKeyChar (), e);
          }
        } // end anonymous class
      ); // end method call

      // Let the canvas panel get the focus.
      canvas.setFocusable (true);
    } // ctor

    /** Display a message in text area about the key
        event. **/
    void saySomething (String eventDescription, KeyEvent e) {
      fTextArea.append (eventDescription + " on "
          + e.getComponent ().getClass ().getName ()
          + "\n");
    }
  } // class KeyTestPanel
```

The `KeyListener` interface also has the `keyPressed()`, `keyReleased()`, and `keyTyped()` methods that fire a `KeyEvent` when a key is pressed, released, and "typed," respectively. Key presses and releases are low level events that depend on the keyboard and platform in use. Key "typed" events are higher level events that occur when a typing action is complete and are the preferred way to find out about character input. Combinations of keys can be detected by testing the key code. The KeyEvent class provides the codes as constants. For example, this code snippet tests if the left arrow and shift keys were simultaneously pressed:

```java
public void keyPressed (KeyEvent e) {
   int keyCode = e.getKeyCode ();
   if (keyCode == VK_LEFT && e.isShiftDown ()) {
      .  .  .
```

The KeyEvent class provides several other useful methods such as the isShiftDown() method used above and isActionKey() which indicates whether the event was triggered by one of the action keys such as HOME, END, etc. See the KeyListener and KeyEvent class descriptions for more information on using key data.

12.7 Audio

In the early versions of Java the audio capabilities in the core language were extremely limited. Only 8 kHz *au* type files could be played. With version 1.2, it became possible to play 22 KHz, 16-bit stereo in the following formats:

- AIFF
- AU
- WAV
- MIDI
- RMF

The new sound engine is a part of the core library. A full featured Java Sound API became available as well in Java 1.3. This includes the packages javax.sound.midi and javax.sound.sampled. These advanced audio capabilities are beyond the scope of this book so we only look at the simple playing of sound clips [2–5].

While audio may have limited applications for scientific programs, they can be useful for such things as warnings and alarms. The AudioClip interface has three methods to implement:

- play ()
- loop ()
- stop ()

The applet method

```
Applet.getAudioClip (URL url)
```

returns an instance of an `AudioClip` object – i.e. an object that implements the
`AudioClip` interface. See the discussion in Section 6.9 on obtaining image files
for information on using the `getResources()` method in the `Class` class for
accessing audio files in a JAR file.

The following applet shows the basics of playing a sound clip. The `loop()`
method continuously repeats the clip.

```java
import javax.swing.*;
import java.awt.*;
import java.awt.event.*;
import java.applet.*;


/** Demonstrate playing an audio clip. **/
public class AudioTestApplet extends JApplet
{
  public void init () {
    Container content_pane = getContentPane ();
    // Create an instance of a JPanel sub-class
    AudioPanel audio_panel =
      new AudioPanel (getClip (false));
    //And add one or more panels to the JApplet panel.
    content_pane.add (audio_panel);
  } // init


  AudioClip getClip (boolean file_in_jar) {
    if (file_in_jar) {
      // Use getResource () to search directory or a
      // jar file.
      return (getAudioClip (
              getClass ().getResource (
              getParameter ("AudioClip")))));
    }
    else {
      // Read audio file from the code directory
      return (getAudioClip (
              getCodeBase (), getParameter
              ("AudioClip")));
    }
  } // getClip
} // class AudioTestApplet
```

```java
/** Panel with a button to play/stop a clip in loop
  * mode. **/
class AudioPanel extends JPanel implements ActionListener
{
  AudioClip fAudioClip;
  JButton fButton;
  boolean fPlay = false;

  /** Constructor gets the clip and makes a button for the
    * panel. **/
  AudioPanel (AudioClip audio_clip) {
    fAudioClip = audio_clip;
    fButton = new JButton ("Play Clip");
    fButton.addActionListener (this);
    add (fButton);
  } // ctor

  /** Button will start/stop the clip playing. **/
  public void actionPerformed (ActionEvent e) {
    if (fAudioClip!= null) {
      if (!fPlay) {
      fButton.setText ("Stop clip");
      fAudioClip.loop ();
      fPlay = true;
    }
    else {
        fButton.setText ("Play clip");
        fAudioClip.stop ();
        fPlay = false;
    }
   }
  } // actionPerformed
} // class AudioPanel
```

12.8 Performance and timing

Reducing the execution time in a Java program can be a very important part of making it a useful tool, especially if the program must execute extensive computations like those needed for a complicated mathematical algorithm or an animation of a complex scene. There are various *profiler* tools available that give detailed information on the time taken by various parts of a program, particularly for method calls. The Sun J2SE `java` program, in fact, includes the options `-Xprof` and `-Xrunhprof` to produce time profiles. We discuss here a more basic but often effective approach. (See the Web Course Chapter 12 for more information about profilers.)

There are some optimization steps that you can make to improve the performance of your programs. For example, if you are sure that you will not be overriding a method then you should you use the modifier `final`. The compiler can then *inline* the method and the interpreter will not have to search for possible overriding methods. Inline puts the program code directly in the execution flow and therefore avoids jumping to and from another section of memory.

When you are unsure about the relative speed of different coding techniques, you can test them with the static method `System.currentTimeMillis()`. This method returns a `long` value equal to the number of milliseconds since a standard date in 1970. That number in itself is of little value, but the difference between two millisecond time readings can be meaningful. Thus, you can bracket a code section of interest with this method and find the difference in milliseconds to get an idea of how long that section of code takes to run.

Note that on today's high-performance machines, you often need to loop over an operation a large number of times to produce a time in the millisecond range. Such a timing measurement has come to be known as a microbenchmark and is fraught with difficulties for a variety of reasons, including just-in-time and dynamic compilation techniques, compilation warm-up periods, the use of separate compile threads, background compilation, compiler optimizations, dead code removal, and others. Nevertheless, many people still rely on `System.currentTimeMillis()` to measure execution time differences. We explain here how that is normally done and attempt to ameliorate many of the concerns that make microbenchmarks untrustworthy.

With multiprocessing happening in the operating system and multithreading in the JVM, the measured times will vary from run to run, so a statistical average is more meaningful than a single run. It is wise to unload any unnecessary programs that might steal time slices away from the program you are testing.

As an example, consider the `java.util.Vector` class we discussed in Chapter 10. It is often very useful compared to a regular fixed-length array because of `Vector`'s ability to remove and add elements. However, in situations where high performance is required, you may find that an array is preferable because `Vector` offers noticeably slower performance than an array. One reason that `Vector` is slow is because it is synchronized for thread safety. An alternative that can be used in most situations is the `ArrayList` class which does not have the synchronization overhead. In the test code below we compare timings of object arrays, `Vector`s, and `ArrayList` objects. For consistency, we populate each with `Integer` objects.

In the `main()` method shown below in the class `TimeTest`, we read in some command line parameters and then call `doTests()` several times. Inside `doTests()` we call four methods that test fetching `Integer` objects from four different container types – `Vector`, `ArrayList`, regular object arrays, and an `ArrayList<Integer>` collection object that takes advantage of the new generics feature in Java 5.0 (see Chapter 10). If you want to run this code on a pre-5.0 Java system, you'll need to comment out the generics version.

```
import java.util.*;

public class TimeTest
{
   private int fWarmup, fNum, fOuter, fSleep;

   public TimeTest (int warmup, int sleep, int num, int outer) {
      fWarmup = warmup;
      fSleep = sleep;
      fNum = num;
      fOuter = outer;
   } // ctor

   public static void main (String[] args) {
      int warmup=0, sleep=0, num=0, outer=0, total=0;
      if (args.length == 5) {
         warmup = Integer.parseInt (args[0]);
         sleep  = Integer.parseInt (args[1]);
         num    = Integer.parseInt (args[2]);
         outer  = Integer.parseInt (args[3]);
         total  = Integer.parseInt (args[4]);
         System.out.println ("Testing with warmup = " +
            warmup + ", sleep = " + sleep + ", num = " + num +
            ", outer = " + outer + ", total = " + total);
      }
      else {
         System.err.println ("Usage: java TimeTest warmup " +
            "sleep loop outer total");
         System.exit (1);
      }

      System.out.println ("V\tAL\tAL<I>\tarray");
      TimeTest tt = new TimeTest (warmup, sleep, num, outer);
      for (int i = 0; i < total; i++) {
         tt.doTests ();
      }
   } // main

   public void doTests () {
      long vectime    = testv ();
      long arlisttime = testal ();
      long alitime    = testali ();
      long arraytime  = testa ();
```

```
      System.out.println (vectime + "\t" + arlisttime +
        "\t" + alitime + "\t" + arraytime);
    } // doTests
    . . .
```

The `testv()` method appears below. The others are similar and appear in the Web Course material. In each case, the fetches from the containers are in a loop in order to make the total time large enough to measure. We have to do something with the `Integer` objects fetched from the containers or else the compiler will notice that nothing is changing inside the loop and remove the loop altogether. So for our timing loop we first cast the `Object` type retrieved from the container to an `Integer` and then we accumulate a total sum of all the values after converting from `Integer` type to `int` type using the `intValue()` method on `Integer`.

```
public long testv () {
  // Create Vector and fill elements with Integer objects
  Vector vec = new Vector (fNum);
  for (int i=0; i < fNum; i++) {
    vec.add (new Integer (i));
  }

  // Now test access times by looping through the Vector
  // to access each Integer element.

  // First warmup the hotspot compiler.
  long sum = 0;
  for (int i = 0; i < fWarmup; i++) {
    sum += ((Integer)(vec.elementAt (i))).intValue ();
  }
  // And give it time to finish the JIT compile
  // in the background.
  if (fSleep > 0)
    try {Thread.sleep (fSleep);
    } catch (InterruptedException e) {}

  // Then do the loop for real and time it.
  long t1 = System.currentTimeMillis ();
  for (int j = 0; j < fOuter; j++) {
    for (int i = 0; i < fNum; i++) {
    sum += ((Integer)(vec.elementAt (i))).intValue ();
    }
  }
  long t2 = System.currentTimeMillis ();
```

```
    long elapsed = t2 - t1;
    vec.clear ();
    vec = null;
    System.gc ();
    return elapsed;
} // testv
```

In this method we first populate a `Vector` with `Integer` objects. Then we begin to fetch from the `Vector` and accumulate the sum. In order to allow the Hotspot compiler time to warm up, we first perform a short summation loop without timing. The number of warm-up loops is specified by one of the command line parameters. After warming up we sleep a short while. The purpose of the sleep is to relinquish the processor so that the Hotspot compiler can finish a background compile of the loop if needed.

Finally we begin the timing loop by calling `System.currentTime-Millis()`, followed by the loop itself (actually an inner and outer loop) and then another call to `System.currentTimeMillis()` so we can calculate the elapsed time. We then clear the `Vector`, set the reference to `null`, and request garbage collection (`System.gc ()`). The elapsed time is returned to `doTests()` where it is printed to the console.

The command-line parameters adjust the number of warm-up loops, the sleep time, the number of inner and outer loops to be timed, and the total number of times to run the tests. If you download this code you can experiment with the various parameters to see what effect they have on the timing results. The results can vary widely based on the platform in use.

On one system we find that `Vector` is consistently the slowest, as expected. The `ArrayList` is slightly faster (typically 80% of the `Vector` time or less) and the fastest of all, as expected, is the plain object array, which is nearly twice as fast as `Vector`. Of special interest is the Java 5.0 parameterized type `ArrayList<Integer>`, which is consistently slightly faster than the unparameterized `ArrayList`, perhaps due to removing the need for casting from `Object` type to `Integer` type. On a different platform, we observed about the same relative performance between `ArrayList` and `Vector` but a factor of nearly nine between a plain array and a `Vector`. On yet another platform, we've seen a factor of about three between an array and a `Vector`. On one other platform the array is only about 20% faster than the `ArrayList` with both about twice as fast as the `Vector`. The moral of this story is that the container type chosen can lead to an important performance difference. Arrays are almost certainly always the fastest and the synchronized collection objects are the slowest with the newer unsynchronized collection objects somewhere in between. Just where in between is platform dependent. Perhaps a second moral is to be sure to test your code on the platform or platforms on which it will be deployed.

Whenever coding for performance, it is a good idea to first write the code in a natural way and then profile it with the built-in profiling tools to see where the problem areas are. Finally, tune those problem areas. In general, extra care in coding should be taken in loops with lots of iterations and in methods that are called frequently.

For the problem areas, here is a list of some basic performance tips:

- Local variables run faster than instance and static variables.
- The `long` and `double` type variables typically require extra time to access and modify since they contain additional bytes compared to `int` and `float`. However, significantly slower performance is not true of all JVMs on all platforms since some take advantage of particular processor capabilities.
- The JVM bytecode is weighted towards `int` operations so use `int` type except where you specifically need to use one of the other types. Math operations with the shorter integer types are widened in the JVM to `int`. For example, a sum of two `byte` values results in an `int` value (that may be cast back to a `byte` if the result is to be stored into a `byte` variable).
- Use the `x += A` type of commands rather than `x = x + A`. The first requires one instruction and the latter four.
- If you are concatenating lots of strings, use `StringBuffer` and its `append()` method instead of the `String` "+" operator. In Java 5.0 and above, use `StringBuilder` instead.
- As demonstrated in the `TimeTest` example above, the original Java utility classes like `Vector`, `Enumeration`, and `Hashtable` are slow because they use synchronization for thread safety. Although synchronization overhead has been significantly reduced in modern versions of Java, if you do not require synchronization, then the newer utilities from the Java Collections Framework are normally superior. The new classes `ArrayList` and `HashMap` generally replace `Vector` and `Hashtable`, respectively, and `Iterator` is preferred over `Enumeration`.
- Avoid creating lots of objects if possible, especially in loops. Creating an object takes up considerable time and uses up memory resources and requires the Java garbage collector to do more work to reclaim abandoned memory.
- Avoid method calls in a loop as much as possible. That is, do your loops inside of methods rather than calling methods inside of loops.
- Performance can vary significantly among different JVMs for the same platform. Do timing measurements for all the JVMs that would potentially be used with your program.

See the book by Shirazi [6] for a complete discourse on performance enhancements.

12.9 Lifelong Java learning

Part I introduced the basics of Java programming and you can now create applets and applications with graphical user interfaces, threads, I/O, image processing, and other capabilities. In Parts II and III you will learn how to use Java for network programming, distributing computing, running native code, and other tasks.

For most of the topics that we discussed in this book, we only had room to present those essential elements of a class, or package of classes, that would allow you to begin using it in your programs. Most of these classes contain lots of other methods with which you should become familiar besides the ones we talked about here. Furthermore, with the ever-expanding capabilities of Java there will always be new tools, classes, APIs, and new versions of the language appearing on the scene (like J2SE 5.0). It is now common to find several books available that focus only on some narrow aspect of the language. Rather than trying to master all of Java from the start, which really is no longer feasible, more likely you will seek out information on particular concepts and classes as you need them for a particular programming task.

As we have often suggested, you should always go first to the Java API Specification to examine the description of a class or a particular method with which you are unfamiliar. If that doesn't suffice, then check for tutorials and articles on the web such as those on the `http://java.sun.com` site. In addition to the supplements on our Web Course, we provide an extensive set of web resource links. For major topics like Swing graphics, you will want to invest in some specialized books.

12.10   Web Course materials

The Chapter 12 Web Course: *Supplements* section gives a brief introduction to Java Beans and some other APIs such as the Media Framework and Java Audio. It also looks at the Web Start system for distributing Java applications. We include more information about performance issues and benchmarking and include links to a number of useful web pages.

The *Tech* section gives more examples of technical applications with Java such as pattern recognition tasks. The *Physics* section looks at developing a set of experiment analysis tools with Java.

References

[1] *Lesson: Printing in 2D Graphics – The Java Tutorial,* Sun Microsystems, `http://java.sun.com/docs/books/tutorial/2d/printing/`.

[2] *Playing Sounds – The Java Tutorial*, Sun Microsystems, `http://java.sun.com/docs/books/tutorial/sound/playing.html`.

[3] *Java Sound Programmer's Guide*, Sun Microsystems, 2002, `http://java.sun.com/j2se/1.4/docs/guide/sound/programmer_guide/`.

[4] Java Sound API, `http://java.sun.com/products/java-media/sound/`.

[5] Java Media Framework API, `http://java.sun.com/products/java-media/jmf/`.

[6] Jack Schirazi, *Java Performance Tuning*, 2nd edition, O'Reilly, 2003.

Part II
Java and the network

Chapter 13
Java networking basics

13.1 Introduction

Java arrived on the scene just as computer networking was expanding from isolated local area networks outward to the whole world via the Internet. The developers of Java quickly realized that exploiting the vast potential of networks would become a major activity for programmers in this new interconnected world, so they built a wide array of networking capabilities into the language. This capability grew with each new version of Java and became one of the primary reasons for its popularity.

In this chapter we review the basics of TCP/IP (Internet) networking and some of the tools that Java provides to exploit it [1,2]. In the rest of Part II we examine many of the more sophisticated networking capabilities of Java with an emphasis on how they could benefit scientific and engineering applications.

13.2 Internet basics

As shown in Figure 13.1, networking architecture is based on the concept of layers of protocols. (The more formal OSI – Open System Interconnection – model has seven layers but this one shows the essential layer definitions.) Each layer has its own standardized protocol and standardized application programming interface (API), which allows the next higher layer to communicate with it. Internally, the layers can be implemented in different ways as long as they provide the standard API. For example, the Network layer does not know if the physical layer is Ethernet or a wireless system because the software device drivers respond to the function calls the same way.

The term "Internet" refers primarily to the Network layer protocol known as the Internet Protocol (IP) and the Transport layer protocol known as the Transmission Control Protocol (TCP) forming the familiar TCP/IP acronym. The application layer includes various web protocols, such as the Hypertext Transfer Protocol (HTTP), which rely on the Internet sub-layers. Most users never see below the application layer.

When you send an email or a file over the Internet, the TCP and IP protocols split the message into groups of bytes called packets. Each packet holds a *header*

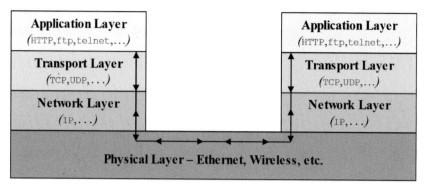

Figure 13.1 Networks use a layer architecture. Highest level applications talk to the Transport layer, which in turn talks to the Network layer and this in turn talks to the Physical layer. To the application user, however, the communications path will appear to be directly from one application to another. [2]

containing its destination and source addresses and other miscellaneous information such as error correction bytes. The body of the packet is called the data payload. These packets travel through the network via routers that lie at nodes (intersections) in the network. The routers read the destination addresses on the packets and, just as with mail in the postal system, send the packet to the next node closest to the final destination. If such a node doesn't respond or if the traffic load needs balancing, the transmitting node looks for alternate routes. When the packets reach their final destination, the original message is rebuilt. The packets may arrive out of order so the rebuilding must wait for all packets to arrive.

13.2.1 IP datagrams

The IP layer communicates via packets called datagrams. Datagrams have headers of 20–60 bytes and data payloads of up to 65K bytes. The headers contain the source and destination addresses. An IP address consists of four bytes displayed as four values separated by periods as in 130.237.217.62. The left-most byte is the highest-order address and so can represent a region or country. The lower-order bytes narrow the address down until the final byte indicates, typically, a particular computer on a WAN (Wide Area Network) or LAN (Local Area Network).

Datagrams from the same message can travel completely different paths as the routers dynamically choose paths for the same destination address so as to avoid loading down any one link. Thus datagrams may become lost or arrive out of order from how they were sent. When TCP/IP is used, the TCP layer is responsible for putting the packets back together in the proper order.

There are a number of special IP addresses. For example, any address beginning with 127, as in 127.0.0.1, translates as a *loop back* address. This means that any packets sent with this destination address automatically return to the source computer.

Users typically deal with text addresses called *hostnames* that are easier to remember than the numerical addresses. For example, `java.sun.com` is a hostname. Special nodes on the Internet called *Domain Name Servers* (DNS) translate hostnames into the corresponding numerical IP addresses. A hostname is composed of a top-level domain such as `com`, `se`, or `edu`. These domains are then divided into second level subdomains such as `sun.com`, `kth.se`, or `ucr.edu`. The systems at these locations can divide a domain further such as `java.sun.com`, `gluon.particle.kth.se`, or `physics.ucr.edu`.

13.2.2   TCP and UDP

Above the IP layer resides the Transport layer. It includes both the Transmission Control Protocol (TCP) mentioned previously and the User Datagram Protocol (UDP). The Transport layer attempts to smooth over the problems of the IP layer. The Transport layer can rearrange packets into their proper order and request retransmission of missing packets.

TCP guarantees that all bytes are provided in the correct order or else an error condition is reported. For text messages and files, for example, this is obviously a requirement. UDP, on the other hand, does not guarantee all the bytes in the correct order, or even that all bytes are received. As such, it has a lower overhead than TCP. For some applications, such as audio and video transmission, the loss of a few bytes is not significant and the use of UDP provides better performance.

13.2.3   Application layer

The Application layer involves all those user programs that we are so familiar with such as web browsers. These use protocols such as:

- `HTTP` – Hypertext Transfer Protocol – web page transmission
- `FTP` – File Transfer Protocol – for sending and obtaining files
- `SMTP` – Simple Mail Transport Protocol – mail
- `Telnet` – console sessions

These protocols require TCP since the programs cannot allow for randomly dropped bytes from source files and web pages. Other application layer programs, such as the `ping` program that sends test packets to a given IP address, can use the simpler UDP. Most Java network programs only deal with the application layer.

13.3   Ports

The IP address is used to get a packet to the right computer, but how does the packet get to the right program on that computer? A computer may have:

- several programs running at the same time
- several programs trying to communicate via the Internet
- all the programs communicating over the same physical Ethernet cable

Figure 13.2 A packet
arrives from the network
and goes to the port to
which it is addressed.

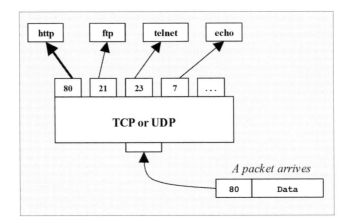

Figure 13.2 A packet arrives from the network and goes to the port to which it is addressed.

The *port* number is the key for organizing all this. The packets are guided by the operating system to the correct program according to the 16-bit port number. Casual internet users occasionally encounter port numbers when they are appended to web addresses, as in:

```
http://www.myschool.edu:80/
```

Port 80 is the default port for the HTTP server so nowadays it is seldom included in the URL. Some popular web servers, notably the open source Tomcat server, commonly use port 8080 by default. So a URL address might appear as

```
http://www.myschool.edu:8080/
```

Various other applications for particular protocols use standard port values as shown in Figure 13.2. Unix machines reserve ports 1–1023 for privileged services (i.e. owned by the Unix *root* account). Windows machines do not restrict these ports but in order to make your Java programs portable it is wise to choose port values above 1023.

One type of firewall assigns port numbers to the machines behind its shield. Incoming packets all go to the same IP address but with different port numbers according to the machine they are destined for. This both reduces the exposure and accessibility to the machines and reduces the need for universally unique IP numbers. Behind the firewall, the local addresses can be the same as in other LANs since the LANs are isolated from each other. This is one of the reasons that 4 bytes remain sufficient for the Internet despite the explosion in the number of devices with IP addresses. (Nevertheless, a 16-byte version of IP addressing called IPv6 will gradually take over.)

13.4   Java networking

Java was designed from the start with networking in mind. Java first became famous because of applets, which were invented to provide dynamic content to web pages. Many other features of Java, however, extend Java network programming far beyond just applets. Some of these network-related capabilities include the following:

- `java.net` contains numerous network-related classes.
- *Remote Method Invocation* (RMI) packages allow a Java program running on one machine to invoke Java methods in a program running on another machine.
- The *Streaming I/O* architecture in Java provides for a uniform structure in that I/O over a network is treated the same as I/O to a disk or the console.
- *Serialization* allows objects to be deconstructed, sent over a network, and then reconstructed on another machine (see Section 9.8).
- *Threading* helps to build server programs since they can create separate threads to tend to new clients as they connect.
- `java.security` and other packages allow for secure interactions over networks.
- *Portability* of java means that Java networking programs developed on one type of platform can also run and communicate with each other when used on other platforms.

In the rest of this chapter and the other chapters of Part II, we discuss these and other features that make Java a powerful tool for network programming.

13.5   The URL class

Browsers obtain resources from the web by specifying web addresses, which are officially called *Uniform Resource Locators* (URLs). The formal description of the URL components goes as follows:

```
Protocol_ID://Host_IP_address:Port/Filename#Target
```

The components of the URL include:

- `Protocol_ID` – HTTP, FTP, etc.
- `Host_IP_address` – host name in either numerical IP or hostname format
- `Port` – port number. Default is 80 if omitted
- `Filename` – name of a hypertext web page or other type of file. Default is `index.html` or `index.htm`. The file name can include a directory path such as `/data/run5.html`
- `Target` – optional reference address within a web page

In Java a URL can be represented by an instance of the `java.net.URL` class. For example, in an applet you can obtain the URL address of the location of the applet's code with the method

```
public URL getCodeBase ()
```

If you want to download an image from the same directory where the applet code is located, you can use

```
getImage (getCodeBase (), filename);
```

This method combines the URL returned by `getCodeBase()` with the image file name to create a URL to locate and download the image file.

The URL class provides several methods for obtaining URL information. The following application (based on a program in Sun's Java Tutorial on networking) breaks apart and displays the components of a complete URL string:

```java
import java.net.*;
import java.io.*;
/** Parse a URL address into its components. **/
public class ParseAddress
{
  public static void main (String[] args) {
    if (args.length!=1) {
      System.out.println (
        "Error: missing url parameter");
      System.out.println (
        "Usage: java ParseAddress <url>");
      System.exit (0);
    }

    try {
      URL url = new URL (args[0]);
      System.out.println ("Protocol = " +
                          url.getProtocol ());
      System.out.println ("Host = " + url.getHost ());
      System.out.println ("Port = " + url.getPort ());
      System.out.println ("File name = " +
                          url.getFile ());
      System.out.println ("Target = " + url.getRef ());
    }
      catch (MalformedURLException e) {
        System.out.println ("Bad URL = " + arg[0]);
    }
  }
} // class ParseAddress
```

Running the program with a URL string on the command line produces the following:

```
c:\>java ParseAddress
  http://www.myschool.edu:80/data/x.html#d

Protocol = http
Host = www.myschool.edu
```

```
Port = 80
File name = /data/x.html
Target = d
```

You can construct a URL object from a URL string:

```
URL kth = new URL ("http://www.kth.se/index.html");
```

or from individual string components as in:

```
URL kth = new URL ("http", "www.kth.se", "/index.html");
```

When you attempt to create an instance of URL, the constructor checks the components for proper form and value. If the URL specification is invalid, the constructor throws a MalformedURLException, which is a checked exception and must be caught.

Typically one of the first tasks that new Java programmers want to do for their own work is to read a file from an applet. This might involve a simple data file in text format such as a list or a table of values and labels. We have learned to read image files and audio files from applets and applications but have not yet read a text file. This is in fact fairly easy to do. However, keep in mind that for applets the SecurityManager in the browser's JVM restricts access to only those files on the host system of the applet. (We discuss security managers further in Chapter 14.)

Below we show the applet ReadFileViaURL, which reads a text file located with an instance of the URL class. It uses a method from the URL class that provides file access via an instance of InputStream (see Chapter 9). We wrap this stream with an InputStreamReader to provide proper character handling. We in turn wrap this class with BufferedReader, which provides buffering to smooth out the stream flow and also contains the convenient readLine() method that grabs an entire line and returns it as a string.

The program can also run as a standalone application. We see in the readFile() method how to obtain a URL object from a File object for local files. Figure 13.3 shows the interface in application mode after reading the default file.

Figure 13.3 The ReadFileViaURL applet displays the text file that it read via a URL address.

```java
import javax.swing.*;
import java.awt.*;
import java.awt.event.*;
import java.util.*;
import java.io.*;
import java.net.*;

/** Demonstration of reading a local file with a URL. **/
public class ReadFileViaURL extends JApplet
            implements ActionListener
{
  // A Swing text area for display of string info
  JTextArea fTextArea;
  String fFileToRead = "data.txt";
  StringBuffer fBuf;

  // Flag for whether the applet is in a browser
  // or running via the main () below.
  boolean fInBrowser = true;

  /** Build the GUI. **/
  public void init () {

    // Create a User Interface with a text area with
    // scroll bars and a Go button to initiate processing
    // and a Clear button to clear the text area.
    Container content_pane = getContentPane ();

    JPanel panel = new JPanel (new BorderLayout ());

    // Create a text area to display file contents.
    fTextArea = new JTextArea ();
    fTextArea.setEditable (false);

    // Add to a scroll pane so that a long list of
    // computations can be seen.
    JScrollPane area_scroll_pane = new JScrollPane
      (fTextArea);

    panel.add (area_scroll_pane, BorderLayout.CENTER);

    JButton go_button = new JButton ("Go");
    go_button.addActionListener (this);

    JButton clear_button = new JButton ("Clear");
    clear_button.addActionListener (this);
```

```
    JButton exit_button = new JButton ("Exit");
    exit_button.addActionListener (this);

    JPanel control_panel = new JPanel ();
    control_panel.add (go_button);
    control_panel.add (clear_button);
    control_panel.add (exit_button);

    panel.add (control_panel, BorderLayout.SOUTH);

    // Add text area with scrolling to the content pane.
    content_pane.add (panel);

    // If running in a browser, read file name from
    // applet tag param value. Else use the default.
    if (fInBrowser) {
      // Get setup parameters from applet html
      String param = getParameter ("FileToRead");
      if (param != null) {
        fFileToRead = new String (param);
      }
    }
  } // init

  /** Use a URL object to read the file. **/
  public void readFile () {
    String line;
    URL url = null;

    // Get the URL for the file.
    try {
      if (fInBrowser)
        url = new URL (getCodeBase (), fFileToRead);
      else {
        File file = new File (fFileToRead);
        if (file.exists ())
          url = file.toURL ();
        else {
          fTextArea.append ("No file found");
          System.out.println ("No file found");
          System.exit (0);
        }
      }
    }
```

```java
    catch (MalformedURLException e) {
      fTextArea.append ("Malformed URL = " + e);
      System.out.println ("Malformed URL = " + e);
      return;
    }

    // Now open a stream to the file using the URL.
    try {
      InputStream in = url.openStream ();
      BufferedReader dis = new BufferedReader (
                           new InputStreamReader (in));
      fBuf = new StringBuffer ();

      while ((line = dis.readLine ())!= null) {
        fBuf.append (line + "\n");
      }
      in.close ();
    }
    catch (IOException e) {
      fTextArea.append ("IO Exception = " + e);
      System.out.println ("IO Exception = " + e);
      return;
    }

    // Load the file into the TextArea.
    fTextArea.append (fBuf.toString ());
} // readFile

/**
  * Can use the start() method, which is called after
  * init() and the display has been created.
  **/
public void start () {
  // Now read the file.
  readFile ();
} // start

/** Respond to the buttons **/
public void actionPerformed (ActionEvent e) {
  string source = e. getActionCommand ();
  if (source.equals ("Go"))
    start ();
  else if (source.equals ("Clear"))
    fTextArea.setText (null);
```

```
      else
          System.exit (0);
  } // actionPerformed

  /** Display the println string on the text area **/
  public void println (String str) {
    fTextArea.append (str + "\n");
  } // println

  /** Display the print string on the text area **/
  public void print (String str) {
    fTextArea.append (str);
  } // print

  /** Create the frame and add the applet to it **/
  public static void main (String[] args) {

    int frame_width = 200;
    int frame_height = 300;

    // Create ReadFileViaURL object and add to the frame.
    ReadFileViaURL applet = new ReadFileViaURL ();
    applet.fInBrowser = false;
    applet.init ();

    // Create frame and then set its exit mode.
    JFrame f = new JFrame ("Read file from URL Demo");
    f.setDefaultCloseOperation (JFrame.EXIT_ON_CLOSE);

    // Add applet to the frame
    f.getContentPane ().add (applet);
    f.setSize (new Dimension (frame_width, frame_height));
    f.setVisible (true);
  } // main
} // class ReadFileViaURL
```

13.6 InetAddress

The java.net.InetAddress class represents IP addresses. It works with either a string host name such as java.sun.com or a numerical IP address such as 64.124.81.56. The InetAddress class has a number of useful methods for dealing with host names and IP addresses. The demonstration programs below illustrate some of the capabilities of the InetAddress class [2].

The applet `LocalAddress` displays the local host and IP address. Note that the `SecurityManager` in the browser JVM may block access to this information. You can try it in the `appletviewer` tool or run it as an application.

```java
import java.applet.*;
import java.awt.*;
import java.net.*;

/** Show how InetAddress can provide the local
  * IP address. **/
public class LocalAddress extends Applet
{
  String fMessage = "";

  /**
    * Create an instance of InetAddress for the
    * local host and display the local IP address.
    **/
  public void init () {
    try {
      InetAddress local_Address =
        InetAddress.getLocalHost ();
      fMessage = "Local IP address = "
                  + local_Address.toString ();
    }
    catch (UnknownHostException e) {
      fMessage = "Unable to obtain local IP address";
    }
    System.out.println (fMessage);
  } // init

  /** Paint IP info in Applet window. **/
  public void paint (Graphics g) {
    g.drawString (fMessage, 20, 20);
  } // paint

  /** Print out the local address in app mode. **/
  public static void main (String [] args) {
    LocalAddress applet = new LocalAddress ();
    applet.init ();
  } // main
} // class LocalAddress
```

The application `TranslateAddress`, shown next, returns the IP address if given a host name, or returns the host name if given an IP address:

```java
import java.net.*;

/** Translate IP to a host name or host name to IP
  * address. **/
public class TranslateAddress
{
   public static void main (String[] args) {

      // Look for command line argument
      if (args.length != 1) {
        System.out.println ("Error! No IP or host name
                            address");
        System.out.println (
           "Usage: java TranslateAddress java.sun.com");
        System.out.println (
           " or java TranslateAddress 209.249.116.143");
        System.exit (0);
      }

      try {
        // When the argument passed is a host name (e.g.
        // sun.com), the corresponding IP address is
        // returned. If passed an IP address, then only the
        // IP address is returned.
        InetAddress address = InetAddress.getByName
                              (args[0]);
        System.out.println ("Address " + args[0] +
                            " = " + address);

        // To get the hostname when passed an IP address
        // use getHostName(), which will return the host
        // name string.
        System.out.println ("Name of " + args[0] + " = " +
                            address.getHostName ());

      }
      catch (UnknownHostException e) {
        System.out.println ("Unable to translate the
                            address.");
      }
   } // main
} // class TranslateAddress
```

Here are two example runs of this program:

```
c:\> java TranslateAddress gluon.particle.kth.com
Address gluon.particle.kth.se = gluon.particle.kth.se/
130.237.34.133
Name of gluon.particle.kth.se = gluon.particle.kth.se

c:\> java TranslateAddress 130.237.34.133
Address 130.237.34.133 = /130.237.34.133
Name of 130.237.34.133 = gluon.particle.kth.se
```

Note that the method `getByName (String str)` returns an instance of
`InetAddress` when given a host name. Then when the `toString()` method of
this `InetAddress` object is invoked by the string concatenation, it displays the
`HostName/IP_Address` pair. However, when given an IP address, the resulting `InetAddress` object does not show the host name. Instead, you can use the
`getHostName()` method to obtain the host name.

13.7 Sockets

Sockets provide connections between applications that allow streams of data
to flow. Sockets in Java are straightforward to set up. The package `java.net`
provides two kinds of socket classes:

1. **Socket** – provides a connection-oriented service that behaves like telnet or ftp. The
   connection remains active, even with no communications occurring, until explicitly
   broken.
2. **DatagramSocket** – involves the following:
   (a) connectionless protocol
   (b) transfer of datagram (i.e. UDP) packets
   (c) no fixed connection
   (d) packets can arrive out of order
   (e) no guarantee a packet will arrive.

We discuss sockets further in Chapters 14 and 15. The demonstration program
below illustrates how to use a socket connection to run the *whois* internet operation
[2]. The `whois` service returns information about a given domain name. The
registry site `whois.internic.net` provides this service.

The code snippet below shows how to create an instance of the `Socket` class
that connects to port 43 at host name `whois.internic.net`. An output stream
is obtained from the socket and a `PrintWriter` wrapped around it. This is then
used to send the address of interest to the `whois` service.

Similarly, an input stream is obtained from the socket and wrapped first
with an `InputStreamReader` and then a `BufferedReader` so we can use
`readLine()` to obtain a whole line of text at one time from the `whois` output.
You can enter other domain names in the text field.

```
. . . GUI code from the WhoisApplet program . . .

  /** Connect to the whois service via a socket. Write the
   * address with a PrintWriter object and read the
   * output via a BufferedReader. Send the output to the
   * text area. **/
  public void whoisConnect (String query_address) {

    int port = 43; // Standard whois port
    String reply;

    try {
      Socket whois_socket =
        new Socket ("whois.internic.net", port);

      PrintWriter print_writer =
        new PrintWriter (whois_socket.getOutputStream (),
                         true);
      print_writer.println (query_address);

      InputStreamReader input_reader =
        new InputStreamReader
        (whois_socket.getInputStream ());
      BufferedReader buf_reader =
        new BufferedReader (input_reader);

      while ((reply = buf_reader.readLine ()) != null) {
        // Write the whois info to the textarea.
        fTextArea.append (reply + '\n');
      }
      fTextArea.append ("\n\n");
        // Add space between queries
    }
    catch (IOException e) {
      fTextArea.append ("IO Exception = " + e);
      System.out.println ("IO Exception = "+ e);
      return;
    }
  }
. . .
```

Figure 13.4 shows the program interface with the output of a typical case. Though written in applet form, the security manager in most browser JVMs blocks

Figure 13.4 The interface for the `WhoisApplet` shows results in a text area for the `whois` operation on the domain name entered into the text field at the lower left.

accesses to IP addresses other than the source of the applet. In the figure the program is running as an application.

13.8   The client/server model

The *client/server* paradigm has become a dominant one for the Internet. In this model the clients are programs running on remote machines that communicate with a program called the server that runs at a single site and responds to requests from many clients. The server provides the clients with, say, web pages or database information. Much of the World Wide Web is built on the client/server paradigm. The clients are web browsers run by many millions of individual users, and the servers are the many web-hosting systems running at the many host sites on the web. A single server at a single host can support many hundreds or thousands or more of clients from around the world. Large systems that serve hundreds of thousands of clients balance the server load over multiple machines in an arrangement called "server farms."

With Java you can build client/server systems with sockets or with RMI (Remote Method Invocation). In the following chapters of Part II we come back to the client/server model repeatedly, though we will keep things simple by considering a server to be a program running on a single server computer.

In a socket-based client/server system, a server listens to a particular port for client applications sending requests for connections. A `ServerSocket` class is provided in Java that allows for a server to monitor and answer such requests for

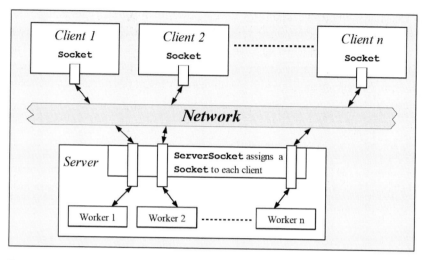

Figure 13.5 Schematic of a socket based client/server system. The `ServerSocket` in the server program monitors a particular port for requests for connections from clients on the *network*, which could range from a local area network to the global Internet. The client programs connect via an instance of `Socket` using the server's IP address and port. When the `ServerSocket` makes a connection with the client, it returns an instance of `Socket` that is given to a thread process, which we call a `Worker`, to attend to the requests of the client. The `ServerSocket` returns to monitoring the port for new requests for connections and spins off a worker process for each client that connects.

connections. The client sends the request for a connection by creating a socket with the host name and port for that server as discussed in the previous section.

Figure 13.5 shows a diagram illustrating the basics of a socket-based client/ server system. The `ServerSocket` instance listens for a client to connect to the particular port. When a client request arrives, the `ServerSocket` object sets up a `Socket` instance for the connection and then spins off a new thread to interact with the client via that socket. Many clients can therefore be served since each client has an independent thread dedicated to it.

A Java web server would be built with this kind of socket-oriented client/server approach. In Chapters 14–15 we discuss this approach further and show how to build custom servers for applications such as providing access to data at a remote device. We also use such a server in Chapter 24 on an embedded Java processor.

In Chapters 16–20 we look at a more direct approach to communications between clients and servers. Using tools like RMI and CORBA, a Java program on one machine can invoke a method in an object on another machine just as if the code was running on the local platform. A client, for example, could call a method on a server program to obtain some parameter of interest. The server, in turn, can invoke methods in the client.

Remote method invocation provides for powerful distributed computing capabilities. Much of the complicated machinery to make this happen is hidden from the application programmer who can instead concentrate on the task at hand. The portability of Java is a further advantage since a distributed computing application can be developed that will run with many different platforms with the same code.

13.9   Web Course materials

The code for the above examples and several other networking demonstrations are available in the Web Course pages. Web resource links to supporting information are also provided.

References

[1] *Trail: Custom Networking – The Java Tutorial*, Sun Microsystems,
    `http://java.sun.com/docs/books/tutorial/networking/`.
[2] E. R. Harold, *Java Network Programming*, 2nd edn, O'Reilly, 2000.

Resources

Ian F. Darwin, *Java Cookbook*, 2nd edition, O'Reilly, 2004.
P. Niemeyer and Jonathan Knudsen, *Learning Java*, 2nd edition, O'Reilly, 2002.

Chapter 14
A Java web server

14.1 Introduction

A web server program runs continuously while waiting for and answering requests it receives over the Internet from browsers. Typically the requestor asks for the transmission of a web page in HTML (Hypertext Markup Language) format or asks for some other HTTP (Hypertext Transmission Protocol) service such as the running of a CGI (Common Gateway Interface) program.

Developing web servers and server applications such as online stores became the first big money-making business area that used Java extensively. Sun offers additional packages with the Java 2 Platform Enterprise Edition (J2EE) to support server development for applications such as database access, shopping cart systems for web stores, and other elaborate *middleware* services that can scale to large numbers of client users. Companies like IBM and BEA have been quite successful in selling their own middleware Java software.

In this chapter and the next we look at a simple socket-based approach to building web servers for specialized applications [1–4]. This can be done with the classes available in J2SE. In Chapters 16–20 we focus on RMI (Remote Method Invocation) clients and servers and other distributed computing techniques. In Chapter 21 we return to web-based networking with a discussion of web services.

We show here how to create a simple web server that could run on any platform that implements a JVM with the `java.net` and `java.io` packages. Such a *micro-server* could, for example, run on remote devices in a scientific experiment and provide data and status information to client programs. It could also accept instructions for operating the devices, installing calibration data, and so forth.

These devices are not theoretical. In Chapter 24 we discuss commercially available processor cards that provide both network access and a JVM (or a hardware implementation) with which to run such a server program. The cards are intended for embedded applications such as monitoring and controlling sensors distributed in a scientific experiment, on a factory assembly line, or in power plant equipment.

14.2 Designing a web server

When you click on a URL link in a browser page, a request is sent to the computer at the host address given in the URL. If the port is not specified, the request goes to the default port 80. The web server program monitors this port and, when a request arrives, the server opens a socket for the client and sends the web page or other data requested.

Different types of servers monitor different ports to provide services such as database access, email, audio/video streaming, and so forth. Web serving is a stateless interaction in that the connection ends after answering the request and the server does not usually maintain any further information about the client. If the same client returns later, then a completely new session is created with no knowledge about the previous session. (An online store might use a *cookie* file on the browser platform and the IP address of the requestor to maintain information about an interaction for a given period of time.) There is no continuous connection as in a `telnet` session, for example.

With Java you can create and run your own web server that follows the HTTP protocols. On Unix machines, the ports numbered 0–1023 are restricted but otherwise your server is free to choose whatever port it wants to monitor for clients requesting a connection.

Note that since you customize the server as you wish, you don't have to break off the connection (that is, close the socket) immediately as with standard web servers. We discuss in Chapter 15 a client server system using sockets that can maintain a continual connection.

In this chapter we develop `MicroServer`, a basic HTTP server that illustrates all the basic components of a web server. (It is similar to an example in the book by Niemeyer and Knudsen [3].) To serve its clients, our little web server needs to:

- create a `ServerSocket` instance that watches for incoming client requests
- create a `Socket` instance to connect with a client
- spin off a thread to handle the client's request
- set up streams to carry out the I/O with the client

The following sections discuss each of these tasks.

14.2.1 `ServerSocket`

In Chapter 13 we briefly discussed communications with the `Socket` and `ServerSocket` classes. In the code snippet below, the `ServerSocket` object waits for a client to request a connection. The `accept()` method blocks (i.e. doesn't return) until the client connects, and then it returns an instance of `Socket`.

```
. . . In the main() method in MicroServer . . .

  // Create a ServerSocket object to watch port for clients
  ServerSocket server_socket = new ServerSocket (port);
  System.out.println ("Server started");
  // Loop indefinitely while waiting for clients to connect
  while (true) {
    // accept () does not return until a client
    // requests a connection
    Socket client_socket = server_socket.accept ();
    // Now that a client has arrived, create an instance
    // of our thread subclass to respond to it.
    Worker worker = new Worker (client_socket);
    worker.start ();
    System.out.println ("New client connected");
  }
  . . .
```

The program passes the socket to an instance of a thread class called `Worker` that handles all further communication with this particular client. The main thread loop then returns to wait for another client connection. When one arrives it will create a new `Worker` to handle it. The details of the `Worker` class are discussed next.

14.2.2   Worker threads for clients

Our `Worker` class is a subclass of `Thread`. There will be one `Worker` object per client:

```
/** Here is our thread class to handle clients.**/
public class Worker extends Thread {
  Socket fClient;
  // Pass the socket as an argument to the constructor
  Worker (Socket client) throws SocketException {
    fClient = client;
    // Set the thread priority down so that the
    // ServerSocket will be responsive to new clients.
    setPriority (MIN_PRIORITY);
  } // ctor

  public void run () {
    . . . Do the client interaction here
  }
  . . .
} // class Worker
```

We set the priority lower on our thread so that the client processing does not dominate other tasks such as the `ServerSocket` monitoring of new incoming clients and the serving of other clients by other `Worker` objects.

The `run()` method is, of course, the heart of the thread class and is where the interaction with the client occurs. Before we discuss possible tasks for the `Worker` we look at setting up I/O streams with the client.

14.2.3 I/O with the client

Our thread communicates with the client via the I/O streams made available by the `Socket` object. The code snippet from the `run()` method here in our `ClientProcess` shows how to obtain an `InputStream` from the socket and then wrap it with an `InputStreamReader` using the 8859_1 character encoding and then wrap that with a `BufferedReader`. (See Chapter 9 for information about character encodings such as 8859_1.)

```
. . . The run() method in Worker . . .

public void run () {
  try {
    // Use the client socket to obtain an input stream.
    InputStream socket_in = fClient.getInputStream ();
    // For text input we wrap an InputStreamReader around
    // the raw input stream and set ASCII character encoding.
    InputStreamReader isr =
              new InputStreamReader (socket_in, "8859_1");

    // Finally, use a BufferReader wrapper to obtain
    // buffering and higher order read methods.
    BufferedReader client_in = new BufferedReader (isr);
    . . .
```

We then obtain an `OutputStream` from the socket as shown in the next code snippet. We wrap this with an `OutputStreamWriter` and then with a `PrintWriter`.

```
. . . Continue in the run() method in Worker . . .


  . . .
    BufferedReader client_in = new BufferedReader (isr);

    // Now get an output stream to the client.
    OutputStream out = fClient.getOutputStream ();
```

```
// For text output we wrap an OutputStreamWriter around
// the raw output stream and set ASCII character encoding.
OutputStreamWriter osr =
                new OutputStreamWriter (out, "8859_1");

// Finally, we use a PrintWriter wrapper to obtain its
// higher level output methods. Use autoflush mode.
// (Autoflush occurs only with println().)
PrintWriter pw_client_out = new PrintWriter (osr, true);
. . .
```

At this point in the code we have streams for both input and output communications with the client, but we haven't said anything about just *what* they communicate. The client can place arbitrary bytes on the stream and the server will see them, but unless some agreement is made about what form those bytes should take and what they mean, the communication that happens is rather meaningless. In other words, we need to develop some sort of *protocol* so that the server and the client can understand each other.

14.3 Hypertext Transfer Protocol (HTTP)

For network communications to work correctly, a common format or protocol must be established. Many standard protocols have already been defined – for example, HTTP (Hypertext Transfer Protocol). In an HTTP request, a line such as the following must be sent from the client to the server:

```
GET /index.html HTTP/1.0 \r\n\r\n
```

Here GET is the request keyword, /index.html is the file requested, and HTTP/1.0 indicates the protocol and version number of the protocol to be used. Finally, the characters \r\n\r\n indicate the two carriage return/linefeed pairs that terminate the line.

The next code snippet from run() obtains the request line sent from the client by invoking the readLine() method of BufferedReader. Then the request text is broken into tokens with the split() method in the String class (see Chapter 10). The tokens are checked to determine if the client sent the "GET" command and, if so, to obtain the name of the file that the client is requesting.

```
. . . In the run() method in Worker . . .

    // First read the message line from the client
    String client_str = client_in.readLine ();
```

```java
System.out.println ("Client message: " + client_str);

// Split the message into substrings.
String [] tokens = client_str.split(" ");

// Check that the message has a minimun number of words
// and that the first word is the GET command.
if ((tokens.length >= 2) &&
    tokens[0].equals ("GET")) {
  String file_name = tokens[1];

  // Ignore the leading "/" on the file name.
  if (file_name.startsWith ("/"))
    file_name = file_name.substring (1);

  // If no file name is there, use index.html default.
  if (file_name.endsWith ("/") || file_name.equals (""))
    file_name = file_name + "index.html";

  // Check if the file is hypertext or plain text
  String content_type;
  if (file_name.endsWith (".html") ||
      file_name.endsWith (".htm")) {
    content_type = "text/html";
  }
  . . .
```

If the request from the client is valid, we then read the local file that the client is requesting and return it to the client. As shown in the following code, to read the file we first obtain a `FileInputStream` for the file. We send text messages back to the client using the `PrintWriter` methods. We note that the `PrintWriter` methods don't throw `IOException` and instead the class offers the `checkError()` method, which returns true if an `IOException` occurred. For the sake of brevity, we did not check for errors after every print invocation. (In the `DataWorker` class in Chapter 15, we place the print invocations in utility methods that check for errors and throw exceptions when they occur.)

The line "`HTTP/1.0 200 OK\r\n`" gives the protocol and version number of the return message followed by the 200 code number that indicates that the file has been found successfully and will be sent. This is followed by the file's modification date, an identification of the server, the length of the file and the file type (e.g. "`text/html`"). Finally, the whole file is sent in one big byte array via the `write()` method of `OutputStream`.

If the file is not available, the program sends the "`404 Object Not Found`" error message, which is a common one for web users. If the request line had problems, a "`400 Bad Request`" error message is sent.

```
. . . Continue in run() method in Worker . . .

     // Now read the file from the disk and write it to
     // the output stream to the client.
     try {
        // Open a stream to the file.
        FileInputStream file_in =
          new FileInputStream (file_name);

        // Send the header.
        pw_client_out.print ("HTTP/1.0 200 OK\r\n");
        File file = new File (file_name);
        Date date = new Date (file.lastModified ());
        pw_client_out.print ("Date: " + date + "\r\n");
        pw_client_out.print ("Server: MicroServer 1.0\r\n");
        pw_client_out.print ("Content-length: " +
                             file_in.available () + "\r\n");
        pw_client_out.print ("Content-type: " +
                             content_type + "\r\n\r\n");

        // Create a byte array to hold the file.
        byte [] data = new byte [file_in.available ()];

        file_in.read (data);     // Read file into byte array
        client_out.write (data);// Write it to client stream
        client_out.flush ();     // Flush output buffer
        file_in.close ();        // Close file input stream

     } catch (FileNotFoundException e) {
        // If no such file, then send the famous 404
        message. pw_client_out.println ("404 Object Not
                                        Found");
     }
   } else {
     pw_client_out.println ("400 Bad Request");
   }
} catch (IOException e) {
  System.out.println ("I/O error " + e);
}

// Close client socket.
try {
  fClient.close ();
} catch (IOException e) {
  System.out.println ("I/O error " + e);
}
// On return from run () the thread process dies.
} // run
```

Figure 14.1 Running `MicroServer` from the command line. The display here shows the output to the console after a request for a file arrived.

This short program provides a basic web server that returns web files to browsers and any other client program that connects to the port and uses the proper HTTP protocol. The `Socket` and `ServerSocket` classes, along with the I/O stream classes, do most of the work. The complete code listing appears in the Web Course for Chapter 14.

14.4 Running the server

We run our server application from the command line with

```
> java MicroServer 1234
```

The program takes a port number in the command line. (On a Unix platform you should pick a port number above 1023.) Figure 14.1 shows an example of running `MicroServer` from the console with the port 1234. The server prints to the console when a client connects to it.

In our directory with `MicroServer` we put a simple web page file named `message.html` containing:

```
<html>
  <head>
    <TITLE>Message Test</TITLE>
    <meta http-equiv="Content-Type"
          content="text/html; charset=iso-8859-1">
  </HEAD>
  <BODY BGCOLOR="#FFFFFF" >
       A file sent by the MicroServer.
  </body>
</html>
```

Figure 14.2 shows the file displayed in a browser that connected to our `MicroServer`. For testing with both the server and browser running on the same machine you can use the special loopback IP address 127.0.0.1. The URL must explicitly include the port number since the server uses a different number than the default HTML port 80.

Figure 14.2   A web page with a single text line in delivered by `MicroServer` to a browser.

14.5   A more secure server

The `MicroServer` has a serious security problem. It allows access to almost any file on the server's host computer. For example, suppose the server runs on the site `www.myschool.edu` with port 2222 and the client connects with a URL such as:

```
http://www.myschool.edu:2222/../restricted.html
```

The "`..`" refers to the directory above the directory where the server code is located. In that case, the file `restricted.html` would be delivered to our inquisitive client.

To control access to resources in Java programs, an elaborate system involving the `SecurityManager` class, policy files, permissions, and other tools are available. Security is a big issue in Java so we can only touch here on a few aspects of it. In the following sections we show the basics of how to make our server program more secure.

14.5.1   The security manager

To restrict a Java program's access to system resources, you can load a `SecurityManager` object that controls access to external resources. An instance of the `SecurityManager` class must be installed when an application first begins. If not, then the default `null` security manager puts the server into a completely unrestricted state. (For applets running in a browser JVM, the security manager severely restricts the actions allowed.)

Before Java 1.2 you needed to create a subclass of `SecurityManager` and customize it for your security requirements. The `SecurityManager` class holds many methods of the form `checkX (params)`, such as `checkDelete (String file)`, that throw an instance of `SecurityException` when they wish to block an attempt to execute the particular action X. The subclass must override those methods for the actions that you want to allow or to block only in particular circumstances. For example, in the case of deleting a file, your overriding version of the `checkDelete()` method could examine the file name and path to determine if a deletion should be permitted.

This approach to security, however, requires hard-coding and is very clumsy considering the large number of different types of system access operations now available to Java programs. With Java 1.2 a much more flexible *permissions* based security system was introduced. In this approach, a security *policy file*, which is a text file and therefore easily modifiable, is checked by the security manager to determine if particular actions can be granted. Anything that isn't explicitly granted is forbidden.

For example, suppose the policy file `myRules.policy` contains the entry

```
grant codeBase "file:C:/Java/apps/" {
   permission java.io.FilePermission "*.tmp", "delete";
};
```

This policy specification allows the application to delete those files in the directory `C:\Java\apps\` that end with the "`.tmp`" suffix but no others. We discuss more about the details of the policy file and permissions in Section 14.5.3.

With this new security design, an application specifies the security manager parameters from the command line using the `-D` option rather than in program code:

```
c:> java -Djava.security.manager -Djava.security.policy
=myRules.policy MyApp
```

(The continuous line is broken here to fit within page margins.) This approach to configuring the security for access to the system provides for much greater flexibility and clarity than customizing a `SecurityManager` subclass that must be recompiled after every modification.

14.5.2 Policy file for the server

We use our `MicroServer` to illustrate how to set up the security permissions for file access. Without these changes, there are no protections that prevent the client from requesting any file on the system. To control access to files, we create the policy file `microServer.policy` in which we put the following code:

```
grant codeBase "file:C:/Java/MicroServer/Secure/"
{
   permission java.net.SocketPermission "localhost:1024-",
                                 "accept,connect,listen";
   permission java.io.FilePermission "/-", "read";
};
```

The server must access sockets and, since this is an external resource, a permission statement is required. So in the above `grant` statement, we permit the program to listen for, accept, and connect with any socket on a port numbered "`1024`" and higher (by using the symbol "1024-"). Similarly, we give permission for file access to the server's directory and to its subdirectories with the "/-" parameter. We put this policy into the directory c:\Java\MicroServer\Secure\ where we also put the server code. (The grant statement must always use forward slashes regardless of the platform.)

To catch the instances of SecurityException that can now be thrown, we create a new version of the server called MicroServerSecure that is identical to MicroServer except that its Worker class adds a new catch statement in the run() method as shown below in bold:

```
. . . Method run() in modified Worker class for MicroServerSecure
              . . .
        }
      catch (FileNotFoundException e) {
        // If no such file, then send the famous 404 message.
        pw_client_out.println ("404 Object Not Found");

      }
      catch (SecurityException se) {
          // An attempt was made to read a file
          // in a forbidden location.
          pw_client_out.println ("403 Forbidden");
      }
    }
    else {
      pw_client_out.println ("400 Bad Request");
    }
  }
  catch (IOException e) {
    System.out.println ("I/O error " + e);
  }
  . . .
```

Now we run this server with

```
c:> java -Djava.security.manager -Djava.security.policy
=microServer.policy MicroServerSecure
```

(This should be one continuous line on a Windows platform or entered with line-continuation characters on Unix or Linux.) When a client browser attempts to access the file in the restricted area, the server now sends the "403 Forbidden" message.

14.5.3   More Java security

Java security is a huge topic that involves not only the security manager but many other issues such as cryptography, public/private keys, certificates, etc. Even a thorough discussion of permissions and the policy file is beyond the scope of this book. (See references [5, 6] for more information and tutorials about Java security capabilities.) However, as we see above, the basics of setting permissions are fairly straightforward.

Besides the `java.io.FilePermission` class, there are a number of permission classes that represent the various types of access to external resources that the security manager controls. They include

```
java.security.AllPermission
java.security.SecurityPermission
java.awt.AWTPermission
java.io.FilePermission
java.io.SerializablePermission
java.lang.reflect.ReflectPermission
java.lang.RuntimePermission
java.net.NetPermission
java.net.SocketPermission
java.util.PropertyPermission
```

and several others. For a complete listing, see the reference for the *Permissions in the Java SDK* document at `http://java.sun.com` [6]. These classes are all subclasses of `java.security.Permission`.

The basic format of the `grant` statement goes as

```
grant codeBase "URL" {
    permission permission_class_name1 "target_name", "action";
    permission permission_class_name2 "target_name", "action";
    . . .
};
```

A more elaborate form of the `grant` statement includes information on where to find certificates with the public keys needed to decode programs signed with private keys.

The codebase item indicates the location of the code to which you are granting a permission. If you load a class from a different location, then the permission does not apply to it. If the codebase is empty then the permissions apply to code from any location. In the example in the previous section, we used the `"file:"` type of URL for referencing a local file.

Some of the permissions require a target name and a listing of the particular actions allowed. See, for example, the `java.io.FilePermission` case mentioned in the previous section.

You can make the policy files by hand, but an alternative is to use the `policytool` program supplied with the SDK. It provides a graphical interface in which you specify the details of the policy file. The program makes it fairly easy to set the codebase information, to choose what permissions you need, to set the target names and action settings, and to select where to put the policy file. See the Web Course Chapter 14 for more information about using this program.

A program may need to check if a particular action is allowed before attempting it. Perhaps you create a class that will be used with other people's programs who might be using different security settings. It might waste a lot of execution time in your code if an action in the final step is not allowed.

So you can check if an action is allowed by first creating an instance of the particular permission and passing it to checkPermission (java. security.Permission) in the SecurityManager class. If it throws a SecurityException, you will then know the action is not allowed. For example, you could check on access to files with the following:

```
java.security.Permission permit =
   new java.io.FilePermission ("/-", "read");
SecurityManager security = System.getSecurityManager ();
if (security!= null) {
   try {
      security.checkPermission (permit);
   }
   catch (SecurityException se) {
      System.out.println ("File access blocked");
   }
}
```

14.6   A client application

We might want to communicate with our server using a custom client program rather than just with a browser. With Java we can create a standalone client application to request and read files from the server just like a web browser but it could also interact in other ways. For example, maybe the client sends a command to the server to instruct it to carry out some action on the local platform such as running a diagnostic program. (See Chapter 23 for a discussion of running external programs from within a Java program.) As we see in the next chapter, we can even maintain a connection with the client indefinitely and avoid the stateless condition of the usual browser–server interaction.

Here we show the run() method in client application ClientApp. This method sends a request to the server for a file. It connects to the server by the creation of a Socket instance with the server's address and port number. Using the input and output streams obtained from the socket, the client sends a request to the server and then reads the data returned by the server one line at a time. The client then displays this data in a text area.

```
. . . In class ClientApp . . .
public void run () {
  // Clear the text area
  fTextArea.setText ("Downloading . . .");

  try {
    // Connect to the server via the given IP address
    // and port number
    fSocket = new Socket (fIpAddr, fPort);

    // Assemble the message line.
    String message = "GET /" + fFilename;

    // Now get an output stream to the server.
    OutputStream server_out = fSocket.getOutputStream ();

    // Wrap in writer classes
    PrintWriter pw_server_out = new PrintWriter (
     new OutputStreamWriter (server_out, "8859_1"), true);
    // Send the message to the server
    pw_server_out.println (message);

    // Get the input stream from the server and then
    // wrap the stream in two wrappers.
    BufferedReader server_reader = new BufferedReader (
       new InputStreamReader (socket.getInputStream ()));

    fTextArea.setText ("");
    String line;
    // Add the text one line at a time from the server
    // to the text area.
    while ((line = server_reader.readLine ())!= null)
      fTextArea.append (line + '\n');

  }
  catch (UnknownHostException uhe) {
    fTextArea.setText ("Unknown host");

  }
  catch (IOException ioe) {
    fTextArea.setText ("I/O Exception");
  }
  finally {
    try {
      // End the connection
      fSocket.close ();
```

```
          fThread = null;
      }
   catch (IOException ioe) {
      fTextArea.append ("IO Exception while closing socket");
      }
   }
}
. . .
```

Figure 14.3 shows the client talking to a server running on the same Windows platform. The client displays the file in a text area and provides text fields to enter the server host address and port and the file that the user wants to download.

14.7 Server applications

We now know how to make a simple HTTP web server. What can we do with it? Several possibilities come to mind:

- **Custom server** – if you don't want to install a full-function web server for your PC, you can develop your own small, customized server.
- **Client input** – a server could record input from clients.

Figure 14.3 The client application – `ClientApp` – displays a file that it obtained from the `MicroServer`. The fields in the lower control panel indicate the host name of the server, the port on which to contact the server, and the name of the file to request from the server. Here the server and client ran on the same machine.

- **Data monitoring** – your custom server could provide recent data files written by an ongoing experiment or sensor reading. (See Chapter 15 for an example of a data-monitoring system.)
- **Embedded server** – in Chapter 24 we discus how a server can run on a Java processor in an embedded application such as monitoring a sensor or other instruments.
- **Applet interaction** – you can set up a two-way link between your client-side applet and your server to carry out special tasks.
- **Run external programs** – a server could start an external program to do some task such as making a measurement or running diagnostic tests. In Chapter 23 we discuss the `Runtime` class that provides a means to run external programs.
- **Secure interaction** – you can customize your server to respond only to clients with allowed usernames and passwords.

While one can certainly do such things with other programming languages, the Java code is quite compact and straightforward because of the core networking and threading capabilities. Furthermore, the portability of Java means that you can run the server and client programs on a wide range of platforms with little or no modification needed.

Note that the client and server roles are not absolute. A server can switch roles and become a client when necessary. For example, a server at a remote station could monitor some system and provide information to clients seeking status reports on the system. However, the server might periodically contact a central server to download data and status information, thus acting as client to the central server.

14.8   Servers, servlets and JSP

We briefly mention here a couple of other popular Java web server tools: *servlets* and *Java Server Pages* (JSP). We have seen how applets are Java programs that run inside the browser on the client's machine. Servlets provide an analogous approach on the server side in that they are Java programs run by a web server to provide specialized services. They can do many of the same tasks that CGI (Common Gateway Interface) programs perform such as processing input from web page forms. However, they offer several advantages over CGI:

- CGI programs run once and go away. Servlets can remain active and respond to another request without additional startup costs.
- Servlets are run in a Java thread by the server and are generally much faster to start up than CGI programs, which require an operating system process.
- A CGI program receives one input request, sends a response, and then dies. Servlets can carry on a two-way conversation for an indefinite period.
- A servlet can communicate with multiple clients simultaneously. An example is serving the players in a multi-player game.

We briefly return to the subject of servlets in Chapter 21 and their role in the area of web services.

Another tool in the Java toolbox is the Java Server Page (JSP). HTML pages at a server are typically static in that the text is fixed until it is edited by hand. In a JSP page, however, the hypertext contains specially tagged areas that signal to the server or servlet where it can insert dynamically created data. For example, the servlet could enter the latest price for a product in a catalog page.

Though servlets and JSP are powerful tools for large enterprise web servers, our emphasis is in small, specialized servers for custom applications such as monitoring remote devices. Such servers can usually suffice with the techniques discussed in this and the following chapters. See the Web Course and the references for more information about servlets and JSP.

14.9   Web Course materials

The Web Course Chapter 14 provides the code for the client/server demonstration programs discussed here along with additional demos and resources. See also the code listings for Chapter 24 where we created a version of `MicroServer` to run on a Java hardware processor. That platform does not provide the `String` `split()` method so we had to create a version of our own. Chapter 21 provides more information about servlets and Web Course Chapter 21 provides an example servlet.

References

[1] *Trail: Custom Networking – The Java Tutorial*, Sun Microsystems,
    `http://java.sun.com/docs/books/tutorial/networking/`.
[2] E.R. Harold, *Java Network Programming*, 2nd edition, O'Reilly, 2000.
[3] P. Niemeyer and Jonathan Knudsen, *Learning Java*, 2nd edition, O'Reilly, 2002.
[4] Budi Kurniawan, *How Java Web Servers Work*, Aug. 23, O'Reilly's OnJava.com, 2003,
    `www.onjava.com/pub/a/onjava/2003/04/23/java_webserver.html`.
[5] *Security in Java 2 SDK 1.2 – The Java Tutorial*, Sun Microsystems,
    `http://java.sun.com/docs/books/tutorial/security1.2/`.
[6] *Permissions in the Java 2SDK Guide*, Sun Microsystems, 2002,
    `http://java.sun.com/j2se/1.4.2/docs/guide/security/`
    `permissions.html`.

Chapter 15
Client/server with sockets

15.1 Introduction

In Chapter 14 we showed how to build a basic web server that sends files to browsers or to browser-like clients. In this chapter we present a more interesting socket-based client/server demonstration system that goes beyond just transmission of web pages. This new server sends data to a client, which then displays the data in histograms. This type of client/server system could be quite useful in various applications such as transmitting data from a remote experiment, running diagnostics under the direction of a client, installing calibration settings, and controlling an instrument remotely. For demonstration purposes, our server generates simulated data.

As in Chapter 14 we use sockets for our client/server communications [1–3]. Later chapters present RMI and CORBA based approaches. In a step-by-step manner we describe the concepts and the code techniques used in the client and server demonstration programs.

15.2 The client/server design

For the web server discussed in Chapter 14 we used socket communications. The server monitors a port with a `ServerSocket`, which returns a socket for a client whenever one requests a connection. The socket is passed to a thread that receives a request from the client for a file and then transmits that file if it is available. The server then breaks the connection and the session ends.

We can, however, create a client/server system in which the server maintains a connection for as long as the client desires. To demonstrate the benefits of such a system, we create a server program that sends simulated data to clients. A client program displays the data that it obtains from the server, and its graphical interface lets the user send requests to update the data.

A client session begins with a log-in procedure for the client with the server. Then the server and client set up I/O streams for their communications. As in

Chapter 14, multiple clients can connect to the server because it can create a new thread for each new client. For the log-in, for the exchange of commands, and for the transmission of data, the two programs need a simple protocol so that they can understand one another.

Both our server and client programs provide graphical user interfaces. The server's GUI displays the status of the communications with the clients in a text area and offers some basic controls. The client's GUI includes its own text area for communications monitoring and a set of controls, plus it includes two histograms to display the data.

15.3 The client/server interaction

In this section we present the main steps involved in our data client/server system. The information flow is similar to that described in Section 13.8 and in Chapter 14. We first start the `DataServer` application program, which uses a `ServerSocket` to wait for clients to request connections at the appropriate port. The client, an instance of `DataClient`, starts on another machine (or on the same machine for convenience during testing) and connects to the server using a socket.

The server initiates a simple log-in procedure by sending the string `"Username:"` to the client, to which the client responds with a name string. Here we do nothing with this string but you could easily add code to compare the name to a list of authorized users; you could require a password as well. On the client-side, you could use `JPassworldField`, which is a subclass of `JTextField` designed for password input, to protect what is typed on the screen. However, you would need to take additional measures to encrypt the data flowing over the Internet if you want to ensure that the password remains secret.

For each client, the server creates an instance of the `Thread` subclass called `DataWorker` whose job is to communicate with the client over the socket and perform the requested tasks. The `DataServer` passes the socket for the client to `DataWorker`, which sets up the I/O streams with the client and carries out all of the communications with the client.

Similarly, `DataClient` creates an instance of `DataClientWorker`, a thread that is in charge of communications with the server. When the server sends data to a `DataClient`, it is the `DataClientWorker` that receives and processes the data. The `DataClient` then displays the results on its graphical interface.

Figure 15.1 shows a diagram of possible data-taking scenarios such as the `DataWorker` reading data from an instrument via a serial port connection. It could also run an external program and read its output. In Chapter 23 we discuss how Java programs can operate in the world outside of the JVM.

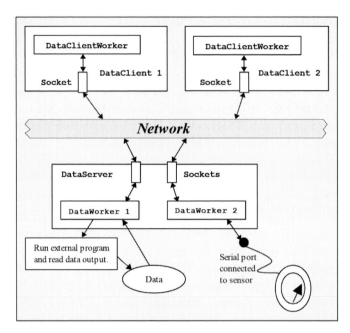

Figure 15.1 The `DataClient`/`DataServer` example demonstrates how remote clients communicate through a network (e.g. a local network or the Internet) with a server offering specialized services. When a client connects to `DataServer`, it assigns an instance of `DataWorker` to attend to the client. Similarly, the `DataClient` application uses a `DataClientWorker` to communicate with server while the `DataClient` provides the user interface. In the example here, the server provides simulated data but it could be modified for other tasks such as sending data files, running external programs, or communicating with a sensor via a serial port (see Chapter 23).

15.4 The `DataServer`

Figure 15.2 shows the interface for the `DataServer` application. A text field allows the user to set the port number that the clients must use to connect with the server. A text area displays the status of the server and messages indicating the various actions taken as connections to the clients occur and services are provided. The utility method `println()` sends messages to the text area:

```
public void println (String str) {
   fTextArea.append (str + "\n");
   repaint ();
}
```

The `"Start"` button puts the server program into a server state by starting a thread where a loop in the `run()` method connects to clients and assigns instances of `DataWorker` to handle them. As with the web server in Chapter 14, the program uses a `ServerSocket` object to wait for clients to attempt to make a socket connection with it on a given port.

Figure 15.2 The interface for `DataServer` allows the user to select a port number. It also holds a Start button and a text area to display information on the status of the connections with the clients.

```
. . . The run() method in class DataServer . . .

/** Create a ServerSocket and loop waiting for clients. **/
public void run () {

   // The server_socket is used to make connections to
   // DataClients at this port number
   try {
      fServerSocket = new ServerSocket (fDataServerPort);
   }
   catch (IOException e) {
      println ("Error in server socket");
      return;
   }
   println ("Waiting for users . . .");

   // Loop here to grab clients
   while (fKeepServing) {
      try {
         // accept() blocks until a connection is made
         Socket socket = fServerSocket.accept ();
```

```
        // Do the setup for this socket and then loop
        // back around to wait for the next DataClient.
        DataWorker worker = new DataWorker (this, socket);
        worker.start ();
      }
      catch (IOException ioe) {
        println ("IOException: <" + ioe + ">");
        break;
      }
      catch (Exception e) {
        println ("Exception: <" + e + ">");
        break;
      }
    }
  } // run
```

DataServer uses a Vector to keep a list of all the clients, and a limit on
the number of clients is set with the fMaxClients variable. A DataWorker
calls back to the clientPermit() method, shown below, to determine if it can
join the list. If so, then it invokes clientConnected(), which adds the client
to the list. When the client disconnects, the DataWorker invokes the server's
clientDisconnected() method to remove itself from the worker list. These
methods are synchronized to avoid any interference if two or more threads are
connecting/disconnecting at the same time.

```
    . . . Other code in DataServer . . .
    // Use a Vector to keep track of the DataWorker list.
    Vector fWorkerList;
    . . .
    /** Before adding a client, check if there is room for it. **/
    public synchronized boolean clientPermit () {
      if(fWorkerList.size () < fMaxClients_)
        return true;
      else
        return false;
    }

    /** Add a new client to the list. **/
    public synchronized void clientConnected (DataWorker worker) {
      fWorkerList.add (worker);
      fClientCounter_++;
    }
```

```
/** Before a client breaks off, remove it from the list. **/
public synchronized void clientDisconnected (String user,
   DataWorker worker) {
   println ("Client: " + user + "disconneced");
   fWorkerList.remove (worker);
   fClientCounter__-;
}
```

The `DataServer` hands off the client socket to `DataWorker` and then the worker begins its job of communicating with the client and providing it the requested services.

15.5 The `DataWorker`

The `DataWorker` is a thread that tends to the needs of its client. It maintains the connection until the client breaks it. The worker calls back to the `DataServer` to add or subtract itself to the list of workers and to send messages for display in the text area in the server's user interface. The worker follows a simple protocol with the client so that each knows when to send a message and when to wait for a message (and when to send or receive numerical values). The server initially carries out a simple log-in procedure, which here just means a request for a user name. As we mentioned earlier, you could easily expand this to include a password exchange as well.

As shown in the following code snippet, the first act by the `run()` method is to invoke the `serviceSetup()` method. This method sets up the streams for I/O with the client. The `PrintWriter` and `BufferedReader` wrappers are used to send and receive text to and from the client. A `DataOutputStream` wrapper is used to send numerical values. The read/write methods for these streams are put into some utility methods discussed later.

If the maximum number of clients has been reached, the worker sends a warning message to its client and breaks off the connection. That worker thread itself then signs off from the server and dies.

If there is room for the client, the `serviceSetup()` method performs a simple log-in procedure with the client that consists of sending the string "Username: " to the client and waiting for a string in return:

```
. . . The run() and serviceSetup() methods in the class
DataWorker . . .

   public void run () {
      // If service setup fails, end thread processing.
      if (!serviceSetup ()) return;
      . . .
```

```
} // run

public boolean serviceSetup () {
  fDataServer.println ("Client setup . . .");

  // First get the in/out streams from the socket to the
  // client
  try {
      fNetInputStream = fSocket.getInputStream ();
      fNetOutputStream = fSocket.getOutputStream ();
  }
  catch (IOException e) {
    fDataServer.println (
      "Unable to get input/output streams");
    return false;
  }

  // Create a PrintWriter class for sending text to the
  // client. The writeNetOutputLine method will use this
  // class.
  try {
    fPrintWriter = new PrintWriter (
      new OutputStreamWriter (fNetOutputStream,
                              "8859_1"), true);
  }
  catch (Exception e) {
    fDataServer.println (
      "Fails to open PrintWriter to client!");
    return false;
  }

  // Check if the server has room for this client.
  // If not, then send a message to this client to
  // tell it the bad news.
  if (!fDataServer.clientPermit ()) {
    try {
      String msg =
        "Sorry, we've reached the maximum number of
        clients";
      writeNetOutputLine (msg);
      fDataServer.println (msg);
      return false;
    }
    catch (IOException e) {
        fDataServer.println (
          "Connection fails during login");
```

```
      return false;
    }
}

// Get a DataInputStream wrapper so we can use
// its readLine() method.
fNetInputReader = new BufferedReader (new
                  InputStreamReader (fNetInputStream));

// Do a simple log-in protocol. Send a request for the
// users name. Note that a password check could
// be added here.
try {
  writeNetOutputLine ("Username: ");
}
catch (IOException e) {
  fDataServer.println (
    "Connection fails during login");
  return false;
}

// Read the user name.
fUser = readNetInputLine ();
if (fUser == null) {
  fDataServer.println (
    "Connection fails during login");
  return false;
}

// Send a message that the login is OK.
try {
  writeNetOutputLine ("Login successful");
  fDataServer.println (
    "Login successful for " + fUser);
}
catch (IOException e) {
  fDataServer.println (
    "Connection fails during login for " + fUser
  );
  return false;
}
fDataServer.println (fUser + " connected!");
fDataServer.println (fSocket.toString ());

// The log-in is successful so now add this DataWorker
// to the DataServer's list of workers.
```

```
// Read a request from the DataClient
String client_msg = readNetInputLine ();
if (client_msg == null) break;

// Only print message if it changes. Avoids printing
// same message for each data set.
if (!client_msg.equals (client_last_msg))
  fDataServer.println (
      "Message from " + fUser + ": " + client_msg);
client_last_msg = client_msg;

// Could interpret the request and do something
// accordingly here, but for this example we will just
// send a set of data values. Send the number of data
// values.
try {
    writeNetOutputInt (DataServer.fNumDataVals_);
}
catch (IOException e) {
  break;
}

// Create dummy data values and send them to the
// DataClient.
for (int i=0; i < DataServer.fNumDataVals_; i++) {

  // Select the range of Gaussian widths for the data
  // values for each channel of the data set. Add an
  // offset to get most negative values above zero.
  int i_std_dev = i%6;
  double dat = 3.0 * fStdDev[i_std_dev] +
              fStdDev[i_std_dev] *
              fRan.nextGaussian ();
  // Force any remaining negative value to zero.
  if (dat < 0.0) dat = 0.0;

  // Pass only integer values;
  int idat = (int) dat;
  try {
    writeNetOutputInt (idat);
  }
  catch (IOException e) {
    break;
  }
```

```
        }
    }

    // Send message back to the text area in the frame.
    fDataServer.println (fUser + " has disconnected.");

    // Do any other tasks for ending the worker.
    signoff ();

} // run
```

The `DataWorker` class could be modified to obtain real data in various ways. For example, a sensor might generate a file that the worker could read and forward to the client.

15.6   The `DataClient`

The `DataClient`, shown running in Figure 15.3, is an applet that makes a socket connection to `DataServer`, grabs data from the server, and displays it in two histograms. Such a client program could serve as a data monitoring tool. From a remote location you could examine samples of data as it is produced during an experiment in a manner that helps to spot anomalies or malfunctions. The data would normally be saved to disk files on the server machine for full analysis later but the monitor helps to prevent the taking of flawed data.

The main job of `DataClient` is to create an interface for the user to initiate and control communications with the server. It provides fields to specify the server's host address, a user name, and which channel in the data set to plot in the lower histogram. Here we are assuming that the data comes in as a set of readings or channels. For example, perhaps there are readings from 20 sensors in each data set. We refer here to a particular reading in the set of 20 as a channel.

The top histogram displays the values for each data set as it arrives. (That is, bin 0 in the histogram corresponds to the value in element 0 of the data array, bin 1 to element 1, and so forth.) At a glance this histogram indicates whether some channels are missing data. The user can also select a particular channel to display its distribution of values in the bottom histogram. A set of buttons starts the process, clears the histograms, and exits the program.

Here we show a large snippet from the `DataClient` applet, skipping the `init()` method that creates the interface. A click on the "Start" button on the interface leads to the invocation of the `start()` method, which in turn invokes the `connect()` method that makes the socket connection to the server.

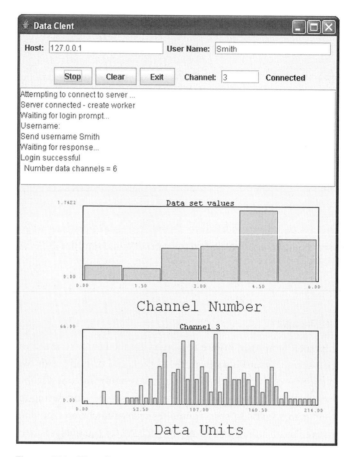

Figure 15.3 The client interface allows the user to initiate a connection with the server at the host address given in a text field. A user name is given to the server in a log in procedure. The data obtained from the server is displayed in two histograms. The top one shows the values in each data channel for a reading of a set of data. The lower histograms displays the distribution of values over a number of data set readings for a channel selected by the user via the text field labeled "Channel".

```
. . . Skip GUI setup in the DataClient . . .

public void actionPerformed (ActionEvent e) {
   Object source = e.getSource ();
   if (source == fStartButton) {
      if (fStartButton.getText ().equals ("Start"))
         start ();
      else
         stop ();
   }
```

```
   else if (source == fClearButton) {
     fHistData.clear ();
     // For adaptable histogram, clear bins
     // and also clear internal data array.
     fHistChan.reset ();
     repaint ();

   }
   else if (!fInBrowser) // Exit button
     System.exit (0);
} // actionPerformed

/**
  * Make the connection to the server. Set up the
  * DataReader and begin recording the data from the
  * server. **/
public void start () {
   // If already connected, then stop the current
   // connection before continuing to set up this new
   // connection.
   if (fConnected) stop ();

   // Clear the histograms
   fHistData.clear ();
   fHistData.clear ();

   // Get the current values of the host IP address and
   // and the username
   fHost = fHostField.getText ();
   fUserName = fUserNameField.getText ();
   try {
     fChannelToMonitor =
       Integer.parseInt (fChanField.getText ());
   }
   catch (NumberFormatException ex) {
     println ("Bad channel value");
     return;
   }

   // Now try to connect to the DataServer
   try {
     if (connect ()) {
       // Successful so set flags and change button text
       fConnected = true;
       fStartButton.setText ("Stop");
       fStatusLabel.setText ("Connected");
```

```
              fStatusLabel.setForeground (Color.BLUE);
          }
          else {
              Println ("* NOT CONNECTED *");
              fStatusLabel.setText ("Disconnected");
              fStatusLabel.setForeground (Color.RED);
          }
      }
      catch (IOException e) {
        println ("* NOT CONNECTED *");
        fStatusLabel.setText ("Disconnected");
        fStatusLabel.setForeground (Color.RED);
      }
  } // start

  /**
    * Connects to the server via a socket. Throws
    * IOException if socket connection fails.
    **/
  boolean connect () throws IOException {
    println ("Attempting to connect to server . . .");
    try {
        // Connect to the server using the host IP address
        // and the port at the server location
        fServer = new Socket (fHost, fDataServerPort);
    }
    catch (SecurityException se) {
        println ("Security Exception:\n"+se);
        return false;
    }

    println ("Server connected - create worker");
    // Create the worker to tend to this server
    fDataClientWorker =
        new DataClientWorker (this, fServer, fUserName);
    fDataClientWorker.start ();

    return true;
  } // connect

  /** Stop the worker thread. **/
  public void stop () {
    // Disconnect and kill the fDataClientWorker thread
    fDataClientWorker.finish ();
    setDisconnected ();
  } // stop
```

```
/** Set buttons for restart. **/
void setDisconnected () {
  fStartButton.setText ("Start");
  fStatusLabel.setText ("Disconnected");
  fStatusLabel.setForeground (Color.RED);
} // setDisconnected

/**
  * The DataClientWorker passes the data array from the
  * server here. Display the data set by packing a
  * histogram.
  *
  * Also, plot the distribution of one of
  * the channels of the data. The channel number is
  * given in the text field.
  **/
void setData (int[] data) {

  // Display each data set
  fHistData.pack (data, 0, 0, 0.0, (double)
                  (data.length));
  fHistDataPanel.getScaling ();

  // Plot the distribution of one of the channels in the
  // data.
  if (fChannelToMonitor >= 0 && fChannelToMonitor <
      data.length) {
    fHistChan.setTitle ("Channel " +
                          fChannelToMonitor);
    fHistChan.add ((double)data[fChannelToMonitor]);
    // Adapt since data varies from channel to channel.
    fHistChan.rebin ();
    // Now rescale and draw.
    fHistChanPanel.getScaling ();
  }
  repaint ();
} // setData

/** Convenience method for sending messages to the text
  * area. **/
public void println (String str){
    fMessageArea.append (str +"\n");
    repaint ();
} // println

. . . Continue class DataClient . . .
```

If the client successfully connects to the server, it spins off the socket to an instance of the Thread subclass DataClientWorker, which handles communications with the server. After the connection is made, the text on the "Start" button becomes "Stop". If the user clicks on the "Stop" button, the stop() method is invoked, via the actionPerformed() method, and this tells the DataClientWorker to break the connection and die.

When the DataClientWorker receives data from the server, it calls back to the DataClient via the setData (int[] data) method to pass the data on. This method uses the pack() method in the Histogram class (see Section 6.11) to display the values in the array for the top histogram in the user interface. For the data channel (i.e. the data array index) given in the interface text field, the distribution of values are displayed in the second histogram.

15.7   The DataClientWorker

As we indicated in Figure 15.1, communication with the server is mostly handled by the DataClientWorker object. This Thread subclass first opens the streams to the server and then carries out a simple log-in procedure with the server by passing the user name to it. See the doConnection() and login() methods in the code snippet shown below. Note that, as with DataServer, the class holds some utility methods for the I/O operations.

```
. . . . In the class DataClientWorker . . . .


/** Remain in a loop to monitor the I/O from the server.
 * Display the data.
 **/
public void run () {

   // The socket connection was made by the caller, now
   // set up the streams and do a login
   try {
      if (!doConnection ()) {
         fDataClient.println (" Connection/login failed");
         return;
      }
   }
   catch (IOException ioe) {
      fDataClient.println (" I/O exception with serve:" +
                           ioe);
   }

   int num_channels = -1;
```

```
// This loops until either the connection is broken or
// the stop button or stop key is hit
while (fKeepRunning) {
  // Ask the server to send data.
  try {
      writeNetOutputLine (" send data");
  }
  catch (IOException e) {
    break;
  }


  // First number sent from server is an integer that
  // gives the number of data values to be sent.
  try {
    num_channels = readNetInputInt ();
  }
  catch (IOException e) {
    break;
  }

  if (num_channels!= fNumChannels) {
    fNumChannels = num_channels;
    fDataClient.println (
      " Number data channels = " + fNumChannels);
  }

  if (fNumChannels < 1) {
    fDataClient.println (" no data");
    break;
  }

  // Create an array to hold the data if not available
  if (fData == null || fNumChannels != fData.length)
    fData = new int[fNumChannels];

  for (int i=0; i < fNumChannels; i++) {
    try {
      fData[i] = readNetInputInt ();

      // Pass the data to the parent program to do
      // with as it wants
      fDataClient.setData (fData);
    }
    catch (IOException e) {
```

```
          fDataClient.println ("IO Exception while
                             reading data");
          break;
        }
      }

      // Ask for data after every fTimeUpdate long period.
      try {
        Thread.sleep (fDataClient.fTimeUpdate);
      }
      catch (InterruptedException e) {}
    }

    if (fServer != null) closeServer ();
    fDataClient.println ("disconnected");
    fDataClient.setDisconnected ();
  } // run

  /** Set up the streams with the server and then login. **/
  boolean doConnection () throws IOException {

    // Get the input and output streams from the socket
    InputStream in = fServer.getInputStream ();

    // Use the reader for obtaining text
    fNetInputReader = new BufferedReader (
      new InputStreamReader (in));

    // User the DataInputStream for getting numerical
    // values.
    fNetInputDataStream = new DataInputStream (in);

    // Output stream for sending messages to the server.
    fNetOutputDataStream = fServer.getOutputStream ();

    // Write with a PrintWriter for sending text to the
    // server.
    fPrintWriter= new PrintWriter (
      new OutputStreamWriter (fNetOutputDataStream,
        "8859_1"), true);

    // Now try the login procedure.
    if (!login ())
      return false;

    return true;
```

```
} // doConnection
/**
  * Here is a homemade login protocol. A password
  * exchange could also be added.
 **/
boolean login() {
  fDataClient.println ("Waiting for login prompt . . .");

  String msg_line = readNetInputLine ();
  if (msg_line == null) return false;
  fDataClient.println (msg_line);
  if (!msg_line.startsWith ("Username:")) return false;

  fDataClient.println ("Send username " + fUserName);
  try {
    writeNetOutputLine (fUserName);
  }
  catch (IOException e) {
    return false;
  }
  catch (Exception e) {
    fDataClient.println ("Error occurred in sending
                         username!");
    return false;
  }

  fDataClient.println ("Waiting for response . . .");

  msg_line = readNetInputLine ();
  if (msg_line == null) return false;
  fDataClient.println (msg_line);

  return true;
} // login

/** Do all of the steps needed to stop the
  * connection. **/
public void finish () {
  // Kill the thread and stop the server
  fKeepRunning = false;
  closeServer ();
} // finish

/** Close the socket to the server. **/
void closeServer () {
  if (fServer == null) return;
```

```
    try {
      fServer.close ();
      fServer = null;
    }
    catch (IOException e) {}
  } // closeServer

  /**
    * The net input stream is wrappped in a
    * DataInputStream so we can use readLine, readInt
    * and readFloat methods.
    **/
  String readNetInputLine () {
    try {
      return fNetInputReader.readLine ();
    }
    catch (IOException e) {
      return null;
    }
  } // readNetInputLine

  /** Read an integer value from the socket stream **/
  int readNetInputInt () throws IOException {
    return fNetInputDataStream.readInt ();
  } // readNetInputInt

  /** Read float value from the socket stream. **/
  float readNetInputFloat () throws IOException {
    return fNetInputDataStream.readFloat ();
  } // readNetInputFloat

  /**
    * The net output is a PrintWriter class which doesn't
    * throw IOException itself. Instead we have to use
    * the PrintWriter checkError() method and throw an
    * exception ourselves if there was an output error.
    **/
  void writeNetOutputLine (String string) throws
  IOException {
    fPrintWriter.println (string);
    if (fPrintWriter.checkError ())
        throw new IOException ();
    fPrintWriter.flush();
    if (fPrintWriter.checkError ())
        throw new IOException ();
  } // writeNetOutputLine
} // class DataClientWorker
```

The loop in the `run()` method sends a request to the server for the data. The server first returns a value indicating how many data values it plans to send. The worker then reads that number of values from the input stream connected to the server. This set of data is then passed to the parent `DataClient` via its `setData(int[])` method. This program only expects integer data but you could easily modify it to obtain floating-point data from the server. The loop in `run()` pauses for a given period and then repeats the process.

15.8 Benefits and shortcomings of sockets

Socket communications work well for the exchange of files as seen with the simple web server in Chapter 14 and for the downloading of data and messages as discussed in this chapter. However, for more ambitious distributed computing tasks, such as multiprocessing on several hardware processors, other tools provide significant advantages. For example, the RMI (Remote Method Invocation) system allows a program on one machine to invoke a method in an object on another machine just as if it were a local object. The user does not need to create any sort of custom protocol as we did for the socket I/O. Object serialization also becomes a powerful feature in such a system. RMI and CORBA procedures rely on serialization to pass objects back and forth between platforms within a distributed program. In Chapters 16–20 we discuss these approaches to distributed computing.

In Chapter 21 we return to a web-based style of distributing computing with an introduction to web services. This involves the exchange of data in the form of XML text documents.

15.9 Web Course materials

Chapter 15 in the Web Course includes the complete code for the `DataServer` and `DataClient` programs discussed above. The `DataClient` runs as applet and communicates with the `DataServer` if it is running on the same host.

The Web Course chapter also includes another example of a scientific client/server application. The `SimServer` and `SimClient` programs demonstrate how a central server could provide a physics simulation service to clients. The server assigns to each client its own simulator process and the client can set up the simulator parameters and run it remotely. Such a simulation service could be useful, for example, in an experimental collaboration in which a simulator resides on a central site where it is continually updated. Remote users then run the simulator with the assurance that they are always using the latest version.

Chapter 16
Distributed computing

16.1 Introduction

As has been demonstrated, Java is a very capable platform for many scientific computing tasks. Yet Java is still sometimes perceived as slow. While this often is a no-longer-deserved reputation, especially in Java 1.4 and later, there are definitely times when the nature of the scientific calculation is so demanding that typical desktop computing resources are insufficient. In such cases, moving portions of the calculation to a heavy-duty remote server machine, perhaps even a "supercomputer," makes good sense. In this chapter, we introduce the concept of distributed computing.

We continue with the client/server paradigm discussed in the previous chapters, but rather than simply passing messages via socket connections, the client and server objects directly invoke methods in each other over the network. This allows for much more elaborate and productive interactions. The Java Remote Method Invocation (RMI) or Common Object Request Broker Architecture (CORBA) frameworks take care of the communications, and we do not need to create our own low-level protocols as we did with sockets.

We first discuss just what distributed computing is and what form of distributed computing is of value to scientific calculations. We introduce just a little Unified Modeling Language (UML) as a visual aid to understanding the various components in a distributed application. This use of UML and the Design Pattern approach allows us to describe client/server programs in a more formal manner than in the previous chapters. We then lay out the design of a simple distributed application, concentrating on the server side in this chapter. Later chapters discuss the client side and RMI, CORBA, and web services as tools to implement a distributed application.

16.2 Distributed computing for scientific applications

There are several viewpoints as to just what constitutes distributed computing. One well-known concept might better be called "distributed data processing," in which calculations are spread over many different computers distributed over a wide area, each working on parts of the problem. SETI@home [1] is an example. Millions of users worldwide have downloaded a client application that processes radio satellite data searching for repetitive signals that might indicate intelligent life elsewhere in the universe. Each client application obtains a small chunk of data, processes it, and returns the results to the central SETI@home computer for compilation. In this concept each client is nearly completely independent, not interacting with, or even aware of, the existence of other clients. There have been other similar distributed applications ranging from encryption/decryption applications to climate study applications (see [2]).

Another distributed computing concept is massively parallel computing on a parallel computer such as the NEC Earth Simulator computer in Japan [3], and, in the USA, the IBM Blue Gene/L at IBM's Thomas Watson Research Center [4], the Apple "BigMac" at Virginia Tech University [5], or the National Leadership Computing Facility being built at Oak Ridge National Laboratory [6], which is targeted to be the world's fastest scientific research computer when completed [7, 8]. All of these systems are clusters of 1000 or more processors. In fact, the biannual Top 500 ranking of the world's fastest supercomputers is dominated by massively parallel machines [9]. A related idea is the Parallel Virtual Machine (PVM) system in which many disparate computers, large and small, and possibly even using multiple operating systems, are linked together via the Internet to create a virtual parallel machine [10]. On these parallel systems, the different parts of the calculation typically interact with each other in some way. While Java code is portable to any platform on which a JVM is available, there may be no JVM on the most exotic supercomputer designs. However, several of the Top 500 supercomputers are Linux clusters, upon which one could probably install the Java Runtime Environment for Linux. Whether or not a JVM is available on these massively parallel supercomputing systems, such is not the topic of this chapter.

Distributed computing in an object-oriented view involves distributing the software objects over multiple nodes or hosts, perhaps utilizing mobile objects that move from node to node as needed instead of pre-configuring the work to be done at each node. Intelligent mobile agent software is an example. Intelligent agents are also not the subject of this chapter.

For general scientific computing, as opposed to state-of-the-art supercomputing, the distributed computing techniques one needs to know are much less grandiose, though still very useful. In this chapter, we discuss a simple two-node distributed computing concept (a paradigm that should now be familiar from the previous chapters where it was known as client/server computing). As we have discussed, in a client/server arrangement, the client typically is a GUI in which the user prepares input and views output in a graphical interface. Heavy duty computations are routed to a remote server machine. There might be many reasons to separate the heavy computations to a server. One obvious reason is to improve performance when the calculations to be done are so intensive that client computers would be too slow. Another valid reason is the protection of intellectual property contained in the server-side code or server-side databases. Yet another reason might be the existence of legacy software written in another language, typically Fortran, that one does not wish to port to Java. Often, there have been dozens of man-years poured into the development of a scientific code. Rewriting the entire code in a modern language like Java would be extremely time-consuming and expensive. Instead, the calculation can often be split into a client portion and a server portion with most of the heavy calculations done on the server with only minor modifications to the original code.

As mentioned in the introduction, for the distributing computing systems discussed here and the rest of Part II, the interactions between a client and server across the network do not involve the programming of low-level socket communications as in Chapters 14 and 15. Instead, objects on one system invoke methods in objects on other computers just as if they all were running in the same program on the same local platform. This magic is accomplished by building the programs on top of the Java RMI or CORBA frameworks, which do all the low level communications work. We discuss how to set up these frameworks in Chapters 18 and 19.

16.3 Minimalist UML

The Unified Modeling Language (UML) has emerged as a powerful tool for object-oriented analysis and design (OOAD). Unlike a traditional programming language, UML is not really a "language" but rather a notation system for modeling systems that use object-oriented concepts. The UML provides a way to visualize the attributes and methods of, and interactions among, objects in an object-oriented system. The full body of UML is very complete, with the capability to describe almost any set of objects and interactions imaginable. Because of this rich capability, UML is also rather complex. Entire books are devoted to the subject, and the full capabilities are well beyond the scope or needs of this chapter (see [11, 12]).

UML may not be an absolute necessity to produce a good object design, but it is an invaluable aid in doing so. UML also serves in a collaborative environment as an excellent vehicle for communicating an object design among the various architects and developers who will be implementing the design. UML also provides an excellent way to illustrate and document designs. For a lone scientist designing a system and writing his own code, perhaps a tool like UML is not important. However, we feel that once a base level of proficiency with UML is achieved, the language provides an excellent mechanism for organizing one's thoughts. Often, illustrating a design in UML terms makes clear a deficiency in the design that can be corrected early at the ground level with a simple design change. The alternative of waiting until the design problem is discovered during implementation will undoubtedly result in much greater repair effort. For these reasons, we aim to arm the reader with a basic UML proficiency in the balance of this chapter. UML also provides the best method we know of to describe the object design used in the rest of this chapter. So we must introduce a basic level of familiarity with UML just to convey the distributed computing ideas discussed here.

Perhaps the most important concept in object-oriented analysis and design is the skillful assignment of responsibilities to software components, and UML is arguably the best tool to describe and visualize that assignment. We introduce only the essential UML features needed to describe a relatively simple distributed application. The description below is not even intended to be rigorously accurate with UML terms. Instead, we freely switch between the common Java names and the generic UML names and notations for certain items. Furthermore, we do not intend to produce a formal or even informal tutorial on UML. Instead, we introduce some simple UML diagrams for our distributed application and explain what they mean. A hands-on minimalist UML training session, if you will, or on-the-job training.

16.3.1   UML interaction diagrams

One of the most commonly seen features of UML is the class diagram, though it is not necessarily the most useful and probably should not be the first diagram that one thinks of when analyzing a problem and designing a solution. A class diagram depicts the software classes and the relationships among them (e.g. the Java "extends" and "implements" relationships) in an object-oriented system. A class diagram also shows the *operations* and *attributes* of a class, or, in Java terminology, the *methods* and *variables*, respectively. We explain class diagrams further in Section 16.6.3, below.

Before designing classes, though, one should create a conceptual model of the problem at hand. A conceptual model represents real-world concepts, not software components. For complex systems, a disciplined OOAD approach has proven to be a very useful technique to create good models and good solutions to real-world problems. Formal OOAD is beyond the scope of this book and is typically not needed for most scientific problems. For simple systems, OOAD begins with a conceptual model that can be sketched out on paper without a lot of rigor. A block diagram of the parts of the real-world system along with some lines connecting them with perhaps some notations about how the various parts interact is a good starting point. Creating such a block diagram forces you to identify explicitly the various parts of a system and the roles that they play. Once a conceptual model is in hand, you can begin designing a software solution.

UML *interaction diagrams* are useful for depicting the software implementation of a real-world conceptual model. As is obvious from the name, these diagrams depict the interactions among objects. There are two types of interaction

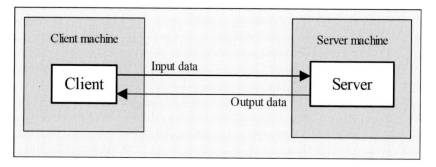

Figure 16.1 A conceptual model for a simple client/server distributed application.

diagrams in UML. One is the *sequence* diagram, which shows the order in which messages are sent between objects as well as an object's life cycle – i.e. creation and termination.

The other UML interaction diagram is the *collaboration* diagram. (The collaboration diagram has been renamed *communications* diagram in UML 2.0.) Collaboration diagrams are useful for visualizing the relationship between objects collaborating to perform a particular task. They illustrate the order and flow of messages between objects or, using Java terminology, the invocation of methods. By thinking about how objects collaborate, one learns the behavior required by each object – i.e., the methods that each class must provide. That is, collaboration diagrams help one design the classes that will appear in the class diagram.

Sequence diagrams seem to be utilized far more often than collaboration diagrams by most UML practitioners. For our purpose here, however, we find the collaboration diagram to be the more useful of the two. In actuality, sequence diagrams and collaboration diagrams show nearly the same information; they just present it differently. The two are so closely related that some modeling tools, such as Rational Rose, can automatically create a collaboration diagram from a sequence diagram and vice versa [13]. We use collaboration diagrams below and say nothing more about sequence diagrams.

16.4 A conceptual model for a simple distributed application

We consider a simple distributed application in which a client sends input data to and receives output data from a server object. A conceptual model of this client/server application would be simple, as shown in Figure 16.1.

This system is considered to be a distributed application since the client and server are on different machines although it is possible to run both client and server applications on the same physical machine in two separate JVMs. One might think that this diagram is almost too simple to be useful, but we build on it below.

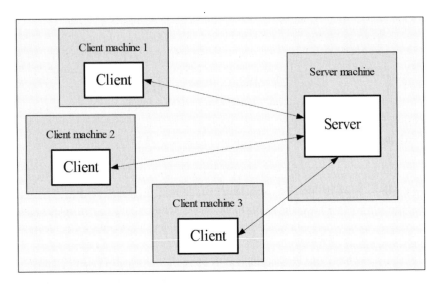

Figure 16.2 Expanded conceptual model showing several clients interacting with a single server machine.

Before building a UML collaboration diagram for this conceptual model, we first introduce a useful *design pattern*. Design patterns came into vogue in 1995 with the publication of the famous *Design Patterns* book [14]. While important, that book is not particularly easy to understand for the non-computer scientist. Yet design patterns are immensely useful, so it is important to understand some simple patterns. Design patterns need not be mysterious. A design pattern can be defined simply as a proven good solution for a commonly occurring software design issue – in other words, a "template" or "best practice" for how to solve a common problem.

The common software design issue in the case of a client/server system is how to support multiple clients with one server. One does not wish to lock out all other clients when one client is using the server. A web server is a common example. Popular web servers such as search engines, auction sites, or online shopping sites might have to support thousands of simultaneous clients. Thus the conceptual model could be expanded to that shown in Figure 16.2.

We can imagine that the multiple clients are created as needed by multiple users, but how does one server handle multiple clients? In an object-oriented solution, the "server" is actually made up of several server objects. Because we're still at the conceptual model phase, remember that the boxes labeled *client* and *server* in the figure do not represent *objects*. They are simply concepts at this point. Later we draw diagrams with actual objects in them. Where do all of the server objects come from, how many are needed, and what happens if the server machine runs out of server objects?

One might recall from Chapters 14 and 15 that we handled the multiple client problem by using multiple threads on the server – one thread per client. That solution is, in fact, very close to the more formal design pattern technique presented here. The relationship between the obvious multithreaded solution used earlier and the design pattern used below will become obvious as we proceed.

The need to handle multiple clients is manifestly a common design problem, and a useful design pattern for this common situation is the *Factory Pattern*, which specifies that multiple server objects are obtained from a server *factory*. A factory is simply an object that creates other objects. Therefore, when we move from conceptual model diagrams to object diagrams, the client object connects first not to a server but to a server *factory* object. The server factory knows how to create server objects. From the factory object, the client obtains a new server object. Further interaction then occurs between the client and its own private server. Other clients, typically on other machines, obtain additional server objects from the factory. Thus the Factory Pattern is a technique to support multiple simultaneous clients. This pattern is very common in distributed programming, and its relationship to the thread-per-client solution used previous is obvious.

It is illuminating to consider attempting to implement the Factory Pattern in some language other than Java. It might be straightforward to have the factory create new server objects as needed, but if one server object has control of the CPU, then the other servers will never get a chance to execute. A necessary condition for properly implementing the Factory Pattern is the ability for each server object to get a portion of the CPU. One particularly heavyweight solution is to create a new process on the server machine for each new server object, letting the operating system take care of CPU allocation among the processes. Often a higher performance solution is to use multiple threads of execution. Since Java has an easy-to-use multithreading API, each new server object can run in its own separate thread (see Chapter 8). If a limit on the number of simultaneous clients is required, for example to keep from overloading the server machine, then the factory can easily keep track of the number of server objects in use and throw a "too busy" exception if that limit is exceeded.

16.5   Collaboration diagram for a simple distributed application

A collaboration diagram illustrating the factory pattern is shown in Figure 16.3. Here the arrows indicate "messages" from one object to another. The arrowhead points to the object that receives the message – i.e. to the object that provides the method. The *sequence numbers* indicate the order of the operations and may be nested. Thus, in sequence number 1, the client sends a `getInstance()` message to the `ServerFactory` object. That is, the client, which already has

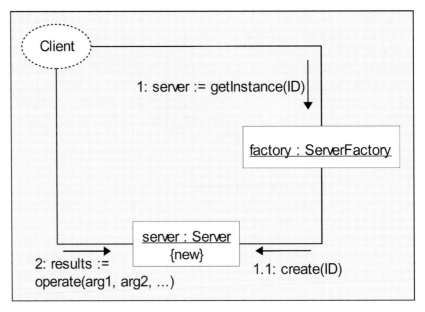

Figure 16.3   UML Collaboration Diagram illustrating the Factory Pattern.

a reference to the `ServerFactory` object, invokes the `ServerFactory.getInstance()` method. (Just how that first reference to `ServerFactory` was obtained will be explained later.) In this diagram the client is shown as an oval *collaborator* instead of an object because we are concentrating on the server side objects in this collaboration diagram. The format of the message labels is

```
sequence_number: return_value := message (arguments)
```

where almost everything except the message name may be omitted.

The first step is sequence number 1, with the client passing an ID argument to `ServerFactory` in the `getInstance()` method. The meaning and type of the ID argument is not rigorously indicated here. It is merely some means by which the factory can identify and distinguish the client making the request from other clients. The ID might be a username, for instance, or a username/password pair, or anything else that the factory requires in order to identify the clients that connect to it. The solid arrowheads indicate synchronous operations. Since sequence number 1 is synchronous, it must be completed before sequence number 2 can occur. But within sequence number 1 is the nested sequence 1.1, which must be completed before sequence 1 is finished. Sequence numbers may be nested to any depth.

There are two objects shown in the diagram – an instance of the `ServerFactory` class on the right and an instance of the `Server` class at the bottom of the diagram. Objects in UML are indicated with an object name (which may be omitted), a colon, and the class name. So the instance name of the

ServerFactory object is *factory* and the instance name of the Server object is *server*. In the first message (i.e. sequence number 1), the client already holds a reference to *factory* and calls its getInstance() method. The return is an object of type Server, though that is not explicitly shown in the diagram. The name of the returned value is *server*, which is explicitly shown. Since the diagrams may be confusing at first we emphasize that the sequence numbered messages are just method calls in Java. In other words, what sequence number 1 really means in Java code executed on the client is something like

```
Server server = factory.getInstance (ID);
```

In order to obtain the Server object to return, the ServerFactory creates a new Server instance in sequence number 1.1, forwarding the ID parameter received from the client. The notation {new} means that a new Server instance is created during this step. The name of the Server object created is *server*, as indicated in the diagram. This name is the same as that returned in sequence 1 because it is the same object. The factory then returns this server to the client.

Once the client has a reference to its own private Server object, all further interaction is between the client and that particular Server instance. All of this interaction is indicated by the operate() method which takes various unspecified arguments and returns unspecified results. In practice, there will be numerous methods on Server that the client invokes to obtain numerous different results. The details are highly dependent on the nature of the actual services provided by the server and are not important here.

We have so far glossed over how the client got a reference to the ServerFactory to begin with. Recall that in a distributed application, the ServerFactory must be constantly running on a server machine somewhere on the Internet, waiting on clients to connect to it to obtain their private Server objects. But how does a client object obtain the original reference to the ServerFactory? The answer is through a *naming* service, which is a well-known service that must be running somewhere accessible to the client. The naming service functions somewhat like a factory, but instead of creating new instances of the objects requested, it returns a reference to an existing object – in this case, to the one single instance of the ServerFactory that already exists. Since there is only one ServerFactory object, it is referred to as a *singleton*, another very common design pattern.

In order to begin this bootstrap process, the client must know how to contact the naming service and must know the proper protocol to request a reference to the ServerFactory. In the Common Object Request Broker Architecture (CORBA, the subject of Chapter 19) the naming service is known as *CosNaming*. There is a certain, well-known protocol for contacting the *CosNaming* service and requesting an object reference. In Java Remote Method Invocation (RMI, the subject of Chapter 18), there is an rmiregistry running on the server machine from which clients obtain server references. In both of these, clients request a

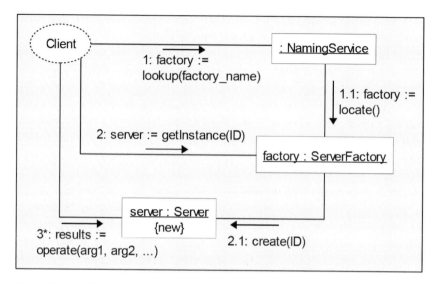

Figure 16.4 Collaboration diagram illustrating the use of the naming service.

lookup of the requested service by name. The registered name of the desired factory server must be known by the client seeking a server, much like you must know the URL address of a web server to which you wish to connect with your browser.

We can now expand our collaboration diagram to illustrate the presence of a naming service as in Figure 16.4. Here the first sequence number is the client requesting a lookup of the named factory server from the naming service. At sequence 1.1, the naming service locates the requested factory (a singleton), which is then returned to the requesting client. This diagram makes clear how the client obtained the reference to the factory server used in the simplified collaboration diagram in Figure 16.4. Then sequence numbers 2 and 2.1 duplicate what we called sequences 1 and 1.1 in the earlier diagram.

In this more complete diagram we've also introduced the *iteration* notation on sequence number 3. The asterisk means that this step can be performed multiple times. For example, the client probably makes several calls to the server to obtain multiple results. A good example is when the server performs an iterative or time-dependent calculation. Each call to the `operate()` method might return status information or the current results of the time-dependent calculation.

Here, too, we have glossed over some details. How, for instance, does the naming service actually find the `ServerFactory`? Those details are private to the naming service and differ for the different naming services available. The important thing to note is that when the `ServerFactory` is created it must *register* with the naming service in some way. Just how this registration is done is dependent on the naming service, but once an object has successfully registered with the naming service, the naming service then knows how to locate that object by name when a `lookup()` request is received.

16.6 Server details

So far we have learned enough UML to illustrate the Factory Pattern and the naming service. We understand the value of the Factory Pattern as a technique to support multiple clients wanting to access the same server, and we know how a client object obtains a reference to its own private server object. But exactly what the server does and how the client interacts with it is yet to be determined. Of course, the behavior of the client and server is highly dependent on the problem at hand. Below we develop a fairly generic client/server interaction that can be used for typical scientific computing problems in which the main computational portion of a problem is performed on a server. As explained previously, the reason for moving the computation to a server might be because of performance concerns, it might be because the computation involves the use of legacy (non-Java) code, or it might be a way to protect intellectual property when a code's services are made available over the Internet. The reason for using distributed computing is not really important to the discussion at hand. We're just going to learn what a server does in a generic client/server application.

In the simplest possible client/server interaction, the client merely prepares input, calls on the server to perform some calculation, and displays the results. This technique might be called "batch mode" and, because it is quite uninteresting, we do not discuss it further.

More interesting is an interactive simulation in which the client prepares the input and receives output on a continuous basis as the calculation is running, perhaps with graphical display of the results. A time-dependent calculation is a good example. Such an arrangement requires the server to output the results in a periodic fashion and the client to retrieve those results and display them somehow.

The server comprises all the code necessary to perform the simulation and communicate with the client. In a simple simulation, the server class typically exposes just a few public methods to the client – methods such as `initialize()`, `receiveInput()`, and `retrieveData()`. There are, of course, other methods internal to the server that are not publicly exposed. In a more complex example, the server would expose additional methods to support the client. These methods might involve storing and retrieving user preferences on the server, retrieving data needed in the simulation from a server-side database, merging user input with static data on the server, etc. The purpose of all these publicly exposed methods is to permit the client to cause the required operations on the server to occur when needed. It is convenient to group all of the publicly exposed remote methods into a Java interface that the actual server class implements. We call this interface `ServerInterface`. Methods internal to the server that are not exposed to the remote client appear in the server class itself but not in `ServerInterface`. This grouping makes explicit the separation of the publicly exposed client-callable methods and the internal methods. We illustrate this separation graphically later when we introduce the UML class diagram.

What we have so far been calling the server class is really just the front end to everything that the server does. In reality, in an object-oriented solution, the "server" consists of multiple objects interacting in such a way to provide the services advertised to remote clients in `ServerInterface`. Internally, the workings of the server are necessarily more complex, as described next.

16.6.1   Model-View-Controller design pattern

Another useful design pattern that we employ is the Model-View-Controller (MVC) design pattern [15]. The Java Foundation Classes (aka Swing) use the MVC pattern. As a reminder, in MVC the *model* is the data. In our distributed computing problem, it is the data generator – i.e. the part of the calculation that creates the data – or what we've previously called the server. The *view* is the user, or consumer, of that data – generally thought of as the presentation to the human user, or the graphical user interface seen by the user. The MVC *controller* is the component that manipulates the data and the view. For example, the user controls the calculation by providing input through the GUI and also, perhaps, by adjusting the graphical output – choosing the variables to observe, for instance. This latter choice may involve communicating to the server which variables to calculate. In any case, it is clear that the user actually does *control* the server.

From the client's point of view, the client provides both the view and the controller to the user, and the server is the model. That is, the client both consumes the data that the server generates and also controls the behavior of the server. We consider the client's implementation in the next chapter. For now, we concentrate on the server.

Even though the server represents just the MVC model to the client, it is also convenient to design the server's internal architecture using an MVC design pattern. The entire server is composed of various parts. The most important part, of course, is the scientific calculation that the server provides. If we break the server down into its various components, the calculation component can be called the "compute engine." Since this is a client/server system, we must assume that the client application is remote, connected via the Internet. Because of network delays, or because of client-side presentation costs, the client may not be able to request and process data as fast as the compute engine generates it. Therefore, the data that is generated must be stored somewhere until the client requests it. The component that stores the calculation data is obviously the MVC data model.

The ultimate consumer of the data is the client, but recall that the client obtains the data by calling a method on the server – referred to as `retrieveData()` above. When the client calls `retrieveData()`, the front-end server object that

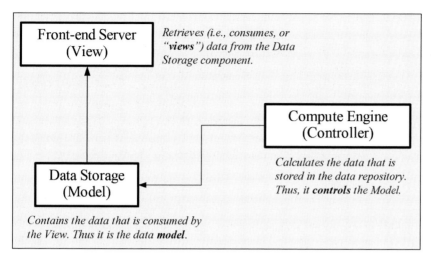

Figure 16.5 Conceptual model showing the model, view, and controller relationships that are internal to the server architecture.

implements that method must fetch the data from the storage location. Thus, from the point of view of the data storage component, the front-end server object is the consumer, or the MVC view. These "view" and "model" relationships are illustrated in the conceptual model shown in Figure 16.5.

The MVC controller is the component that controls the data. From this point of view, controlling the model means controlling what is stored there – i.e. in the data storage component. It is the compute engine that calculates the data and provides it to the data storage component. The data storage component only contains data placed there by the compute engine. Thus the compute engine can be thought of as the controller. In this description, the controller controls only the data model; it has no direct control over the view component.

16.6.2 Internal server collaborations

The conceptual model developed so far is not complete for at least two reasons. First, recall that the client must have some way to tell the compute engine what to compute. The client does so by providing input data via the front-end server object, which is the only server-side object that the client has access to or even knows about. Thus there must be some association between the front-end server and the compute engine. That is, the front-end server must somehow communicate the input data to the compute engine.

The second shortcoming in the conceptual model is more subtle. Notice that there must be two threads of execution occurring inside the server. One thread is the compute engine, calculating data and storing it into the data storage component. But any response to the client's request to retrieve data must occur on a

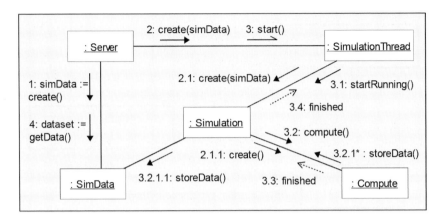

Figure 16.6 Collaboration diagram illustrating the internal server architecture.

separate thread. The client's data request comes to the front-end server component. If the compute engine has control of the CPU, then there is no way for the front-end server to handle that request unless the front-end server is running on a separate thread. (In a single-threaded world, the compute engine would have to periodically give up the calculation and poll the front-end server to see if any new data request had been made.) While we're considering multithreaded issues, also notice that both the compute engine and the front-end server have access to the data in the data storage component. Therefore there will be synchronization issues to deal with to ensure the integrity of the data in the data storage component since that data is stored and retrieved by separate threads.

With these concerns in mind, we can now create a collaboration diagram showing the interactions among the various server-side components. We identify the front-end server object merely as `Server`. It is the object with which the client interacts. Though it contains other methods, recall that the only publicly exposed methods that the client knows about are defined in `ServerInterface`, which `Server` must implement.

Let's call the object that embodies the compute engine `Simulation` since the client/server application we are designing is best thought of as an interactive simulation rather than a batch-mode client/server interaction. We call the data storage component `SimData`, since it stores the results of the simulation. To explicitly show the multithreaded nature of the architecture, we also introduce a `SimulationThread` object. These objects are shown in the collaboration diagram in Figure 16.6, which also introduces a bit more UML notation.

As this is more complicated than the previous diagrams, we'll walk through the diagram by sequence number and point out the new features. In sequence number 1, the `Server` instantiates a `SimData` object and retains a reference to it called `simData`. (We have dispensed with the "`{new}`" notation here to reduce clutter and because it provides no additional information. Instead, we indicate the creation of the new `simData` object with the `create()` method. In practice, in Java, the `create()` is actually implemented as a Java new operation.) In sequence 2, a `SimulationThread` object is created. This is the thread that runs the calculation. During the `SimulationThread` constructor (i.e. during sequence 2), a new `Simulation` object is created at sequence 2.1 and then a new `Compute` object at sequence 2.1.1. The `Compute` object is broken out separately for convenience, although its operations could be rolled into `Simulation`. We explain `Compute` in more detail shortly. In practice, the construction of `Simulation` and `Compute` may be done during the `SimulationThread` constructor as shown, or it may be postponed until the thread starts running during sequence 3. There is no particular advantage either way.

Observe that `Simulation` has an association with the `SimData` object – i.e. it must call a `SimData` method. Thus `Simulation` needs a reference to the `SimData` instance. By using the parameter list features of the UML messages, we've explicitly shown how the `simData` reference gets communicated to `Simulation`. Eventually, though not shown here, the client instructs `Server` to start running the simulation. At this time, in sequence 3, the `Server` sends a `start()` message to `SimulationThread`, which begins the thread's `run()` method in the normal Java fashion. That is, the `run()` method in `SimulationThread` begins execution in another thread of control, and the `start()` message returns immediately. Because `start()` returns immediately while the other thread is still running, this is an asynchronous method invocation. Asynchronous messages are indicated in UML with half arrowheads as shown.

Sequence 3 is fairly complicated, with several sub-sequences as shown. At 3.1, the `Simulation`'s `startRunning()` method is called, which calls on the `Compute` object to do the actual calculation at 3.2. During the computation, the `Compute` object periodically calls `storeData()` at 3.2.1, marked as an iteration since there are several such calls over the course of the calculation. When `Simulation` receives the `storeData()` call, it is forwarded to `SimData` at sequence 3.2.1.1, and the data is stored in the `SimData` object. Finally, when the calculation is complete, we show an explicit "finished" message from `Compute` to `Simulation`. This is actually a return from successful completion of sequence 3.2, but it is shown explicitly because of the complexity of sequence 3. Return messages are shown in UML as dashed arrows. When `Simulation` receives the return message, it is finished as well, and another finished message is shown returning to `SimulationThread` in sequence 3.4. That finished message corresponds to the end of the thread's `run()` method.

Much simpler is sequence 4 in which the client has requested data from `Server`, and `Server` requests data from `SimData` with a `getData()` message. At this point, the data stored in `SimData` is returned to `Server` and is then forwarded to the client.

The reason that `Compute` is broken out as a separate object is because the calculation is often performed in legacy code, and that legacy code usually amounts

to a separate object. In the case of a Fortran, C or C++ legacy code, it is implemented as a set of one or more "native" methods, the subject of Chapter 22.

Recall that we may want to provide a way for the client to send control messages
to the simulation – i.e. messages that affect the calculation while it is running.
For simplicity, we have omitted these from Figure 16.6. If control messages are
present, they would be analogous to collaborations 3.2.1.1 and 4 but in reverse
order. That is, a control message would be sent from `Server` to `SimData`
and stored until retrieved by `Simulation` at a convenient time. In practice,
a convenient time to retrieve the control message is as a return value from each
`storeData()` call. Obviously the methods for storing and fetching both control
and output data must be synchronized to prevent thread collisions.

Notice that in this scheme the computation is running in its own thread at its
own pace, providing current simulation results to the `SimData` object as they
are calculated. The client, on a remote machine, is also running at its own pace,
receiving simulation data with each request. The server has no control over how
fast the client requests data, or how fast the network connection between client
and server is. If the computation is time dependent, generating a data set at each
time step, it is very possible for the computation engine to take multiple time
steps between client data requests. On a fast server machine and a slow client or
a slow network, the server could calculate many time steps and produce many
data sets between client data requests. It is up to the designer to decide how to
deal with these "extra" data sets. The simplest solution is to ignore them. The
client always gets the most recent data set, with any intervening data sets lost. If
the client is unhappy about losing some data, then the client needs to request data
more often.

A slightly better approach is to design one of the control messages to permit
the client to tell the server to slow down, perhaps by sleeping a short while
between time steps. A more complicated solution is for `SimData` to store all
data sets received until a data request is received and then return the entire saved
set at that time. All these details are unimportant to the generic design being
presented here, but they are extremely important once an implementation is begun.
In the sample code provided in Chapter 20, we use the simplest approach of
ignoring the extra data sets. This approach generally seems to work well when
both the client and server are on a fast LAN, but might be unsuitable on a slow
network.

Let us now consider the code that does the actual computation, i.e. the
`Simulation` and `Compute` objects in Figure 16.6. First, recall that, using the
Factory Pattern, we are able to support multiple clients, each with its own private server object. We see from the discussion above that the "server object" is
really several interacting objects, but the client only sees the interface exposed by
`ServerInterface`. When there are several clients, there will be several
`Server` objects, all in the same JVM but running on multiple threads. If the entire
code is written in Java from the ground up, then maintaining thread safety in the

Simulation and Compute objects is straightforward. In the case of a legacy code, however, particularly codes written in Fortran, ensuring thread safety is more difficult because Fortran codes are almost never reentrant due to their use of global variables in the form of Fortran common blocks. This situation is one reason we broke out Compute as a separate object in the collaboration diagram above. If several Server objects exist, each of which creates Simulation objects, each of which uses its own instance of the Compute object to perform the legacy calculations, then the legacy Fortran code will almost surely cause data corruption when it attempts to access global variables. The reason is that multiple clients accessing the single Fortran image will overwrite each other's global variables, resulting in incorrect answers for at least one of the clients at the very best. More likely, an outright core dump could result.

The best solution for such legacy codes is often to run the legacy code in a separate OS process. Separation between OS processes is maintained by the operating system, with each process getting its own memory map. In that way, the Fortran images are isolated from one another, preventing data corruption as they internally access their global variables. Some kind of inter-process communication (IPC) is required in order for each Simulation object to communicate with its per-client Compute process. Since the spawning of a new process is an asynchronous task, special care must be taken to enforce the synchronous nature of the interaction. Notice that the compute() call (i.e. sequence 3.2 in Figure 16.6) is synchronous. It does not return until the computation is complete. But if the computation is to run in a new process, the method call that spawns the child process returns immediately. This issue can be solved through careful use of IPC.

Another option to protect legacy code is for ServerFactory to create each Server in a new per-client OS process (i.e. the entire Server is in a new process, not just the Compute object). Then the object returned to the requesting client during the getInstance() method (i.e. sequence 2 in Figure 16.4) will in fact be running in a separate process instead of as a separate thread within the same JVM.

16.6.3 UML class diagrams

We described earlier the grouping of the server's publicly exposed methods into a Java interface. To illustrate this grouping, we now introduce the UML class diagram. A class diagram shows classes along with their variables and methods (called *attributes* and *operations*, respectively, in UML) and how those classes are related to one another. The collaboration diagrams seen previously deal with objects, not classes, and the sequence-ordered messages among them. By contrast, class diagrams deal with static relationships among classes. A class in a UML class diagram is shown as a box divided horizontally into three sections. The top section is the class name, the middle section contains the attributes, and

Figure 16.7 UML class
diagram showing the
client's view of the
`Server`.

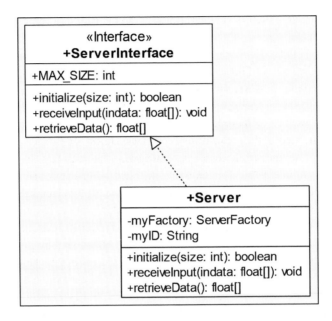

the bottom section contains the operations. The middle and bottom sections are optional. See the example in Figure 16.7.

One of the most important static relationships between classes is inheritance, discussed in Chapter 4. Inheritance, or the Java *extends* relationship (which is known as *generalization* in UML terminology), is shown with a solid line and a hollow arrowhead pointing from the child to the parent – i.e. from the subclass to the superclass. Chapter 17 shows an example of generalization.

A class diagram is a view of the classes from which the objects used in the collaboration diagram are instantiated. Doing the collaboration diagram first helps us understand some of the methods needed on the server-side objects. For example, from the server collaboration diagram, we see that the `SimData` class must have `getData()` and `storeData()` methods.

What we show here is the client's view of the server, specifically the grouping of the publicly exposed methods into a Java interface. A Java interface is really a special kind of class (see Chapter 4) so the UML notation for an interface is similar to a plain class. An interface is denoted in UML with the `<<interface>>` label. Strictly speaking, a Java interface is not quite the same thing as a UML interface, but the differences are subtle enough as to be unimportant for our purposes. So we freely use the `<<interface>>` notation in our diagrams. In addition to the `<<interface>>` label, a common practice is to use a naming convention that appends the word "Interface" (or sometime just "Ifc") to an interface class name. Thus the name of our interface class is `ServerInterface`. Since the `Server` class implements the server interface, the Java code for the `Server` class begins like this:

```
public class Server implements ServerInterface {. . .}
```

To illustrate this relationship in UML, there is one box for the `ServerInterface` class and another box for the `Server`. The Java *implements* relationship (called *realization* in UML terminology) is shown as a dashed line with a hollow arrowhead. As with inheritance, the arrowhead points to the parent.

The client obtains a reference to an object that implements `ServerInterface`. The only methods on the `Server` known to the client are those defined in `ServerInterface`, a simplified version of which is shown in the class diagram in Figure 16.7. Here we see an attribute named `MAX_SIZE` in the central section of the `ServerInterface` box. `MAX_SIZE` is an `int`, as indicated, and the + sign is UML notation signifying public access (see Chapter 5). Following standard Java coding conventions, `MAX_SIZE` is all uppercase, indicating that it is a constant. The idea here is that `MAX_SIZE` will be used to define the size of the data arrays passed between client and server. An actual working example would probably use dynamic array allocation instead of fixed array sizes, but the usage here serves as a good example of the use of UML attributes.

We also see the three publicly exposed methods, `initialize()`, `receiveInput()`, and `retrieveData()`. The `initialize()` method receives a size parameter which specifies the size of the data arrays (up to a pre-compiled maximum of `MAX_SIZE` in this example) and returns a `boolean` indicating success or failure of the initialization operation. The client provides a `float` array of input data to the server in the `receiveInput()` method, which returns `void`. We assume that the calculation begins (i.e. the "start" message, sequence 3 in Figure 16.6) when the server receives and processes the `receiveInput()` call. As the server is running the simulation, the client must periodically call the `retrieveData()` method, obtaining a `float` array of results in return. There can be a special sentinel value in one of the array elements when the simulation is complete. For the control data sent from the client to the server, we could define a `receiveControlData()` method on the server that would receive an array of control data of some kind. However, it is usually more convenient simply to add a parameter to the `retrieveData()` method containing the control data as shown. That is, the current set of control data could be sent to the server during each `retrieveData()` operation rather than creating a new method just to receive the control data. If no new control data is available when the client calls `retrieveData()`, then `null` or the previous control set could be sent, a convention that must be agreed to by both client and server.

The technique for obtaining the server's data illustrated here is called "polling," in which the client periodically polls the server for new results. An alternative solution is to define methods on the client callable by the server so the server can "push" the new data to the client when it has new data available. The client and server roles are reversed temporarily while what we have been calling the server acts as a client by making a call to the real client, which plays the role of a server for that call. In this technique, the real server is said to be "calling back" to the client, and the technique is generally referred to as a "callback" scheme. Callbacks are a good technique in some situations, and they reduce network overhead by eliminating unnecessary polling calls. Unfortunately, callbacks seldom work through firewalls. So, if the server is designed to be called from outside its internal network, then using callbacks is generally not a successful approach.

We have left out many details in the description above. What we have provided is a general outline of how to produce a distributed object application, and we've learned enough UML notation to describe the behavior of the server. While fairly complex, all of the above should become clear when we produce actual working code in Chapter 20 after we've covered a few more theoretical issues. In the general description above, the main thing missing is any mention of client behavior, the subject of the next chapter.

Before proceeding to the client though, let's finish the discussion of the UML class diagram. Notice that the `Server` implements `ServerInterface`, as shown by the dashed line and hollow arrowhead. Since an implementation must provide concrete methods for the methods defined in the interface, we show `Server` with the same three public methods that appear in `ServerInterface`. We also indicate two attributes on `Server`, marked with a "`-`" sign to indicate private access. One is a reference to `ServerFactory`. This reference can be used by `Server` to communicate with its factory. An example of such use is for diagnostic logging in which the factory keeps a log of server behavior. Another use is for synchronization with respect to the factory object for short-running pieces of native code that are not inherently thread-safe. By synchronizing with respect to the singleton factory, we can ensure that no two clients access a non-thread-safe native method at the same time. As long as the non-thread-safe native methods are not time consuming, this simple solution provides thread safety with very little effort. Obviously this solution is not adequate for long-running or frequently accessed native methods because too much thread blocking would occur. The other private `Server` attribute is a copy of the ID that was received when the factory instantiated the server during sequence 2.1 in Figure 16.4.

16.7 Web Course materials

The Chapter 16 in the Web Course provides further discussion and resources regarding distributed computing for technical applications, client/server design, and UML.

References

[1] SETI@Home, `http://setiathome.ssl.berkeley.edu`.

[2] Distributed computing project examples include: the DataGrid Project (`http://eudatagrid.web.cern.ch`), which seeks to develop scientific data analysis techniques with distributed computing; a decryption project at `www.distributed.net`; and a climate study at `www.climateprediction.net`.

[3] The Earth Simulator, `www.nec.co.jp/press/en/0203/0801.html`.

[4] Blue Gene, `www.research.ibm.com/bluegene/`.

[5] Terascale Cluster, `http://computing.vt.edu/research_computing/terascale/`.

[6] ORNL – The National Leadership Computing Facility project, `www.ccs.ornl.gov/nlcf/`.

[7] "Department of energy awards $25 million to Oak Ridge National lab to lead effort in building world's largest computer," ORNL press release, `www.ornl.gov/info/press_releases/get_press_release.cfm?ReleaseNumber=mr20040512-00`.

[8] "DOE leadership-class computing capability for science will be developed at Oak Ridge National Laboratory," DOE press release, `www.energy.gov/engine/content.do?PUBLIC_ID=15871&BT_CODE=PR_PRESSRELEASES&TT_CODE=PRESSRELEASE`.

[9] The Top 500 List, `www.top500.org`.

[10] PVM (Parallel Virtual Machine) project, `www.epm.ornl.gov/pvm`.

[11] UML Resource Page, `www.omg.org/uml/`.

[12] J. Rumbaugh, I. Jacobson and G. Booch, *The Unified Modeling Language Reference Manual*, Addison-Wesley, 1998.

[13] IBM Rational Software, `www.rational.com`.

[14] E. Gamma, R. Helm, R. Johnson and J. Vlissides, *Design Patterns*, Addison-Wesley, 1995.

[15] S. J. Metsker, *Design Patterns Java Workbook*, Addison-Wesley, 2002.

Chapter 17
Distributed computing – the client

17.1 Introduction

Chapter 16 introduced distributed computing and enough UML to describe server-side interactions. In this chapter we describe the design of the client for a distributed scientific application in which the computationally intense calculations are performed on a remote server. Like the server, the client details necessarily depend heavily on the calculation being performed and the data that is to be presented to the user. Nevertheless we can provide some general guidelines that should apply to many scientific applications.

17.2 Multithreaded client

Recall that the server in Chapter 16 is running on a remote machine, at least conceptually, and is generating results continuously as the simulation is running. To avoid problems with intervening firewalls, we designed the server to be polled by the client rather than using a callback from the server to the client when new data is available. That is, the client must poll the server periodically to retrieve the results being calculated by the server. Assuming that the server calculation is generating results somewhat uniformly, we clearly would prefer for the client to poll the server on a regular basis. Meanwhile, once the client receives the current set of results, the client must display them to the user in some fashion and allow the user to interact with the displayed data. Whenever human users are involved, one can be assured that the user's actions will not be uniform and regular. The user may pause to closely examine the data being displayed or may even take a coffee break while the calculation runs. It is clear that the polling loop cannot rely on user behavior. The time required for the client to display the data may depend on the user's actions and choices and may also depend on the data itself. So it is also clear that the polling loop should not rely on a uniform processing time for the most recent set of data. What is needed is a multithreaded client with one thread handling a regular and uniform polling of the server for new data and one or more other threads handling the display of the data and the user interactions. As always in a multithreaded design, care must be taken to prevent thread collisions. In this case, we must ensure that the polling thread does not attempt to modify the data while the data-display thread is using it.

As mentioned in the previous chapter, some decisions must be made regarding how to deal with the relative speeds of the client and server machines and the network between them. That is, should the `SimData` object in Chapter 16 store all simulation data until retrieved by the client or discard old data if new data arrives before the client has requested it? Or perhaps the client should attempt to retrieve data as fast as possible and store that data on the client side. As before, the details are not important to a general discussion but become very important when an implementation is begun. For simplicity, we assume that the client and the network are fast enough to collect the data from the server as often as needed. This assumption essentially means that if the server calculates data faster than the client's polling thread retrieves it, then `SimData` merely discards the old data and always returns the most recently calculated simulation data to the client.

17.3 Model-View-Controller for the client

Recall from the general discussion in Chapter 16 that we desire a way for the user to control the simulation in some fashion. We explain how the control data is collected and used in more detail below. For now, consider that we have a server generating data to be displayed, a client displaying that data, and some means for the user to control the simulation. This arrangement leads quite naturally to the Model-View-Controller design pattern for the client, this time a more obvious application of MVC. The client's graphical display of the data is clearly the view, and the control data portion of the user interface is the controller. The server is obviously the data model. From a client-design point of view, it is better to view the client-side object that fetches data from the server as the data model. How it generates the data (which happens to be by contacting the server) is immaterial to the client design.

The Java class library conveniently provides the `java.util.Observable` class and the related `java.util.Observer` interface as perfect aids for implementing the data and view components, respectively, of the MVC design pattern. By extending `Observable`, we obtain a class that our client application wants to have observed. This technique provides a perfect data model since data can reside in the `Observable` subclass, and the view component is the observer of that data. When the data contained in the `Observable` changes, the view component can be notified to update the view. In practice, there can be more than one view component. For example, we may wish to display the data in multiple ways – tables of numbers, graphs of different types, etc. Each type of display is a part of the MVC view but will be implemented as a different view object. The most straightforward way to implement these view objects is to create classes that implement the `Observer` interface, which requires that the `update()` method be implemented. When the `Observable` obtains new data, the `Observer` objects are notified by having their `update()` methods called, permitting them to update their displays.

These notifications and responses don't happen quite automatically, of course. The observers must first register with the observable object, and the observable's `notifyObservers()` method must be called in order to initiate the notification. The `notifyObservers()` method sees to it that each observer's `update()` method gets called, during which the updated data must be retrieved and the view modified accordingly.

We create a class called `DataManager` that serves as our `Observable`. This class contains the polling thread, which fetches data from the server on a regular basis and then calls `notifyObservers()`. We also need a top-level class instantiated by the `main()` method of the client application that puts everything together. We call this class `SimClient`. The application's `main()` method instantiates the `SimClient` object, which then instantiates `DataManager`, which then instantiates other objects, some of which will present a GUI. With this much design in mind, we can begin to create a client class diagram. In this case it makes sense to think about the class relationships and create the class diagram first because we're using the `Observer/Observable` classes from the Java class library.

Before turning to the diagram, we need to introduce a little more UML notation. We've already described the UML *realization* relationship (dashed line with hollow arrowhead) and the UML *generalization* relationship (solid line with hollow arrowhead). As a reminder, these relationships in Java are the `implements` and `extends` relationships, respectively. Classes can have relationships other than realization and generalization. One of the most common is known in UML as an *association*, in which objects of one class have some reason to interact with objects of another class. Without getting deep into the finer UML points of composition vs. aggregation associations, we use just two simple types of associations. When an object instantiates another object (which should happen only when the first object needs some services from the instantiated object), we show the class association as a solid line with a simple arrowhead pointing to the instantiated class. Similarly, a dependency on another class is shown with a dashed line and a simple arrowhead pointing to the depended on class. These UML notations are shown in the diagram in Figure 17.1. To reduce clutter, we have not shown the visibility indicators (i.e. the + signs indicating public classes) or any of the attributes or operations in each class. We've also shown the two classes from the Java class library as lightly shaded to distinguish them from the custom classes created for our application.

Here we show two observers – `PlotPane` and `TextPane`. The idea is that the client GUI has both textual and graphical output windows. Both are observers because they implement the `Observer` interface as shown. `DataManager` is an `Observable` because it extends the `Observable` class. Several things are not shown in the diagram, including how the observers' `update()` methods get

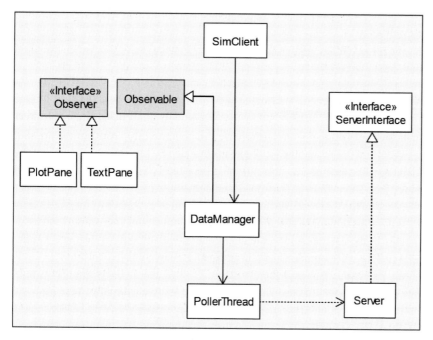

Figure 17.1 A simplified class diagram of the client application.

called by `notifyObservers()`. This action can be assumed to occur behind the scenes, courtesy of the Java class library's implementation of `notifyObservers()`.

`DataManager` also contains `PollerThread`, which has a dependency on the `Server` class, which implements the `ServerInterface`. Also missing from the diagram so far is any indication of how this server object is instantiated. Note that this `Server` isn't the actual server-side server object described in Chapter 16, but rather some client-side construct that knows how to communicate with the real server. Just how the client-side server object actually communicates with the real server object is discussed in Chapters 18 and 19. In practice, the client-side server object is obtained from an analogous client-side representation of the server factory object using the Factory Pattern, as explained in Chapter 16.

For simplicity, we've also omitted from the above discussion any mention of the controller component. Recall that our design includes the capability for the user to control certain aspects of the simulation as it is running by passing an array of control variables to the server in the `retrieveData()` method call. One can imagine some sort of user-adjustable values controlled by GUI elements – buttons, checkboxes, sliders, etc. – in the `PlotPane` and/or `TextPane` windows, or perhaps in a new `InputPane` class. As is common in Swing applications, the view and controller components are often merged, at least in the user interface if not in the design. In any case, these input values must somehow be communicated to

`PollerThread` since that is the object that makes the `retrieveData()` call. Since the observers are shown as standalone classes without any associations to any other classes, that route of communication is not shown in the class diagram.

17.4   More client details

Once the basic design is understood, we can begin to fill in some of the missing details. `SimClient` is the first object instantiated by the `main()` method of the client application, and it creates all the other objects as needed. One of the first things that `SimClient` must do is contact the `ServerFactory` to obtain a reference to the `Server`. We assume that the `ServerFactory` object already pre-exists on the remote server machine. For completeness, we show those associations in the next version of the client class diagram, though they don't really add much to what we must know in order to implement the client classes.

We also explicitly show the `InputPane` class that collects the user's input data. For organizational reasons, it is convenient to collect the input pane class and the two (or more) observer classes as subcomponents of a `DisplayManager` class that manages all user input and display functions. It is this `DisplayManager` and its subcomponents that present the GUI to the user.

It is also convenient to collect all interactions with the server into a central location which we call the `ServerGateway` class. Any class that needs to call a server method does so through the server gateway, meaning that only the server gateway needs a reference to the server class. In that way, any changes to the server interface will need to be dealt with only in the gateway class instead of spread over several client classes. The one exception might be the poller thread class which makes frequent calls to `retrieveData()` on its own schedule. To avoid function-call overhead for such frequent server calls, it may be advisable to permit `PollerThread` to call the server directly rather than going though `ServerGateway`.

While we now have a good handle on the client design, perhaps a collaboration diagram will be helpful to understand the order and flow of messages in the client. Having the collaboration diagram handy also serves as a reference when we start writing actual Java code to implement the client. Therefore we present the diagram in Figure 17.2.

Here we can see that `SimClient` first obtains a server from the server factory, as explained in Chapter 16, and then initializes the server. At first glance, this initialization might be thought unnecessary since initialization could occur when the server factory creates the server. However, a separate initialization step is more general and permits the factory, for instance, to keep a pool of inactive uninitialized servers instead of constructing a brand new server for each client. At sequence 3, `SimClient` instantiates the `DataManager`, passing a reference to the server. `DataManager` then instantiates the `PollerThread`,

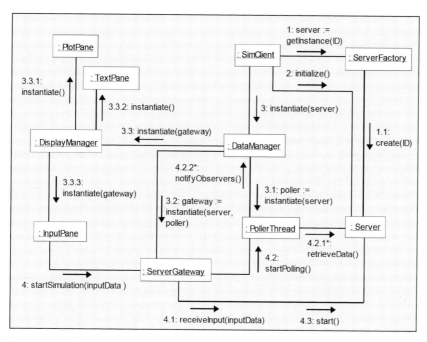

Figure 17.2 Client collaboration diagram.

the `ServerGateway`, and the `DisplayManager`, in that order. Those objects must be created in the order shown because the gateway needs a reference to the poller and the display manager needs a reference to the gateway. We also see that `DataManager` must pass a server reference to `PollerThread` and `ServerGateway`, since both must access the server. Then `DataManager` must provide a `ServerGateway` reference when it instantiates `DisplayManager`, which must provide the reference to `InputPane` since `InputPane` needs to access the gateway. `DisplayManager` also creates the plot and text output panes. The `InputPane` object contains a GUI element permitting the user to start the simulation.

When the user causes the simulation to begin, several things must happen during sequence 4. First, `InputPane` must call the gateway's `startSimulation()` method. At this time `InputPane` provides a set of input data used to initialize the calculation. This data is passed to the server's `receiveInput()` method. It is important to note that this data set is different from the control data that controls the simulation while running. It is also different from any data used to initialize the remote server object as was done in sequence 2. The server initialization is done very early, before `InputPane` even exists. Later, `InputPane` presents its GUI to the user and collects the calculation initialization parameters. These parameters might include, for example, time steps, maximum run time, or other values that set up the desired simulation. They are passed to the gateway in the `startSimulation()` call. The gateway then sends these initialization parameters to the server's `receiveInput()` method.

After initializing the simulation, the gateway tells the `PollerThread` to begin polling and then calls the server's `start()` method which starts the simulation actually running. `PollerThread` repeatedly polls the server by calling `retrieveData()`. Whenever new data is received, `PollerThread` calls the `notifyObservers()` method in the `DataManager` to cause the two observers to update their output displays.

To reduce clutter in the diagram, we've omitted how the control data from `InputPane` is communicated to `PollerThread` in order to pass it to the server during the `retrieveData()` call. One method is for this flow to mimic that of sequence 4's start simulation messages, but bypassing the `receiveInput()` initialization call. Alternatively, a particularly elegant solution can be had by making `InputPane` into an `Observable` and `PollerThread` an `Observer` of `InputPane`. Then, when new control data is entered by the user into `InputPane`, the `PollerThread` observer is notified of the changes, ready to pass the new control set to the server during the next poll.

Through the process of creating this client collaboration diagram, we've learned that the server needs another method that we forgot, or at least glossed over, in Chapter 16. That missing method is the `start()` method that the client uses to tell the server to begin running the simulation. In Chapter 16, we assumed that the simulation would start at the end of the `receiveInput()` method, once the calculation was initialized. However, we need a chance to begin polling after initializing the simulation but before actually starting the simulation. Alternatively, if we started the simulation and then started polling, we might miss the first few data points. The astute reader can think of several other solutions to this timing problem, including some more elegant ideas, but the most obvious technique is to separate simulation initialization and simulation start into the two `receiveInput()` and `start()` methods. We might also wish to change the name of the `receiveInput()` method, first invented back in Chapter 16 while designing the server with only a little thought about the client, to something more descriptive like `initializeSimulation()`.

17.5 Improved client class diagram

By examining the collaboration diagram, we can now add some of the missing pieces to the initial class diagram. This time we show some of the operations identified while creating the collaboration diagram. We won't show the instantiate methods since they're really object instantiations rather than methods anyway. This more complete class diagram appears in Figure 17.3.

The only new UML notation used here is the italicized *notifyObservers()* method in `DataManager`. Italics are used to indicate that a method is inherited from a superclass. We've also switched to the `initializeSimulation()`

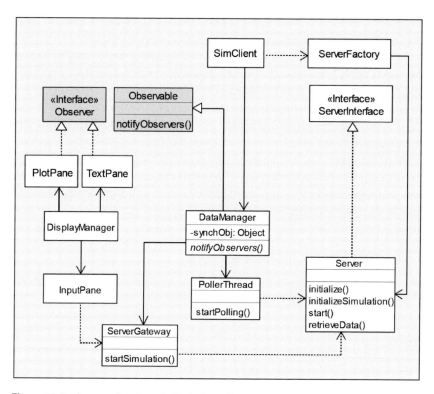

Figure 17.3 A more developed client class diagram.

name in the `Server` object. The same change, as well as the addition of the `start()` method to `Server`, must be propagated to `ServerInterface` and the server-side UML diagrams as well. We won't take up space showing the modified diagrams here, but we must remember to make those changes when implementation is begun.

There are a few more loose ends to be cleaned up during implementation. One important example is protecting against thread collisions between the poller thread and the thread that is displaying the data in `PlotPane` and `TextPane`. A good approach is for the poller to duplicate the received data during a synchronized block. Then the display objects can lock on the same synchronization object while updating their displays with that data. Accordingly, we show a private `synchObj` attribute in `DataManager`. Both `PollerThread` and the output panes could obtain a reference to that synchronization object and use it to protect code that accesses the data copy. (Note that the `synchObj` is declared to be `private` to `DataManager` in order to preserve good encapsulation, but that does not prevent `DataManager` from making it available in a public `getSynchObj()` method.) If the poller obtains another data set from the server while the output panes still have the lock, the poller must wait until the lock is released before making a copy of the new data set. Since we want the poller to poll uniformly, it is obvious that the output panes must not hold the lock for too long. One way to accomplish this goal is for `DataManager` to request the lock and make yet another copy of the data that the output panes use instead of having the output panes lock and copy the data. Since the output panes and `DataManager`'s methods run in the same thread, there is no chance of collisions between `DataManager` and the output panes.

Another issue not shown is that the `Observable` superclass requires that `setChanged()` be called in order to mark the `Observable` object as having been changed before calling `notifyObsevers()`. Otherwise, `notifyObservers()` does not notify anything because it thinks that nothing has changed. For simplicity, this detail is not shown in the class or collaboration diagrams but must be dealt with during implementation.

17.6   Web Course materials

The online materials include more discussion of client design and of the Model-View-Controller design pattern. Also, there is additional information about UML.

Resources

Model-View-Controller, *Design Patterns Catalog*, Sun Microsystems,
  http://java.sun.com/blueprints/patterns/catalog.html.
Model-View-Controller Architecture, *Fundamentals of JFC/Swing*, Sun short course,
  http://java.sun.com/developer/onlineTraining/GUI/Swing/
  shortcourse.html.

Chapter 18
Java Remote Method Invocation (RMI)

18.1 Introduction

Chapters 16 and 17 described the client/server or distributed object paradigm at a somewhat abstract level using the UML notation language. Because of the abstract nature of that discussion there was almost no Java code in either of those chapters. In fact, the only place that any code snippets appeared at all was to illustrate UML concepts in a familiar concrete Java environment.

Now we will begin to explain how to really implement the distributed object paradigm using Java Remote Method Invocation (RMI). By the end of this chapter, we will have some real code that demonstrates simple communication between two distributed Java objects, though we still won't have a running application using the architecture of Chapters 16 and 17. That complete application will be developed in Chapter 20.

Before getting to the real code, we must first describe how distributed computing can be implemented. The concepts are not particularly new, nor are they unique to Java. However, just as in other areas, the Java platform offers advantages and an easy programming style that are absent in other languages. These advantages are discussed as we explain how distributed computing works. While the full capabilities of RMI are beyond the scope of this book, we will learn the basics of RMI, and with these skills we can implement the distributed architecture described in Chapters 16 and 17.

18.2 How distributed computing works

Consider a conventional (non-distributed) computer program, be it written in Java or some other language. Certain program elements, call them "functions" as in C, or "subroutines" as in Fortran, or "methods" as in C++ and Java, will call other program elements. All these program elements (we call them functions for simplicity) are typically loaded somewhere into the machine memory allocated for the program, and each function has a unique memory address. A computer's *program counter* points to the address of the next machine-level instruction to be executed. Through the magic of compilation and linking, a *calling* function knows the memory address of the *called* function. When the call is made, the calling function somehow instructs the computer to change the program counter to point to the called function, and the computer then starts executing the code at the new location. There are other complications having to do with the machine stack used to pass function parameters and to hold the return address, but the point is that *all* the code resides in a single process or memory space in which all the machine addresses of all the program elements are known. In a Java program, all the code resides in a single instance of the JVM.

In a distributed program, some of the code resides in another process, perhaps on another machine. The calling process has no way of knowing the machine addresses of methods in a different process or on a remote machine and certainly has no way to instruct the remote machine to move its program counter to begin executing code for a called method.

If the two machines have different operating systems or different architectures (32-bit vs. 64-bit, for example), there can be differences in floating-point format, integer size, and byte order (big-endian vs. little-endian). For Java programs, which have automatic garbage collection, there is also the messy issue of remote, or distributed, garbage collection when an object is no longer needed.

The Java platform easily provides elegant solutions to some of these problems. For instance, because Java is a cross-platform system, one need not worry about floating and integer formats or endian order. Java defines consistent sizes and formats for all the primitives, and the byte order is the same on all Java platforms regardless of the underlying operating system. Therefore Java application programmers need not worry about reversing the byte order or dealing with different `int` sizes, a common problem with remote procedure calls (RPCs) using C or C++.

As has already been seen in Chapters 14 and 15, Java provides low-level socket-based networking. Socket programming is appropriate for some tasks, but it is far too complicated for the kinds of distributed computing tasks we're dealing with here. No one wants to invent his or her own RPC protocol for each new distributed application.

Fortunately, the Java RMI system handles all the low-level networking tasks. Using RMI, the semantics for making remote calls is *almost* the same as for making normal local method calls. RMI is part of the Java 2 Standard Edition, so it is available on any J2SE platform. Since the programming style is nearly the same as normal Java programming, it is *almost* as easy to work with as normal Java programming. Everyone in the large pool of Java developers should be able to program with Java RMI with little additional training, unlike socket-based programming or proprietary alternatives like Microsoft's DCOM. Java RMI also offers ease of deployment. Because of the cross-platform nature of Java, it is even possible for clients to automatically download the client-side bytecodes as needed, relieving the developer from having to distribute client-side code. Because of all these advantages, it is clear that Java and RMI provide an excellent platform for distributed computing while retaining the object oriented nature of Java.

An alternative to RMI is the Common Object Request Broker Architecture (CORBA). CORBA has some advantages over RMI, particularly because it is language independent, meaning that C or C++ clients, for example, can make CORBA calls on a Java server or vice versa. While Java RMI requires Java on both the client and the server, CORBA permits the use of any programming language that has a CORBA binding. The most popular languages that support CORBA are C++ and Java. Yes, Java fully supports CORBA as well as RMI, and the CORBA support is built into Java, just like the RMI support. Because of the added complexities required to be language independent, CORBA is more difficult to use than RMI, is not as fully object oriented as Java, and suffers some performance penalties compared to RMI. However, CORBA remains an important technology that has a well-deserved place in the distributed computing world. We discuss CORBA in more detail in Chapter 19.

18.3 RMI overview

RMI is the mechanism that allows an object in one Java virtual machine to invoke methods on an object running in another Java virtual machine. Both JVMs may be present on the same physical computer or, as is more likely, the JVMs can be on different computers connected over a network. While RMI works for applets too, we consider two Java applications, rather than applets, running in the two JVMs. One of the Java applications can be thought of as the client and one as the server. Neither Java application is a standalone entity. It is the sum of the client and server applications together that offers a useful whole program to the user.

18.3.1 Remote objects and remote exceptions

An RMI server application typically creates several server-side objects, makes them remotely accessible (i.e. accessible by other JVMs), and then waits for external clients to invoke methods on them. These remotely accessible server-side objects are referred to as *remote* objects. Server-side objects may also include methods that are not made remotely accessible. Such methods are available only locally within the server application. We explain below the simple steps necessary to create remote objects.

Ignoring the degenerate case of running both JVMs on the same host, a network is assumed to be present between the client and server. Because the possible failure modes increase when additional machines and networks are in the mix, RMI client applications must be prepared to deal with additional exceptions. These extra exceptions are grouped under the `java.rmi.RemoteException` class.

One of the design goals for RMI was that it should fit naturally into the Java programming language. As such, the semantics for making a remote call are nearly the same as for making a local method call. The only real differences are the steps needed to obtain the remote reference to begin with and dealing with the additional exceptions that can occur for remote method calls. An RMI client application first obtains references (we explain just how below) to the remote objects and then invokes remote methods. The RMI subsystem handles all the communications between client and server in a way that is nearly transparent to the developer.

Much more detail about the design and operation of RMI can be found in the online API documentation. What we describe here is the basic use of RMI needed to implement the distributed client/server design of Chapters 16 and 17. We also demonstrate the downloadable bytecodes feature.

18.3.2   Stubs and skeletons

As already mentioned, calling objects cannot access the program counter in the remote JVM. So how does RMI work its magic to present the illusion that a remote method call is just like a local call? The answer is through the use of a tried and true mechanism (from RPC technology) known as stubs and skeletons. When a client makes a remote call, it is actually making a local call to a *stub*. The stub then handles the communication to the remote object. Part of this communication is to package up and send the method parameters to the remote side, a process called *marshalling*. On the server side there is a *skeleton* that reads the incoming parameters (*unmarshals* them) and then dispatches the incoming call to the actual server-side implementation of the remote object. For the return, the reverse happens – the skeleton marshals the results and sends them to the stub, which unmarshals the return value or exception and returns the value to the caller. Fortunately, we don't have to write the stub and skeleton. Java provides a tool (the `rmic` compiler) to create them from the remote object class file(s).

Without going into detail that, while interesting, is not important to the casual RMI programmer, Figure 18.1 illustrates how the various players interact in a remote method call. The Java 2 SDK introduced a new stub protocol that eliminated the need for skeletons on the server side as long as both client and server use Java 2, which is now quite likely, at least for applications instead of applets. Instead, generic code is used to carry out the duties previously performed by skeletons in JDK1.1. Nevertheless, it is still useful to think of stubs and skeletons as the components handling the RMI communications between the calling and called objects.

18.3.3   Creating remote objects

We mentioned above that an RMI server must make its objects that have remotely-callable methods into *remote* objects. Just how does an object become a remote

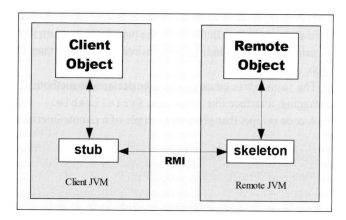

Figure 18.1 This diagram shows the stub/skeleton mechanism that RMI uses to allow methods in the client and remote objects on different platforms to invoke methods in the other object across the network.

object? The simple requirements are summarized here:

- The remote class must be split into both an interface and an implementation class.
- Each remotely callable method must be declared in the interface.
- The interface must extend `java.rmi.remote`.
- The declaration of each remotely callable method in the interface must include `java.rmi.RemoteException` in the `throws` clause.
- If another remote object is included in the parameter list or the return value, it must be declared using the corresponding remote interface rather than the implementation class.
- The implementation class must extend `java.rmi.server.RemoteObject` or, normally, the subclass `java.rmi.server.UnicastRemoteObject`.

That list may sound daunting, but using it is really quite simple. Each of these requirements is explained in more detail in the following sections.

18.3.3.1 The remote interface

Any class that wishes to be remotely accessible must first be split into a Java interface that declares the remotely callable methods and an implementation class that provides the implementation of those methods. Said another way, each remotely callable method must be defined in an interface that the remote class implements.

The interface declares the methods that are to be remotely accessible and is referred to as a *remote interface*. To be a remote interface, the interface class must extend, either directly or indirectly, the `java.rmi.Remote` interface. Also, each method declared in the remote interface must include the `java.rmi.RemoteException` (or one of its superclasses) in its `throws` clause.

All method parameters and return values that are either primitives or regular objects – either standard Java language objects or custom objects the developer defines – are treated just as in a normal non-remote method. However, there is

sequence to run an RMI application is usually:

1. Start the registry, specifying the desired port number.
2. Start the server, which registers itself in the registry.
3. Run the client, which finds the server by querying the registry.

Clients query the registry using a known sever name to obtain a reference to the desired remote object. Of course, the remote objects must first be entered into the registry, using that same known server name, before they can be found. This *binding* into the registry is typically handled by the server code itself at server startup time, and the server is expected to be running essentially all the time, waiting on client connections. Alternatively, RMI supports an activation model in which servers can be activated on demand instead of running all the time. We describe only the standard persistent binding rather than activation. For more information on RMI activation, consult the online Java documentation.

One way to handle binding is to add a `main()` method to the remote method implementation class shown above. This `main()` would instantiate an `RMIExampleImpl` object and then bind the object reference into the registry. We use a slightly more general method and create another class altogether to do the instantiation and binding. In this way, the implementation class remains separate from the class that handles server instantiation and binding. We call this special class `RMIExampleServer`. Its sole job is to create an instance of the `RMIExampleImpl` implementation class and bind it into the RMI registry under a known name so that the remote methods are accessible to clients. All the actual remote methods reside in `RMIExampleImpl` while the action of starting up the server occurs in the `main()` method of `RMIExampleServer`.

Assuming the registry is already running, the registry is accessed programmatically through the `java.rmi.Naming` class, which has methods to bind, unbind, rebind, list, and look up names. The `bind()` and `rebind()` methods are similar except that `bind()` requires that no previous binding exist with the specified name – it throws an `AlreadyBoundException` if the name is already in use. The method `rebind()`, because it permits, but does not require, a preexisting name, is most commonly used – it replaces the object reference previously bound to a name with the new object reference, if necessary. Both `bind()` and `rebind()` take two parameters – the name under which to bind the object and a reference to the `java.rmi.Remote` object to be bound. An example of the use of `rebind()` is shown below:

```
import java.rmi.*;


public class RMIExampleServer
{
  public static void main (String[] args) {
    try {
```

```
      // Instantiate the remote object implementation class.
      RMIExampleImpl impl = new RMIExampleImpl ();
      // Define the "well-known" name to use in the
      // registry.
      String server_name =
        "//localhost:3001/ rmi-example-server";
      // Bind the implementation object under that name.
      Naming.rebind (server_name, impl);
    } // try
    catch (Exception e) {System.err.println (e);}
  } // main
} // class RMIExampleServer
```

Note that we first instantiate and obtain a reference to the `RMIExampleImpl`
object that provides the remote methods. Then we bind it into the registry under
the name "`//localhost:3001/rmi-example-server`". The format of
the name is a URL formatted string with the URL protocol omitted but with the
two forward slashes (`//`) retained. (As discussed in Chapter 13, the URL format
is an industry standard and always uses forward slashes regardless of what the
underlying operating system may use as a file separator.) We use `localhost`
as the host since we bind into the RMI registry running on the same host as
`RMIExampleServer`. In fact, for security reasons, an application can bind or
unbind only to a registry running on the same host. This requirement prevents a
malicious program from removing or overwriting any of the entries in a server's
remote registry. A lookup, however, can be done from any host. The hostname is
followed by an optional colon and port number. If unspecified, the default port
number is the RMI standard port, 1099. The host and optional port are followed
by a slash and a completely arbitrary name for the server. The only requirement
is that the client must know and use the same name. There are two exceptions
that can occur – `java.net.MalformedURLException` if the name is in an
illegal format and `RemoteException` if the registry cannot be contacted for
some reason. For simplicity, we just catch the `Exception` superclass in the code
snippet above.

   To start the server running, we simply execute `RMIExampleServer`. There
are Java security issues that we've glossed over so far that must be addressed
before we have an example that actually works. We discuss those issues after
discussing the client code.

18.4 The RMI client

The client's job is pretty simple. It must first look up the desired server name
in the registry and then make method calls to the remote methods. As explained
above, the only difference in a local call and a remote call on the client side is

the requirement of placing the remote calls within a `try/catch` block. The registry lookup is handled by the following code snippet:

```
// Get a reference to the RMIExampleServer.
RMIExampleInterface server = null;
try {
   // Lookup the server.
   String server_name = "//localhost:3001/rmi-example-server";
   server = (RMIExampleInterface) Naming.lookup (server_name);
} // try
catch (NotBoundException nbe) {. . .}
catch (MalformedURLException mue) {. . .}
catch (RemoteException re) {. . .}
```

Notice that we have used the same name in the `lookup()` call that the server was bound under. This example is set up to run both client and server on the same machine, so `localhost` works for both. If the server were running on a remote machine, then that machine would be specified instead of `localhost`. Of course the proper port number must be specified as well, and the final part of the name must match the original "`rmi-example-server`" name used when the server was bound into the registry. `Naming.lookup()` must be enclosed in a `try/catch` block as well. There are several exceptions that could occur, including `NotBoundException`, which is thrown if the requested server name is not found in the RMI registry running on the indicated host and port number. Other possible exceptions are `MalformedURLException` and the usual `RemoteException`. If the lookup fails, then `null` is returned, so one should always check for `null` before continuing.

`Naming.lookup()` returns a `Remote` interface, which must be cast into the `RMIExampleInterface` that we really want. Notice that it is the *interface*, not the remote `RMIExampleImpl` implementation class that we need. Since every remote method implemented in the implementation class is declared in the interface, we can access all those remote methods through the interface in the normal Java fashion. This interface object reference actually refers to the client-side stub, and the RMI system handles all the actual communication from client, through stub and skeleton, to the remote implementation class. This object reference can be, and often is, thought of as a reference to the remote server implementation. In fact, we refer to it as "server" in the example above.

If a non-null reference to the desired server is received, then we can proceed to making a remote method call as follows:

```
try {
   server.method1 ("hello from client");
```

```
} // try
catch (RemoteException re) {
   System.err.println (re);
}
```

The only exception to deal with here is the general RMI `RemoteException`. If the remote `method1()` had declared any custom exceptions that it might throw, then those would be caught here too.

Obviously the remote `add()` method would be called similarly. And the server-side `doSomethingLocal()` method cannot be called at all from the client, since that method is not a remote method.

18.5 RMI security issues

Since the advent of Java 2, the security issues surrounding RMI have become more stringent. Some of these issues were present in JDK 1.1 and before, but we assume a Java 2 platform in the following discussion. The issues to be dealt with are the need for a *security manager*, the specification of the *codebase* where downloadable bytecodes may be found, and the *policy file* that defines permissions granted to the client and server applications.

18.5.1 The security manager

On both the client and server sides, a security manager typically must be running (see Chapter 14 for a discussion of security managers). Normally, the `java.rmi.RMISecurityManager` is used, though you are free to use one of your own if special requirements must be met. A security manager is required in order to guarantee that the classes that get loaded do not perform operations that they should not be allowed to perform. If no security manager is specified, then Java will not permit any class loading, by either RMI clients or servers, aside from what can be found in the local `CLASSPATH`. For this simple example, the server finds all of its classes in its own `CLASSPATH` so a security manager is not strictly required. However, more complicated servers might need to receive a remote object from the client as a method parameter, possibly requiring the transfer of bytecodes from the client to server, and thus involve the security manager. It is safest to always install a security manager with the following code:

```
if (System.getSecurityManager() == null) {
   System.setSecurityManager (new RMISecurityManager ());
}
```

The same code snippet should be used on the client side as well, since it is normal for clients to download remote bytecodes for the stub objects if nothing else. The one exception is when the RMI client is an applet, in which case the web browser controlling the applet will have already installed a security manager. Therefore it's a good idea to include the if-test shown above to check for the presence of an existing security manager before installing another one.

18.5.2   The codebase

A concept related to the security manager is the *codebase*. The `java.rmi.server.codebase` property is specified when the server starts up and identifies where downloadable bytecodes may be found. The RMI codebase is closely related to an applet codebase, and both are similar to a `CLASSPATH`. The codebase specifies a source from which a remote JVM can load needed classes that cannot be found in the local `CLASSPATH`. When a local JVM is running and loading classes from a local disk-based source, it searches for classes within the list of locations specified in the `CLASSPATH`. When a remote client's JVM needs to obtain RMI stub classes, it looks first in the local `CLASSPATH` and then in the list of locations specified by the codebase.

Note that the codebase is used by the client, but it is specified on the server. When the RMI server registers itself with the registry, the codebase is "remembered" by the registry. In fact, the registry itself uses the server's codebase to find the remote object's stub class during registration. (For this reason, the registry should not be started with a `CLASSPATH` that includes the remote object's stubs.) If the codebase is not properly specified – i.e. if the registry cannot find the needed stub classes in the location specified by the codebase – then an exception is thrown when the server attempts to `bind()` or `rebind()` itself into the registry. The exception is a `RemoteException` nesting an `UnmarshalException` nesting a `ClassNotFoundException`. This chain of exceptions is a common error and is almost always the result of a bad codebase value.

The codebase is a URL or a space-delimited list of URL locations where the needed stub classes, as well as any other classes needed by the stubs, can be found. These URLs must be absolute paths, not relative paths. When running both client and server on the same machine, they may be `"file:"` URLs, though it is more common to use `"http:"` URLs. The URL may point to a directory location or a JAR file. If a directory, it is important that the URL end in a trailing `"/"` character. If the downloadable bytecodes are supplied by an `"http:"` URL, then there must be an http server running on the specified host that can serve those bytecodes.

The codebase property can be set using the `-D` syntax on the command line when the server is started. An example is

```
-Djava.rmi.server.codebase=file:///rmiservers/example1/classes/
```

Since it is a "file:" URL, this codebase is suitable for when the client will be running on the same machine as the server. The URL specifies an absolute path to a directory named `/rmiservers/example1/classes`. The needed class files must appear below the `classes` subdirectory, organized in package-named sub-directories. This example assumes a Unix-like directory system. On a Windows system, an analogous codebase specification would be

```
-Djava.rmi.server.codebase=file:///c/rmi-servers/example1/classes/
```

where the absolute path starts at the C: drive but we continue to use forward slash characters as the separators.

If the needed class files are in a directory served by an HTTP server, then the codebase might look like

```
-Djava.rmi.server.codebase=http://myserver/rmi/example1/classes/
```

If the downloadable classes are in a JAR file, then the specification might be

```
-Djava.rmi.server.codebase=http://myserver/rmi/example1.jar
```

If the needed classes are split across two jar files then the following might be used:

```
-Djava.rmi.server.codebase=
"http://myserver/rmi/example1.jar http://myserver/rmi/more.jar"
```

where we have split the single line into two lines. In practice, the entire quoted and space-delimited string should be all on one line.

18.5.3 The policy file

Even with the codebase properly specified and a security manager in place, both client and server must still navigate through the Java permission system to run correctly. As discussed in Chapter 14, once a security manager is used in Java 2 the policy file is consulted each time certain potentially sensitive operations are performed. The default policy file is quite restrictive, permitting little more than the minimum permissions needed to run the JVM and load classes from the local `CLASSPATH`. Therefore, it is vital that a custom policy file be specified for both client and server. This custom policy file is specified using the `java.security.policy` property.

For the simple example used here the server permission needed is `java.net.SocketPermission`, which controls access to network sockets. The permission needed is a specification of the host to be used and a set of "actions" that identify ways to access that host. For this example, as long as we're running the client and server on the same machine, the server needs network access to localhost, and the actions are *accept, connect*, and *resolve*. The need for this permission may be determined by examining the source code along with a knowledge of which Java API methods require various permissions. Unfortunately, that body of knowledge is difficult to learn and remember. In addition, often the method that requires some permission may not be called directly by the source code you write, but indirectly by some method that your code calls. In practice, the best way to determine the needed permissions is often to run the code without any policy file and see what runtime exceptions are thrown. Then adjust the policy file to add the indicated permissions and run again.

To get started and to be sure that all other requirements are met and that the code is working nominally, one could use a wide open policy – i.e. grant all permissions to everything. The policy file to grant all permissions reads

```
grant {
   permission java.security.AllPermission;
};
```

Obviously, using such a policy file in a production environment is, well, bad policy. But using this policy file during testing is a good way to get both client and server running quickly. The above lines can be created with a text editor and saved in a file named, say, `grantall.policy`. Then, add the following to the launch line when launching both the server and the client:

```
-Djava.security.policy=grantall.policy
```

As discussed in Chapter 14, it is probably easiest to use the SDK `policytool` to create and edit policy files. That tool has a graphical user interface with intelligent pull-down menus that restrict the choices to ones that make sense for the task at hand, thereby reducing errors by minimizing the need to fully understand the policy file syntax. Still, policy files are just text files, and this makes it easy to display policy file snippets here.

Once we are convinced that everything else is working correctly, this permissive policy file should be replaced with a more restrictive one. For the server, we create a file called `local-server.policy` containing these lines:

```
grant {
   permission java.net.SocketPermission "localhost",
   "accept,connect,resolve";
};
```

Later we create another policy file for use with a remote server.

For the client, when running on the same host as the server, the permission required is also a network socket permission, but only for the *connect* and *resolve* actions. The client also needs a `java.io.FilePermission` in order to read the location specified by the server's "`file:`" codebase.

To summarize, when running the client and server on the same machine, we arrange for the client to use a local CLASSPATH that does not include the remote stub classes. That means the client must look in the server's codebase for the downloadable class files. That codebase specification is received from the RMI registry. Since we're using the same machine, that codebase will be a "file:" URL. The client must read the remote stub classes from that URL, and a suitable permission to read that location must be granted in the client's policy file. Therefore, the client's policy file must read as follows:

```
grant {
    permission java.net.SocketPermission "localhost", "connect,
    resolve";
    permission java.io.FilePermission
        "c:\\javatech\\rmi18\\build\\classes\\-", "read";
}
```

The syntax of the read permission shown is for a Windows machine. The path specified must be an absolute path and must correspond to the absolute path specified in the "file:" URL given as the server's codebase. The use of the double "\\" characters is required on Windows. For an RMI client running on a Unix box, the file might read as follows

```
grant {
    permission java.net.SocketPermission "localhost", "connect,
    resolve";
    permission java.io.FilePermission
        "/javatech/rmi18/build/classes/-", "read";
}
```

In both cases, notice the use of the "-" at the end of the directory specification which indicates the named directory and all directories below it. For more details of the policy file syntax, see the Java SDK documentation.

An alternative to using the file permission when client and server are on the same machine is to arrange the client's CLASSPATH to point to where all the class files, including the remote stubs, reside. Since the default policy file always permits reading the CLASSPATH, no special file permission is needed. However, doing so would not test or demonstrate the downloadable bytecodes feature of RMI. Therefore, we use a client-side JAR file that includes only the minimum classes needed to start the client, specifically avoiding the remote stub class files when we create the JAR file. Then we set the server's codebase to include the location of the remote stubs and use the file permission shown above.

18.6   Finally, a working example

We finally have all the pieces in place to run a real, but simple, example RMI client/server application with the restriction that both client and server must be on the same machine. We make minor modifications in the next section to permit the use of a remote machine. For completeness, the entire code is shown here, including full package statements and exception handling which were omitted for brevity in the code snippets shown above. The complete source code is also available on the Web Course site with slightly different formatting than shown here. After listing the code, we also show the step-by-step instructions needed to run this example.

We use a root package name of `javatech.rmi18` for this example. The remote interface file is in package `javatech.rmi18.server`, in a subdirectory named `javatech/rmi18/server`. (We use the Unix standard forward slash as a directory separator. Windows users need to mentally replace the "`/`" with a "`\`".) To recap the role of this interface, it declares each method that is to be remotely callable. It must extend the `java.rmi.Remote` interface, and each remote method must be declared to throw `java.rmi.RemoteException`.

```java
// RMIExampleInterface.java
package javatech.rmi18.server;

import java.rmi.Remote;
import java.rmi.RemoteException;

/** The remote interface for use with the RMI Example
  * in Ch. 18. **/
public interface RMIExampleInterface extends Remote
{
   public void method1 (String s) throws RemoteException;
   public int add (int a, int b) throws RemoteException;
} // interface RMIExampleInterface
```

Next comes the class that implements the remote methods declared by the remote interface. This class obviously must contain an "`implements RMIExampleInterface`" clause. It must also extend `UnicastRemoteObject` and must provide an implementation of each of the remote methods declared in the interface. It may also include other methods, such as the `doSomethingLocal()` method.

We find it convenient to keep our remote interfaces separate from the implementation classes, so we put the implementation classes into an `impl` subdirectory below the directory containing the interface itself. Therefore, the package for the implementation classes is named `javatech.rmi18.server.impl`. Another choice that keeps the interfaces and implementations separate is to put the interfaces into something like a `javatech.rmi18.server.interfaces` package. Alternatively, it is perfectly legal for the interfaces and implementations to be kept together in the same package called, say, `javatech.rmi18.server`.

The choice is rather arbitrary and personal. If a different scheme than that shown here is desired, just be sure to modify the package statements, subdirectories, and import statements accordingly.

```java
// RMIExampleImpl.java
package javatech.rmi18.server.impl;

import java.rmi.Remote;
import java.rmi.RemoteException;
import java.rmi.server.UnicastRemoteObject;
import javatech.rmi18.server.*;
   // location of the interface classes


/** The implementation class that implements the remote
  * methods declared in the <tt>RMIExampleInterface</tt>
  * interface. **/
public class RMIExampleImpl
   extends UnicastRemoteObject
   implements RMIExampleInterface
{
   // The default constructor will not work since
   // UnicastRemoteObject, which we extend,
   // throws RemoteException, which the default
   // constructor provided by the compiler does not.
   // Therefore, we must implement a default constructor
   // that throws RemoteException even if the
   // constructor does nothing at all. It will, of course,
   // automatically call its superclass constructor.

   /** Constructor.Must throw <tt>RemoteException</tt>. **/
   public RMIExampleImpl () throws RemoteException {}

   /** Echoes the input string provided by the client. **/
   public void method1 (String s) {
     System.out.println ("RMIExampleImpl.method1: " + s);
   } // method1

   /** Adds the input parameters and returns the sum. **/
   public int add (int a, int b) {
     System.out.println (
       "RMIExampleImpl.add: computing sum of" +
       a + " and " + b);
     return a + b;
   } // add
```

```
/** Constructor **/
public RMIExampleClient () {
  // Create and install a security manager.
  if (System.getSecurityManager() == null) {
    System.setSecurityManager(new RMISecurityManager());
  }
  // Get a reference to the remote server named
  // rmi-example-server.
  RMIExampleInterface rmi_example_server = null;
  try {
    // Find the server.
    String server_name = "//localhost:3001/rmi-example-server";
    System.err.println (
       "RMIExampleClient: looking up server " + server_name
    );
    rmi_example_server =
       (RMIExampleInterface) Naming.lookup (server_name);
    System.err.println ("RMIExampleClient: found it!");
  } // try
  catch (MalformedURLException mue) {
    System.err.println (mue);
  }
  catch (NotBoundException nbe) {
    System.err.println (nbe);
  }
  catch (RemoteException re) {
    System.err.println (re);
  }
  if (rmi_example_server == null) {
    System.err.println ("\nEXITING BECAUSE OF FAILURE");
    System.exit (1);
  }

  // Test method1.
  System.err.println (
    "Calling remote method1 which should echo the string "
    + "'hello from client'"
  );
  try {
    rmi_example_server.method1 ("hello from client");
  }
  catch (RemoteException re) {
    System.err.println (re);
  }

  // Test the add method.
```

```
      try {
         int sum = rmi_example_server.add (18, 34);
         System.err.println ("According to the remote add(), the "
                             + "sum of 18 and 34 is " + sum);
      }
      catch (RemoteException re) {
         System.err.println (re);
      }
   } // ctor

   public static void main (String[] args) {
      new RMIExampleClient ();
   } // main
} // class RMIExampleClient
```

That finishes all the code. The steps to compile and run are summarized as follows:

1. Compile everything.
2. Run the `rmic` compiler.
3. Package the client code into a client JAR file.
4. Start the RMI registry.
5. Start the server.
6. Run the client.

We detail each of the steps below.

18.6.1 Compile everything

For all the client/server examples in this book, we find it convenient to use the Ant build tool, though describing Ant is not within the scope of this book (see [1]). The Web Course contains Ant build files as well as Windows bat scripts for building each example. For simplicity we show only the Windows bat scripts here. Unix shell scripts are very similar.

Ant strongly encourages a directory structure in which the compiled class files are kept separate from the source files. An advantage of this arrangement is that directory listings in the source directories are not cluttered up with a lot of compiled `.class` files. We use that directory arrangement here, even though we demonstrate building and running the examples with Windows bat scripts only. Accordingly, all our source files and directories appear below a directory named `src`. The name is completely arbitrary but is a common one for Ant users. Therefore, at some top level directory containing everything pertinent to this example, there is a `src` subdirectory. Below `src` appears `javatech/rmi18/server`, for the interface file, `javatech/rmi18/server/impl`, for the implementation files, and `javatech/rmi18/client`, for the client. All the class files are compiled into a separate subdirectory tree. In this way, making a clean rebuild of everything is easy since all the compilation results can be discarded simply by deleting the entire compiled class file tree. It is common for Ant users to use a `build` directory that is parallel to the `src` directory for the purpose of containing all such generated results. Sometimes things other than just class files need to go into `build`, so class files are generally sent to a `classes` subdirectory below `build`. Of course, to maintain Java's required package-named subdirectory arrangement, the entire `javatech/rmi18/server`, `javatech/rmi18/server/impl`, and `javatech/rmi18/client` directory structure is replicated below `build/classes`.

To summarize, the entire directory structure appears as follows, assuming we begin at a directory named `javatech-18`. A trailing "`/`" indicates a directory, and file names are italicized.

```
javatech-18/
   (build and run scripts, policy files, etc.)
   src/
      javatech/
         rmi18/
            client/
               RMIExampleClient.java
            server/
               RMIExampleInterface.java
               impl/
                  RMIExampleImpl.java
                  RMIExampleServer.java
   build/
      classes/
         javatech/
            rmi18/
               client/
               server/
                  impl/
```

For simplicity we do not show the various build and run scripts, policy files, etc. in the diagram above. All those files appear at the Web Course. Initially, the build tree will be empty. In fact, in won't even exist at all. It is created when running Ant or when running the build scripts, and compiled `.class` files will appear in the appropriate places below the `build/classes` tree.

The first task is to compile all the Java source files. In order to use the directory structure above, we need to direct the compiler output to a different directory than the source code directory. The -d switch on the `javac` command line does just that. The compiler also needs to know where to look for already compiled class files that might be needed as new files are compiled. The familiar -classpath switch handles that and should point to the same directory as the -d output directory. First, we create the `build` and `build/classes` directories. For a Windows bat script, the `mkdir` command must use the standard Windows "\" character as a directory separator. Elsewhere we use the conventional "/" between directory names since the `javac` tool knows how to interpret it.

```
rem Create the build and class directories
mkdir build
mkdir build\classes
```

Next we set up some environment variables to make the compilation lines a little shorter:

```
rem setup some environment variables to use as abbreviations
set classdir=build/classes
set cp=%classdir%
set ifcdir=src/javatech/rmi18/server
set impldir=src/javatech/rmi18/server/impl
set clientdir=src/javatech/rmi18/client
```

And finally we compile all the sources:

```
rem Compile everything
javac -classpath% cp% -d %classdir% %ifcdir%/*.java
javac -classpath% cp% -d %classdir% %impldir%/*.java
javac -classpath% cp% -d %classdir% %clientdir%/*.java
```

18.6.2   Run the `rmic` compiler

The next step is to generate the stubs and skeletons using the `rmic` compiler. It operates on the remote implementation class file, not the source file, so the `javac` compilation must be completed first. The input to `rmic` must be the fully qualified implementation class – i.e. with the complete package name prefix, `javatech.rmi18.server.impl.RMIExample`. Of course `rmic` needs to know the local CLASSPATH so it will know where to begin looking for the named class to compile. We also want to direct the `rmic` output to the same `build/classes` tree used for other class files. Just like `javac`, `rmic` uses the -d switch to indicate the output directory. Continuing with the Windows bat script begun above, we run `rmic` as follows (line split to fit page):

```
rmic -classpath %cp% -d %classdir%
   javatech.rmi18.server.impl.RMIExampleImpl
```

18.6.3   Create a client JAR file

When we run the server, we'll set the local CLASSPATH to point to the build/classes directory tree. We could run the client pointing to the same CLASSPATH, but, as described earlier, doing so would not demonstrate the ability of RMI to download bytecodes from a codebase. Instead, we create a JAR file that contains only the pieces needed for the client to start up. It intentionally does not include the stubs just generated by rmic. The only pieces absolutely required for the client are the compiled RMIExampleClient.class file and the compiled interface file. The JAR file must be created carefully so that it contains the directory structure matching the package structure of the compiled class files. Therefore the JAR must begin at the build/classes directory level. One way to do that is to use the -C switch to the jar command. It is often easier to just cd to the required root directory, run the jar command, and then cd back to the starting location, the approach used here:

```
cd build\classes
jar cf client.jar
        javatech/rmi18/server/RMIExampleInterface.class
        javatech/rmi18/client/RMIExampleClient.class
move client.jar ..\..
cd ..\..
```

(The list of files to include in the JAR file is too long to fit on one line. On Unix and Linux platforms, a continuation character ("\") can be used to nicely format the script file, but on Windows both files must appear in one long line. We've split it here only for appearance reasons.)

Note that we've also moved the created client.jar back up two directories, back to the javatech-18 root directory for this example.

18.6.4   Start the RMI registry

The next step is to start the RMI registry. When doing so, it is important that the registry be started without a CLASSPATH that includes the stub files. In general, no CLASSPATH at all should be used. In all the examples shown so far, we've assumed that the command shell window does not have a CLASSPATH environment variable set. We explicitly set the CLASSPATH using the -classpath command line switch whenever necessary. The reason that no CLASSPATH should be set is that the registry is a Java application, and like any Java application, the JVM looks first in its local CLASSPATH for the class files it needs. The files the registry needs are the stub classes. If the registry finds the stub classes in its CLASSPATH, then it won't look in the java.rmi.server.codebase. When a client asks the registry for the codebase, there won't be one, and the client will fail with a nested ClassNotFoundException. Therefore, we intentionally leave off any `-classpath` specification when starting the registry. The RMI registry is started, in Windows, with the `rmiregistry` command as follows:

```
start rmiregistry 3001
```

For Unix, we would use the command

```
rmiregistry 3001 &
```

In both cases, we've specified the port number 3001. If left off, the default RMI port of 1099 is used. Since we've hard coded port 3001 in our source code, we need to be sure to use port 3001 for the RMI registry. If there is already a process using port 3001 on your machine, then you must choose a different unused port number.

For some applications, it is possible and convenient to start the registry from within the server program itself. Doing so saves the separate step of starting the registry and keeps the registry port number and the server that uses the registry at just one place in the code. We learn in Chapter 20 how to use that technique.

18.6.5 Start the server

We are finally ready to run the server. To do so, there are four things that must be specified on the Java command line – the `CLASSPATH`, the code-base, the policy file, and the name of the class to run. The `CLASSPATH` must point to the `build/classes` directory. Alternatively, we could have created a `server.jar` file containing all the classes in `build/classes`. The codebase and policy files are specified as system properties using the `-D` syntax shown earlier, and the class to run is the fully qualified `RMIExampleServer` class. Therefore the command line to start the server is

```
java -classpath build/classes
  -Djava.rmi.server.codebase=file:///c:/javatech-18/
   build/classes/
  -Djava.security.policy=local-server.policy
  javatech.rmi18.server.impl.RMIExampleServer
```

Again, we've broken the command into multiple lines for appearance reasons. On a Windows machine, this entire command must be entered on a single line. Unix and Linux shells can use the line-continuation character. The policy file specified, `local-server.policy` was shown earlier but is repeated here for completeness:

```
grant {
  permission java.net.SocketPermission "localhost","accept,
  connect, resolve";
};
```

The source code listed earlier and appearing on the Web Course includes some diagnostic output written to `System.err`, so one should see the following appearing in the console window in which the command is issued:

```
Constructing an RMIExampleImpl
(re)binding it
RMIExampleServer ready and waiting on clients
```

18.6.6   Run the client

The registry and server are now both running, so we can finally test the client application. The command line for the client needs the following three items – the `CLASSPATH`, the policy file, and the client class name. As explained earlier, the `client.policy` file is

```
grant {
  permission java.net.SocketPermission "localhost", "connect,
  resolve";
  permission java.io.FilePermission
    "c:\\javatech-18\\build\\classes\\-", "read";
}
```

Another console window is required in order to launch the client since the previous console window is occupied running the server. The command line to launch the client is

```
java -classpath client.jar
  -Djava.security.policy=client.policy
  javatech.rmi18.client.RMIExampleClient
```

The expected output in the client console window is

```
RMIExampleClient: looking up server //localhost:3001/rmi-
  example-server
RMIExampleClient: found it!
Calling remote method1 which should echo the string 'hello
from client'
According to the remote add(), the sum of 18 and 34 is 52
```

Meanwhile, the following two lines should be added to the output in the server console window:

```
RMIExampleImpl.method1: hello from client
RMIExampleImpl.add: computing sum of 18 and 34
```

18.7 How to run on two machines

Everything so far in this chapter has used `localhost` as the host name. Doing so has been a convenience to avoid specific host names. Additionally, the reader can download the code from the Web Course and run it on any machine without any manual editing required. To run the client and server as RMI was intended – on two different hosts – requires only a few changes which we detail now.

Let's suppose that we have a host named `myserver.somewhere.com` on which to run the server, and host `client.somewhere.com` for the client. The RMI registry must be running on the server machine because applications are permitted to bind only to a registry running on the same host as the application. Since the server is running on `myserver`, the client must look up the server on that machine instead of `localhost`. Therefore the string passed to `Naming.lookup()` must refer to `myserver.somewhere.com` rather than `localhost`. Of course, the server application must be sure to bind using that address as well – i.e. the server application should bind to its actual machine name instead of `localhost`. The use of `localhost` works fine when the client is on the same machine, but for a remote client, the correct server hostname is required.

The only source code change needed on the server side is to replace the binding to `localhost` in `RMIExampleServer.java` with a binding to `myserver.somewhere.com`. The rebind line must become

```
Naming.rebind (
    "//myserver.somewhere.com:3001/rmi-example-server", impl);
```

where we still assume that port 3001 is free. Similarly, the lookup line in `RMIExampleClient.java` must be changed to

```
rmi_example_server = (RMIExampleInterface) Naming.lookup (
    "//myserver.somewhere.com:3001/rmi-example-server")
```

The server's policy file must permit outgoing network connections from the server itself as well as incoming network connections from the anticipated client machine. Thus the `server.policy` file must be modified to read

```
grant {
  permission java.net.SocketPermission
    "myserver.somewhere.com", "accept, connect, resolve";
  permission java.net.SocketPermission
    "client.somewhere.com", "accept, connect, resolve";
};
```

If being so explicit in one's policy file seems too restrictive, there are ways to write more general policies. Wild card usage is permitted in the host name, for example. And a policy file can include a codebase to which it applies. Using a codebase in the policy file means that only codes from that codebase will be granted the permissions listed in the policy file. In that way, a rather permissive policy file still will not open up your machine to attacks from unknown codebases.

Almost everything is ready now, except for the codebase and a way for the client to download the needed stub classes from the server. Typically, the download is provided with a web server using an "`http:`" URL as the codebase. Setting up a web server such as Apache is not particularly difficult, but is beyond the scope of this book [see 2]. Instead, we package all the needed class files, including the stubs this time, into a `clientall.jar` file that must be copied to the client machine. That way, the client will find all its needed classes in its local `CLASSPATH` and will not need to consult the codebase at all. The server still needs to provide a valid "`file:`" URL codebase so the RMI registry can find the stubs it needs.

To summarize the steps required for running on two machines:

1. Make the source code changes above to specify the actual server hostname instead of `localhost`.
2. Modify the `server.policy` file to permit connections by the server host and the client host.
3. Build a JAR file containing the classes needed by the client, including stubs
4. Copy that JAR file to the client machine.
5. Start `rmiregistry` on the server machine using the port number hard coded into the client and server source codes.
6. Start the server application.
7. Run the client application.

Note that for this JAR file, one can explicitly list the `.class` files needed by the client. For convenience, one can also make a JAR file containing everything in the `build/classes` directory, including server-side classes at the cost of having to transfer a slightly larger JAR file than actually needed on the client machine.

The Web Course contains an `rmi18-2` directory that has these changes as well as the needed build and run scripts. The reader will, of course, have to change `myserver.somehwere.com` and `client.somewhere.com` to the proper host names.

18.8 Conclusion

This completes our simple RMI example. It may not have sounded so simple, since the description was rather long and detailed. As one creates more realistic RMI applications, one will realize that this truly was a simple example. However, other applications are just extensions of the skills learned here. The only critical aspect of RMI not demonstrated here is the use of remote objects as method call parameters and the associated transfer of client-side bytecodes to the server. Doing so is a simple extension of what was learned. There are also examples to be found in the online Java documentation. Before developing a complete example that demonstrates the client/server simulation architecture described in Chapters

16 and 17, we take a brief detour to discuss CORBA as an alternative to RMI. Future discussions of RMI do not go into such laborious detail as shown here, since the basic skills needed to create and compile the sources, create the stubs, start the registry, and run the server and client have already been learned.

18.9 Web Course materials

All of the codes discussed here are available in the Web Course Chapter 18 along with additional examples, resources, and discussion of RMI.

References

[1] Ant build tool, `http://ant.apache.org`.
[2] Apache open source server, `www.apache.org`.

Chapter 19
CORBA

19.1 Introduction

CORBA is an acronym for Common Object Request Broker Architecture, a name that does not really convey to the new user the purpose of the technology. Most people just think of CORBA as the name of an important distributed object technology without really considering what the acronym stands for. For Java developers, RMI is generally the preferred distributed object architecture – especially when it is known that both client and server will be written in Java. However, CORBA has the advantage that it is language independent, meaning that non-Java clients can make CORBA calls to a CORBA server.

CORBA is a standard, so there is an official specification of that standard. The standard is maintained by the Object Management Group, a consortium of over 800 members. See the www.omg.org home page for voluminous information on CORBA and other technologies developed by the OMG. CORBA is also a very broad technology that covers much more than we introduce here. Our point is to demonstrate how CORBA technology can be used to implement straightforward distributed computing solutions analogous to the RMI example developed in Chapter 18.

A typical scientific application, such as the simulation described in Chapters 16 and 17, generally has only a few users. The developer writes both the client and server and, if developing in Java, probably uses RMI. However, if one is designing a public server that will be accessed by multiple clients, including the possibility of non-Java clients written by someone else, then CORBA is a better choice than RMI. CORBA is such an important technology that the Java 2 SDK fully supports CORBA applications. The official name of the CORBA support in the SDK is "Java IDL," but in this book we use the name CORBA instead. Programming in CORBA is similar to RMI in concept, but has some significant differences. Because of the need to be language independent, CORBA is generally somewhat more difficult to use than RMI. And while CORBA is object-based, it generally does not "feel" as object-oriented as RMI.

This chapter develops a simple CORBA client/server application, similar to the simple RMI example developed in Chapter 18. Because many of the basic distributed computing concepts should already be understood from Chapter 18, we do not need to go into as much detail as earlier. For example, CORBA uses the same stubs and skeletons framework as in RMI, so we don't have to rehash that concept. The Java 2 SDK provides tools to automatically generate the CORBA stubs and skeletons, similar to that provided with RMI. CORBA includes its own highly developed language-independent exception mechanism similar to Java exceptions. When using the Java language to implement a CORBA application, the exception mechanism maps very cleanly to Java's exception system. In fact, one will find that almost all CORBA concepts map well to the Java language and much better than other programming languages.

19.2 CORBA IDL

Perhaps the biggest difference between CORBA and RMI is that the CORBA interfaces must be defined in a language independent fashion. This is done using the CORBA Interface Definition Language (IDL) rather than a Java interface file. IDL is a limited C-like language whose only purpose is to define interfaces. There are absolutely no implementation features in the IDL language. The IDL file is processed with an IDL-to-language-of-choice "compiler" for the language chosen to generate the language-specific stubs along with numerous other support files for that language. Then the developer provides the implementation of the server code using those support files. There must be an IDL compiler for whatever programming language is used for the implementations. The Java 2 SDK includes an IDL-to-Java compiler. If one were developing in C++ or Smalltalk, for example, one would need a CORBA environment for that language, including an IDL-to-C++ or IDL-to-Smalltalk compiler. Running such an IDL compiler would produce the required stubs and support files for that language, and, depending on the language, they could look very different from the files created by the IDL-to-Java compiler.

The first step in developing a CORBA application is to define the server interfaces in an IDL file, a task that requires learning the IDL syntax. Learning the syntax is not particularly difficult, as we demonstrate shortly, but doing so does represent an additional step. In fact, Java now provides a way to generate an IDL file directly from a Java interface file, relieving the developer from needing to learn IDL. However, since CORBA cannot support some features that are easily supported in RMI, one must be careful to restrict the Java interface file to only those features supported by CORBA. This subset of RMI is referred to as RMI over IIOP, and is fully documented in the online Java documentation. Further discussion of RMI over IIOP is beyond the scope of this book.

We introduce IDL here with an example that duplicates the functionality of the RMI server developed in Chapter 18, but we do not go into all the features available in IDL. The online Java documentation includes a complete description of IDL for those interested in the details.

CORBA predated the invention of Java, but the Java language structure maps closely to the features provided in CORBA. Java packages, for instance, map closely to IDL *modules*, which, like Java packages, provide a way to avoid namespace collisions. To correspond to the package structure we've already become familiar with in Chapter 18, we create a `javatech` module and nest within it a `cor19` module and a `server` module. Within the `server` module, we define an IDL interface that corresponds to the `RMIExampleInterface` used with RMI. We also included a few other IDL features. The IDL interface is very similar in concept to a Java interface. It declares the remotely callable methods that you wish to provide in the CORBA server, including the method name, parameter types, return types, and any possible exceptions. We emphasize again that the IDL file defines the interfaces only; the implementations of those methods must be provided elsewhere. The IDL file we use for this simple example is shown here. We describe each feature below.

```
// server.idl
module javatech {
  module cor19 {
    const string COR19_CONSTANT = "Chapter 19";
    module server {
      const string SERVER_CONSTANT = "JavaTech Book";

      exception Cor19UserException {
        string message;
      }; // exception Cor19UserException

      struct CustomData {
        float someFloatValue;
        long someIntegerValue;
      }; // struct CustomData

      interface Cor19Example {
        const long COR19_EXAMPLE_CONSTANT = 19;
        attribute long huh;
        readonly attribute float hah;
        void method1 (in string s);
        long add (in long a, in long b);
        void demo (inout CustomData cd) raises
          (Cor19UserException);
      }; // interface Cor19Server
    };
  };
};
```

Notice the use of a semicolon terminating every method definition, similar to a Java interface file, but also after every closing brace, unlike Java. Leaving out the terminating semicolon generates an error from the IDL compiler and is a common mistake. The IDL compiler honors `//` as well as `/* . . . */` comments. The compiler provided with Java 2 SDK even retains javadoc-style comments and forwards them to the generated Java source code where appropriate.

When this IDL file is processed by the IDL-to-Java compiler, several support files are generated in packages named to match the module names. These support files are definitely important, but their contents are rarely of interest. These files simply provide most of the underlying plumbing needed to implement the CORBA programming paradigm. Because the files themselves are normally of little use, and to avoid cluttering up our implementation directories, it is convenient to instruct the IDL compiler to direct the generated files into some directory tree other than our source tree.

Because CORBA is language and platform independent, it must support languages and platforms that are not case sensitive. Therefore, the IDL file is not case-sensitive. One must be careful to avoid names that differ only in case; else name conflicts will occur. For case-sensitive languages such as Java, the IDL compiler is permitted to retain case-sensitive names in the support files that it creates. Then the implementer can retain case-sensitive names for convenience in the implementation files. Even so, the requirement to maintain identifier uniqueness without regard to case must still be met.

In the IDL file above, we created an interface named `Cor19Example`. That interface includes two methods that are similar to the methods defined in the RMI interface from Chapter 18. (In CORBA terminology, the "methods" are actually referred to as "operations," which is a relatively language-neutral term. However, Java programmers will almost always think of them as methods instead of operations.) We also included one new operation, `demo()`, that we describe shortly.

The IDL file also defines a data structure (CORBA keyword `struct`) called `CustomData`. Structures have no direct analog in Java since everything in Java is either a primitive or an object, and all objects inherit from `java.lang.Object` and thus have behavior (methods) associated with them. Structures in CORBA are pure data containers with no behavior. Like structures in other languages, they are used to group together related values. If one is to pass structured data between CORBA clients and servers, those data structures must be defined somewhere, and where they are defined is in the IDL file using the `struct` keyword as shown.

A feature of IDL that may be unfamiliar is that each of the method parameters must be identified as an input only, output only, or input/output parameter. The keyword `in` identifies input-only parameters. They are forced to be read-only parameters in the implementation files. No changes to their values are communicated back to the calling program, no matter what the server-side code does with them. Output only parameters are identified with the keyword `out`, and input/output parameters with the keyword `inout`. Because Java always passes parameters by value, the IDL-to-Java compiler must generate special "Holder" classes to provide support for `out` and `inout` parameters.

For example, the `demo()` operation uses an `inout` parameter of type `CustomData`. The IDL compiler generates a special support class named `CustomDataHolder`. When the implementation class is built, the parameter passed in the `CustomData` position must actually be a `CustomDataHolder` object. Holder classes are created to "hold" an object of the root type. Thus `CustomDataHolder` holds a `CustomData` object. In that way, a Java implementation can set the value of the `CustomData` object within the Holder class for return to the caller. Both `out` and `inout` parameters are handled with Holder classes. In fact, in a Java implementation, there is little difference between `out` and `inout` parameter types. The only difference is that `inout` parameters are expected to contain data upon entry to the method implementation while `out` parameters should not. An attempt to retrieve the inner value of an `out` parameter results in a runtime failure, as does an attempt to pass an `inout` parameter whose inner value has not been set.

Our IDL file also defines an exception called `Cor19UserException`. Exceptions are not "thrown" in IDL terminology, but rather "raised," as shown in the example. However, the `raises` keyword maps into the Java `throws` keyword in the output from the IDL compiler. There are a variety of built-in exception types, but if a custom type is desired, it must be defined in the IDL file.

The `Cor19Example` interface declares that `method1()` returns `void`, while `add()` returns `long`. There is no `int` keyword in IDL, but when IDL types are mapped to Java primitive types, an IDL `long` parameter becomes a Java `int`. One of the first things a novice CORBA programmer must learn is to mentally map everywhere that a Java `int` would be used to a `long` in the IDL file. When the Java implementation is built, the parameters and return values identified as `long` in the IDL file appear as `int` types in the Java implementation.

The parameter for `method1()` is an input (`in`) parameter of type `string`. Note the lowercase "`s`". Lowercase `string` is an IDL type that maps to a Java `String` object. If one accidentally uses uppercase `String` in an IDL file, an error is seen when the file is processed by the IDL compiler. This mistake is common, but the solution is simple – just replace the erroneous `String` with the correct `string`.

It should now be clear that the IDL file shown above defines a CORBA server that provides the same services as the simple RMI example from Chapter 18 plus a few more. Other features of IDL not yet described include the use of two kinds of *attributes* and two kinds of constants (CORBA keyword `const`). Attributes map into instance variables in Java with accessor and mutator (also known as getter and setter) methods (which must be implemented in the implementation class). Attributes with the keyword `readonly` have only an accessor method. Read-only attributes are values that may be changed within the server implementation, and can be read but not set by a client. Regular attributes may be set by a client using the mutator method.

Constants defined within a module but outside an interface in the IDL file map into a public Java interface (with the same name as the constant) in the package corresponding to the module in which the constant is defined. Thus our COR19_CONSTANT, which is in the cor19 module, maps to a Java interface (in a file named COR19_CONSTANT.java) in the javatech.cor19 package, while the SERVER_CONSTANT, being in the server module, maps to a Java interface in the javatech.cor19.server package. These Java interfaces define no methods, just constant values in a field named value. So the constant can be accessed like this:

```
String a_fun_chapter = javatech.cor19.COR19_CONSTANT.value;
```

or, using imports,

```
import javatech.cor19.*;
   . . .
String a_fun_chapter = COR19_CONSTANT.value;
```

Constants defined inside an IDL interface map to public static final fields in the Java interface that corresponds to the IDL interface.

There are more features to IDL that we do not go into here since our aim is only to introduce CORBA rather than give an exhaustive tutorial. Some features that might be of interest include the oneway keyword, which defines a non-blocking method, the typedef keyword, enum and union objects, and the Any type.

19.3 Compiling the IDL file

The IDL-to-Java compiler provided with the Java 2 SDK is known as idlj. There exist CORBA implementations for Java other than the free built-in implementation. Some of these may provide better performance or adherence to newer versions of the CORBA standard than the built-in CORBA support. Each implementation must provide its own IDL compiler, and the name and usage of the compiler is likely to be different from the name and usage of idlj. However, most implementations should produce essentially similar output files, with perhaps slight naming differences. We deal only with the free Java 2 SDK CORBA implementation.

When the above IDL file, with its nested module statements, is compiled, it generates the support classes in the javatech.cor19.server package. The IDL file itself may be kept almost anywhere. We use a directory structure similar to that used in the RMI example in Chapter 18 with an src directory containing all the implementation files in a package-named directory structure and a build/classes directory containing the compiled class files. With this structure, it is convenient to keep the IDL file in the src directory. The IDL compiler supports a -td switch that directs where the output should be generated. In order to keep the IDL-generated files separate from our implementation files, we direct the output to a directory called idlfiles, parallel to the src and build directories. The command line to compile the IDL file is

```
idlj -fall -td idlfiles src/server.idl
```

The idlj compiler is even smart enough to create the idlfiles directory if it doesn't already exist. The -fall switch instructs the compiler to output both client and server support files. If omitted, then -fclient is assumed, generating only the client-side support files, an option that might be desired if one were creating only a CORBA client to connect to an already existing CORBA server. After running the idlj compiler, and assuming that we've already created the empty src directory tree structure, the directory structure appears as shown here:

```
javatech-chapter-19/
   (build and run scripts, etc.)
   src/
      server.idl
      javatech/
         cor19/
            client/
            server/
            impl/
   idlfiles/
      javatech/
      cor19/
         (1 IDL-generated file)
         server/
            (13 IDL-generated files)
   build/
      classes/
```

The build/classes tree will not exist yet either, but we show it here for completeness. When completed, there will be javatech/cor19/server/impl and javatech/cor19/client trees below build/classes.

The several IDL-generated files are of little interest to the developer. They are mainly used internally by the CORBA system. In order to create the implementation files, the programmer needs only to have knowledge of the interfaces defined in server.idl and the standard way in which CORBA works. A brief listing of the files and their purposes is given in Table 19.1. There is also a COR19_CONSTANT.java file in the javatech.cor19 directory one level up, corresponding to the IDL constant COR19_CONSTANT.

Table 19.1 *IDL files list.*

File Name	Description
Cor19Example.java	A Java interface corresponding to the IDL interface, defines constants that were defined within the IDL interface.
Cor19ExampleHelper.java	Static support methods. The most important method here is the narrow() method used to perform object casting in CORBA.
Cor19ExampleHolder.java	Holder class for Cor19Example.
Cor19ExampleOperations.java	A Java interface that defines the methods corresponding to the IDL operations.
Cor19ExamplePOA.java	Abstract class, must be extended by the implementation class. Serves as the server-side skeleton class.
Cor19UserException.java	Java Exception class corresponding to the user exception defined in the IDL.
Cor19UserExceptionHelper.java	Helper for Cor19UserException.
Cor19UserExceptionHolder.java	Holder for Cor19UserException.
CustomData.java	Final Java class corresponding to the user-defined struct in the IDL.
CustomDataHelper.java	Helper for CustomData.
CustomDataHolder.java	Holder for CustomData.
SERVER_CONSTANT.java	Java interface defining the value of SERVER_CONSTANT.
_Cor19ExampleStub.java	Stub class needed by clients.

All of these generated classes are important, but the most commonly seen usage is probably the narrow() method in the Helper classes. All CORBA objects are derived from the org.omg.CORBA.Object class. To cast a generic object into a specific type, the narrow() method must be used. Conceptually, narrowing is nearly identical to the normal Java cast operation, but a Java cast does not work. Each CORBA Object type has an associated Helper class with a narrow() method in order to do the casting. The Helper classes also provide other services that we do not discuss here.

19.4 Creating the server implementation

Similarly to what we did for RMI, we build an implementation file in the javatech.cor19.server.impl package. We can name the implementation class anything we choose, but to be consistent with certain CORBA naming conventions, we use the filename `Cor19Servant.java`. Another name one might see in CORBA examples is `Cor19ExampleImpl.java` since it is the implementation of the interface known as `Cor19Example`. However, we find the "servant" nomenclature easier to keep straight in our minds for reasons that should become clear shortly.

In CORBA terminology, a "servant" is an implementation of an interface. In Java, we can say that a servant is an instance of the Java object that implements the operations declared in the IDL interface. Then a "server" is a process that instantiates the servant objects and makes them available via the CORBA subsystem. In Java, we create another class to be the CORBA server process for this example, somewhat like the `RMIExampleServer` class from Chapter 18, whose role was to create an instance of the implementation class.

19.4.1 Servant implementation

In order to be a CORBA servant, our class must extend `org.omg.PortableServer.Servant`. Actually, this extension is handled by extending `Cor19ExamplePOA`, which itself extends `org.omg.PortableServer.Servant` and also adds other features needed by the CORBA subsystem. Thus our implementation file begins

```
public class Cor19Servant extends Cor19ExamplePOA {
```

Our implementation (servant) class must implement each of the methods defined in the IDL file. Servant methods look just like regular Java methods. The code to handle all of the CORBA communications is provided by the skeleton class in `Cor19ExamplePOA` and its superclasses.

Returning to the IDL file, we first have the attributes huh and hah to implement. Hah is read-only, so we need only provide a getter method of the appropriate signature. That signature is a method named hah with no parameters, returning the type of the attribute, in this case `float`. We create a private instance variable named fHah for the value of the hah attribute. The idea is that the attribute hah, embodied in the implementation code in a variable named fHah is used for some purpose within the implementation. Since the IDL file has identified this as an attribute, we must assume that the client might want to query its value. The client does so by calling the hah() method. Since hah is read-only, there is no way for the client to set its value. Our implementation is simply

```
public float hah () {
    return fHah;
}
```

Next comes the huh attribute, which is not read-only. There must be a getter method, similar to that for hah, along with a private instance variable to contain the value of the attribute. There also must be a setter method. The Java signature of the setter method for this CORBA attribute is a `void` method named `huh(int value)` where the parameter is the type matching the type of the attribute. The purpose of the setter method is to allow the client to pass in a value to assign to `fHuh`, therefore we use

```
public void huh (int huh) {
   fHuh = huh;
}
```

We also need to implement `method1()` and `add()`, similarly to the way they were implemented in the RMI example in Chapter 18. One difference is that here we demonstrate accessing the `COR19_CONSTANT` and `SERVER_CONSTANT` values as follows:

```
public void method1 (String s) {
   String cor19_con = COR19_CONSTANT.value;
   String server_con = SERVER_CONSTANT.value;
   System.out.println (s + "/" + cor19_con + "/" +
   server_con);
}
```

The `add()` method is straightforward as well. The complete code for this example is available on the Web Course.

The `demo()` operation was created to demonstrate the use of an `inout` parameter and the corresponding Holder object. To access the base object contained in a Holder object, we use the public instance variable named `value`. Since `demo()` was declared in the IDL as receiving an `inout` `CustomData` structure, the Java implementation receives a `CustomDataHolder` object, and we obtain the contained `CustomData` object as follows:

```
CustomData cd = custom_data_holder.value;
```

where `custom_data_holder` is the input parameter. Then we can manipulate the fields within `cd` at will. Recall that `CustomData` was declared to have two fields – a `float` and an `int` (well, an IDL `long`, which becomes an `int` in the Java implementation). For demonstration purposes, we multiply the `float` field by two and the `int` field by three:

```
public void demo (CustomDataHolder custom_data_holder)
throws Cor19UserException {
   CustomData cd = custom_data_holder.value;
   cd.someFloatValue *= 2;
   cd.someIntegerValue *= 3;
}
```

Note that once we have a reference to the contained `CustomData` object, we do not need to "put it back" into the Holder object.

19.4.2   Server implementation

We now turn to the server class that instantiates the servant and makes it available to the CORBA subsystem. Our `Cor19Server.java` class has only a `main()` method that initializes CORBA, locates the CORBA naming service, creates a `Cor19Servant` instance, and loads that instance into the naming service. These steps are analogous to the steps followed in the RMI example, but the details are quite different for CORBA than for RMI.

Part of the CORBA subsystem is an Object Request Broker (ORB). The ORB is a library of support code that accomplishes the low-level communications between CORBA clients and servers. The ORB class is in package `org.omg.CORBA` and provides many functions, only a few of which are of interest to the casual CORBA programmer. The only method we use is `ORB.init()` to initialize the ORB.

Another concept important in CORBA is the *Portable Object Adapter*, or POA. We briefly mentioned POA when discussing the abstract `Cor19ExamplePOA` class, which must be extended to create the servant class. It is useful to think of an "object adapter" as the way in which clients, servers, and servants interact with the ORB. Before the OMG adopted the POA standard, each vendor of CORBA software implemented their object adapters in different ways. As a result, source code was not portable between CORBA ORB vendors. To resolve this problem, the OMG created the *portable* object adapter specification, or POA. A POA is a CORBA object, but because of the services it provides, it is a very special kind of object. A POA manages the implementation of a collection of objects and provides a namespace for those objects, including other POAs. It is not necessary to understand all the services provided by the POA, which is a large and complicated subject. We use only the most basic POA features in this example without going into detail about what is really happening behind the scenes.

The server class's `main()` method must perform the following six steps:

1. Create and initialize the ORB.
2. Get the root POA and activate the POAManager.
3. Create an instance of the servant object.
4. Get a reference to the naming service.
5. Bind the servant into the naming service.
6. Tell the ORB to wait for incoming requests from clients.

We discuss each of these six steps in turn. First, though, since any call to a CORBA method could result in a CORBA exception, we wrap everything in a `try/catch` block. Thus,

```
public static void main (String[] args) {
   . . .
   try {
     // All six server steps
   }
   catch (Exception e) {
     System.err.println ("ERROR: " + e);
     e.printStackTrace (System.err);
   }
} // main
```

This `main()` method appears in the `Cor19Server` class, where we have assumed the following package and import statements:

```
package javatech.cor19.server.impl;

import javatech.cor19.server.*;
import org.omg.CORBA.*;
import org.omg.CosNaming.*;
import org.omg.CosNaming.NamingContextPackage.*;
import org.omg.PortableServer.*;
```

19.4.2.1 Create and initialize the ORB

The ORB is created and initialized with the static `ORB.init()` call in the `org.omg.CORBA` package. For a Java application (as opposed to an applet), the `init()` method takes two parameters – a `String` array and a `Properties` object. Either parameter may be null. The idea for the `String` array is that `ORB.init()` is typically called from an application's `main()` method, and since `main()` receives a `String` array of command line arguments, these can be passed directly to the ORB. The `init()` method uses any of these command line arguments that it can interpret and ignores the rest. An example is the argument `-ORBInitialHost`, which identifies the host on which to find the naming service.

The `Properties` parameter provides an alternate way to deliver certain values to the ORB. For example, the initial naming service host can be specified with a `org.omg.CORBA.ORBInitialHost` property instead of `-ORBInitialHost` command line argument. The `Properties` object passed to the ORB can be created and populated within the `main()` program, or you can use the system properties object and load the desired property names and values from the command line with the `-Dprop=value` syntax.

You are free to use either the `args` method or the `Properties` object method, or both. A code snippet showing ORB initialization utilizing both methods is

```
ORB orb = ORB.init (args, System.getProperties ());
```

Here, we passed in the `args` parameter from the command line as well as the system properties object, allowing both methods of ORB parameter specification to be used.

19.4.2.2 Get the root POA and activate the *POAManager*

When there is an ORB, there are certain important services either already available or assumed to be available. An example of the latter is the CORBA naming service, known as *COSNaming*, which we learn about shortly. An example of the former is the root POA. The root POA is the root of a hierarchical chain of POAs. In many cases, including all the examples used in this book, you need use only the root POA. For more complicated cases in which certain POA policies need to be changed, you would have to create one or more child POAs in order to have different behaviors.

The ORB class provides the `resolve_initial_references()` method to obtain references to services such as the naming service and the root POA. As usual, the return from `resolve_initial_references()` is a generic CORBA object that must be narrowed to a POA using the `POAHelper` class. A reference to the root POA is retrieved with the following code snippet;

```
POA rootpoa =
  POAHelper.narrow (orb.resolve_initial_references
  ("RootPOA"));
```

Each POA has an associated `POAManager` object, which may be the manager of more than one POA. The `POAManager` class has several methods, as documented in the `org.omg.PortableServer` package, but the only thing we really need to know about the `POAManager` is that it must be activated, which permits its associated POAs to start processing requests. The code to activate the `POAManager` is

```
rootpoa.the_POAManager().activate ();
```

19.4.2.3 Create an instance of the servant object

Recall that a server is a process that instantiates and makes available one or more servant objects. So our server must create an instance of our servant. Doing so is a simple Java instantiation:

```
Servant servant = Cor19ExampleServant ();
```

At this point, we have a Java servant instance, which is an instance of `org.omg.PortableServer.Servant`, which is a Java object, but not a CORBA object. To create a CORBA object (actually, an object reference) from the servant, we use a method on the POA:

```
org.omg.CORBA.Object obj_ref =
  rootpoa.servant_to_reference (servant);
```

Notice that the type of `obj_ref` is the fully qualified `org.omg.CORBA.Object`. Whenever we refer to a CORBA `Object`, we must fully specify the type to distinguish it from `java.lang.Object`.

Now we have a generic CORBA object reference, but not a reference to the required `Cor19Example` interface that is needed by the naming service and, eventually, the client. To obtain the specific type from the generic CORBA `Object` type, we must narrow the object reference using the `Cor19ExampleHelper` class:

```
Cor19Example cor19 = Cor19ExampleHelper.narrow (obj_ref);
```

19.4.2.4 Get a reference to the naming service

The naming service in CORBA is known as COSNaming and is implemented in the `org.omg.CosNaming` package. COSNaming provides functions somewhat like the RMI naming service in that server-side objects (servants) are bound into the naming service under a specific name and clients perform a lookup on that name to gain access to the servant's operations. An object reference to the naming service is found with the `ORB.resolve_initial_references()` method using the well-known name "`NameService`," which is required to be defined for all CORBA ORB implementations. As usual, this call returns a generic CORBA object which must be narrowed to the type desired using the appropriate Helper class. For the case of the naming service, the desired type is a `NamingContextExt` object:

```
org.omg.CORBA.Object ns_ref =
  orb.resolve_initial_references ("NameService");
NamingContextExt ncRef =
  NamingContextExtHelper.narrow (ns_ref);
```

19.4.2.5 Bind the servant into the naming service

Now we can register our servant with the naming service under a specific name. The name given to `NamingContextExt` is not quite as simple as a plain Java `String`, again a complication owing to the generality and language-neutrality of CORBA. Instead of a plain `String`, we must pass in an `org.omg.CosNaming.NameComponent` array. Fortunately, the `NamingContextExt` class provides a `to_name()` method to translate a plain Java `String` name into a `NameComponent` array:

```
NameComponent[] path = ncRef.to_name ("Cor19");
```

Then, to bind into the naming service, we use

```
ncRef.rebind (path, cor19);
```

Recall that `cor19` is a reference to the `Cor19Example` interface that corresponds to the `Cor19ExampleServant` object that was instantiated above. Now, when the client calls `resolve()` using the name "Cor19", it receives back an object reference that can be used to make remote method calls to our servant implementation.

19.4.2.6 Tell the ORB to wait on incoming requests

This last step is simple. We must notify the ORB that everything is ready and that it should begin listening for requests from clients. The following code should appear at the end of, but within, the `try/catch` block:

```
orb.run ();
```

Generally, this method never returns. It waits until a client invokes a remote method on the servant and then dispatches that incoming request to the servant code. When the invocation is complete, the ORB waits again for more incoming requests.

19.5 Client implementation

The client class begins similarly to the server implementation, by importing the required packages and starting a `main()` method that encloses all CORBA activity within a `try/catch` block to catch any CORBA errors. We put the client application into the `javatech.cor19.client` package under the name `Client`. The client must create and initialize the ORB and obtain a reference to the `NamingContextExt`, similar to what was done in the server class. To perform a lookup, we could use the `resolve()` method by providing a `NameComponent` array, as was done for the server. But `NamingContextExt` also includes a `resolve_str()` convenience method that accepts a Java `String` and internally performs the conversion necessary to a `NameComponent` array. So the `main()` method in our `Client` class goes as follows:

```
public static void main (String[] args) {
  try {
    // Create and initialize the ORB
    ORB orb = ORB.init (args, null);
    // Get the root naming context
    org.omg.CORBA.Object ns_ref =
      orb.resolve_initial_references ("NameService");
    NamingContextExt ncRef =
      NamingContextExtHelper.narrow (ns_ref);
    // Get a Cor19Example from the name server.
    Cor19Example cor19 =
      Cor19ExampleHelper.narrow (ncRef.resolve_str
        ("Cor19"));
```

Notice that we have used the same name, "Cor19", that the servant was bound with and that we have narrowed to a `Cor19Example` interface using the `Cor19ExampleHelper` class. Now we are ready to call methods on the servant:

```
cor19.method1 ("Hello");
int result = cor19.add (4, 9);
System.err.println ("add(4,9) returns " + result);
```

We must, of course, close the `try/catch` block and handle any errors. For this simple example, we just dump the stack trace:

```
    } catch (Exception e) {
        e.printStackTrace (System.err);
    }
} // main
```

19.6   Running the example

To test this example, we need to start the name service, start the server, and run the client. For CORBA, the name service is started with `orbd`, which needs to know the port number and host. For running client and server on the same machine, the host can be left off, allowing it to default to localhost. So we start *orbd* as follows:

```
start orbd -ORBInitialHost localhost -ORBInitialPort 3001
```

To start the server, we run the `javatech.cor19.server.impl.Cor19-Server` class. Notice that we pass in the `ORBInitialPort` property via the system properties object (i.e. the `-Dname=value` syntax) rather than as a command line parameter. Either method may be used, since the server was coded to accept either command line arguments or system properties:

```
java -classpath build/classes
   -Dorg.omg.CORBA.ORBInitialPort=3001
   javatech.cor19.server.impl.Cor19Server
```

(Here, and in later examples, we split command lines to fit within the page margins.) We run the client, in another command shell window, similarly. But notice that the client code above does not pass the system properties object to `ORB.init()`. Therefore we must use the command-line arguments feature to specify the port number on which to lookup the name server:

```
java -classpath build/classes javatech.cor19.client.Client
   -ORBInitialPort 3001
```

The output should be

```
add(4,9) returns 32
```

But wait, the sum of 4 and 9 should be 13, not 32! What went wrong? Actually, nothing went wrong, because we coded the add() method to add the two input parameters *plus* the value of COR19_EXAMPLE_CONSTANT from the IDL file. We didn't show that code above, but it appears here and on the Web Course:

```
public int add (int a, int b) {
   return a + b + Cor19Example.COR19_EXAMPLE_CONSTANT;
}
```

There are other methods defined in the server.idl file that we have not tested, such as the getter and setter methods for the huh and hah attributes. Test code in the client for those methods is:

```
// Try getting and setting the huh and hah attributes.
System.err.println ("hah() = " + cor19.hah ());
System.err.println ("Before setting, huh() = " +
   cor19.huh ());
cor19.huh (19);
System.err.println ("After setting, huh() = " +
   cor19.huh ());
```

To demonstrate the use of the Holder object for the inout CustomData parameter, we first create and populate a CustomData object. CustomData.java is one of the files created by the IDL compiler – its function is to serve as the CustomData struct defined in the IDL file.

There are two ways to populate the CustomData object. One is to supply the values as parameters to the CustomData() constructor since the CustomData class generated by idlj includes an overloaded constructor that receives values for each field in the struct. For example,

```
// Create and populate a CustomData object.
CustomData cd = new CustomData (12.0f, -13);
```

This technique requires only one line of code and is especially useful when the values of all the fields are known at construction time.

If some values are not known until later, then an alternative method is required. There is also a no-arg constructor that merely creates an empty CustomData object. The idlj-generated CustomData class implements all the structure fields as public variables, which means we can directly access the fields. (Implementing fields as public variables is generally frowned upon as poor OO design, but that's the way CORBA does it.) Each field name in CustomData is named with the same name used in the structure definition in the IDL file. Therefore, the alternative way to create and populate a CustomData object is:

```
// Create and populate a CustomData object.
CustomData cd = new CustomData ();
cd.someFloatValue = 12.0f;
cd.someIntegerValue = -13;
```

Once we have a `CustomData` object, we must "wrap" it in a `CustomData-Holder` object as follows. The `CustomDataHolder` constructor accepts a `CustomData` parameter to be wrapped:

```
CustomDataHolder cdh = new CustomDataHolder (cd);
```

Alternatively, the public `value` field in the Holder class could be used to set the value of the wrapped object:

```
CustomDataHolder cdh = new CustomDataHolder ();
cdh.value = cd;
```

Then we call the `demo()` method. We print out the contents of `cd` before and after calling `demo()` to see what happens:

```
System.err.println ("Before calling demo():" +
   "\ncd.someFloatValue = " + cd.someFloatValue +
   "\ncd.someIntegerValue = " + cd.someIntegerValue);
// Call demo() to manipulate the CustomData within the
// (inout) CustomDataHolder parameter.
cor19.demo (cdh);
// Retrieve and print the modified CustomData from the
// holder.
cd = cdh.value;
System.err.println ("After demo():" +
   "\ncd.someFloatValue = " + cd.someFloatValue +
   "\ncd someIntegerValue = " + cd.someIntegerValue);
```

The complete expected output from the client is:

```
add(4,9) returns 32
hah() = 3.1416
Before setting, huh() = 0
After setting, huh() = 19
Before calling demo():
cd.someFloatValue = 12.0
cd.someIntegerValue = −13
After demo():
cd.someFloatValue = 24.0
cd.someIntegerValue = −39
```

Note that if we run the client again without restarting the server, the initial value of `huh()` will be 19, not 0. Once the value in the servant is set, it remains set until changed. For this reason, one must be careful when initializing attributes.

19.7 Running the CORBA example on two machines

The example above ran both client and server on the same machine. To run on two different machines, the changes are very minimal. The name server, `orbd`, must be started on the server machine, and the server machine name must be specified in the command line parameter or system property to both `orbd` and the server application rather than taking the default localhost. Thus we start `orbd` and the server as follows:

```
start orbd -ORBInitialHost myserver.somewhere.com
  -ORBInitialPort 3001
java -classpath build/classes
  -Dorg.omg.CORBA.ORBInitialHost=myserver.somewhere.com
  -Dorg.omg.CORBA.ORBInitialPort=3001
  javatech.cor19.server.impl.Cor19Server
```

And then start the client on a different machine like this:

```
java -classpath build/classes javatech.cor19.client.Client
  -ORBInitialHost myserver.somewhere.com -ORBInitialPort 3001
```

19.8   Conclusion

This completes our introduction to CORBA technology. As we have seen, CORBA can provide much of the same basic functionality as RMI, but with a somewhat different programming paradigm. The general concepts are similar to the RMI programming model – clients and servers, stubs and skeletons, a naming service, and some behind-the-scenes plumbing that makes it easy to make complex remote method calls to another JVM. However, dealing with the CORBA complexities amounts to a significantly different, and generally more difficult, programming effort compared to RMI. As stated at the beginning of the chapter, the main advantage of CORBA is that it is language and platform independent. Non-Java clients can have full access to a server if it is written in CORBA and the interfaces are published in an IDL file.

19.9   Web Course materials

The files for the programs discussed here are available on the Web Course Chapter 19. Also, there are additional discussion, resources, and examples of CORBA.

Resources

Introduction to CORBA, Java Short Course at Sun Microsystems, December 1999,
    http://java.sun.com/developer/onlineTraining/corba/.
Object Management Group, www.omg.org.
"Overview of CORBA," Chapter 11 in: Qusay H. Mahmoud, *Distributed Programming with Java*, Manning Pub., 2001, http://java.sun.com/developer/Books/corba/ch11.pdf.

Chapter 20
Distributed computing – putting it all together

20.1 Introduction

Chapter 16 introduced distributed computing, the UML notation language, and the design of the server side of a client/server application. Chapter 17 described the client side. Before beginning an actual implementation, it was necessary to describe Java RMI in Chapter 18. Then we took a brief detour in Chapter 19 to describe CORBA as an alternative to RMI. Now that we have all the pieces, we can put them together to build a simple example of a distributed computing application in which the calculation engine is implemented completely in Java. If the calculation engine involves legacy code in a language other than Java, then we must use JNI (the Java Native Interface, which is described in Chapter 22).

20.2 The sample application

A real client/server application using the design of Chapters 16 and 17 will be more complex than the sample described here. In fact, if the application were not somewhat complex, there would be little reason to implement it in a client/server design in the first place! However, for the sake of demonstration purposes and to provide a template for the reader's own applications, we provide a simple application that illustrates the important features of the client/server design presented in Chapters 16 and 17.

Recall that the client/server system designed earlier assumed a time-dependent simulation – i.e. one with a solution that varies as a function of time. The same approach could also work with monitoring and controlling a real system such as a remote sensor. We wish for the simulation code to return a periodic "snapshot" of the solution as a function of time as the simulation runs. Alternatively, in the case of monitoring a remote sensor system, we could return periodic data as it is collected from the sensor. For our demonstration, we need either a time-dependent simulation problem to use as an example or a remote sensor or experimental data apparatus that is generating time-dependent data. We choose a software simulation code for the example, but we could just as well use a software mock-up of experimental data. For simplicity we refer to the data-generation source as a simulation code rather than repeatedly mention that it could just as well be a remote sensor device.

When choosing a simulation code to use for the demonstration, we must keep in mind that we need the code to run long enough to permit demonstration of the use of control data that modifies some parameters of the simulation as it runs. If we choose a typical time-dependent calculation that exhibits transient behavior, such as a simple heat-diffusion problem, a steady state is likely to be reached much too quickly to demonstrate the time-dependent behavior of our client/server design. Thus a problem whose solution is a periodic function of time is more appropriate for demonstration purposes. A suitable simulation parameter to be modified can be a time constant that modifies the frequency of the periodic solution.

With these concepts in mind, we choose a simple harmonic motion example – a mass m on a perfect, frictionless spring with spring constant k set in motion with an initial amplitude A. We define the coordinate system with $x = 0$ at the center of the motion – so the mass m oscillates between $x = +A$ and $-A$. This is a simple first-year physics problem with the following solutions:

$$\begin{aligned}
\text{position } x &= A \ \sin(\omega t) \\
\text{velocity } v &= \mathrm{d}x/\mathrm{d}t = \omega A \ \cos(\omega t) \\
\text{acceleration } a &= \mathrm{d}v/\mathrm{d}t = -\omega^2 A \ \sin(\omega t) \\
\text{kinetic energy } K &= \tfrac{1}{2}\, m v^2 \\
\text{potential energy } p &= \tfrac{1}{2}\, k x^2
\end{aligned}$$

where $\omega^2 = k/m$.

While not very interesting as a numerical example, this choice has the advantage that it is very easy to calculate – so easy, in fact, that we have to build in a delay on the server side to approximate what would happen in a truly complex simulation. At least the numerics of this example will not get in the way of explaining the client/server implementation. A control parameter to be varied could be the mass or spring constant or, equivalently, the frequency ω. If we were to plot the position as a function of time, the plot would be a simple sine curve with amplitude A and period $T = 2\pi/\omega$. If we then changed ω we would see the shape of the curve change to a sine curve with increased or decreased period.

We describe the implementation of this example in four phases. First, Section 20.3 describes the factory and server interfaces. Section 20.4 discusses the implementation of the factory interfaces while Section 20.5 describes the server implementation. Finally, the client implementation is built in Section 20.6.

20.3 Server interfaces

We begin by implementing the server design from Chapter 16. Since we are writing both the client and the server, and since we assume that we do not need to support non-Java clients, we implement the server using RMI. Following the pattern of the `rmi18` and `cor19` packages, we use the `javatech.all20` package for this application. The "20" part indicates Chapter 20 while "all" is intended to be a mnemonic abbreviation for "putting it all together," the title of this chapter.

20.3.1 The factory interface

We first need to define the interfaces for the factory and server objects. From Chapter 16, both interfaces are quite simple. The factory has only one method, `getInstance (id)`, which returns a reference to an object that implements the server interface when given an ID parameter.

Recall that the purpose of the factory pattern is that multiple clients can call the factory, each receiving its own private server object. The ID parameter serves to distinguish clients from one another. If the server needs to store any temporary data during runs, then the ID parameter is a good way for the server to keep data from each client separate from other clients, most likely by creating a subdirectory on the server side named with the client's ID. The server might also keep log files in that subdirectory logging the progress of the calculation. If an error occurs on the server, then the log files can be examined during debugging. In this case, the ID might be a project name or a run number – anything that the client can supply as identifying information for the run so that the log files on the server side can be identified. Note that it is not essential to the design that the client supply the ID. The factory could simply make up a unique ID for each client using, say, the millisecond clock time. Debugging, however, is often easier if the client has supplied a meaningful ID of some sort.

If the server uses any proprietary information, then the ID might be a username or even a username/password pair used to authenticate the user and authorize that user's access to the proprietary data. For security purposes, the password should be encrypted since the client/server communication will be occurring over the Internet. Encryption and user authentication and authorization are very important topics but beyond the scope of this book. But not at all beyond the scope of Java! See, for example, the Java Authentication and Authorization Service (JAAS)[1], the Java Cryptography Extension (JCE)[2], and the Java Cryptography Architecture (JCA)[3].

While a general design might require three parameters – a username, an encrypted password, and a project name, we implement this example assuming that a single string specifying a project name is given as the ID. In practice, we won't store any data files on the server side in this example, so we won't actually do anything at all with the ID parameter except log it.

From Chapter 18 on RMI we know that the interface must extend `java.rmi.Remote` and that each method declared in the interface must throw `java.rmi.RemoteException`. Therefore, our `FactoryInterface` looks like:

```
package javatech.all20.server;

import java.rmi.Remote;
import java.rmi.RemoteException;

public interface FactoryInterface extends Remote {
  final static String FACTORY_NAME =
    "all20ServerFactory";
  public ServerInterface getInstance (String id)
    throws RemoteException;
}
```

where we have used package `javatech.all20.server`, following the pattern of Chapter 18. Here we have also defined the constant `FACTORY_NAME`, which contains the name that the factory will be bound under in the RMI registry. Both the client and the server need to know the binding name, and defining it here is convenient since both client and server can see the same name in the same place, which is much preferred to defining the same name in two different places.

20.3.2 The server interface

The server interface designed in Chapter 16 and clarified in Chapter 17 (see the class diagram in Figure 17.3) has four methods:

```
package javatech.all20.server;
...
public interface ServerInterface extends Remote {
  public boolean initialize (String initparam)
    throws RemoteException;
  public void initializeSimulation (float[] indata)
    throws RemoteException;
  public void start ()
    throws RemoteException;
  public float[] retrieveData (float[] indata)
    throws RemoteException;
}
```

The `initialize()` and `initializeSimulation()` methods are almost redundant in this simple example. Recall from Chapter 17 that `initialize()`

is used to initialize the server object while `initializeSimulation()` is used to initialize the simulation code within that server object. In many cases, both duties could be combined into one. However, examining the collaboration diagram in Figure 17.2 (where `initializeSimulation()` was called `receiveInput()`) shows that `initializeSimulation()` at sequence number 4.1 occurs much later in the collaboration than `initialize()` at sequence number 2. In a complex simulation the data known by the client at the two different times in the collaboration sequences could become important, so we keep the two methods separate.

In a technique that is useful to efficiently support multiple simultaneous clients, the factory server keeps a pool of instantiated but uninitialized servers. This pool is created when the factory starts up, long before any client exists. To build the pool, the factory merely creates several instances of the `Server` class and keeps them in a `HashMap`. Using such a pool improves response time by allowing the factory to supply an already constructed server from the pool during a client's call to `getInstance()` rather than incurring the potentially nontrivial cost of a new server construction for each new client. Since the client credentials in the ID parameter provided to the factory cannot be known at server construction time, when the pool is initially created, the `initialize()` method on the server interface provides a good opportunity to supply those client credentials. When that client is finished, the server can be returned to the pool, awaiting use by another client with different credentials.

The server pool technique is useful and straightforward but a bit beyond the scope of this chapter where we want to provide a simple example that demonstrates working client/server concepts unencumbered by concerns of user/client credentials and pool creation and maintenance. Therefore, the example for this chapter passes only a simple `String initparam` to the `initialize()` method. As with the ID passed to the factory, for simplicity we won't actually do anything with `initparam` other than log it.

There are three other factory methods that we have not yet discussed – `initializeSimulation()`, `start()`, and `retrieveData()`. Referring to the UML diagrams in Chapters 16 and 17, we see that `initialize-Simulation()` is where the initial data for the simulation code is provided and `start()` is where the simulation thread is actually begun. Then the client calls `retrieveData()` periodically to retrieve the results from the simulation.

20.4 Server factory implementation

The server factory implementation is rather simple. From the `FactoryInterface` class shown above, its only remotely exposed method is `getInstance()`. The `ServerFactory` class implements `FactoryInterface` and extends `UnicastRemoteObject`. Since we put the implementation into the `impl` subpackage below `javatech.al20.server`, the beginning of `Server-Factory` is:

```
package javatech.al20.server.impl;

// various imports...
import javatech.al20.server.*;

public class ServerFactory
   extends UnicastRemoteObject
   implements FactoryInterface
{
   private static int fServerIndex__ = 0;
   public ServerFactory () throws RemoteException { }
```

We have omitted the various standard `java.rmi` imports for brevity. We do show the `import javatech.al20.server.*` because it is necessary for the compiler to see the `FactoryInterface` and `ServerInterface` definitions. The constructor, as shown, does nothing. However, as explained in Chapter 18, the constructor must be present and must be declared to throw `RemoteException` since `UnicastRemoteObject` throws `RemoteException`. We explain the class variable `fServerIndex__` when we discuss the `getInstance()` method below.

20.4.1 Automatically starting the RMI registry

Recall that in Chapter 18 we manually started the RMI registry using the `rmiregistry` tool from the Java 2 SDK. Then the server implementation's `main()` method instantiated the server and bound it into the registry under a known name. Thus starting the server was a two-step process: start `rmiregistry` and start the server.

Here we demonstrate how to skip the first step by automatically starting the RMI registry from within the server factory's `main()` method. Then all we need to do to start the server running and listening for clients is to run the server factory class.

The `java.rmi.registry.LocateRegistry` class has a `create-Registry()` method that starts the registry on a specified port. Its use is simple:

```
try {
   java.rmi.registry.LocateRegistry.createRegistry (port);
}
catch (RemoteException re) {
   System.err.println ("Could not create rmiregistry:" + re);
   System.exit (1);
}
```

After the registry has been created, we continue as in Chapter 18 by instantiating a `ServerFactory` object and binding it into the registry under the name in the `FACTORY_NAME` constant:

```
try {
  ServerFactory factory = new ServerFactory ();
  Naming.rebind ("//localhost:" + port + "/" + FACTORY_NAME,
    factory);
  System.out.println (
    "\n READY AND WAITING ON CLIENTS ON PORT " + port);
}
```

As usual we must catch the following exceptions:

```
catch (MalformedURLException mue) {
  System.err.println (mue);
  System.exit (1);
}
catch (RemoteException re) {
  System.err.println (re);
  System.err.println ("\n EXITING BECAUSE OF FAILURE");
  System.exit (1);
}
```

20.4.2 Implementing `getInstance()`

Our `getInstance()` implementation receives the ID parameter and returns an object that implements `ServerInterface`:

```
public ServerInterface getInstance (String id)
  throws RemoteException
{...}
```

For this example, we simply create a `Server` object for return. We haven't yet discussed the `Server` implementation, and although we know the signatures of the methods it must contain (since it must implement `ServerInterface`), we don't yet know how to call its constructor. Often, implementations evolve through an iterative process, so let's start by calling the constructor as follows:

```
ServerInterface server = new Server (id);
```

Our `getInstance()` needs to return a `ServerInterface`, not a `Server` object, so we declare the object returned by the `Server` constructor to be of type `ServerInterface`. (It is, in fact, a `Server` object, but since it implements `ServerInterface`, it is also of type `ServerInterface`.)

As shown, we also passed the `id` value received by `getInstance()` to the `Server` constructor. From the discussion above, the ID parameter serves to distinguish clients from one another when there are multiple clients. Obviously it makes sense for the server to know that ID. However, there is no guarantee that the `id` value supplied by the client is unique. If we want to uniquely identify server instances, which is a very good idea, then we need to make certain that the server ID is unique. A simple and successful technique is to append a unique number to the `id` received from the client. Therefore the factory keeps a `static int` field named `fServerIndex__`. By incrementing its value each time `getInstance()` is called and appending it to the `id` supplied by the client, the factory can create a guaranteed unique ID with which to identify the server:

```
System.err.println (
   "\nServerFactory: making a new Server instance");
fServerIndex__++;
String unique_server_id = id + "-" + fServerIndex__;
```

It turns out in a real-world simulation problem that it is often useful for the factory class to provide some "utility" methods of use to servers. An example is when the factory maintains a pool of server objects, in which case some pool management methods are needed on the factory. To support such usage, we also pass the factory's `this` parameter to the `Server` constructor, along with the unique `server_id`:

```
ServerInterface server = new Server (this, unique_server_id);
```

In this way, the server has a reference to the factory, permitting the server to call factory methods. Since none of these utility methods need to be remotely accessible to clients, they do not appear in the `FactoryInterface`.

When the `Server` constructor returns, we simply return that object reference to the client:

```
return server;
```

20.5   Server implementation

The `Server` implementation is considerably more complicated than the factory. Recalling that we spent considerable design effort in Chapter 16, it is a great help to build the server from the collaboration diagram created there and shown in Figure 16.6. For convenience, we have reproduced that diagram in Figure 20.1 with one small change – the `initialize()` invocation has been inserted as sequence number 3, bumping the remaining sequence numbers up by one.

As we develop the code for this example, keep in mind that the example is necessarily a simplified one while the collaboration diagram supports a more general and more complicated implementation. Some of the features that we implement will seem to have little or no value to this simple example. However, those features have been found to be of value in more realistic client/server simulations. All

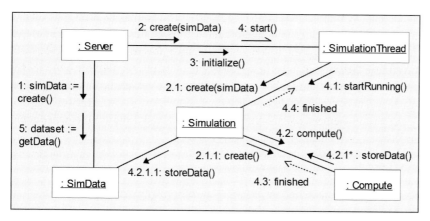

Figure 20.1 The server collaboration diagram. It is the same as in Figure 16.6 except for the insertion of the `initialize ()` invocation as sequence number 3.

the code here is derived from actual working examples of complex simulations. Sometimes the simplifications necessary to produce an example of pedagogical value that can be discussed in a textbook setting result in the removal of features that are quite valuable for real world cases. Rather than reduce the example simulation to an even more limited one, we have chosen to retain certain features as placeholders for more complicated client/server solutions that the reader may wish to put into practice.

20.5.1 The `Server` Class

The `Server` class is the class that implements `ServerInterface` and is the only server-side object with which the client interacts directly. As normal in RMI implementations it extends `UnicastRemoteObject`:

```
public class Server extends UnicastRemoteObject
   implements ServerInterface
{...}
```

20.5.1.1 The Server constructor

The `Server` constructor is called by the factory upon demand from a client during `getInstance ()`. From the factory discussion above, we already know the signature of the constructor. As required by RMI, the constructor, and all methods declared in `ServerInterface`, must be declared as possibly throwing `RemoteException`:

```
public Server (ServerFactory myfac, String id)
   throws RemoteException {
   fFactory = myfac;
   fID = id;
```

Notice that we keep a reference to the factory that created this `Server` instance in the variable `fFactory`. We also keep a copy of this server object's unique ID in the string `fID`, which is useful to identify this server if we should need to call any factory methods. In a real-world example, we might use the ID for user authentication and authorization as described earlier. In this example, we just keep a copy for possible later use.

We can implement collaboration sequences 1 and 2 during the constructor. Sequence 1 is to create an instance of `SimData`:

```
fSimData = new SimData ();
```

Sequence 2 is to create an instance of `SimulationThread`, giving it a copy of the `SimData` reference:

```
fSimulationThread = new SimulationThread (fSimData);
```

That completes the `Server` constructor. To complete the implementation of the `ServerInterface`, we must also implement `initialize()`, `initializeSimulation()`, `start()`, and `retrieveData()`.

20.5.1.2 The `initialize()` method

Our `initialize()` method is quite simple. It receives a single `String` `initparam` and does nothing at all with it. As explained above, `initialize()` is provided to initialize the server object with information that is unknown at `Server` construction time but available later. In this example, no additional information is required. The `initparam` parameter is provided as a placeholder for more complicated client/server problems that the reader may develop. It is a good idea to keep track of whether or not the `Server` object has been initialized. Therefore we set an instance variable `fInitialized` to `true` and then return `true` to indicate a successful initialization:

```
public boolean initialize (String initparam)
   throws RemoteException
{
   // Log initparam here if desired
   fInitialized = true;
   return true;
}
```

20.5.1.3 The `initializeSimulation()` method

Our `initializeSimulation()` method is also quite simple. Recall that it is called by the client to provide initial simulation data. (In the collaboration diagram in Figure 17.2 `initializeSimulation()` was called `receiveInput()`, the name invented early in our OOAD phase. Its name was changed late in Chapter 17 to be more descriptive of its function after our analysis of the problem improved. Such name changes are common during OOAD and should not be eschewed. The proper and sensible naming of objects and operations is important for a clear understanding of a design.) When the client calls `initializeSimulation()`, the server performs sequence 3, which initializes the `SimulationThread` with the input data:

```
public void initializeSimulation (float[] indata)
   throws RemoteException
{
   fSimulationThread.initialize (indata);
}
```

20.5.1.4 The `Server.start()` method

Finally the client calls the server's `start()` method in order to start the simulation running. This call is forwarded immediately to `SimulationThread`'s `start()` method at sequence 4 in the collaboration diagram.

```
public void start() throws RemoteException {
   fSimulationThread.start();
}
```

As usual, calling a thread's `start()` method invokes the `run()` method, which we show later.

20.5.1.5 The `retrieveData()` method

As the simulation runs, simulation data is repeatedly loaded into the `SimData` object as shown in sequence numbers 4.2.1 and 4.2.1.1. We explain those details shortly. For now, we assume that the client periodically polls the server for current simulation data. The client does so by calling `retrieveData()`, which simply obtains and returns the latest available data from `SimData` using its (yet to be discussed) `getResults()` method (shown as sequence 5 in the collaboration diagram of Figure 20.1):

```
public float[] retrieveData (float[] indata)
   throws RemoteException
{
   return fSimData.getResults (indata);
}
```

Here we have made the simplifying assumption that the set of data to be returned is a simple `float[]` array. More complicated structured data could be returned in a Java object with minor modifications. We have also taken the opportunity to allow the client to provide a new set of input data during each `retrieveData()` poll. If the input data has not changed, then the original `indata` array can be passed each time.

As mentioned earlier, for this example we are also using the simple technique of returning only the most recent data stored in `SimData`, allowing any intervening data generated by the computation to be lost if the client does not poll often enough. More elaborate schemes can be devised, such as storing intermediate data in `SimData` and returning a collection of results, but doing so now would distract from the basic client/server design being demonstrated here.

20.5.2   The `SimulationThread` class

Continuing around the collaboration diagram in a clockwise direction, which is roughly the order in which the ojects are encountered, the next class to discuss is the `SimulationThread` class. It is a thread so it extends `java.lang.Thread`:

```
public class SimulationThread extends Thread {...}
```

When `SimulationThread` is constructed (at sequence 2 in the collaboration diagram), we see that sequence 2.1 also must be performed to construct a `Simulation` object. So the `SimulationThread` constructor looks like:

```
public SimulationThread (SimData sim_data) {
   fSimulation = new Simulation (sim_data);
}
```

where we have passed the `SimData` object on to the `Simulation` constructor as shown in the collaboration diagram.

This thread's `run()` method starts the simulation running by calling the `startRunning()` method on the `Simulation` object. We can note from the collaboration diagram that `startRunning()` is a synchronous call. That is, it blocks until the calculation is complete, at which time `run()` exits.

```
public void run () {
   fSimulation.startRunning ();
}
```

Recall that the best and proper way to end a thread is to set a `boolean` value that the running thread checks periodically. In this case, our `run()` method is not in a loop itself, but has called `Simulation.startRunning()` instead. If the server ever needs to halt the simulation, perhaps on command from the client (through additional remote methods that we do not show here), it is a good idea to provide a `halt()` method that passes the halt command on to `Simulation`:

```
void halt () {
   fSimulation.halt ();
}
```

20.5.3 The `Simulation` class

The computational work is done in the `Simulation` class. The constructor does nothing except store a reference to the `SimData` object:

```
public Simulation (SimData sd) {
   fSimData = sd;
}
```

The `initialize()` method is called in sequence 3.1 after the client has called `Server.initializeSimulation()`. Recall that `initialize-Simulation()` is used by the client to provide initial conditions to the simulation. In our case, we receive a `float[]` array of values. Each element in the array represents a known physical quantity – initial amplitude, spring constant, etc. Both the client and server must agree upon the proper order of values in the array or else the calculation will be incorrect. A less error-prone technique is to define a data-only class, perhaps named `SpringParameters` with named fields for each desired parameter:

```
class SpringParameters {
   float amplitude;
   float springConstant;
   ...
}
```

This class would best appear in the `javatech.all20.server` package where the interfaces appear, giving it visibility to both the client and server. The client would instantiate the class, populate each element, and pass it to the server during `initializeSimulation()`. (If you have read Chapter 19, then this class will be familiar since it essentially mimics the role played by a CORBA `struct`.) Since RMI is adept at moving objects across the network, such a technique would be entirely satisfactory. We have chosen not to do that here because reading data from a `float` array is slightly faster than dereferencing an object. For initialization, the speed difference is negligible, but when the client starts polling the server for data updates and providing new input data with each poll, then performance becomes more important. In a complex simulation, the input array could be substantially larger than the example here. There must be a close coupling between the client and server anyway – the server is computing a known simulation on behalf of the client, after all – so it's not much of a burden to require that the client and server agree on the order of elements in the array, at the slight cost of increased risk of error.

Therefore, instead of a `SpringParameters` object, we receive a simple `float[]` array. Let's now decide on the order of elements. We certainly need the amplitude A, spring constant K, and mass m. From K and m we can calculate the frequency ω. However, it is more convenient to specify A, K, and ω and derive the mass. Since this is a discrete simulation, we also need to know the time step to use. It is also convenient to specify a maximum run time at which point the simulation will end. We arbitrarily choose the order of elements to be A, ω, K, ΔT, and T_{max}. From K and ω we calculate m. So the `initialize()` method becomes

```
public void initialize (float[] indata) {
    fAmplitude  = indata[0];
    fOmega      = indata[1];
    fSpringCons = indata[2];
    fDeltaT     = indata[3];
    fMaxTime    = indata[4];
    fMass       = fSpringCons/(fOmega*fOmega);
}
```

Notice that the size of the array received in the parameter list must be at least five. If it is longer, the server just ignores the rest of the elements. Our agreement is only that the first five elements represent the values of the physical parameters as chosen. We use this fact shortly. (If the input array is shorter than five, there will be an `ArrayIndexOutOfBoundsException` on the server.) Next let's implement the simple `halt()` method to be called by `SimulationThread.halt()`. It merely sets a `boolean` that is checked periodically by the running simulation. In a real-world example, there may be other clean-up tasks that should be performed here, such as closing a connection to a database, etc.

```
public void halt () {
    fKillMe = true;
    // Additional clean-up tasks...
}
```

At sequence 4, upon command from the client, the `Server` object calls `SimulationThread.start()`, which begins the thread's `run()` method, which was implemented above to call `Simulation.startRunning()` at sequence 4.1. Our collaboration diagram shows that `startRunning()` actually passes the call on to the `Compute` object in sequence 4.2. The purpose of the `Compute` object is to separate the actual numerical computation from `SimulationThread`, thereby permitting the use of legacy code via the Java Native Interface (JNI). We discuss JNI in Chapter 22 where we show how to call native methods in an external library. This example uses no legacy code, plus it is simple enough that the entire code can easily fit inside `SimulationThread`, so we won't actually be using a `Compute` object. Therefore, all our numerical work appears inside `startRunning()`.

20.5.3.1 The `Simulation.startRunning()` method

Our calculation loops over time, increasing t by ΔT each step. This loop is a convenient place to check the `fKillMe` instance variable set by the `halt()` method. That is,

```
while (!fKillMe) {
   // calculate everything...
}
```

Before the loop begins, we need an array to hold our results. There are six results returned – current simulation time, position, velocity, acceleration, and kinetic and potential energy. As with the input data array, the client and server must agree on the order of the array elements in the output data array. Again, a better solution might be to define a data-only results class that would make the order unambiguous. However, we stick with the array approach for this example:

```
float[] data = new float[6];
```

The calculations are straightforward. First, we increment an iteration counter and the time value and store the current time in the output array:

```
iter++;
time = iter * fDeltaT;
data[0] = time;
```

The position of the oscillating mass is $x = A \sin(\omega t)$ so

```
data[1] = fAmplitude * (float) Math.sin (fOmega * time);
```

The velocity is $v = \omega A \cos(\omega t)$

```
data[2] = fOmega * fAmplitude *
   (float) Math.cos (fOmega * time);
```

Acceleration is $a = -\omega^2 A \sin(\omega t) = -\omega^2 x$

```
data[3] = -fOmega * fOmega * data[1];
```

We also calculate and store the kinetic and potential energy as a function of time. For this example, they could be calculated on the client side from the returned position and velocity, but we include them here anyway. One is often faced with a decision about how to divide the work load between client and server. In general, if there are good reasons for splitting a calculation into client and server pieces, it usually makes sense to perform as much work on the server as possible, the thinking being that a server machine is often a high-powered machine with more resources than a client.

The kinetic energy is $K = \frac{1}{2}mv^2$ and the potential energy is $P = \frac{1}{2}kx^2$. So

```
data[4] = 0.5f * fMass * data[2]*data[2];
data[5] = 0.5f * fSpringCons * data[1]*data[1];
```

Once these values are calculated, we only need to store them in the `SimData` object:

```
fSimData.storeData (data);
```

Then we return to the top of the `while()` loop and continue with the next iteration.

There are a few minor details to deal with before we close out the `start-Simulation()` method. First, recall that the simulation loads its data into `fSimData` each time step and that the client periodically polls the server for new data. During each client poll, we have arranged for the client to send a new array of input data. For example, the client may wish to change the frequency ω used in the simulation, and we need a mechanism to communicate that change to the running simulation. We haven't yet discussed the `storeData()` method, but the most convenient way for the simulation to obtain a new input data array from the client is as the return value when the simulation calls `storeData()`. Therefore, we modify the previous line of code to make use of that return value. That is, the code becomes

```
new_indata = fSimData.storeData (data);
```

where `new_indata` is a `float[]` array. To avoid confusion, we use the same order of input data array elements as used during `initializeSimulation()`. Perhaps some of those values should not be varied, in which case a completely different array could be used. Instead of defining another array and another agreement about the order of the data values, we simply agree by convention that only some of the array values are subject to be changed. If any of the other values are changed, the server is free to ignore them.

Some of the parameters that could be changed in this simulation are the amplitude, the spring constant, or the frequency. Obviously a real spring would not change its spring constant or natural frequency, but a more realistic simulation might. One application for which this technique has been used simulates blood flow in a human heart. Depending upon external stimuli, the heart rate can change, corresponding to a change in frequency. In that case, the input array of data contains a representation of the external stimuli that the server-side calculation uses. When the input data changes, the simulation responds and the heart rate changes.

For illustration purposes we allow resetting A, ω, and K but do not permit changing ΔT or T_{max}. Therefore we read only the first three values of the `new_indata[]` array. If the `new_indata[]` array is actually longer than three elements, we simply ignore the elements in index positions 3 and 4:

```
// Obtain new amplitude, omega, and spring constant values
fAmplitude  = new_indata[0];
fOmega      = new_indata[1];
fSpringCons = new_indata[2];

// new_indata[3] and new_indata[4] are ignored
```

In this way, we permit the client to use the very same data array in both `initializeSimulation()` and when providing new input data during the running of the simulation.

Suppose the client wishes to cause the simulation to halt before reaching T_{max}. We already have the `halt()` method, but it is not callable by the client. Neither have we provided a remote method in `ServerInterface` that the client can use to instruct the simulation to halt early. One way to add this functionality is to assign special meaning to one of the input data elements. For example, we could interpret a negative spring constant, which makes no physical sense, to mean the simulation should be halted. Such overloading of meaning might make sense in a tightly resource-constrained environment, but doing so is sure to be a source of confusion.

We find it more convenient and less potentially confusing to add one additional element to the new input data array. This additional element is a control parameter that, when set, can be used to terminate the simulation. It must appear in array index 5 since elements 3 and 4 are already defined even though we don't use them. So, by convention, whenever `new_indata[5]` is set to 1, we terminate the loop by calling `halt()`. Note that this convention means that the client must be sure to set the fifth element to 0 when it wants the simulation to continue normally.

```
if ((int)indata[5] == 1) halt ();
```

In this example, we could just as well set `fKillMe` to true directly. However, it is best to call the `halt()` method in order to be sure that any additional clean-up tasks performed by `halt()` are run.

Unless the client takes the special action needed to halt the simulation early, then the simulation should run until `fMaxTime` is reached. Therefore we need to check that our simulation time has not exceeded `fMaxTime` before continuing with the simulation loop.

Our entire simulation code is extremely simple. A real scientific simulation will be much more complex and will require substantial CPU time to calculate. In order to approximate the effects of a more complex simulation, we build in a fake time delay of 100 ms per loop. Thus

```
... calculation section of startRunning()...

  float[] data = new float[6];
  float[] new_indata = new float[6];
  if (time <= fMaxTime) {
    try {
      Thread.sleep (100);
    }
    catch (InterruptedException ie) {/*ignore*/}
    // Continue with calculations ...
    data[0] = time;
    // position x = A * sin (omega*t)
    data[1] = fAmplitude * (float) Math.sin (fOmega * time);
    // velocity v = omega * A * cos (omega*t)
    data[2] = fOmega * fAmp * (float) Math.cos (fOmega *
      time);
    // acceleration = -omega**2 * A * sin (omega*t)
    //              = -omega**2 * x
    data[3] = -fOmega * fOmega * data[1];
    // kinetic energy = 1/2 * m * v**2
    data[4] =.5f * fMass * data[2] * data[2];
    // potential energy = 1/2 * k * x**2
    data[5] =.5f * fSpringCons * data[1] * data[1];
    new_indata = fSimData.storeData (data);
    // Obtain new amplitude, omega, and spring constant
    // values
    fAmplitude  = new_indata[0];
    fOmega      = new_indata[1];
    fSpringCons = new_indata[2];
    // new_indata[3] and new_indata[4] are ignored
    if ((int)new_indata[5] == 1) halt ();
  }
  else {
    // Set time negative indicating that the loop has
    // finished
    data[0] = -1.0f;
    fSimData.storeData (data);
    // Break out of the while loop
    break;
  }
```

Here we have set the time negative when the loop has finished as a signal to the client that the simulation has completed normally.

20.5.4 The `SimData` class

`SimData` is a small but important class. Simulation data is stored in it whenever `storeData()` is called by the `Simulation` object as the calculation runs. `SimData` must also be ready to provide the most recent set of results whenever the server calls `getResults()` responding to a client call to `Server.retrieveData()`.

`SimData` methods can be called by two different threads – the simulation thread, which calls `storeData()` periodically, and the main server thread, which calls `getResults()` upon request from the client. Therefore `SimData` must be careful to avoid thread collisions. In particular, requests to retrieve data must not overlap requests to store data. This situation is easily handled by Java's synchronization features.

`SimData` internally maintains two private arrays – one for input data and one for output data. These are created at construction time.

```
public SimData () {
    fInputData = new float [INSIZE];
    fOutputData = new float [OUTSIZE];
}
```

where `INSIZE` and `OUTSIZE` are constants that give the required sizes of the input and output arrays.

During `storeData()`, the supplied data array from the calculation is copied to the private `fOutputData` array and the latest array of input data is returned:

```
public synchronized float[] storeData (float[] data) {
    System.arraycopy (data, 0, fOutputData, 0, OUTSIZE);
    return fInputData;
}
```

When the server calls `getResults()`, the `fOutputData` array is returned. At the same time, a new input array is received and copied to the private `fInputData` array:

```
public synchronized float[] getResults (float[] indata) {
    System.arraycopy (indata, 0, fInputData, 0, INSIZE);
    return fOutputData;
}
```

In these methods we see another advantage of using arrays of `float`s as data containers rather than custom classes. The use of the optimized `System.arraycopy()` method to copy array elements is very fast. Cloning objects containing structured data is not nearly as fast. Since the getting and setting of the data arrays is a frequently occurring event, the time saved here by using float arrays instead of custom classes can be significant, especially in a simulation that involves hundreds or more data elements.

20.6 Client implementation

One of the purposes of splitting the calculation into client and server pieces is
for the server to do most of the work. For this example, our client can be pretty
simple. We do not need the full client design shown in the client collaboration
diagram of Figure 17.2, but we do need to perform the following steps:

1. Lookup the factory.
2. Get a server instance from the factory.
3. Initialize the server.
4. Initialize the simulation.
5. Start the simulation running.
6. Poll for results.
7. Display the results.

Referring to the client collaboration diagram, we see that these steps comprise
sequences 1, 2, and 4. The additional interactions in sequence 3 are needed to
set up a complex client with output plotting frames and input controls. To avoid
clutter in the client description, we simply print the results to standard output and
dispense with the complicated steps in sequence 3.

The first two steps are straightforward:

```
String factory_server = "//" + host + ":" + port +
  "/" + FactoryInterface.FACTORY_NAME;
fFactory = (FactoryInterface) Naming.lookup
  (factory_server);

// Get a server instance from the server factory.
fServer = fFactory.getInstance (id);
```

where `host` and `port` are pre-defined with the factory server hostname and the
port number, respectively, and `id` is the ID parameter discussed in Section 20.3.1.
Notice that we use the normal RMI-style URL when performing the `Naming.`
`lookup()` and that we use the `FactoryInterface.FACTORY_NAME`
constant to name the factory, as planned earlier in Section 20.3.1. Of
course this code must appear in a `try/catch` block to catch the possible exceptions `MalformedURLException`, `NotBoundException`, and
`RemoteException`.

Next we initialize the server by calling `fServer.initialize`
`(initparam)` where `initparam` is a `String` described in the server implementation above. Since this example does not actually use `initparam`, we
can pass anything at all in its place. As usual we must be prepared to catch
`RemoteException` any time we call any remote method:

```
try {
  fServer.initialize ("doesn't matter");
}
```

```
catch (RemoteException re) {
   ...
}
```

In the complete client collaboration developed in Chapter 17, sequence 3 would involve setting up the client user interface. Since this example has no user interface, we can proceed to sequence 4 where we call `initializeSimulation()` and `start()`. The parameters passed to `initializeSimulation()` do matter since they define the spring constant, amplitude, etc. of the simulation we wish to run. We must follow the convention established on the server for the order of parameters – i.e. A, ω, K, ΔT, and T_{max}. We also defined one additional parameter that, when set to 1, ends the simulation. So we need an input array of six floats where the final value is anything except 1:

```
fIndata = new float[] {
   fAmplitude, fOmega, fSpringCons, fDeltaT, fMaxTime, 0.f
};
fServer.initializeSimulation (fIndata);
```

The actual values of the parameters can be hard coded or can come from command-line parameters passed to `main()`.

Finally we start the simulation with

```
fServer.start ();
```

and begin a polling loop that repeatedly calls `fServer.retrieveData()` until a negative time value is seen.

The reader will recall that the collaboration diagram in Chapter 17 took special pains to start the polling loop before starting the simulation in order to not miss any simulation results. For this simple example that lacks a user interface, we have avoided that complexity. Instead we simply start polling immediately after calling `fServer.start()` and print the results to standard output.

20.7 Enhanced client using the histogram class

For an improved client with a simple user interface, we can use the `Histogram` class developed in Chapter 6. Let's make a histogram of the time the oscillating mass spends in each of a number of bins between $\pm A$. Since this is a discreet simulation, the histogram will not be perfectly smooth, though the smoothness can be increased by choosing a small ΔT so that all time bins are visited nearly equally.

Setting up the histogram is easy:

```
fPositionHist = new Histogram ("position (x)", "", fNumBins,
   −fAmp, fAmp);
fPositionHPan = new HistPanel (fPositionHist);
```

```
JFrame position_frame = new JFrame ("Position Histogram");
position_frame.getContentPane ().add (fPositionHPan);
```

Let's add a `Quit` button to demonstrate the use of the fifth element of the input data array to control the server-side simulation:

```
fReallyQuit = false;
fQuitButton = new JButton ("Quit");
fQuitButton.addActionListener (new ActionListener() {
   public void actionPerformed (ActionEvent ev) {
      if (fReallyQuit)
         System.exit (0);
      fIndata[5] = 1.f;
      fReallyQuit = true;
      fQuitButton.setText ("One more time to exit");
   }
});
position_frame.getContentPane().add (fQuitButton,
BorderLayout.SOUTH);
```

When `fReallyQuit` is false, as it will be the first time the `Quit` button is clicked, then we set the value of `fIndata[5]` to be 1, the special sentinel value that the server uses to halt the simulation. Upon the next poll for new data, `fIndata` is sent to the server, and the server sees the 1 in the fifth position and halts the simulation. We also set `fReallyQuit` to true and change the text of the `Quit` button. The second time the button is clicked causes the client application to exit.

During polling, we load the histogram with new data and cause it to be repainted:

```
outdata       = fServer.retrieveData (fIndata);
time          = outdata[0];
xposition     = outdata[1];
velocity      = outdata[2];
acceleration  = outdata[3];
kinetic       = outdata[4];
potential     = outdata[5];
total_energy  = kinetic + potential;
fPositionHist.add (xposition);
fPositionHPan.repaint ();
```

We can just as easily make histograms of the other quantities as well. They all look similar. Using $A = \omega = K = 1$, $\Delta T = 0.1$ and $T_{\max} = 2\pi$ we obtain the following position histogram when using 21 bins (see Figure 20.2). A better approximation is obtained using $\Delta T = 0.001$ and 51 bins (see Figure 20.3).

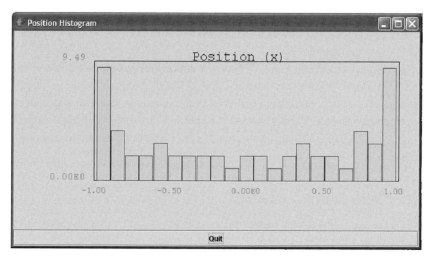

Figure 20.2 Histogram of the position of the spring mass in the simulation for $\Delta T = 0.1$ and 21 bins.

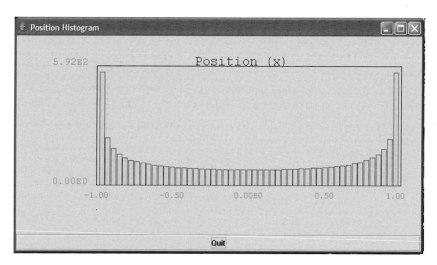

Figure 20.3 Histogram of the position of the spring mass in the simulation for $\Delta T = 0.001$ and 51 bins.

20.8   Conclusion

This chapter has drawn together the client/server design of Chapters 16 and 17 and the power of RMI described in Chapter 18 with a simple example. The code snippets shown here give a good starting point for implementing a server that calculates a general simulation. For a real-world example, the details will vary. The arrays of input and output variables will differ and will probably be much larger. The CPU time on the server to calculate the simulation will certainly be larger. However, the overall structure of the server-side code can remain close to that developed here. The same technique has been used on a variety of time-dependent simulations.

The biggest change in a reader's implementation will surely come on the client side. The clients shown here are very simple, serving only as examples to call the server side code. For a large real-world example, the collaboration diagrams developed in Chapter 17 give a good starting point of how to implement a complete client with a graphical user interface complete with graphical output and input areas.

20.9   Web Course materials

All of the code files for the client, server and simulation classes discussed here are available on the Web Course.

References

[1] JAAS, http://java.sun.com/j2se/1.4.2/docs/guide/security/jaas/
    JAASRefGuide.html.
[2] JCE, http://java.sun.com/j2se/1.4.2/docs/guide/security/jce/
    JCERefGuide.html.
[3] JCA, http://java.sun.com/j2se/1.4.2/docs/guide/security/
    CryptoSpec.html.

Chapter 21
Introduction to web services and XML

21.1 Introduction

Over the last few years, a new distributed computing technology based on the now-ubiquitous World Wide Web and known as *web services* has become very popular. In this chapter we introduce web services and briefly discuss how the technology can be used in a scientific application. We also introduce the closely-related Extensible Markup Language (XML). These are both large subjects. A full treatment is outside the scope of this book, but this chapter is designed to give the interested reader enough basic information to get started and pointers on where to look for more.

21.2 Introducing web services for distributed computing

We have already learned about distributed computing in the abstract sense using UML in Chapters 16 and 17. And we've learned concrete implementations using Java RMI in Chapters 18 and 20 and CORBA in Chapter 19. Both RMI and CORBA are technologies invented to implement distributed computing. One can think of both as transport mechanisms used to move data and, at least in the case of RMI, objects from clients to servers and back. There are other technologies for distributed computing as well, such as the low-level socket based networking technologies discussed in Chapters 14 and 15, and higher-level proprietary single-platform technologies, particularly Microsoft® products, that are not of interest to this book. Web services offer another alternative technology for distributed computing.

The idea is simple. When human users view web pages over the Internet, they are using a computer program – the web browser – that "talks to" a remote web server computer. A human controls the browser by entering URLs or clicking on HTML links, and the browser forwards the human's directions to the web server, which responds accordingly by sending a new web page. In the web services paradigm, a computer program on a client machine "talks to" a remote computer program using the same web technology used by browsers and web servers, but no humans are involved. There is no magic here; both ends of the interaction – the client program and the web server program – must be carefully

```
    <LastName>Turner</LastName>
  </Name>
  <AgeInYears>49</AgeInYears>
  <Sex>Male</Sex>
  <Smoker>No</Smoker>
</LifeExpectancyParameters>
```

Even without understanding the XML syntax, the meaning of each value is manifest. The fact that each value is associated with a "tag" that identifies the data value's meaning is what is meant by the phrase "self-describing data."

The XML tags are reminiscent of HTML tags, but, unlike HTML, the tag names may be anything you want! The <Smoker> tag, for instance, could be changed to <Nonsmoker>, with a corresponding change of the value to Yes, and the meaning would still be clear. The fact that programmers are free to make up their own tag names is the "extensible" part of XML.

Let us now return to the "self-describing" aspect of XML data. As demonstrated above, the XML syntax permits, and even strongly encourages, data to be associated with human-readable tags that identify the meaning of the data values. This tagging is done for the benefit of humans, not computers. Computers are just as happy to receive a string of ASCII or even binary data as long as they are programmed to know the order and meaning of each parameter. There is nothing, for example, in XML that prevents us creating a document like this:

```
<LifeExpectancyParams>Jackson Turner 49 Male
No</LifeExpectancyParams>
```

The server can receive and decode this document and perform exactly the same calculation as with the previous document. The disadvantage of this second XML example is that it is not as meaningful to *humans*, not that it is less meaningful to a computer or more difficult to process. In fact, parsing this second document is probably much easier for a computer than the previous document. And the second document is certainly much shorter, requiring considerably less bandwidth than the first XML document. Shorter still, and even easier for a computer to "parse" is the binary format that would be used by RMI or CORBA. The self-describing aspect of XML documents is for humans, not computers.

However, just seeing the XML document shown above – the good one, with meaningful tags – does little to define what is to be done with the data. In that sense, the data may be somewhat self-describing, but the XML data alone provides no description of how it will be used. There still needs to be some documentation that describes how a server uses the data for there to be true understanding of the meaning of the data itself. Just as important, both client and server must agree on the XML tags to be used. It would not be successful for a client to use the <Nonsmoker> tag while the server is expecting a <Smoker> tag. So there still must be close cooperation among clients and servers.

21.3.1 Tag definitions, DTD, Schema, and XML namespaces

The need for universal agreement between client and servers on the naming and meaning of XML tags is obvious. A server requires an XML document that uses the "correct" tags and that has "proper" values for each tag. Everything in XML is encoded as strings, but a string with a decimal point in it would be inappropriate in a field that is expected to be an integer. Some elements may be required, some optional. As in the example above, some tags can appear nested within others. The nesting can go to any depth, but knowing just what is nested within what is important. It is clear that there must be some way of communicating among users of XML which tags are to be used and how they should nested, what kinds of data values are expected, etc.

There are two standards in use that address the definition of legal and illegal XML tags – the Document Type Definition (DTD) and the newer and more general XML Schema notation. While discussing these is well beyond the scope of this book, both provide techniques to declare which XML tags should appear in a document, which elements are required and which are optional, how they should be nested, etc.

Many industries have created their own specifications of valid XML documents to be used in their particular industry. Some non-profit standards bodies have provided XML definitions suited to various tasks. These "standard" XML formats are defined using either DTD or XML Schema.

A closely related concept is that of XML namespaces, which become important in complex XML documents [4]. To avoid name collisions, particularly when common tag names are used, it is possible to prefix a tag name with the name of the namespace to which it belongs. For a contrived example, suppose there is a namespace for the life expectancy calculation XML document. Then we would prefix each tag name with the namespace name. In practice, to avoid lengthening tag names too much, an abbreviation of the namespace is used. If the abbreviation `life:` is used, then the XML document introduced above would look like this:

```
<life:LifeExpectancyParameters>
  <life:Name>
    <life:FirstName>Jackson</FirstName>
    <life:LastName>Turner</LastName>
  </life:Name>
  <life:Age>49</life:Age>
  <life:Sex>Male</life:Sex>
  <life:Smoker>No</life:Smoker>
</life:LifeExpectancyParameters>
```

Each tag name is prefixed with `life:` which avoids any collision with another tag named, for example, `<FirstName>` elsewhere in the document.

21.3.2 XML parsers

We've only touched the surface of XML syntax. XML documents can become very long and complex. In fact, long XML documents become almost anything but human readable. Only the most determined (and perhaps demented) human will want to read through a complex XML document. The good thing about XML is that the syntax is very well defined and strict, which makes parsing by computer straightforward. Obviously, the casual programmer does not want to have to write code to parse long and ugly XML documents. Thankfully, the computing industry has already provided many XML parsing libraries for a variety of languages, including Java.

There are two basic ways to do XML parsing. One is to read an XML document into the parser and set up the parser so that it makes callbacks to certain handlers each time a tag is seen. This technique is known as the Simple API for XML, or SAX. The other method is to read in the entire document and build a structure in memory that represents the structure of the document. Then the values associated with the various tags in the document can be accessed in a structured way, reminiscent of the way Java handles instance variables in a deeply nested object hierarchy. This scheme is known as the Document Object Model, or DOM [5–7].

There are a variety of implementations of both the SAX and DOM models. Java includes both SAX and DOM parsers as part of the standard Java installation (in the `org.xml.sax` and `org.w3c.dom` packages, respectively). These and other Java XML technologies are grouped together under the term Java API for XML Processing, or JAXP. Space does not permit discussing these tools, but there are good online tutorials [3]. There you will also find optional Java technologies that support web services development. These are discussed in more detail below.

XML parsers provide many benefits, not the least of which is relieving you of having to parse a complex XML document yourself. Parsers can also check a document for conformance to the DTD or Schema that the document should be using. Documents that adhere to the DTD or Schema are said to be conforming documents. Invalid or non-conforming documents can be rejected at the outset of processing by the parser rather than waiting for some hard-to-find error much later while a server is performing operations on the data.

21.4 Java web services

So far we've explained, briefly, what web services are and how XML documents are used. But we have not described how to actually implement a client/server application using the web services paradigm. This section provides an introduction to doing just that. Again, the subject is large and encompasses many technologies that we do not have the space to discuss here. So we just introduce the important features and tools and provide pointers to more information.

Web services are so important that Sun now provides many Java libraries to support web services development with Java. The entire package is available as one (large) add-on to the standard Java installation. The package is called the Java Web Services Developer Pack, or JWSDP, and includes a variety of technologies, only a few of which we discuss here [8, 9].

21.4.1 Java servlet technology

Before web services, there was the World Wide Web, which delivered static HTML documents to web browsers over the Internet using the hypertext transport protocol (HTTP) to send data back and forth between client (i.e. browser) and web server. Soon it was learned that people needed to provide information to web servers, not just read static web pages. HTTP includes the GET and POST methods that provide information to web servers. When you do a web search for instance, you're probably using, behind the scenes, a GET operation. When you fill out a form at an online merchant providing your name, shipping address, billing information, etc. you're almost certainly using POST. All this happens completely transparently to the user.

Then it was realized that web servers needed to be able to generate dynamic content, and thus was born the common gateway interface (CGI), which provided a way for web servers to send data to an external program – often written in perl or C – to perform some calculations and generate HTML pages dynamically.

To resolve some performance problems with CGI, Sun introduced the Java servlet technology with which a web server can quickly execute Java code to generate dynamic content (see also Section 14.8). The Java code that is executed by the web server is called a Java servlet [10]. It must be written in a special way to be a servlet, much like Java applets must be written to extend `java.applet.Applet`. The part of the web server that deals with Java servlets is called a "servlet container." The reference implementation of a servlet container is an open-source product called Apache Tomcat [11]. Tomcat can be used in conjunction with the standard Apache web server or it can be run standalone as both a web server and a servlet container. JWSDP includes support for Tomcat, but Tomcat itself is a separate download from `http://jakarta.apache.org`.

21.4.2 Java servlets for web services

So what do Java servlets have to do with web services? The answer is that perhaps the simplest way to implement a web service is with Java servlets! Servlets process HTTP GET and POST operations. In the most basic web services paradigm, XML documents are sent from the client as the payload of a GET or POST operation, and Java servlets running on the server receive that XML payload. Then, within the servlet, an XML parser processes the payload, feeding the various parameters to calculations to be performed by the server. In a standard servlet, the servlet might generate dynamic HTML content for display as a web page, but there certainly is no requirement to do so. In a web services distributed computing application, the servlet might very well reply to the server with another XML document containing the results of the calculation. No web browser and no human may ever see the output.

21.4.3   Sin (ωt) as a web service

In practice, of course, the calculations are being done for the benefit of some humans somewhere. We could, for example, implement the sin (ωt) calculation of Chapter 20 as a web service. The client would be much the same, except it would send data to the remote servlet as the XML payload of a HTTP GET or POST rather than in an RMI remote method call. Part of that payload would include the name of the function to perform – `initialize`, `receiveInput`, or `retrieveData`. The servlet would use some XML parser to read the input data and perform the calculations. The servlet would also use XML tools to package the results as a reply XML document to be sent back to the client. The client would still use its graphical interface to plot the results for human consumption. Note that no web browser entered the discussion above at all. We simply used XML and the technology behind the Web – i.e. HTTP – to communicate between a client and a server, which, in this case, is implemented as a Java servlet.

21.4.4   Web services for scientific applications

The astute reader might have noticed that XML documents tend to be large, and that passing large documents around the Internet will require high bandwidth, and that building and parsing XML documents will take time, time that might better be spent on the calculation itself. For these reasons, Web services are typically not appropriate for fine-grained calculations such as the sin (ωt) example where there is a large amount of data passed between client and server.

In a large-grained scientific application in which the calculation time is large compared to the data transmission time, web services might be an excellent solution. Web services are straightforward to implement, and the technology is so popular that it is understood by many people.

21.5   Other web services technologies

We do not want to leave the impression that web services are no more complicated than the relatively simple Java servlet technology. In fact, web services are usually thought of as more than Java servlets. There are many other technologies in use in the Web services arena, including wrapping XML documents in Simple Object Access Protocol (SOAP) wrappers, the use of the Universal Description,

Discovery and Integration (UDDI) directory service to register and look up web services, and the Web Services Description Language (WSDL) to define and describe a web service [12–14].

Another very popular technology for implementing web services in Java is the Java API for XML-based RPC (JAX-RPC) [15]. In JAX-RPC, calling a web service is handled much like a remote method call in RMI except the transport mechanism used is HTTP.

We've only scratched the surface of web services. In fact, we've only scratched the surface of Java servlets. The technologies involved are not difficult to understand or use, but they are many. We do not have the space in this book to provide real working examples, partly because setting up and configuring a web server and servlet container such as Tomcat and loading servlets into it is a task that requires much explanation.

The JWSDP provides many of these technologies in one convenient download. There is also a very complete JWSDP Tutorial available. See the references for further information about all these technologies.

21.6   Conclusion

Java clearly offers a diversity of resources and techniques for distributed computing. In Chapter 14 we showed how to create a basic web server with low-level socket code to connect with and send HTML data to browsers and custom clients. Chapter 15 looked at how to use sockets to pass data back and forth directly between custom built servers and clients. In Chapters 16–20 we switched to building programs on the RMI and CORBA frameworks in which objects can invoke methods in other objects across a network just as if the objects were local. With web services we return to a web server type scheme but at the higher level of servlets and XML documents. Each of these approaches to distributed computing has its advantages and disadvantages, and you can choose the one that best serves a particular application.

21.7   Web Course materials

The Web Course provides further discussion of web services techniques and includes an example that illustrates development of a servlet and a custom XML format.

References

[1] Web services information available at www.w3.org/2002/ws, http://java.sun.com/webservices, and http://msdn.microsoft.com/webservices.

[2] XML, www.w3.org/XML.

[3] Java and XML Tutorial, http://java.sun.com/xml.

[4] XML Namespaces, www.w3.org/TR/xml-names11/.

[5] SAX, `http://sax.sourceforge.net`.

[6] DOM, `www.w3.org/DOM`.

[7] JDOM, `www.jdom.org`.

[8] Java Web Services Developer Pack (JWSDP), `http://java.sun.com/webservices/jwsdp`.

[9] JWSDP Tutorial, `http://java.sun.com/webservices/downloads/webservicespack.html`.

[10] Mark Andrews, *Story of a Servlet: An Instant Tutorial*, Sun Microsystems, `http://java.sun.com/products/servlet/articles/tutorial/`.

[11] Apache Tomcat, `http://jakarta.apache.org/tomcat`.

[12] UDDI, `www.uddi.org` and `www.uddicentral.com`.

[13] SOAP, `www.w3.org/TR/soap/` and `http://ws.apache.org/soap`.

[14] WSDL, `www.w3.org/TR/wsdl`.

[15] JAX-RPC, `http://java.sun.com/xml/jaxrpc/index.jsp`.

Part III
Out of the sandbox

Chapter 22
The Java Native Interface (JNI)

22.1 Introduction

Chapter 20 developed a complete client/server application in which both the client code and the server-side code were implemented completely in Java. Pure Java on the client side is desirable for many reasons – platform portability, rich user interface, object-oriented programming environment, ability to run as an application, an applet or a Java Web Start application, etc. For many of the same reasons, a pure Java server offers obvious advantages too. However, for some calculation-intensive processing tasks, particularly if legacy code in another language already exists, it can be advantageous for a Java program to gain access to code written in another language such as C or Fortran.

Java permits calls to code written in languages other than Java. Such external languages are referred to as "native" languages, and the API for accessing them is called the Java Native Interface, or JNI [1]. The decision to use JNI should be made with great care. JNI is designed for use when it is necessary to take advantage of platform-specific functionality that is not available within the Java Virtual Machine. There are two key concepts in that previous sentence – *necessary* and *platform-specific functionality*. To utilize platform-specific functionality via JNI obviously removes Java's platform portability. In fact, any use of JNI at all renders a Java application no longer platform portable since the Java application requires a native shared object library for each platform to which it is targeted. In some cases, such as when the native code is very generic or easily portable, a simple recompile of the native code on a new platform can produce a working shared object library, and thus a working JNI application for that platform. In many cases, though, the native code could be difficult to port, particularly when the purpose of utilizing native code is to gain access to platform-specific features of a particular platform. Porting among various versions of Unix or Linux is normally easy, but porting a library that uses OS-specific functionality to or from a Microsoft Windows platform is often quite difficult.

Because of the loss of platform portability, one should carefully consider whether the desired platform-specific functionality is really necessary to the task at hand and if that functionality is really unavailable in the Java API. As the API has grown, access to almost everything that might be needed from the underlying operating system has become available through standard Java API classes. Another important consideration militating against the use of JNI is that accessing native language features voids many of the security and safety features of Java. Using a native language that permits pointer manipulation, buffer or array overruns, and memory leaks carries many risks.

One oft-cited reason for using JNI is performance – other languages are thought to be "faster." The perceived lack of performance of modern-day Java is often just that – perception – rather than reality. We urge any developer to demonstrate to himself that Java's platform independent performance is insufficient to the task at hand before making the leap to platform *de*pendence and JNI.

A more-defensible reason for using JNI is to gain access to legacy code. Some legacy applications, particularly large scientific calculations, represent many years of effort, development, and debugging. Rewriting and re-debugging such codes in Java could be prohibitively time consuming and expensive. Since legacy codes are likely already tied to a specific platform or operating system, once a design decision is made to use the legacy code in the first place, then the use of JNI probably does not add to the platform dependency of the legacy code. In such cases, using JNI to access a legacy calculation engine in a distributed application makes good sense and can provide the best of both worlds – a rich and portable Java user interface for the client, and a highly developed and debugged calculation engine on the server.

22.2   What is JNI?

Just what is "native" about the Java Native Interface? When using JNI, a Java program makes a method call to a method whose implementation is written in a native language. Such a method is called a *native* method. To the calling Java program, it's just a regular method call, complete with Java primitive or object parameters and/or return types. In fact, the calling code has very little evidence that the method is actually implemented in another language, and the calling code doesn't really care. Nor should it care. Someday the native method could be re-implemented in Java, and the calling program should not have to make extensive changes to use the new implementation.

When a Java program makes a native method call, some special libraries have to get involved to route the native method call to the native language implementation. Those libraries are necessarily both system dependent and language dependent – i.e. they have to be written for the particular operating system and native language in use. Sun's J2SE implementations for Solaris, Linux, and Windows include the JNI libraries for those platforms for the C and C++ languages. Other implementations of J2SE on other platforms must also provide C and C++ JNI libraries for their platform in order for the implementation to be J2SE compliant. So a developer can be sure that any compliant J2SE implementation on any platform will include the required JNI libraries for C and C++.

This chapter explains how JNI is used to call C and C++ code. The examples given are necessarily simple and somewhat contrived since it is not the purpose of this book to describe C/C++ programming, but we provide enough C/C++ boilerplate code to produce working examples. (See references [2,3] for additional examples.)

JNI is often thought of as complicated and difficult. Actually, "verbose" is a more apt description of JNI. While it can be argued that JNI code is "ugly," which makes it appear complicated, the concepts are straightforward, not difficult. Much of the ugliness can be attributed to the fact that JNI is designed to be general enough to work with almost any native language. It is possible, at least in concept, that some vendor could implement JNI libraries for languages other than C and C++, say Smalltalk or Fortran. To our knowledge, no vendor has done so. To gain access to codes written in Fortran, for example, one must use a C or C++ layer as an intermediary. Such C or C++ code is often called *glue code* since it is the "glue" that holds the Java and Fortran code together. While calling any native code from Java using JNI is always done the same way, there is no such standard for calling Fortran from C/C++. Therefore, we do not address here how to write such glue code.

A JNI native method written in C/C++ has full access to Java objects, although the way in which those objects are accessed and modified is necessarily different than how they are used in the Java language. Native methods can obviously call other methods written in that native language and can use any other feature that is available in that language. What may not be so obvious is that native methods can also call Java methods, including standard Java API methods or methods that are part of a custom written Java class. Native methods can create Java objects, including arrays. Native methods can cause Java classes to be loaded and can discover and gain access to fields and methods in those classes. Native methods can throw Java exceptions that can be caught in the Java code. The rest of this chapter describes how to use JNI to take advantage of all these features and more.

While JNI is most often used to call from Java into a native language, it is even possible for a regular non-Java program running on the native operating system to invoke a JVM to gain access to Java classes and features. This use of JNI is referred to as the Invocation API, but we do not discuss it further.

22.3 Hello World in JNI

We begin our study of JNI with a simple example that calls a native method that prints "Hello World" on the console. We first write the Java program that makes the call and then implement the native method in C++. To be callable by the JVM, that C++ code must be written in a very specific way with a very specific signature, about which we learn in this example. To complete this example, we need the following:

- a Java class that calls the native method
- a native method implementation in the proper form
- a runtime-loadable library containing the native method

The native method must reside in a library that can be loaded at runtime by the JVM. For Linux/Unix operating systems, the library must be a shared object library (typically with the .so filename extension). For Windows, the library must be a Windows DLL (.dll extension). We need to know how to write the native method and how to create the shared library in such a way that the JVM can load the library and call the method. The Java class that calls the native method looks much like any other Java class except for two additions: it must declare the native methods it uses as `native`, and it must load the library that contains the native code.

The Java source code that calls a native method looks just like source code that calls any other method, but when the JVM routes that native call to the native implementation, the JVM requires a very specific native function signature – i.e. the function name, the list of parameters and types, and the return type. You must know that signature in order to properly implement the native method in the native language. The signature is derived from the native method declaration in the calling Java class using a very specific set of rules. We describe those rules here but do so through examples rather than a boring listing of the rules. After you have used JNI for a while, you will understand those rules and should be able to derive the native language signature simply by inspecting the native method declaration in the Java source file. For beginners, rather than memorizing and applying the rules, it is far easier to use a tool that generates the required signature. The Java 2 SDK includes the `javah` tool that generates a C-style header file from a compiled Java class that contains native method declarations. Therefore the steps to create and run a JNI application are:

1. Create a Java class that declares the native method and loads the native library that contains the native implementation.
2. Compile the Java source file.
3. Run the `javah` tool to generate a header file for the native method.
4. Create an implementation of the native method using the generated header file.
5. Compile the native implementation.
6. Create a shared object library containing the native implementation.
7. Run the Java class.

These steps must be performed in the order shown. The following sections demonstrate how to perform each step for the simple Hello World example.

22.3.1   Create the Java class

The Java class must declare the native methods it uses. A native method declaration looks much like an abstract method declaration except that it uses the
`native` keyword instead of `abstract`. Like an abstract declaration, a native
declaration ends in a semicolon and has no body. This example:

```
public native void nativeHelloWorld ();
```

declares a public native method named `nativeHelloWorld()` that returns
`void` and takes no parameters. Such a declaration appears inside a class definition
but outside of any constructors or method definitions. Native declarations generally appear at the top of a class definition, along with instance variables, though
there is no requirement that they appear there. There is not even a requirement
that the native declaration appear before it is used, but it must appear somewhere
within the class in which it is used.

Much like `abstract`, the `native` keyword informs the Java compiler that
the method is implemented elsewhere, which prevents the compiler from emitting
an error when it cannot find the implementation. A Java source file that declares
and uses a native method compiles without errors, even though the native method
is not implemented anywhere. In fact, being able to compile a class containing
a not-yet-implemented native method declaration is a good thing, as it allows us
to generate the required header file. (If one attempts to run such a Java class, a
runtime `UnsatisfiedLinkError` occurs at the point where the native method
is called.)

For this example, we create a simple Java class that declares two native methods
and then uses them. We arrange to call one of the two native methods from within
`main()` and the other native method from within the constructor. Recall that
the `main()` method is static, and that static methods cannot call instance (i.e.
non-static) methods. Attempting to do so results in a compile-time error like
this:

```
non-static method nativeHelloWorld() cannot be
referenced from a static context
```

Therefore, the native method that we call from `main()` must be declared to be
`static`. Static methods require a slightly different implementation in the native
language than non-static methods.

Once we create an object instance – i.e. once we call the constructor – then
we can call instance methods. So we also declare a non-static native method
to be called from the constructor. Of course, it is also possible to call static
methods from inside the constructor, but we do not demonstrate that in this
example.

Our Java class is shown below. Ignore the `System.loadLibrary()` call
for now. Its purpose is explained shortly.

```java
package javatech.jni22;
public class JNIHelloWorld
{
   static {System.loadLibrary ("NativeHelloWorld");}

// Declare the two native methods.
   public native void nativeHelloWorld ();
   public static native void nativeHelloWorldStatic ();

   public static void main (String[] args) {
      // Call the static native method.
      System.out.println (
         "main: calling nativeHelloWorldStatic()");
      nativeHelloWorldStatic ();

      // Call the constructor, which will call the
      // non-static method.
      System.out.println (
         "main: instantiating JNIHelloWorld");
      new JNIHelloWorld ();

      // Exit.
      System.out.println ("main: exiting");
   } // main

   // Constructor
   public JNIHelloWorld () {
      // Call the non-static native method.
      System.out.println (
         "ctor: calling nativeHelloWorld()");
      nativeHelloWorld ();
   } // ctor

} // JNIHelloWorld
```

We have named the two native methods with the prefix "native" to make it explicit that they are native methods. We've also named the static native method with the suffix "Static" to distinguish it from the non-static method.

The only other new feature in this source code is the static initializer that calls System.loadLibrary(). The purpose of System.loadLibrary() should be obvious – to load the native library that contains the native method implementations. Recall that a static initializer runs when the class is first initialized. (The details of class initialization are a bit esoteric, as well as almost always unimportant to most Java programmers. It is sufficient to know that the JVM performs initialization before a class is actually used. Therefore, loading the native library in a static initializer ensures that the library is available when needed.) Alternatively, one could load the library later, as long as it is loaded before the first call to a native method is encountered. For this example, the library could be loaded at the beginning of `main()`, just before calling `nativeHelloWorld-Static()`. However, `System.loadLibrary()` is most commonly seen in a static initializer.

The parameter passed to `System.loadLibrary()` is the shared library name, which must correspond to the name of the actual shared library that contains the native code. `System.loadLibrary()` performs the same function on all platforms, but the way in which it does that function is platform-specific. In particular, the name provided is converted to a platform-specific standard shared library name. On a Solaris or Linux system, the input name "NativeHelloWorld" is converted to a library named `libNativeHelloWorld.so`, while a Windows system converts the same input name to `NativeHelloWorld.dll`. Capitalization is preserved in the name conversion on all platforms, which is important to keep note of on a case-sensitive operating system but doesn't really matter on a Windows system.

22.3.2 Compile the Java source file

The compilation of the source file above is simple:

```
javac -classpath build/classes -d build/classes
    src/javatech/jni22/JNIHelloWorld.java
```

As in some previous examples, we separate the generated `.class` files from the source files by using a `build/classes` directory.

22.3.3 Generate the header file

In order to implement the method in a native language, a header file for that native language method is required. As described above, the header gives us the function signature for the native implementation. Because it does a bit more than just that, the header file must be included (`#include`) in C or C++ implementations. The Java 2 SDK supplies the `javah` tool to generate the C header file from a compiled Java class file.

The name of the header file is formed from the fully qualified class name of the class on which it is based. As with other tools, the `-d` switch directs the output to a named directory. Since it is generated output, rather than source code that we write, we put the output into a `headers` subdirectory below the `build` directory:

```
javah -classpath build/classes -jni -d build/headers
    javatech.jni22.JNIHelloWorld
```

This command produces a file named `javatech_jni22_JNIHelloWorld.h` in the `build/headers` directory. Notice the naming scheme in which the package name (`javatech.jni22`) and class name (`JNIHelloWorld`) are used with underscore characters replacing the dots.

Let's examine the generated header file (the formatting has been modified slightly from the actual generated source to better fit the page):

```
/* DO NOT EDIT THIS FILE - it is machine generated */
#include <jni.h>
/* Header for class javatech_jni22_JNIHelloWorld */

#ifndef _Included_javatech_jni22_JNIHelloWorld
#define _Included_javatech_jni22_JNIHelloWorld
#ifdef _cplusplus
extern "C" {
#endif
/*
 * Class: javatech_jni22_JNIHelloWorld
 * Method: nativeHelloWorld
 * Signature: ()V
 */
JNIEXPORT void JNICALL
Java_javatech_jni22_JNIHelloWorld_nativeHelloWorld (
  JNIEnv *, jobject);
  /*
   * Class: javatech_jni22_JNIHelloWorld
   * Method: nativeHelloWorldStatic
   * Signature: ()V
   */
JNIEXPORT void JNICALL
Java_javatech_jni22_JNIHelloWorld_nativeHelloWorldStatic (
  JNIEnv *, jclass);
#ifdef _cplusplus
}
#endif
#endif
```

This may look unwieldy at first (like we said, JNI code is "ugly"), but there are two important lines here – the two that begin with `JNIEXPORT void JNICALL`. These give the signatures of the two native C functions that must be used in the implementation source. We see that each method name declared in the Java source code (`nativeHelloWorld` and `nativeHelloWorldStatic`) gets converted into a long function name beginning with `Java_` followed by the fully-qualified method name, again with dots replaced by underscores. This lengthy naming scheme is followed for all native methods. For a deep package structure, the native function name can grow quite long.

This process of creating a long C-side function name from the Java-side package, class, and method name is referred to as *name mangling* and is used to avoid namespace collisions. The generated long names are called *mangled* names. If one uses overloaded native method declarations, then the mangled native names are longer still, with the overloaded names augmented by a mangled argument signature to distinguish the multiple overloaded function names. Again, the rules for determining the mangled function names are straightforward, but it is easiest to just compile the Java class and run `javah` to generate the header file containing the exact mangled function names needed, especially since the header file is always required anyway.

Even though our native method declarations declared no parameters, the implementation functions actually receive two arguments. This pattern is true of all native methods. There are always two "extra" arguments in the native implementation function followed by the actual parameters declared in the Java class file.

The first argument is always a `JNIEnv` pointer, and the second argument is always either a `jclass` type for static methods or a `jobject` type for instance (non-static) methods. The `JNIEnv` pointer is very important to JNI. (For those unfamiliar with C/C++, the presence of the "*" following the `JNIEnv` type indicates that it is a pointer.) It is through the `JNIEnv` data type that native code implementations gain access to Java objects. We explain how this works in more detail later.

The data types seen above – `JNIEnv`, `jobject`, and `jclass` – along with several others, are defined in the `jni.h` file that is included at the top of the generated header file. The `jni.h` file can be found in the Java 2 SDK installation directory in the *include* subdirectory. Examining the `jni.h` file can be daunting, but if you choose to examine it, you will see that it contains different definitions for C and C++. Therefore the same `jni.h` file can be used for both C and C++.

In addition to definitions of the `jobject` and `jclass` data types, there are also definitions of various other "j" data types that represent C or C++ versions of Java data types – for example, `jint` for Java's `int`, `jfloat` for Java's `float`, and `jstring` for `String`. There is a mapping from each Java primitive type and a few important Java objects (like `String`, `Class`, and `Throwable`) to corresponding "j" data types in C or C++. These "j" types are needed on the C-side to ensure a match between primitive types. For example, a Java `float` is always 32 bits, but a C `float` could be 32 or 64 bits, depending on the underlying platform. Therefore it is unsafe to map a Java `float` directly to a C `float`, but a C-side `jfloat` is guaranteed to always be 32 bits in length, just like a Java `float`. Similarly, a `jint` is always 32 bits long, never 64. Because C doesn't have a primitive Boolean type, Java `boolean` types are mapped to `jboolean` types in `jni.h`, along with the constants `JNI_TRUE` and `JNI_FALSE`. We discuss these data type mappings later, but for this example, we won't need any of the mappings. In fact we won't even need the `jobject` and `jclass` data types.

22.3.4   Create the native implementation

Our fourth step is to create the native implementation. From the generated header file we now know the required signature of the native implementation function. Our implementation must include the header file, and we must provide native function definitions that exactly match the signatures in the header file. Let's implement the non-static method first. Recall that the function signature, obtained from the generated header file, is

```
JNIEXPORT void JNICALL
Java_javatech_jni22_JNIHelloWorld_nativeHelloWorld (
    JNIEnv *, jobject);
```

A C implementation of the method is quite simple. We just use the C `stdio` library method `printf` to print "Hello World" on standard output. Actually, we prefix the string with a few spaces and the word "`<native>`" to distinguish this output from the Java `System.out.println` statements. We also include the phrase "`(non-static)`" to distinguish this output from the output to be generated later by the static method.

```
#include "javatech_jni22_JNIHelloWorld.h"
#include <stdio.h>

JNIEXPORT void JNICALL
Java_javatech_jni22_JNIHelloWorld_nativeHelloWorld (
    JNIEnv *, jobject)
{
    printf ("    <native> Hello World (non-static)\n");
    return;
}
```

The first line `#include`s the generated header file, as is required.

The static method implementation is very similar to the non-static method. Since we don't do anything with either the `jclass` or `jobject` input parameters, the only difference is the function signature. We can put this function definition in the same C/C++ file as the previous function.

```
JNIEXPORT void JNICALL
Java_javatech_jni22_JNIHelloWorld_nativeHelloWorldStatic (
    JNIEnv *, jclass)
{
    printf ("    <native> Hello World (static)\n");
    return;
}
```

We place the source code for these two methods in a file called `NativeHelloWorld.cpp` in our `src/javatech/jni22` directory.

22.3.5 Compile the native implementation

Recall that we instructed `javah` to direct its output to the `build/headers` directory. Therefore we must arrange the C/C++ compiler command line arguments to specify our `build/headers` directory as a compiler include directory. Unfortunately, since C is not platform portable, compiling instructions are different for different platforms. We begin with a Windows example, since most readers probably have access to a Windows system.

We assume that Microsoft Visual C++ is installed on the user's Windows machine. There are, however, other alternatives, including the excellent and free CYGWIN library (see [4]) which includes the GNU C++ compiler as well as many other Unix-like tools.

First, we must enable command-line usage of the C/C++ compiler by running the `vcvars32.bat` script. In a standard Visual C++ installation, that script appears in `C:\program files\microsoft visual studio\vc98\bin\vcvars32.bat`. The command line compiler tool is `cl`.

When compiling, we must specify the location of the needed include files. The `cl` tool uses `/I<dir>` (or `-I<dir>`) to specify directories to add to the include file search path. There are three such include files – the header file generated by `javah`, the `jni.h` file that is included by the generated header file, and a machine-dependent header file that is included by `jni.h`.

We already know that the generated header file is in the `build/headers` directory (known as `build\headers`, with a backslash, on Windows). The `jni.h` file appears, on Windows, in `%JAVA_HOME%\include` where `JAVA_HOME` is the Java 2 SDK installation directory. The machine-dependent file appears in `%JAVA_HOME%\include\win32`. So our command line to compile the implementation source becomes

```
cl -c -Ibuild\headers -I%JAVA_HOME%\include I%JAVA_HOME%\
    include\win32 -Fobuild\objs\NativeHelloWorld.obj
    src\javatech\jni22\NativeHelloWorld.cpp
```

We have used the `-c` switch to specify compile-only and `-Fo<file>` to name the output directory and file name of the compiled `.obj` file. Notice that we have continued the pattern of placing compiled output in the `build` directory by placing the `.obj` file into a `build\objs` directory. That way, it is easy to clean up all generated and compiled output simply by deleting the *build* directory. For this choice to work, we must create the `build\objs` directory before running the above command. (Although the line above is shown as three lines to fit on the page, it must be entered as all one line in a Windows command shell or batch script.)

The result of the command line above is a file named `NativeHelloWorld.obj` in the `build\objs` directory which must be linked to create a Windows DLL file.

On a Sun Solaris system, the compilation command line is

```
cc -c -Ibuild/headers -I$JAVA_HOME/include \
  -I$JAVA_HOME/include/sparc \
  -obuild/objs/NativeHellowWorld.o \
  src/javatech/jni22/NativeHelloWorld.cpp
```

where we have used the "`\`" continuation character to break the long line into four lines.

22.3.6   Create a shared library

To create a Windows DLL file, we use the Windows `link` command line tool, which is part of the Visual C++ installation:

```
link -dll build\objs\NativeHelloWorld.obj
  -out:build\NativeHelloWorld.dll
```

Alternatively, the `cl` tool can be used to do the linking as follows:

```
cl -LD build\objs\NativeHelloWorld.obj -link
  -out:build\NativeHelloWorld.dll
```

Both of these commands place the DLL directly into the `build` directory. The former also puts the associated but unneeded EXP and LIB files in the build directory, while the latter puts the EXP and LIB files in the directory from which it is run.

22.3.7   Run the Java class

The final step is to run the Java class. We start Java in the normal way and specify the `CLASSPATH` and the `javatech.jni22.JNIHelloWorld` class to be run. We also must tell the operating system where to find the shared library file. In Windows, the only search path used is `PATH`, so it is important to add the `build` directory (where we created the DLL file) to the `PATH` environment variable:

```
set PATH=build;%PATH%
java -classpath build/classes javatech.jni22.JNIHelloWorld
```

Alternatively, instead of modifying the `PATH` variable, we can utilize the `java.library.path` system property:

```
java -classpath build/classes -Djava.library.path=build
  javatech.jni22.JNIHelloWorld
```

Both of these commands should produce the following output:

```
main: calling nativeHelloWorldStatic()
    <native> Hello World (static)
main: instantiating JNIHelloWorld
ctor: calling nativeHelloWorld()
    <native> Hello World (non-static)
main: exiting
```

If you try this example and do not see the output shown above, then you've made some mistake. The most common runtime error when running a simple example like this is an `UnsatisfiedLinkError`, which can occur because of two situations. If an error message like this

```
Exception in thread "main" java.lang.UnsatisfiedLinkError:
    no NativeHelloWorld in java.library.path
```

appears, then the system could not find the shared library file, almost surely because of an incorrect `PATH` or `java.library.path` setting. The stack trace should lead to the Java source code line that calls `System.loadLibrary()`.

The other possible reason for an `UnsatisfiedLinkError` is that the specified library was found but a required native method could not be found in that library. This problem can sometimes happen if one has declared many native methods and simply forgotten to implement one of them. Perhaps more likely, the native method signature is not quite letter perfect. In both cases, the error message simply includes the name of the native method that cannot be found. An example would be mistakenly using `jclass` instead of `jobject` when implementing the `nativeHelloWorld` method. Recall that `jclass` is used for static implementations but that `jobject` is required for non-static methods like `nativeHelloWorld`. Since there is no (correct) implementation of the required non-static `nativeHelloWorld` method, the JVM is unable to locate that method when needed and the error message reads

```
Exception in thread "main" java.lang.UnsatisfiedLinkError:
    nativeHelloWorld
```

The stack trace should lead to where the Java source first attempted to call `nativeHelloWorld`. If you're sure the native method being called is present, be sure to check that the signature in your implementation exactly matches the signature in the generated header file.

22.4 Deeper into JNI

The example above served as a useful introduction to JNI but omitted many features. It did not pass any parameters from the Java side to the C side, did not return a value back to Java, did not use the important `JNIEnv` pointer, and made no attempt to access fields or methods of Java objects. It did not do any error checking and did not throw any exceptions. In this section, we go deeper in the details of JNI and explain how to use JNI for more realistic examples. JNI supports all the features just mentioned, and we need to know how to use them.

First, as we explain those features, one may think that JNI is unnecessarily complicated. A somewhat complicated JNI function call is needed, for example, to gain access to Java object fields. Why couldn't the Java designers simply permit the native implementation to access the fields directly as, say, members of a C data structure? In fact, an early (Java 1.0) native method interface did just that but was abandoned in favor of the more general JNI as Java matured. The modern JNI appears complicated because it is important to avoid exposing the JVM's internal layout of Java objects to native code. Doing so would preclude ever changing the internal implementation and also make efficient garbage collection difficult. (See the JNI Specification [1] for more details about the design decisions.)

It is because of this generality that JNI appears complicated. Again, we claim that JNI is not so much complex as it is verbose. Learning the concepts of JNI is not difficult. Applying them requires lots of typing of verbose (and ugly) code. That verbosity tends to make any discussion of JNI equally verbose. We apologize in advance for the apparent complexity of the discussion to follow.

When parameters are passed from Java to a native language, they are not received as Java objects, nor are they received as objects or structures in the receiving native language. The former would be impossible, since a native language has no facilities to deal with Java objects. The latter is somewhat conceivable for an object-oriented native language such as C++, but is impossible for a general JNI design that supports non-object-oriented languages such as C or assembly language. There are two key things to remember throughout the discussion to follow:

- Java objects passed to a native language are not received as normal objects or structures in that language but rather as some intermediate data types.
- There are special JNI functions provided to access fields within those intermediate types and even to convert some intermediate types to native data types.

Because of this need to convert to and from the intermediate data types, the code grows verbose (and ugly).

22.4.1 The Java interface pointer

The access and conversion functions are provided in the JNI library itself. There are many JNI functions – `FindClass()`, `GetMethodID()`, `CallInt-Method()`, `CallStaticFloatMethod()`, to name just a few. These are C and C++ *functions*, not Java methods. The function naming convention does not use the standard Java method naming convention of initial lower case, but rather uses an initial upper case letter.

These C and C++ functions appear in the JNI library. C and C++ native code gains access to these JNI library functions via the `JNIEnv` pointer, which is a pointer into a C/C++ function table. You can, however, think of the `JNIEnv` variable as a pointer into the Java "environment." It is automatically provided to your C/C++ code by the JNI system when the JVM routes a method call from Java to your native code. It is always the first of the two "extra" function arguments that appear in the native function signature.

As an example, let's see how a C implementation calls the `GetMethodID()` function. If we added a call to `GetMethodID()` to our nativeHelloWorld implementation, this JNI function would be accessed as follows:

```
JNIEXPORT void JNICALL
Java_javatech_jni22_JNIHelloWorld_nativeHelloWorld
    (JNIEnv *jenv, jobject jo) {
      jmethodID mid =
        (*jenv)->GetMethodID (jenv, <actual arguments>);
      ...
}
```

Here the Java interface pointer is named `jenv` in the function signature. This `jenv` pointer is used to call the `GetMethodID()` function as shown. Note the double dereferencing (i.e. the use of the construct `(*jenv)->`) and note that `jenv` is also passed as the first argument to the function. (We discuss the `<actual arguments>` later when we describe `GetMethodID()` in more detail.) This code is, admittedly, a little ugly, but it demonstrates how the Java interface pointer is used to call JNI functions from the C language. Any call to any JNI function in C looks similar.

The C++ syntax is somewhat cleaner. The extra level of indirection is unneeded and the interface pointer is not passed as the first argument in the function call:

```
jmethodID mid = jenv->GetMethodID (<actual arguments>);
```

This cleaner way of calling JNI functions is possible because of some magic worked by the `jni.h` header file when using C++. We use C++ in the rest of the examples in this book for convenience.

The interface pointer received in a native call is valid only for the current Java thread. The JVM always passes a valid interface pointer when it makes a native method call, and if there are multiple calls to a native method, the JVM always passes the same interface pointer as long as all those calls are from the same Java thread. However, if a native method is called from different Java threads, the native implementation may receive different JNI interface pointers. A native method implementation must not pass the interface pointer from one Java thread to another. In general, one should just use the interface pointer received from the

JVM whenever inside any native method rather that attempting to save a copy of the interface pointer.

22.4.2 Calling conventions

The JVM always uses the standard library calling convention for a given platform, though that calling convention is likely to be different on different underlying platforms. On Unix platforms, the C calling convention is used; on Windows, the __stdcall convention. Fortunately, this situation is completely transparent to the programmer. The only requirement is that the programmer write the implementation to match the signatures in the generated header file and compile the implementation on the target native platform. A native implementation can be moved to another platform and recompiled without changes to the source code on either the Java side or the native side. The jni.h file on each platform transparently handles the differences in calling conventions between platforms. (Of course, if the native implementation uses platform-specific features, then those features have to be ported to the new native platform.)

Primitive type values – int, float, char, etc. – are copied (passed by value) between the Java and native sides, automatically accounting for field size and byte order through the defined jint, jfloat, etc. types. Java objects, including arrays, are passed by object references, but the object references seen on the C/C++ side are the intermediate data types described in Section 22.4. Therefore, the JNI library provides special JNI functions to access array elements and the fields internal to a Java object. The next two sections deals with Java String objects and arrays in more detail.

22.5 Java string objects

Java String objects are quite different from C strings. For one thing, Java Strings are full-fledged objects with both state and behavior (i.e. data and methods) while C strings are simply char arrays. When the JVM passes a Java String to a native C or C++ implementation, a jstring is received in its place. Just what is a jstring? It is certainly not a C string, and attempting to treat it as one inside the native code likely results in a crash of the JVM. Actually, a jstring is typedef'ed in jni.h to be a jobject and could be manipulated like any other jobject (more about that appears below). However, because strings are so important, JNI provides special functions for converting jstring types into C strings. A similar situation applies to Java arrays, which we discuss after taking a look at the JNI string-related functions.

Recall that Java Strings are stored as Unicode characters, and that C strings are arrays of C char types, which are 8-bit fields. Normally, C strings are composed of just the 7-bit ASCII characters with the eighth bit of the char empty. To convert Java strings into C strings, it is necessary to convert the Unicode string into UTF-8 encoding (see Chapter 9), something that C and C++ can use. UTF-8 maps, and fits, into C's 8-bit `char` representation. Generally, the results are as expected as long as the original string contains only characters in the ASCII range of UTF-8, which is a good idea for anyone using a C or C++ native implementation since those languages don't provide native support for non-ASCII characters anyway. Special care must be taken in the native implementation to deal with non-ASCII strings. The examples in this book always assume ASCII strings.

The main JNI function for working with `jstring` types is `GetString-UTFChars()`, which converts the internal Unicode Java string into a UTF-8 representation. Suppose we have a native method declared as follows:

```
package javatech.jni22;
public class StringExample {
   ...
   public native String nativeProcessString (String s);
   ...
}
```

This native method receives a Java `String`, processes it somehow, and returns a Java `String`. The native method signature is obtained from the generated header file after running `javah`. Thus the beginning of the C/C++ implementation looks like this:

```
JNIEXPORT jstring JNICALL
Java_javatech_jni22_StringExample_nativeProcessString
(JNIEnv *jenv, jobject jo, jstring js) {
   // code to do the processing
   ...
}
```

Note here that the single `String` parameter on the Java side appears as a `jstring` in the third argument on the C side, after the `JNIEnv` pointer and the `jobject`. Also note that the C function is declared to return a `jstring`.

The first thing we must do is convert the received `jstring` to a C string. To do so, the C++ calling sequence is

```
const char* cs = jenv->GetStringUTFChars (js, NULL);
if (cs == NULL) {
   // Handle the error and return immediately
   return NULL;
}
// Use cs in some way ...
...
jenv->ReleaseStringUTFChars (js, cs);
...
```

Here, `js` is the `jstring` received as a function argument that is to be converted. The return value from `GetStringUTFChars()` is `cs`, which is a pointer to an array of UTF-8 characters that can be treated within C/C++ code as a C string. We explain the `NULL` argument passed to `GetStringUTFChars()` later.

It is very important to call `ReleaseStringUTFChars()` when finished with the returned array of characters so the JVM can clean up the storage allocated for `cs`. Otherwise, a memory leak occurs. One cannot assume that the memory allocated for `cs` is freed automatically when the native method goes out of scope as would happen in a true Java method.

It is also important to check the return value from `GetStringUTFChars()`. If sufficient memory could not be allocated to contain the C string, then the function returns `NULL`. The function also throws an `OutOfMemoryError`, but because of the way exceptions are handled in JNI, you must still check for an error return and then return to the calling Java method before the exception is seen on the Java side. We learn more about exceptions in JNI later.

If buffer space to contain the converted string is already pre-allocated, or if it is known that only a substring of the original Java string is needed, then the `GetStringUTFRegion()` function may be used. This function is much like `GetStringUTFChars()` except that its arguments include a beginning index into the string and the number of characters that should be converted. It also requires a pointer to a sufficiently-sized pre-existing buffer in which to place the converted UTF-8 characters. A short code snippet illustrating `GetString-UTFRegion()` is

```
char buffer[21];
jenv->GetStringUTFRegion (js, 5, 20, buffer);
```

This code extracts 20 UTF-8 characters beginning at index 5 (counting from the first character at index 0). The buffer is sized at 21 in order to contain the standard C terminating null character. If the input string does not contain enough characters, then a `StringIndexOutOfBoundsException` is thrown because the function attempts to extract more characters from the `jstring` than exist. If the buffer is not sized large enough, on the other hand, then data is written to memory locations past the end of the buffer, producing unpredictable (and probably bad) results. It is your responsibility to write error-checking code to prevent and/or handle any such abnormal situations. If you are certain that no index overflow can occur, then `GetStringUTFRegion()` is easier to use than `GetString-UTFChars()` because the former does not allocate memory, thereby removing the necessity to call `ReleaseStringUTFChars()` and never raising unexpected out-of-memory exceptions. In practice, `GetStringUTFRegion()` is preferred for small fixed-size strings since the required buffer can be allocated on the C stack very cheaply. Another useful string handling function is `GetString-UTFLength()`, which returns the length in bytes needed to contain the UTF-8 version of a `jstring`.

We note that `GetStringUTFRegion()` was not in the original JNI specification. It was added for Java 2 SDK version 1.2 (commonly called JDK 1.2) meaning it has been around for several years now. Unfortunately, it can be easy to miss the documentation about `GetStringUTFRegion()` and a few other new JNI functions because of the way the JNI documentation is organized. Most of the JNI functions are described in the online docs [1] in the section titled "JNI 1.1 Specification". The enhancements added since Java 1.1 appear in the section labeled JNI Enhancements, below which are links to "JNI Enhancements in version 1.2", and "JNI Enhancements in version 1.4". In order to see all the JNI functions, you need to follow the links to all three documents. (Fortunately, this situation has improved with the J2SE 5.0 document set where all the JNI functions are combined into one document.)

One might also come across some (scant) documentation (see the *JNI Programmer's Guide* [3]) referring to a `SetStringUTFRegion()` function that permits directly setting the characters in a region of a `jstring`. This function would be analogous to the "`Set`" functions for array elements, to be discussed later. However, to the best of our knowledge, such a function does not really exist in JNI.

The Web Course contains a simple string handling example in files called `StringExample.java` and `NativeString.cpp`. The string handling example takes an input string and uses C code to reverse the order of the characters. The reversed string is returned as the return value of the native method. Of course the same thing could be done purely in Java. The example is simply meant to illustrate the use of the JNI string handling functions.

When the native method declaration declares the method return to be `String`, as our `nativeProcessString` example does, the value returned to the calling Java code must be a Java `String`. To create one on the native side, the native code uses the `NewStringUTF()` function which takes a C array of UTF-8 characters as an input argument and creates and returns a `jstring`. When the `jstring` is returned to the calling Java code, it is converted to a Java `String` by the JNI subsystem. The code to create the new `jstring` appears as follows, assuming that the C string `reverse` has already been populated with the reversed-order version of the input string:

```
jstring jreverse = jenv->NewStringUTF (reverse);
return jreverse;
```

Like `GetStringUTFChars()`, `NewStringUTF()` must allocate memory, so it is written to return `NULL` and throw an `OutOfMemoryError` if memory for the new `jstring` cannot be allocated. In general, the returned value should be checked and appropriate action should be taken if it is `NULL`. Here, however, since we immediately return the value obtained from `NewStringUTF()` anyway, there is no explicit need to check for a `NULL` return.

There are a few other JNI string handling functions, particularly `GetStringChars()` and `ReleaseStringChars()`. These are useful for getting string characters in Unicode format instead of UTF-8 characters. Doing so is not relevant to this book and is not discussed further.

22.6 Java primitive arrays

Arrays in Java can be arrays of Java primitives (e.g. `float[]`) or arrays of Java objects. Arrays of Java primitive types are treated similarly to Java strings in JNI. Both are objects on the Java side, both are passed to native methods as object references, and both have special types defined in `jni.h` on the C side for convenience. Object arrays are a bit more complicated than primitive arrays, and we discuss them in the next section.

A Java primitive array of Java `floats` is passed to a JNI method as a `jfloatArray`. There are also `jintArray`, `jbyteArray`, `jdoubleArray`, etc. types for the other Java primitives. Just like a `jstring` is not a C string, neither is a `jfloatArray` an array of C/C++ `floats`. And, just like with strings, JNI provides special functions for converting the various `jxxxArray` types into corresponding C/C++ arrays. One of the special functions for converting `jfloatArray` types is `getFloatArrayElements()`, which returns a pointer to an array of C `floats`. Given a `jfloatArray` called `the_jfloatarray`, we obtain a C array as follows:

```
float* c_array = jenv->GetFloatArrayElements (
  the_jfloatarray, NULL);
if (c_array == NULL) {
  return 0.;
}
// Use c_array in some way, and then
ReleaseFloatArrayElements (the_jfloatarray, c_array, 0);
  ...
```

Here we have converted `the_jfloatarray` into an array of C `floats`. The JNI function `GetFloatArrayElements()` allocates the space for the C array, performs the conversion, and returns a pointer to the new C array. Since the function must allocate enough memory for the C array we must be sure to check the return value for `NULL` and respond accordingly, just like we did with `GetStringUTFChars()` above. After `c_array` has been used, we must be certain to clean up by calling `ReleaseFloatArrayElements()`. Similar code applies for all the other primitive array types.

During garbage collection on the Java side, Java arrays may be moved in memory without warning. The JVM ensures that a garbage collection event does not impact behavior on the Java side. For native code, the JVM also guarantees that the C side array does not move in memory unexpectedly. It does so by either making a copy of the Java array elements for use on the C side, where the copy is not subject to Java garbage collection, or by "pinning" the actual Java array elements in memory. The choice to copy or pin is completely up to the JVM; the programmer has no control over which method is used. If pinned, then calling `ReleaseFloatArrayElements()` is important in order to unpin the array elements in memory so that future runs of the garbage collector are able to move the array as needed. If copied, then calling `ReleaseFloatArrayElements()` is important in order to copy the changed array back to the Java side and avoid a memory leak. In other words, *always* be sure to call the corresponding `ReleaseXxxArrayElements()` function after using one of the `GetXxxArrayElements()` functions.

The careful reader may have noticed that we passed NULL as the second parameter to both `GetStringUTFChars()` and `GetFloatArrayElements()`. That parameter may optionally be a `jboolean` type. Upon return from the array functions, the `jboolean`, if present, is set to `JNI_TRUE` if a copy of the actual Java array elements is returned and `JNI_FALSE` if a pointer to the actual elements themselves is returned – i.e. if the array is pinned in memory. If `JNI_FALSE`, then changes made to array elements appear instantly on the Java side. If `JNI_TRUE`, then any changes to the array do not appear on the Java side until `ReleaseFloatArrayElements()` is called. The same concept applies to the string functions as well, although a copy is almost always made for strings since the native platform is unlikely to have direct support for Java's Unicode character format.

When calling `ReleaseFloatArrayElements()` from C++, there is a `jint` "mode" parameter in the third argument position. The mode may be 0, as in the example above, or one of the constants `JNI_COMMIT` or `JNI_ABORT`. Mode 0, almost always the proper choice, means to copy back the content, if necessary, and release or unpin the memory for the `c_array`. `JNI_COMMIT` means to copy the content but not release the memory, and `JNI_ABORT` means to free the buffer without copying back the possible changes. Both these special values should be used with great care, if at all.

22.6.1 Handling subsets of arrays with `Get` and `Set` region functions

In practice, getting an entire array with one of the `GetXxxArrayElements()` functions can be expensive, especially if a copy is made (and that choice is out of the developer's hands). If only a subset of an array is needed, then JNI provides `GetXxxArrayRegion()` methods, with "Xxx" replaced with `Float`, `Int`, `Double`, etc. Suppose we have a Java `int[]` array that is 1000 elements long. It becomes a `jintArray` on the C side. If we only need to access elements 5 to 14, we can get access to that subset of the large array as follows:

```
jint region[10];
jenv->GetIntArrayRegion (the_jarray, 5, 10, region);
```

Here it is necessary to have a pre-allocated buffer of the required size. Over-running the buffer results in unpredictable results. The function performs array bounds checking on the_jarray, throwing an ArrayIndexOutOfBounds-Exception if one attempts to extract a region that is beyond the range of the original Java array. The Web Course provides an example program that uses Get-IntArrayRegion() to sum elements 5 to 14 of a long integer array. In general the "region" methods are preferred, especially for short arrays, because they do no memory allocation.

Recall that Java arrays contain length information in addition to the array elements. In Java, the length is obtained directly from the length field of the array (i.e. int len = array.length). A useful function for getting the size of a jxxxArray type on the C side is the GetArrayLength() function. This function can be used to ensure that the region buffer for a GetXxxAr-rayRegion() function call is sized large enough. GetArrayLength() takes a jarray as an argument. The jarray type can be thought of as a "super-type" of all the individual jxxxArray types, so any jxxxArray type can be passed to GetArrayLength(), including a jobjectArray, to be discussed below.

There are also SetXxxArrayRegion() functions for each primitive xxx type. These "set" functions permit copying a C-side buffer directly into a region of a jarray. The Web Course provides an example program that uses SetIntArrayRegion() to set elements 5 to 14 of a long Java int array.

22.6.2   Creating new primitive arrays

If the native implementation needs to create a new Java primitive array, then it does so using one of the NewXxxArray() functions. For example to create a new jintArray ten elements long, we use

```
jintArray my_jarray = jenv->NewIntArray (10);
if (my_jarray == NULL)
   return NULL;
```

As this function allocates memory, one should be sure to check for a NULL return value and react accordingly. About the only reason to create a new "j" array on the C side is for return to Java.

22.7   Java object arrays and multidimensional primitive arrays

While it is easy for JNI to provide the jintArray, jfloatArray, etc. types for convenience, there is no way to provide types for arbitrary Java objects on the C side. An array of objects, such as MyCustomObject[], on the Java side gets passed to the C side as a jobjectArray. Unlike with the primitive arrays, there is no way to obtain a C version of the entire object array or even a sub-region of the object array. A language like C or even C++ has no idea what the objects that make up the elements of a Java object array look like or how they behave. Any knowledge of the structure and behavior of Java objects must be written into the native code by the programmer.

Individual array elements from an object array can be accessed and modified with the `GetObjectArrayElement()` and `SetObjectArrayElement()` functions. The return from a "Get" is a `jobject`, and the input to a "Set" is a `jobject` of the proper type. For example, for an object array named `the_jo_array`, we can obtain element number 7 with

```
jsize index = 7;
jobject jo = jenv->GetObjectArrayElement (the_jo_array, 7);
```

This method throws an `ArrayIndexOutOfBoundsException` if the index specified is out of bounds. The return is a `jobject`, a C reference to a single instance of one of the Java objects in the Java-side object array. To set a `jobject` into an object array, use

```
jenv->SetObjectArrayElement (the_jo_array,
    index, the_jo_value);
```

where `the_jo_value` is the `jobject` being inserted into `the_jo_array`.

Before discussing more about `jobjects` and `jobject` arrays, we need to point out that multidimensional primitive arrays in Java are not really primitive arrays. They are implemented in Java as arrays of arrays. Since arrays are really objects, then an array of arrays is really an array of objects – i.e. an object array. If a Java `int[][]` array is passed to C/C++, then a `jobjectArray` is received. To access the array elements on the C side, one must first use the `GetObjectArrayElement()` function to extract one of the underlying single-dimensional arrays. That single-dimensional array is returned as a `jobject` that then can be cast into a `jintArray` and manipulated like any other `jintArray`.

The Web Course includes the `ArrayExample.java` and `NativeArray.cpp` codes that demonstrate handling of 2D `int` arrays.

22.8 Java objects on the C side

We mentioned earlier that native codes have full access to Java objects. This section explains how C/C++ can deal with Java objects. Obviously, JNI cannot have a-priori knowledge of any custom objects that a programmer creates, so those objects cannot possibly be handled the way Java `String` objects are handled with a special `jstring` data type. In theory, JNI could perhaps provide special data types for all the myriad objects in the Java API, though doing so would make JNI unnecessarily huge. Except for a few special cases like `jstring`, `jclass`, and jthrowable, all other Java objects appear on the C side as jobject data types.

A Java object has both state and behavior, or data and methods, and JNI must provide a way to gain access to the data inside an object for manipulation on the C side. In addition, JNI also makes it possible to make method calls to the methods in a Java object. Said another way, not only can Java call C/C++ code using JNI, but C/C++ code can also call Java code using JNI. We look first at accessing the data fields within an object.

22.8.1 The field ID

JNI provides a way for the native language programmer to get and set the values of member variables inside a Java object or class. Both class (static) and instance variables are available. On the Java side, a member variable, also known as a *field*, is identified by its name using the object.field_name syntax. As might be imagined, things are not so simple on the native side. Like everything else in JNI, the getting and setting of fields is done through the use of JNI functions. Doing so is a two-step procedure. First, you must obtain an identifier for the desired field within the Java class. This identifier then serves as a kind of index used to locate the member variable within the class or object.

The identifier is called the *field ID* and is of type jfieldID. Java fields can be of any type, and it is necessary to know both the field name and type signature in the Java class to obtain the corresponding field ID. Once the field ID is had, it is used to get or set the value of the corresponding field. JNI factors out the process of obtaining the field IDs from the process of getting and setting the actual field values so that the field IDs do not have to be recalculated each time they are used, saving time when the same field is accessed multiple times.

Field IDs are obtained with the GetFieldID() and GetStaticField-ID() functions as illustrated here:

```
jfieldID fid = jenv->GetFieldID (cls, <field-name>,
    <type-signature>);
if (fid == NULL)
    return NULL;
```

where <field-name> is a C string naming the desired field, <type-signature> is a C string containing the type signature, and cls is an argument of type jclass that identifies the class the desired field is found in. We explain the cls argument in more detail shortly.

As usual, we check for a NULL return in case of any errors. The exceptional conditions that can be thrown by GetFieldID() are NoSuchFieldError, ExceptionInInitializerError, and OutOfMemoryError. The most common is NoSuchFieldError, usually due to a misspelled <field-name> or incorrect <type-signature>, both of which must be exactly correct. A

`NoSuchFieldError` can also result from passing the wrong `jclass` reference, resulting in a lookup of the correct field name and signature but in a class that does not contain that field.

Like most everything else in JNI, type signatures can appear messy, but they are well-defined and straightforward. It is easiest to introduce the type signature notation by way of an example. Suppose we have a Java class named `JNIDemo` like this:

```
package javatech.jni22;
class JNIDemo
{
    static float a_static_float;
    int some_int;
    int[] array;
    int[][] array2d;
    String some_string;
    MyCustomObject my_custom;

    ...
}
```

The type signatures follow a well-defined and deterministic formula. All the Java primitive types have a single-letter designation – `I` for `int`, `F` for `float`, `D` for `double`, etc. So the first two fields are simple. The type signature of `a_static_float` is "F" and for `some_int`, the type signature is "I".

With this information, we can get the field IDs of the `float` and `int` fields as follows:

```
jfieldID a_static_float_fid =
    jenv->GetStaticFieldID (cls, "a_static_float", "F");
if (a_static_float_fid == NULL) return NULL;
```

and

```
jfieldID some_int_fid =
    jenv->GetFieldID (cls, "some_int", "I");
if (some_int_fid == NULL) return NULL;
```

Arrays of primitives use the same single-letter mapping as the primitive prefixed with a left square bracket. So the signature of the `int[]` array is "[I". You can think of this "[I" notation meaning "array of type I." Thus,

```
jfieldID array_fid = jenv->GetFieldID (cls, "array", "[I");
```

A `float[]` array would obviously be "[F" and the 2D int array (`int[][]` on the Java side) has a type signature of "[[I".

The type signature of an object field is more complicated. It begins with the letter `L` followed by the fully qualified name of the object type with slashes (`/`)

Table 22.1 *JNI Type Signatures.*

Java type	Signature
boolean	Z
byte	B
char	C
short	S
int	I
long	J
float	F
double	D
type[]	[*type*
Class	L*fully-qualified-class;*

replacing dots, and ends with a semicolon. So the `String` field's type signature is "`Ljava/lang/String;`". The type signature for a custom object is similar. It just specifies the custom path to the object instead of a path inside the `java.lang` package. So the type signature of `MyCustomObject` is "`Ljavatech/jni22/MyCustomObject;`". An array of object types is merely prefixed with a left square bracket, just like arrays of primitives.

The complete mapping of Java types to JNI type signatures is given in Table 22.1. Notice that the type signature does not differ for static and non-static fields. That difference is handled by using either the `GetStaticFieldID()` or `GetFieldID()` function, respectively. With this table and the rules given above, one can generate the type signature of any field in any class. However, rather than manually using the table and rules, it is often easier to use the `javap` tool to display the signatures automatically. This tool operates on a compiled class file, and the `-s` option is used to generate type signatures. By default, `javap` generates only the signatures for the default, protected, and public access scope classes and fields. Use the `-private` option to show all classes and fields. With our `javatech.jni22.JNIDemo` class described above compiled into `build/classes`, we launch `javap` as follows:

```
javap -s -private -classpath build/classes
    javatech.jni22.JNIDemo
```

where we fully specify the class name (`javatech.jni22.JNIDemo`) on the command line and also set the classpath to point to the `build/classes` directory. The output from `javap` is

```
1   Compiled from "JNIDemo.java"
2   public class javatech.jni22.JNIDemo extends
    java.lang.Object{
```

```
3   private static float a_static_float;
4      Signature: F
5   private int some_int;
6      Signature: I
7   private int[] array;
8      Signature: [I
9   private int[][] array2d;
10      Signature: [[I
11   private java.lang.String some_string;
12      Signature: Ljava/lang/String;
13   private javatech.jni22.MyCustomObject my_custom;
14      Signature: Ljavatech/jni22/MyCustomObject;
15   public javatech.jni22.JNIDemo();
16      Signature: ()V
17   }
```

where we have added line numbers to aid the discussion. Lines 3, 5, 7, etc. echo
the original lines declaring the fields in the source file. Lines 4, 6, 8, etc. give the
type signatures for each type. This pattern continues for each field until we get
to lines 15 and 16, which give the signature of the class constructor. We discuss
more about method signatures shortly.

22.8.2   The `jclass`

We used a variable called `cls` above when calling `GetFieldID()`. To obtain
a field ID for a field in a particular class, the JNI subsystem obviously needs to
have information about that class. That information is embodied in the `jclass`
type. For static native methods called from Java, recall that the native function
signature includes a `jclass` parameter. Therefore, to access fields within the
class that declares the native method, the `cls` variable is already available.

Suppose our `JNIDemo` class used in the examples above declared and called
a static native method `doStatic()` and a non-static method `doNonStatic()`
as shown here:

```
package javatech.jni22;
public class JNIDemo
{
   // native methods
   public static native void doStatic ();
   public native void doNonStatic ();
   // a class variable
   private static float a_static_float;
   // instance variables
   private int some_int;
   private int[] array;
```

```
    private int[][] array2d;
    private String some_string;
}
```

The native implementation of doStatic() begins

```
JNIEXPORT void JNICALL
Java_javatech_jni22_JNIDemo_doStatic (JNIEnv *jenv,
    jclass cls) {
    ...
}
```

Since the jclass is received in the parameter list, it is already available for any calls to GetStaticFieldID() or GetFieldID().

However, the native implementation of the doNonStatic() method receives a jobject parameter instead of a jclass. The reason should be obvious. Static methods are class methods. They are available even if the class has never been instantiated. Said another way, static methods are independent of any particular instantiation (i.e. object) of the class in which they appear. They have no this variable. When a static method is called, it is called without any object of that class necessarily having been instantiated. But instance methods (i.e. non-static methods) *require* an instantiation of the class (i.e. an object). For this reason, instance methods are sometimes called *object* methods. Each object method is associated with a particular object, and object methods always have an implied this variable. So, while a native implementation of a static method receives a reference to the class in which the native method is declared on the Java side (the jclass parameter), the native implementation of an object method receives a reference to the object from which it was called (the jobject parameter).

Since a jclass reference is needed to call GetFieldID(), receiving the jclass directly in the argument list of the native function that implements a static native method is convenient for finding field IDs. Non-static native methods are not so lucky. Obviously a jobject is closely related to a jclass (the former is an instantiation of the latter), so it should be easy to find the jclass that is associated with the jobject. In fact, it is easy. For a jobject known as jo, the JNI function to get the corresponding jclass is

```
jclass cls = jenv->GetObjectClass (jo);
```

Using this JNI library function, non-static native implementations can find their jclass to use to find field IDs.

It is even possible to find field IDs for classes for which we have neither a jclass nor a jobject reference. We later discuss the FindClass() function that provides this service.

Table 22.2 *JNI functions to get Java field values.*

Java field type	JNI function	Return type
boolean	GetBooleanField()	jboolean
byte	GetByteField()	jbyte
char	GetCharField()	jchar
short	GetShortField()	jshort
int	GetIntField()	jint
long	GetLongField()	jlong
float	GetFloatField()	jfloat
double	GetDoubleField()	jdouble

22.8.3 Getting and setting non-static field values with the field ID

Getting the field ID is only half the battle, since what we really want is the field itself. With a field ID in hand, we can then get and set the values of static fields in a class or non-static fields in an object. We discuss non-static fields first. Let us consider the some_int field of our JNIDemo class. Since it is an int, we get its value with the GetIntField() method:

```
jint my_int = jenv->GetIntField (jo, some_int_fid);
```

The jo parameter is a reference to the object that contains the desired field. Since there may be multiple object instantiations, each with different values for the fields, we must pass in a reference to the object that we're interested in. GetIntField() also requires the field ID of the field we desire. Here we have used the some_int_fid obtained earlier in Section 22.8.1. There are analogous GetXxxField() functions for the other primitives as shown in Table 22.2. Each "Get" method takes a jobject parameter, which must not be NULL, and a valid field ID. Each returns a corresponding "j" data type to the C side as shown in the table.

To set the value of a primitive field, we use the field ID and a "Set" method. For example, if we want to change the value of the JNIDemo.some_int field from a native method implementation, we use code like the following:

```
jint new_value = 7;
jenv->SetIntField (jo, some_int_fid, new_value);
```

There are, of course, "Set" methods for each primitive type.

These built-in JNI methods handle all the Java primitive types. For object types, there are general GetObjectField() and SetObjectField() functions. They apply for standard objects, like the some_string String field of JNIDemo, and for any custom objects the programmer might define. The return from `GetObjectField()` is a `jobject`, which could be cast into a `jstring`, if appropriate.

For most objects, including all custom objects, there is no built-in "`j`" type to cast to. The only way to handle such objects is a nesting of the above procedure. We give a short example here using the `MyCustomObject` type in the `JNIDemo` class. Let's define a very simple `MyCustomObject` class that contains only one `int` field, initialized to the value 13 (the constructor is implied):

```
class MyCustomObject {
   int val = 13;
}
```

Recall that the `my_custom` field of `JNIDemo` is a `MyCustomObject` type. If `jo` is a `jobject` reference to an instance of `JNIDemo`, then we first need to find a `jclass` for `JNIDemo`:

```
// Find the jclass corresponding to the jobject
jclass cls = jenv->GetObjectClass (jo);
```

Then we need the field ID of the `my_custom` field of `JNIDemo`:

```
// Find the field ID of the 'my_custom' field
jfieldID my_custom_fid = jenv->GetFieldID (
   cls, "my_custom", "Ljavatech/jni22/MyCustomObject;"
);
```

With the field ID in hand, we can get a `jobject` reference to the `my_custom` object:

```
// Get the jobject reference to 'my_custom'
jobject my_custom_jo = jenv->GetObjectField (jo,
     my_custom_fid);
```

About the only meaningful thing we can do with this `jobject` is use it to look into the `MyCustomObject` class. First, we get the corresponding `jclass` and then get the field ID of the `val` field:

```
// Find the field ID of the 'val' field of MyCustomObject.
jclass my_custom_cls = jenv->GetObjectClass (my_custom_jo);
jfieldID val_fid = jenv->GetFieldID (my_custom_cls, "val", "I");
```

Now we can obtain the value of the `val` field itself:

```
// Get the 'val' field
jint val = jenv->GetIntField (my_custom_jo, val_fid);
```

and manipulate it somehow:

```
// Manipulate it somehow
val *= 2;
```

We then use `SetIntField()` to set the new value back into `my_custom_jo`:

```
// Set it back into custom_jo
jenv->SetIntField (my_custom_jo, val_fid, val);
```

At this point, the object referenced by `my_custom_jo` has been modified such that the `val` field has a new value. If the original Java class prints out the value of `my_custom.val`, its value will have changed to 26 from the original 13.

The Web Course contains a complete working example of the code snippets shown above in the `JMIDemo.java` and `NativeJNIDemo.cpp` files. Obviously the nesting procedure used above can be used to any depth.

22.8.4   Getting and setting static field values with the field ID

The procedure for dealing with static fields is very nearly the same as for non-static fields. The difference is that there is never a `jobject` variable in use. We only know about the `jclass`. Corresponding to each `GetXxxField()` function for non-static fields there is a `GetStaticXxxField()` function for static fields. The `GetStaticXxxField()` functions take a `jclass` parameter instead of a `jobject`. Otherwise, they are used just like the `GetXxxField()` functions.

For example, to access the static `a_static_float` field, we would do the following:

```
jfieldID fid = jenv->GetStaticFieldID (cls,
   "a_static_float", "F");
if (fid == NULL) return NULL;
jfloat a_static_float = jenv->GetStaticFloatField (cls,
   fid);
```

Of course there are `SetStaticXxxField()` functions as well to set the value of static fields.

22.9   Calling Java methods from native code

We have demonstrated how C/C++ code can use the function arguments received in the native function call and how C/C++ code can locate, get, and set the values of fields within Java objects and classes. It is also possible for native code to call Java methods.

Since Java methods exist only inside of classes, it is obviously necessary to have a reference to the class or object that contains the desired method. Also, the method within that class must be identified somehow. Identification of the desired method is handled with a *method ID*, much like a field ID is used to identify fields. JNI has `GetMethodID()` and `GetStaticMethodID()` functions to locate method IDs of non-static and static methods, respectively. Those functions are used much like the analogous `GetFieldID()` and `GetStaticFieldID()` functions. Instead of a field type signature, they are passed a method signature.

22.9.1   The method ID and method signature

Method signatures use an extension of the notation described in Section 22.8.1 for type signatures. A method signature begins with a parenthesized list of its parameters, each using the type signature notation. After the closing parenthesis appears the return type of the method (in type signature form). There are no commas or other separators between the parameter types. For example, the signature "`(II)F`" describes a method that takes two `int` parameters and returns a `float` – such as `float some_method (int a, int b)`. The letter `V` is used for a void return type. If there are no parameters, then the method descriptor begins with "`()`", not "`(V)`", which is an error since there is no void type in Java. The same "`L`" notation is used for object types and "`[`" notation for arrays. For example, a method that takes an array of Java `String` types as a parameter and returns a `String` is described as "`([Ljava/lang/String;)Ljava/lang/String;`". As with type signatures, the easiest and most error-free way to determine method signatures is to use `javap -s` as described above. If we add the following method to our `JNIDemo` class:

```
int callback (int x) {
   return 2*x;
}
```

that method's signature is "`(I)I`" either by inspection or by examining the output of `javap`. To get the method ID, we use

```
jmethodID callback_mid = jenv->GetMethodID (cls,
   "callback", "(I)I");
```

The function arguments are a `jclass` reference to the class where the method resides, the name of the method, and the method's signature. There is a corresponding `GetStaticMethodID()` function for static methods.

`GetMethodID()` can fail in three ways, the most likely being `NoSuchMethodError` in which no method with the given name and signature can be found in the class referenced by `cls`, most likely due to a mispelled name or a bad signature. The other two possible error conditions are `ExceptionInInitializerError` and `OutOfMemoryError`. If any of these occur, the function returns `NULL` which we must handle in the usual way:

```
if (callback_mid == NULL)
   return NULL;
```

Table 22.3 *JNI functions to call Java methods.*

Java return type	C/C++ return type	JNI function name
boolean	jboolean	CallBooleanMethod()
byte	Jbyte	CallByteMethod()
char	Jchar	CallCharMethod()
short	jshort	CallShortMethod()
int	Jint	CallIntMethod()
long	Jlong	CallLongMethod()
float	jfloat	CallFloatMethod()
double	jdouble	CallDoubleMethod()
Object	jobject	CallObjectMethod()

22.9.2 Calling Java methods using the method ID

With a method ID in hand, we finally can call the Java method from native code. Since the Java `callback()` method we wish to call returns a Java `int`, we use the JNI function `CallIntMethod()` as follows:

```
// Create a jint to pass to the 'callback' method
jint param = 8;

// Call the 'callback' method, placing the return in a jint
// named ret
jint ret = jenv->CallIntMethod (jo, callback_mid, param);

/**** NOTE: exception handling code needed here ****/

// Print the returned value
printf (" return from 'callback' =%d\n", ret);
```

The arguments to `CallIntMethod()` are the `jobject` reference for the object whose method is to be called, the method ID, and then the actual parameters required by the Java method being called. This function can throw any exception that the called Java method can throw. We have omitted the exception handling code until we discuss exceptions in the next section. The `JNIDemo` application on the Web Course demonstrates this technique in a complete working example.

There are corresponding `CallXxxMethod()` functions for each primitive return type and a `CallObjectMethod()` for methods that return Java objects (seen as `jobjects` on the C side). The complete list of "`Call`" functions is given in Table 22.3.

To call static Java methods, there are analogous `CallStaticXxxMethod()` functions. Like the `Get/SetStaticXxxField()` methods, they differ from the non-static versions only in that they require a `jclass` reference to the class in which the static method resides rather than a `jobject` reference.

For each `CallXxxMethod()` function there are two additional functions available that utilize different mechanisms for passing parameters to the Java methods they call. The `CallXxxMethodV()` family of functions pass the parameters in an argument of type `va_list`, and the `CallXxxMethodA()` family of functions pass the parameters in an array of `jvalues`. Type `va_list` is a special type used within the C/C++ language to pass a variable number of arguments. The `printf` family of functions is a well-known example using a variable number of arguments. Type `jvalue` is defined as a union in `jni.h`. In general, the plain functions are the easiest to use. Readers familiar with C/C++ unions and `va_list` arguments might find the "A" and "V" styles useful. The `CallXxxStaticMethod()` functions exist in "A" and "V" styles as well.

22.9.3   Finding classes

In the examples above, we assumed that the `jclass` identifying the class containing the desired method was already available. In practice, we must often first obtain the `jclass` for the desired class. Suppose, for example, that we would like to call a method on the Java `String` class. We first need a `jclass` for `java.lang.String` so we can find the required method ID. The JNI `FindClass()` function serves that purpose, and using it is quite simple. `FindClass()` takes a fully qualified name of a class and searches the `CLASSPATH` to find, and load if necessary, the named class. To find the `jclass` for `java.lang.String` class, we use

```
jclass cls = jenv->FindClass ("java/lang/String");
```

As usual, we need to check for a `NULL` return in case of exceptions, the most likely being `NoClassDefFoundError`. `FindClass()` can find the `jclass` for any class on the `CLASSPATH`, not just Java system classes.

22.10   Exceptions in JNI

We already briefly touched upon Java exceptions in the discussions above. In those cases, the JNI system itself was responsible for raising the exception due to unexpected events such as memory allocation errors or the inability to find field or method IDs. We discuss these system-generated exceptions below. (For simplicity, we refer to all these exceptional conditions as *exceptions* rather than distinguishing between exceptions and errors.) It is also possible for custom code that you write to throw and catch Java exceptions. We describe throwing and catching your own exceptions after we discuss exceptions raised by the JNI system in more detail. In all cases, if an exception is not handled within native code, it eventually propagates back to the JVM once the native code returns.

However, it is not safe to ignore exceptions, expecting the JVM to handle them. It is possible that an unhandled exception in native code could lead to corruption in the native code or in the JNI subsystem itself, leading to a system crash before the native code has a chance to return. Properly dealing with exceptions inside the native code is very important.

22.10.1 Exceptions raised by JNI functions

Whenever a JNI function must allocate memory, there is always the chance that insufficient memory is available from the operating system, resulting in an `OutOfMemoryError` condition. We have also seen an `ArrayIndexOutOf-BoundsException` raised by the array functions when a region or element is requested that is beyond the bounds of the array being accessed, and a few other exceptions from the `GetFieldID()` and `GetMethodID()` functions. In all these cases, the JNI system itself raises the exception.

In the case of the array functions, recall that Java keeps track of array lengths inside an array object. When a native method calls one of the JNI array handling functions, the Java array objects are accessed by (opaque) JNI system code that implements the JNI function. That code takes care of checking the array bounds and throwing the exception, if necessary. Similarly, if the `GetFieldID()` or `GetMethodID()` method cannot find the requested field or method, then JNI system code raises the exception. (When referring to exceptions in JNI we have been using the terms *raises* and *throws* almost interchangeably. There is, however, a subtle difference, which is explained below.)

Most JNI functions are declared to return values of some type. When an exceptional situation arises, most functions can return a special value, usually `NULL`, known as an error code. (This is quite different from Java-side programming in which error code returns are almost never used.) In fact, most JNI functions report an error condition by returning an error code *and* raising an exception. Therefore you can check for the error code value to know whether or not an exception has been raised. The error code for each JNI function is documented in the JNI Specification [1]. If the documented error code is not received – i.e. if a valid value is returned rather than the special error code value – then one can be certain that no exception has occurred.

However, there are a few situations where the JNI system does not return a special error code, meaning that if an exception occurs, the only way you can know about it is by explicitly checking for the presence of an exception. One such situation is when a JNI function is used to call a Java method. In that case, the JNI function must return the result of the Java method. If that Java method throws an exception, there is no special error code value for which to check. Another case is with certain JNI array functions that do not return error codes but instead raise exceptions such as `ArrayIndexOutOfBoundsException` or `ArrayStoreException`.

Native code you write can also raise Java exceptions – either standard exceptions like `java.lang.IOException` or custom exceptions of your own design. Of course the Java definition of the custom exception must already exist on the Java side – i.e. if you want to throw `MyCustomException` from native code, there must already exist a Java `MyCustomException` class somewhere in the `CLASSPATH` on the Java side.

There are two ways to handle exceptions in native code. In the examples shown so far, after detecting an error code return, we have chosen to return immediately to the calling Java method. This return causes the exception to be thrown and handled on the Java side like any Java exception. Note that the exception isn't really thrown to Java until the native code returns. The exception has been *raised* on the native side by JNI, but it is said to be *pending* until the native code returns to the Java side. Once the native code returns, then the JNI system arranges for the pending exception to be *thrown* on the Java side. In that way, returning immediately to Java when we detect the presence of an error code allows the exception to be handled on the Java side.

The second way to handle exceptions in native code is to explicitly check for and handle a pending exception within the native code itself. We describe this technique next.

We emphasize that it is *extremely* important to handle all exceptions by one of these two methods. Continuing to make additional JNI function calls after an exception has been raised can lead to unpredictable results. In fact, the only JNI functions that are safe to call after an exception has been raised are the four special exception handling functions, `ExceptionOccurred()`, `ExceptionCheck()`, `ExceptionClear()`, and `ExceptionDescribe()`. The main JNI exception handling function is `ExceptionOccurred()`, which checks to see if any exception is pending and returns a `jthrowable` if so. It is used as follows:

```
// ...
// some code that might raise an exception
jthrowable jth = jenv->ExceptionOccurred ();
if (jth) {
  jenv->ExceptionClear ();
  // handle the exception...
}
```

The `ExceptionOccurred()` function returns `NULL` if there is no exception. If a `jthrowable` is returned, then there is a pending exception of some kind. If you plan to handle the exception within native code, the JNI system should be told about your plans by calling the `ExceptionClear()` function. Otherwise the exception is left pending and will be thrown on the Java side when the native code eventually returns to Java. The `ExceptionDescribe()` function is used to print a debugging message about the pending exception.

If you want only to check on the existence of a pending exception without creating a local reference to the `jthrowable` exception object, the `ExceptionCheck()` function may be used. It returns `JNI_TRUE` if an exception is pending and `JNI_FALSE` if not.

Often, after detecting an exception in native code, what you really want to do is handle the exception in Java code by throwing your own exception, perhaps with a bit more description about what went wrong. Or sometimes, by careful coding, you might detect an error condition in your own native code even before calling a JNI function that would raise an exception. In these cases, you need to throw your own exceptions back to Java, the subject of the next section.

22.10.2 Throwing exceptions in native code

Suppose we determine, either by detecting an exception thrown by a JNI function we called, or by our own error checking of the arguments received from the Java side, that some of the arguments are invalid. In that case we might want to throw a `java.lang.IllegalArgumentException` back to the Java code.

On the Java side, throwing an `IllegalArgumentException` is simple:

```
throw new IllegalArgumentException ("...");
```

Throwing an exception on the native side is only slightly more complicated. The `ThrowNew()` function is used to throw an exception from native code to Java. This function takes a `jclass` specifying the exception class to be thrown and a string message to include with the exception. Therefore we must first find the `jclass` representing the `IllegalArgumentException`. We can find that `jclass` using `FindClass()`, just like finding any other class:

```
jclass iae_cls =
jenv->FindClass ("java/lang/IllegalArgumentException");
if (iae_cls == NULL)
  // just give up if we can't even find the
  // IllegalArgumentException class
  return;
```

With the desired exception `jclass` in hand, we throw it as follows:

```
jenv->ThrowNew (iae_cls,
  "illegal argument detected in native code");
```

The second parameter to `ThrowNew()` is the message to be used when constructing the `java.lang.Exception` or `java.lang.Throwable` object. On the Java side, J2SE 1.4 added the ability to chain exceptions by specifying a `Throwable` as a *cause* parameter to the `Exception` and `Throwable` constructors. As of this writing, there is no simple way to provide a `Throwable` cause from native code. (The not-so-simple way is to explicitly construct a `Throwable` object and provide it a cause. Then use the JNI `Throw()` function.)

For special needs, you can define custom `Exception` classes in Java that extend `java.lang.Exception`. If a custom exception can be found in the `CLASSPATH`, then `FindClass()` can find it and native code can throw the custom exception just as well as the standard Java library exceptions.

22.11   Local and global references

The JVM must keep track of references to all Java objects passed to native code, primarily so the Java garbage collector does not arbitrarily free an object while it is in use in the native code. There are two basic types of references to objects used by native code: *local* and *global* references.

Local references are valid only during the duration of the native method call. They are freed automatically upon return from native code to Java. Global references exist even after the native method goes out of scope. Global references must be freed explicitly by the programmer when no longer needed.

All objects are passed to native code as local references, and all objects created within the native code by JNI functions such as `NewStringUTF()` are created as local references. This arrangement is normal and expected. When the native method returns, the local references are deleted, permitting the garbage collector to free the memory associated with those objects if needed (and, of course, if there are no outstanding references still in use on the Java side).

For special needs, JNI permits you to create global references from local references with the `NewGlobalRef()` function. If a global reference is created, then it is vital that `DeleteGlobalRef()` be called when the global reference is no longer needed. Otherwise, a memory leak and/or heap fragmentation can occur as the Java garbage collector is never able to free or move the memory associated with the global reference.

A tempting but mistaken tactic is to attempt to cache a `jclass` across native method invocations in order to save the cost of the call to `GetObjectClass()` or `FindClass()`. Under normal circumstances, the `jclass` is a local reference and so it becomes no longer valid after the first native method returns. An attempt to use the cached value on a subsequent call to the native method produces unpredicatable results, possibly including a JVM crash.

A related mistake is to attempt to cache method or field IDs, probably in an attempt to save the cost of calling `GetMethodID()` or `GetFieldID()`. It turns out that method and field IDs are valid only as long as the class from which the ID is derived is not unloaded. After a native method returns to Java code, the Java garbage collector could possibly unload the class to which the IDs refer. If so, then subsequent use of the cached IDs can result in unpredictable behavior.

To solve the latter problem it is safest to re-compute the field or method IDs when needed again. Another solution is to create a global reference to the `jclass` that remains valid even after the native method goes out of scope. Since a global reference prevents the garbage collector from unloading the class, the field and/or method IDs remains valid for subsequent use. Again, we emphasize that it is important to call `DeleteGlobalRef()` when the global reference is no longer needed.

22.12   Threads and synchronization in JNI

As we discussed in Chapter 8, Java is a multithreaded system. Therefore, native methods can conceivably be called by multiple threads. As such, it is your responsibility to be certain that native methods are thread safe. In some situations, you may have special knowledge that there will always be only one thread of control calling a native method. For example, the Java method that calls a native method might be synchronized. Or the native method itself can be declared to be synchronized.

Without such special knowledge, you must ensure that native methods do not modify sensitive global variables in unprotected ways. A few simple rules apply:

- The JNI environment pointer is valid only in the current thread. It must not be passed between threads or cached and used in multiple threads. The JVM always passes the same environment pointer in consecutive invocations of a native method from the same thread. However, different threads pass different interface pointers to native methods.
- Local references are valid only in the current thread. They should not be passed between threads. If different threads need references to the same Java object, use global references.
- Global variables in native code have no intrinsic thread safety. If multiple threads access global variables, you must be very careful to protect such access.

Critical sections of native code can be protected much like critical sections of Java code. In Java, an entire method may be declared to be `synchronized` or, for finer-grained control, a critical block of code can be protected with the `synchronized()` statement:

```
synchronized (some_obj) {
   ... // critical code
}
```

The JVM permits only one thread to enter the synchronized block at a time.

For native code, the analogous JNI functions are `MonitorEnter()` and `MonitorExit()`, used as follows

```
jenv->MonitorEnter (some_jobject);
... // critical code
jenv->MonitorExit (some_jobject);
```

Another option is to use the Java `wait()`, `notify()` and `notifyAll()` mechanisms. JNI does not provide functions to directly support these operations, but, since they are normal Java methods on `java.lang.Object`, they

may be called just like any other Java method using the techniques described in
Section 22.9.

22.13   Conclusion

The Java Native Interface is a vital component of Java. The JVM itself calls native
methods to implement much of the functionality of Java. This has been a long and
dense chapter because JNI is a large subject area. One can do almost anything
in JNI that one can do in Java. Partly because of the generality of the design of
JNI and partly because many native languages are substantially less capable than
Java, the API for accessing JNI functionality is messy. In addition to writing the
native code itself, one must also know how to compile and link the native code
into a shared library object and how to gain access to that object from Java code.

We have described almost all of JNI, though a few advanced features that
you can read about in the JNI Specification have been omitted. The only subject
we have avoided completely is the Invocation API, not because it is any more
difficult to use than the rest of JNI but because it is less likely to be used in typical
scientific programs than the techniques described here for Java code to call native
methods.

The decision to use native methods should be made with great care. As can be
deduced from the length of this chapter, using JNI properly can require significant effort. In addition, as emphasized in the introduction, using JNI necessarily
renders a Java application no longer platform portable. Recent versions of Java
often remove performance concerns as a valid justification for using JNI, while
accessing large legacy computer codes remains a valid reason. We urge you to
be sure that the use of native methods is really necessary before embarking on a
journey into JNI that can often be verbose, messy (and ugly).

22.14   Web Course materials

The code files for the various programs discussed above are available on the Web
Course, as well as some additional examples. We also give an example of linking
Java to a Fortran program via JNI and an intermediate C code.

References

[1] JNI documentation, http://java.sun.com/j2se/1.5.0/docs/guide/jni/.
[2] *Trail: Java Native Interface – The Java Tutorial*, Sun Microsystems,
     http://java.sun.com/docs/books/tutorial/native1.1/.
[3] Sheng Liang, *The Java™ Native Interface Programmer's Guide and Specification*,
     Addison-Wesley. 1999. Available online at http://java.sun.com/docs/
     books/jni/.
[4] CWYGIN, www.cwygin.com.

Chapter 23
Accessing the platform

23.1 Escaping the sandbox

The Java Virtual Machine (JVM) is often said to provide a safe and self-contained *sandbox* where programs such as applets can play without accidentally or deliberately entering restricted areas of the platform. That is sufficient if all a program does is interact with the user via a graphical interface but many programs need to reach out and access the world beyond the JVM to obtain information and interact with external hardware.

In Chapter 22 we showed how Java classes can link to native codes, which possess none of Java's security restrictions. This is the ultimate form of local platform access but it involves a lot of inelegant coding and violates the portability of Java. In this chapter we look at less drastic ways that a program can access the platform. We first show how a program can obtain properties describing the platform such as the operating system, the Java version, and screen size. We then explain how to run a non-Java program from within a Java program. Next we discuss how to use serial ports to communicate with external devices. We include a demonstration program in which a Java application communicates via a serial port with a temperature sensor.

We note that the security restrictions put into place by a browser JVM place severe limits on the access that applets have to the platform and the network. However, as we saw in Chapter 14, the security restrictions on Java applications can be easily customized as needed. (We don't have space here to discuss *trusted* applets that use digital signatures to authenticate their identity. Trusted applets can be given access similar to that for applications.)

23.2 Accessing system properties

The class `java.util.Properties` is a subclass of `Hashtable`, which we discussed in Chapter 10. Java provides `Properties` tables with various system settings in *key/value* string pairs. See the Java API Specification for the `System.getProperties()` method for a listing of the keys, some of which include `java.version`, `java.home`, `os.name`, `user.name`, `user.home`, etc.

Applets are restricted by the `SecurityManager` from accessing many of the properties such as the user's directory information. Here we show a section of code from the program `SysProperties`, which can run both as an applet and as an application. When run as an applet it displays only a fixed subset of the properties.

```
. . . In the class SysProperties . . .

public void start () {
try {
   Properties sysProps = System.getProperties ();
   int i = 0;
   Enumeration names = sysProps.propertyNames ();
   while (names.hasMoreElements ()) {
      String key = (String) names.nextElement ();
      i++;
      print (i +". "+key+ " = " +
      sysProps.getProperty (key) + "\n");
   }
}
catch (Exception e) {
   // If browser security manager throws an exception,
   // then just ask for the following property values:
   String[] key = {
      "java.version",
      "java.vendor",
      "java.class.version",
      "os.name",
      "os.arch",
      "os.version",
      "file.separator",
      "path.separator",
      "line.separator",
   };
   for (int i=1; i<=key.length; i++) {
      print (i+". "+key[i] + " = " +
         System.getProperty (key[i]) + "\n");
   }
}
   . . .
```

If you run this program as an application, you will see a system properties listing similar to the output below (line breaks were added in some items to fit the output within the page here):

1. java.runtime.name = Java(TM) 2 Runtime Environment,
    Standard Edition
2. sun.boot.library.path = C:\ProgramFiles\Java\
    jdk1.5.0\jre\bin
3. java.vm.version = 1.5.0-beta2-b51
4. java.vm.vendor = Sun Microsystems Inc.
5. java.vendor.url = http://java.sun.com/
6. path.separator =;
7. java.vm.name = Java HotSpot(TM) Client VM
8. file.encoding.pkg = sun.io
9. user.country = US
10. sun.os.patch.level = Service Pack 1
11. java.vm.specification.name = Java Virtual Machine
    Specification
12. user.dir = C:\Java\Book\WebCourse\Course\Code\P3\
    Properties
13. java.runtime.version = 1.5.0-beta2-b51
14. java.awt.graphicsenv = sun.awt.Win32GraphicsEnvironment
15. java.endorsed.dirs = C:\ProgramFiles\Java\jdk1.5.0\
    jre\lib\endorsed
16. os.arch = x86
17. java.io.tmpdir = C:\DOCUME~1\User\LOCALS~1\Temp\
18. line.separator =
19. java.vm.specification.vendor = Sun Microsystems Inc.
20. user.variant =
21. os.name = Windows XP
22. sun.jnu.encoding = Cp1252
23. java.library.path = C:\Program Files\Java\
    jdk1.5.0\bin;.;
    C:\WINDOWS\System32;C:\WINDOWS;C:\WINDOWS\system32;
    C:\WINDOWS;C:\WINDOWS\System32\Wbem;
24. java.specification.name = Java Platform API
    Specification
25. java.class.version = 49.0
26. sun.management.compiler = HotSpot Client Compiler
27. java.util.prefs.PreferencesFactory =
    java.util.prefs.WindowsPreferencesFactory
28. os.version = 5.1
29. user.home = C:\Documents and Settings\User
30. user.timezone =
31. java.awt.printerjob = sun.awt.windows.WPrinterJob
32. file.encoding = Cp1252
33. java.specification.version = 1.5
34. user.name = User
35. java.class.path =.

```
36. java.vm.specification.version = 1.0
37. sun.arch.data.model = 32
38. java.home = C:\Program Files\Java\jdk1.5.0\jre
39. java.specification.vendor = Sun Microsystems Inc.
40. user.language = en
41. awt.toolkit = sun.awt.windows.WToolkit
42. java.vm.info = mixed mode, sharing
43. java.version = 1.5.0-beta2
44. java.ext.dirs = C:\Program Files\Java\jdk1.5.0\jre\
    lib\ext
45. sun.boot.class.path = C:\Program Files\Java\jdk1.5.0\
    jre\lib\rt.jar;
    C:\Program Files\Java\jdk1.5.0\jre\lib\i18n.jar;
    C:\Program Files\Java\jdk1.5.0\jre\lib\sunrsasign.jar;
    C:\Program Files\Java\jdk1.5.0\jre\lib\jsse.jar;
    C:\Program Files\Java\jdk1.5.0\jre\lib\jce.jar;
    C:\Program Files\Java\jdk1.5.0\jre\lib\charsets.jar;
    C:\Program Files\Java\jdk1.5.0\jre\classes
46. java.vendor = Sun Microsystems Inc.
47. file.separator = \
48. java.vendor.url.bug =
    http://java.sun.com/cgi-bin/bugreport.cgi
49. sun.cpu.endian = little
50. sun.io.unicode.encoding = UnicodeLittle
51. sun.desktop = windows
52. sun.cpu.isalist =
```

From this output, you can see that the list of system properties includes many useful and interesting items. Everything you might need or want to know is not included there, however. One thing that is missing is a list of operating system or command shell environment variables. Alas, unlike C or C++, there is no getEnv() method in Java. Part of the reason is that Java could conceivably be run on a platform that does not support the concept of environment variables. The expected way to pass environment-variable-like values to a Java application is with the -Dname=value syntax seen a few times in earlier chapters. Using the -D syntax on the java command line effectively adds the specified name and value to the list of system properties. Therefore, if you need to send a system environment variable named SomeEnvVar to your Java code, you can include it on the command line like this:

```
java -Dsome.env.variable=$SomeEnvVar YourClass (Unix/Linux)
or
java -Dsome.env.variable=%SomeEnvVar% YourClass (Windows)
```

Then you access the new system property as follows:

```
String some_value =
   System.getProperty ("some.env.variable");
```

Obviously, you can name the system property anything you want.

Information on the platform display is available from the `java.awt.Toolkit`, which we used in Section 6.9 to obtain images and in Section 12.2 to obtain instances of `PrintJob`. The method

```
int getScreenResolution ()
```

returns the resolution in dots-per-inch. (Unfortunately, on Windows systems the value returned is actually the font size setting rather than the screen resolution setting.) The method

```
Dimension getScreenSize ()
```

returns the width and height of the screen in pixels. For example, you can use this to set the location of a frame at the center of the screen with a method like this:

```
public void center () {
   Dimension screenSize =
      Toolkit.getDefaultToolkit ().getScreenSize ();
   Dimension frameSize = getSize ();
   int x = (screenSize.width - frameSize.width) / 2;
   int y = (screenSize.height - frameSize.height) / 2;
   setLocation (x, y);
}
```

Finally, the colors of the systems GUI are available from the class `java.awt.SystemColor`. For example, you can use it to coordinate the colors in your interface to those of the underlying platform system colors with commands like `myCanvas.setBackground(SystemColor.window)`.

23.3 Running external programs

There are situations in which a Java application needs to run non-Java programs on the platform where it resides. A common example involves opening a browser when the user selects a web link in the interface as might occur, for example, by clicking on a help button. Another example is the case of a server running on a remote device (see Chapters 14 and 15) that starts a non-Java program, e.g. a diagnostic test, at the request of a client.

The `Runtime` and `Process` classes (both part of `java.lang`) are available for launching and communicating with external programs. The following example for a Java program on a Windows system runs the MS-DOS batch file doDir.bat that includes the line

```
dir *.java
```

to produce a directory listing. (A slightly different command is needed on a Unix or Linux system. There are platform portable ways to get a directory listing purely with Java code – see the java.io.File.list() method – but this example serves to demonstrate how to call external programs.) The program RuntTimeApp shown below launches doDir.bat and then reads the output and prints it:

```java
import java.io.*;

/** Demonstrates how to run an external program. **/
public class RunTimeApp
{
  public static void main (String[] args) {
    try {
      Runtime rt = Runtime.getRuntime ();          // step 1

      // Run the external program doDir.bat
      Process process = rt.exec ("doDir.bat");  // step 2

      InputStreamReader reader =                  // step 3
        new InputStreamReader (process.getInputStream ());

      BufferedReader buf_reader =
        new BufferedReader (reader);              // step 4

      String line;
      while ((line = buf_reader.readLine ())!= null)
        System.out.println (line);
    }
    catch (IOException e) {
      System.out.println (e);
    }
  }
}
```

First an instance of the Runtime class is obtained with the factory method:

```
Runtime rt = Runtime.getRuntime ();
```

With the Runtime instance the program is run with the exec() method:

```
Process process = rt.exec ("doDir.bat");
```

This method returns an instance of `Process` that represents the external program. This class provides access to the standard `in`, `out`, and `err` streams with which the Java program can communicate with the external program. In steps 3 and 4, the `Process` object is used to read the output from the external process. Another useful method on `Process` is the `waitFor()` method, which can be used to force the calling Java thread to wait until the external process terminates.

Note that running external programs clearly involves the details of the particular host platform and OS. For example, the `exec()` method does not use a shell. If a shell is needed, then it can be run directly [1].

Running an external program also obviously limits portability. However, in cases where the external program is available on multiple platforms, such as for web browsers, the application could use the system properties (Section 23.2) to determine the platform and then use this information to select a platform specific command.

23.4 Port communications

Many platforms provide a standard serial communications port. A Java application can use a serial port to monitor, control, and record data from external devices. For example, a remote weather station might have several sensors connected to a PC via serial ports. A control program on the PC could collect the data and then use a serial line connection to a modem to dial up a central station and transmit the latest sensor data (or answer a central station that polls remote stations by phone). The `javax.comm` package provides the essential tools to do these tasks. It also supports the parallel port, though this type of port is becoming much less common.

Unfortunately, `javax.comm` does not come as a standard part of the J2SE but as an *optional package*. This means that it is available for several platforms but not for every platform for which a JVM exists. Sun offers versions for Windows and Solaris, and some independent sources provide it for Linux and other platforms [2,3]. Currently there is no standard extension package for other types of communications ports, though some independent sources provide classes to work with the USB port on some platforms [4].

To run the programs discussed here, you will need to download the `javax.comm` set of files to your computer and install them with you JDK files. Instructions are included and we also offer some tips in the Web Course Chapter 23.

We first give an overview of port communications in general and then look specifically at serial port I/O. Although the USB and other faster ports are becoming more popular for desktop peripherals than parallel and serial ports, we expect that the RS-232 serial port will remain common on many devices for many years.

The books *Java I/O* by Elliotte R. Harold [5] and *Java Cookbook* by Ian F. Darwin [6] are popular references for the `javax.comm` package. We refer the reader to these sources for more details about using the parallel port and for aspects of serial port I/O not covered here.

23.4.1   Port classes

There are two types of port classes in `javax.comm`. The abstract class `CommPort` holds methods to control and perform I/O over a specific port. The class `CommPortIdentifier` gives information about the ports on the system and can create an instance of `CommPort` for a given port.

23.4.1.1   CommPort

This abstract class for port representation includes various methods such as `getInputStream()` and `getOutputStream()` to obtain streams to transmit and receive data over a port. The subclasses of `CommPort`, also abstract, include:

- **SerialPort** – this class works with RS-232 ports. Methods in this class provide for control, monitoring, transmission and reception. You can set parameters such as the baud rate, parity, numbers of stop bits and data bits and can choose flow control protocols. Individual control pins, such as DTR (Data Terminal Ready) and CTS (Clear to Send), can be set directly.
- **ParallelPort** – this class works with the 8-bit IEEE-1284 parallel (or printer) port. Methods allow for setting the port mode such as the extended and enhanced modes. Also, the transmission can be suspended and restarted (useful for pausing and restarting a printout), and several status messages can be read such as *paper out* and *printer busy*.

Concrete subclasses of `SerialPort` and `ParallelPort` are available in `javax.comm` for each particular platform. You do not create instances of these classes directly from their constructors but from factory methods in the `CommPortIdentifier` class.

23.4.1.2   CommPortIdentifier

An instance of this class, discussed further in the next section, provides information about a specific port, such as the port's name and type, but it does not set ownership of the port or allow for any control or I/O over the port. Instead, you use the `open()` method of an instance of `CommPortIdentifier` for a specific port to obtain an instance of a `CommPort` subclass for that port.

   As with the `SerialPort` and `ParallelPort` classes, you do not normally obtain instances of `CommPortIdentifier` directly from a constructor but with a factory method in `CommPortIdentifier`.

23.4.2   Finding ports

The `CommPortIdentifier` class acts as both a source of information about ports on a system and also as a descriptor of a particular port. The static method

getPortIdentifers () provides a list of CommPortIdentifier objects for
each port, serial and parallel, on the platform:

```
Enumeration portList =
   CommPortIdentifier.getPortIdentifiers ();
```

This enumeration lists the instances of CommPortIdentifier, one for each
port. The methods of this class provide information about the particular port such
as its name and type via getName () and getPortType (), respectively, as the
PortList example below illustrates:

```java
import javax.comm.*;
import java.util.*;

/** List all the ports available on the local machine. **/
public class PortList
{
  public static void main (String[] args) {

    Enumeration port_list = CommPortIdentifier.getPortIdentifiers ();

    while (port_list.hasMoreElements ()) {
      CommPortIdentifier port_id = (CommPortIdentifier) port_list.nextElement ();

      if (port_id.getPortType () == CommPortIdentifier.PORT_SERIAL) {
        System.out.println ("Serial port: " + port_id.getName ());

      }
      else if (port_id.getPortType () == CommPortIdentifier.PORT_PARALLEL) {
        System.out.println ("Parallel port: " + port_id.getName ());
      }
      else
        System.out.println ("Other port: " + port_id.getName ());
    }
  } // main
} // class PortList
```

We used the constants PORT_SERIAL and PORT_PARALLEL from the
CommPortIdentifier class to test for the port type. On a desktop machine
with four serial ports and two parallel ports, the output would go as:

```
Serial port: COM1
Serial port: COM2
Serial port: COM3
Serial port: COM4
Parallel port: LPT1
Parallel port: LPT2
```

Conversely, you can test for the presence of a port with a particular name, as follows:

```java
import javax.comm.*;
import java.util.*;

/** Look for COM# ports on the local machine. **/
public class PortTest
{
  public static void main (String[] args) {
    String port_name;
    int i = 0;
    while (true) {
      i++;
      port_name = "COM" + i;
      try {
        CommPortIdentifier port_id =
          CommPortIdentifier.getPortIdentifier
            (port_name);
        System.out.println ("Port " + port_name +
            " exists");
      }
      catch (NoSuchPortException e) {
        System.out.println ("No port " + port_name);
        break;
      }
    }
  } // main
} // class PortTest
```

The output of this program might go as:

```
Port COM1 exists
Port COM2 exists
Port COM3 exists
Port COM4 exists
No port COM5
```

The following methods in `CommPortIdentifier` provide information about the status of a port:

- `boolean isCurrentlyOwned()` – indicates if another Java application owns the port
- `String getCurrentOwner()` – a description of the Java application that owns a port

Unfortunately, these methods only work properly if the port is owned by a Java application and not by some other non-Java program. However, if a port is already in use by a non-Java program, then an attempt to open it reveals that situation, as explained next.

23.4.3 Opening ports

If a port is available, you take exclusive possession of it via one of the two overloaded `open()` methods and then use the port for reading and writing to the external device connected to that port. The `CommPortIdentifier` method

```
CommPort open (String ownerName, int timeout)
   throws PortInUseException
```

takes possession of the port and passes it the name of the owning application. The timeout parameter determines how long in milliseconds the method will block while waiting for the port to become available.

For systems such as Unix where ports can be assigned a `FileDescriptor`, the following overloaded `open()` method is provided:

```
CommPort open (java.io.FileDescriptor fd)
   throws UnsupportedCommOperationException
```

Below we show an example where ports are opened. If they are already owned by some other application, a `PortInUseException` is caught. If the owner is a Java program, it can be identified by the name given in the `open()` method; otherwise there is no name available.

```java
import javax.comm.*;
import java.util.*;

/** Check each port to see if it is open. **/
public class PortListOpen
{
  public static void main (String[] args) {

    Enumeration port_list = CommPortIdentifier.getPortIdentifiers ();

    while (port_list.hasMoreElements ()) {
      // Get the list of ports
      CommPortIdentifier port_id = (CommPortIdentifier) port_list.nextElement ();

      // Find each ports type and name
```

```
        if (port_id.getPortType () == CommPortIdentifier.PORT_SERIAL) {
          System.out.println ("Serial port: " + port_id.getName ());
        }
        else if (port_id.getPortType () == CommPortIdentifier.PORT_PARALLEL) {
          System.out.println ("Parallel port: " + port_id.getName ());
        } else
          System.out.println ("Other port: " + port_id.getName ());

        // Attempt to open it
        try {
          CommPort port = port_id.open ("PortListOpen",20);
          System.out.println (" Opened successfully");
          port.close ();

        }
        catch (PortInUseException pe) {
          System.out.println (" Open failed");
          String owner_name = port_id.getCurrentOwner ();
          if (owner_name == null)
            System.out.println (" Port Owned by unidentified app");
          else
            // The owner name not returned correctly unless it is a Java program.
            System.out.println (" " + owner_name);
        }
      }
  } // main
} // class PortListOpen
```

Output from this application might go as:

```
      Serial port: COM1
        Opened successfully
      Serial port: COM2
        Opened successfully
      Serial port: COM3
        Open failed
        Port currently not owned
      Serial port: COM4
        Open failed
        Port currently not owned
      Parallel port: LPT1
        Opened successfully
      Parallel port: LPT2
        Open failed
        Port currently not owned
```

To monitor a port for changes in its ownership, you can implement the `CommPortOwnershipListener` interface. You must override the method `ownerShipChange (int typeChange)`, which can detect three types of ownership changes: ownership attained, port now available, and ownership requested.

23.4.4 Port communications

The procedure to communicate over a port involves the following operations:

1. `open()` in the port's `CommPortIdentifier` provides a `CommPort` object for the port.
2. `getInputStream()` and `getOutputStream()` in `CommPort` provide streams for reading data from and writing data to the port.
3. Use these streams to carry out the desired I/O.
4. `close()` in `CommPort` releases the port for other applications to use.

Other `CommPort` methods can set parameters for the port communications such as the I/O buffer sizes and how long the read operation will wait for data before returning.

23.4.5 Serial port I/O

The `javax.comm` serial port classes assume the system has one or more ports following the RS-232 (or EIA232) standard. Table 23.1 lists the pins for the DB9 connector that follows this standard. RS-232 originated in the 1960s and dealt with a computer talking to a display terminal. This historical basis explains some of the names for the 9 pins on the connector. (See [7] for a description of the less common 25-pin DB25 connector.)

Serial ports send and receive one bit at a time using a positive voltage (between 3 and 25V) to indicate a 0 bit and a negative voltage (between -3 to -25V) to indicate a 1 bit. The duration of a voltage depends inversely on the *baud rate*.

An asynchronous serial protocol (not part of the RS-232 standard) is required to determine how to decode the bits into bytes. The standard protocol groups the bits into a *standard data unit* (SDU) consisting typically of either 7 data bits (for ASCII) or 8 data bits. To indicate where a SDU begins, a start bit value of 0 is sent. To indicate the end of a SDU, the group ends with one or two *stop bits*, each of value 1. To combat noise and bit errors, a *parity bit* is usually included. For *even* parity the bit value is 1 if the number of one bits in the SDU is even and zero if the number is odd. For *odd* parity the bit value is 1 if the number of one bits is odd and zero if even. So, depending on the protocol settings, the total number of bits sent for an SDU can vary from 9 bits (7 data bits, 1 stop bit and no parity) up to 12 bits (8 data bits, 2 stop bits and a parity bit).

Table 23.1   *Pin assignments for the DB9 serial connector.*

Pin	Name	Abbreviation	Direction	Function[1]
1	Carrier Detect	CD	In	Modem & destination modem connected
2	Receive Data	RD	In	Data from modem
3	Transmit Data	TD	Out	Data to modem
4	Data Terminal Ready	DTR	Out	Computer ready to send & receive
5	Ground	GND	Common	–
6	Data Set Ready	DSR	In	Modem ready to send & receive
7	Request to Send	RTS	Out	Computer waiting to send
8	Clear to Send	CTS	In	Modem ready to receive
9	Ring Indicator	RI	In	Modem says phone is ringing

[1] The pin function descriptions here are for the case of a computer connected to a modem.

The `SerialPort` class provides a single method to set the baud rate, number of data bits, number of stop bits, and the parity:

```
void SetSerialPortParams (int baud, int dataBits, int
    stopBits, int parity)
```

If a port does not support any of the values passed in the parameters, the method throws the `UnsupportedCommOperationException`.

The `SerialPort` class includes a set of constants to use for these parameters, such as:

```
DATABITS_7
DATABITS_8
STOPBITS_1
STOPBITS_2
PARITY_NONE
PARITY_EVEN
PARITY_ODD
```

The two devices connected via the serial line need *flow control* settings to determine who is sending and who is receiving and when to switch between the two states. The `XON/XOFF` and `RTS/CTS` are the two primary protocols for this. The former is a software protocol while the latter is implemented in hardware. The methods

```
void setFlowControlMode (int protocol)
int getFlowControlMode ()
```

provide for setting and getting the flow control mode. The protocol value sets both input and output protocols with a bitwise AND of the constants in the SerialPort class:

```
FLOWCONTROL_NONE
FLOWCONTROL_XONXOFF_IN
FLOWCONTROL_XONXOFF_OUT
FLOWCONTROL_RTSCTS_IN
FLOWCONTROL_RTSCTS_OUT
```

The "set" method throws the UnsupportedCommOperationException if the protocol value is invalid.

23.4.6 Serial line connections

In Table 23.1 the *Function* column indicates the purpose of each pin for the case of a computer connected to a modem. The serial line sends and receives only one bit at a time but it uses separate lines for transmission and reception and it uses six other lines (not counting the ground line) for setting up the protocol for the communications.

The class CommPort provides methods to access the six control lines. For the two output control wires, DTR and RTS, there are methods both to set the line and to find the current setting:

```
void setDTR (boolean val)
boolean isDTR ()
void setRTS (boolean val)
boolean isRTS ()
```

For the other four input control lines, there are methods to find their current state:

```
boolean isCTS ()
boolean isDSR ()
boolean isRI ()
boolean isCD ()
```

For connecting to devices other than modems, it is often unnecessary to use all of these control lines. In the serial line demonstration program below only the CTS and RTS (plus the ground line) lines are active.

Note that when you start to set up serial connections, you will find that the cables and connectors vary according to whether you connect a computer to a device like a modem or to another computer. Table 23.1 shows the pins for the connector at the computer. There is a one-to-one correspondence in the numbering of the pins on the computer's male connector and the modem's female connector. That is, pin 2 on the computer connects to receptor number 2 on the modem's connector, pin 3 connects to receptor 3 on the modem, and so forth. This obviously cannot hold for connecting two computers that both have male connectors as in the table. Instead a so-called *null modem* cable is required. It connects pin 2 (RD) on computer A to pin 3 (TD) on computer B, connects pin 3 (TD) on computer A to pin 2 (TD), and so forth. See the book by Strangio for the specifications and diagrams of various types of connectors and cables [7].

23.4.7   Serial port demo

The demonstration application uses a serial connection to obtain temperature readings from a device connected to the local platform via a serial port. The program uses the classes and methods discussed above for communications over the serial port. This program can provide a template for obtaining data from any device connected to a serial line. The program can be adapted to work with the client/server system discussed in Chapter 15 and so provide a complete system for accessing a remote device over the Internet.

The device is in fact a Java hardware processor board that we discuss in Chapter 24. We have programmed it to follow a simple protocol for sending data. This protocol requires that the application on the desktop first send a "password" (actually a 2-byte numerical value) to the device. If this is accepted, the device program then sends a 2-byte integer data value for the temperature in units of 0.5 degrees Celsius.

The desktop (or any platform that can handle J2SE and `javax.comm`) program consists of two primary classes:

- **GetJavelinData** – this class is in charge of sending and receiving data over the serial port with the device (a Javelin Stamp board discussed in Chapter 24). This class implements the `Runnable` interface and its `run()` method contains a loop that first sends the password and then reads the value sent to it from the sensor. It converts each byte pair received into a 4-byte `int` value using the techniques discussed in Chapter 9 to convert byte array values to primitive types. Before repeating the data read operation, the loop pauses for a period whose length is passed via the constructor parameter.
- **SerialToJavelin** – this class provides a graphical user interface with a dropdown menu on the menu bar that lists the serial ports on the platform. Selecting one of these causes the program to attempt to open the port. If it succeeds, it passes the port's `SerialPort` object to an instance of `GetJavelinData`. Hitting the `Go` button invokes the `start()` method in the `GetJavelinData` object, which begins its data taking loop. Hitting `Stop` invokes `stop()` in that object. A text area displays the temperature values which are printed from the `GetJavelinData` loop via print methods of the `Outputable` interface. A file menu provides for saving this data to file.

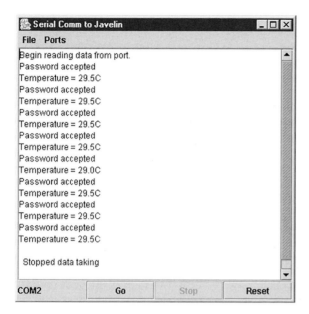

Figure 23.1   The graphical user interface for the `SerialToJavelin` program after connecting via port `COM2` to the Javelin Stamp card (see Chapter 24) and making several temperature readings.

The full code for these two classes can be found in the Web Course Chapter 23. The essential code sections are discussed here. Note that we use the technique discussed in Chapter 6 in which `SerialToJavelin` implements the `Outputable` interface and provides a text area to display strings from the `print()` and `println()` methods rather than on the console. Figure 23.1 shows the user interface after the program has received several temperature readings from the Javelin.

When the program `SerialToJavelin` first begins, it builds a menu with the names of the serial ports on the platform. It invokes the method `getPorts()`, shown below, which uses the `CommPortIdentifier` static method `getPortIdentifiers()` to obtain the `CommPortIdentifier` object for each port on the platform. Those that are serial port types are saved in a `Hashtable`.

```
. . . The getPorts() method in the class SerialToJavelin
      . . .

   /**
    * Use the CommPortIdentifier static method to obtain
    * a list of ports on the platform. Pick out the
    * serial ports from the list and save in a Hash
    * table. Use the port names as keys.
    */
```

```
   static void getPorts () {
     // First get the list of ports on this platform
     Enumeration port_list =
        CommPortIdentifier.getPortIdentifiers ();

     // Scan through the list and get the serial ports
     while (port_list.hasMoreElements ()) {
       CommPortIdentifier port_id = (CommPortIdentifier)
         port_list.nextElement ();

       if (port_id.getPortType () ==
           CommPortIdentifier.PORT_SERIAL) {
         fNumSerialPorts__++;
         fSerialTable__.put (port_id.getName (), port_id);
       }
     }
   } // getPorts
```

When the user selects a serial port from the menu, the `actionPerformed()`
method is invoked. The following code shows the actions taken to open the port:

```
. . . The actionPerformed() method in the class
    SerialToJavelin . . .

public void actionPerformed (ActionEvent e) {

    . . . tests for other commands . . .

    // Scan the serial ports names to look for a match
    // to the menu items.
    Enumeration enum_ports = fSerialTable__.keys ();
    while (enum_ports.hasMoreElements ()) {
        String port_name = (String) enum_ports.nextElement ();

        if (command.equals (port_name)) {
          fSelectedPortID__ =
             (CommPortIdentifier) fSerialTable__.get
               (port_name);
          if (fCurrentPort__ != null) fCurrentPort__.close ();

          // Open port. Allow for 20 seconds block.
          try {
            fCurrentPort__ =
               (SerialPort) fSelectedPortID__.open (
```

```
                              "Serial to Javelin", 20000);
            }
            catch (PortInUseException pie) {
               println ("Error: Port in use");
               fCurrentPort__ = null;
               fSelectedPortID__ = null;
               return;
            }

            // Set up the serial port
            try {
               fCurrentPort__.setSerialPortParams (
                  BAUD_RATE,
                  SerialPort.DATABITS_8,
                  SerialPort.STOPBITS_1,
                  SerialPort.PARITY_NONE
               );
            }
            catch (UnsupportedCommOperationException uce) {
               // This error shouldn't happen
            }

            try {
               getStreams ();
            }
            catch (IOException ioe) {
               println ("Error: Cannot open streams to port");
               fCurrentPortLabel.setText ("No Port Open");
               fCurrentPort__.close ();
               return;
            }
            fCurrentPortLabel.setText (port_name);
            fGetJavelinData.setStreams (fPortInStream,
                                        fPortOutStream);
            fGoButton.setEnabled (true);
         }
      }
   } // actionPerformed
```

The code first looks for a match to the port name. It then attempts to open the
port. If it succeeds it sets the serial port parameters to a baud rate of 9600, eight
data bits, one stop bit, and no parity. The serial line device must, of course, use
the same settings (we show how to program the Javelin Stamp evaluation card in
Chapter 24).

The input and output streams for the serial port are then obtained. The input stream is wrapped as a `DataInputStream` and the output stream is wrapped with a `PrintStream`.

```
. . . The getStreams() method in the class SerialToJavelin
     . . .


/** Open the input and output streams **/
void getStreams () throws IOException {
  fPortInStream =
    new DataInputStream (fCurrentPort_.getInputStream ());
  fPortOutStream =
    new PrintStream (fCurrentPort_.getOutputStream (),
      true);
}
```

When the user hits the `"Go"` button, the `actionPerformed()` method is invoked again and the code section for the `"Go"` command invokes the `start()` method in the `GetJavelinData` class. The complete listing of this class is given here:

```
import java.io.*;
import javax.comm.*;
import java.util.*;

/**
 * This program communicates over a serial port with the
 * Javelin evaluation card. It obtains data values from
 * the device following a simple protocol that involves
 * first sending a "password" (a two byte number) and
 * then reading two bytes at a time that are converted
 * to an int primitive type value.
 *
 * The Javelin has been programmed to respond
 * appropriately to the protocol used here. See
 * Chapter 24 for information about the Javelin.
 *
 * The class is runnable so that the data taking loop
 * runs in a thread.
 */
public class GetJavelinData implements Runnable
{
  // These are the input and output streams over the port
  DataInputStream fPortInStream;
```

```
    PrintStream fPortOutStream;


    // This array and streams are used to convert a stream
    // of bytes to primitive type values.
    byte[] fByteArray;
    ByteArrayInputStream fByteIn;
    DataInputStream fDataIn;


    // Values are returned as text to the fParent that
    // implements the Outputable interface.
    Outputable fParent;


    // Use for text from device.
    StringBuffer fStrBuf = new StringBuffer ();


    // Data taking parameters
    // flag for the data taking thread
    boolean fTakeData = false;
    int fPauseTime = 1000;


    // Data description and info
    String fDataDescription = "Temperature";
    double fSlope = 0.5;
    double fOffset = 0.0;
    String fDataUnit = "C";

   /**
    * Constructor
    * @param fParent implements the Outputable interface
    * for print output.
    * @param rate is the data taking rate in number of
    * values per second.
    */
   public GetJavelinData (Outputable parent, int rate) {
     fParent = parent;
     fPauseTime = 1000/rate; // Pause time in milliseconds
     // Set up the byte array and streams for converting
     // bytes to an int primitive value.
     fByteArray = new byte[4];
     fByteIn = new ByteArrayInputStream (fByteArray);
     fDataIn = new DataInputStream (fByteIn);

   } // ctor

   /** Obtain streams for the serial lines. **/
```

```java
void setStreams (DataInputStream inStream,
                 PrintStream outStream) {
  fPortInStream = inStream;
  fPortOutStream= outStream;
}

/** Start data taking. **/
public void start () {
  if (fPortInStream!= null && fPortOutStream!= null) {
    fTakeData = true;
    Thread thread = new Thread (this);
    thread.start ();
  }
}

/** Stop data taking. **/
public void stop () {
  fTakeData = false;
}

/**
 * In this method data is obtained in a loop until the
 * fTakeData flag goes false.
 * A simple protocol requires that it first sends a
 * password value to the device using the sendPW ()
 * method. If that value is accepted, then receiveInt ()
 * gets two bytes from the data stream and returns a int
 * value created from them. The loop pauses for a time
 * set by fPauseTime parameter before getting the next
 * value;
 */
public void run () {
  fParent.println ("Begin reading data from port.");

  // Now get data
  do {
    try {
      // The "login"
      if (!sendPW ()) return;
      // Get the raw data from the serial connection.
      int data = receiveInt ();
      // Calibrate the data.
      double correctedData = fSlope*data - fOffset;
      // Print the corrected data on the Outputable
      // parent.
      fParent.println (
```

```
          fDataDescription +" = "+ correctedData +
            fDataUnit);
      }
      catch (IOException e) {
        fParent.println ("Input or output error: " + e);
        break;
      }
      try {
        // Pause before next data request.
        Thread.sleep (fPauseTime);
      }
      catch (InterruptedException ie) {}
  } while (fTakeData);
} // run


/**
 * For our simple protocol, a value (0x2201) must first
 * be sent to the device before it will send back data.
 * A text response is returned from the device.
 */
public boolean sendPW () {
  // Send 2 byte password.
  byte[] out = new byte[2];
  out[0] = 0x22;
  out[1] = 0x01;
  fPortOutStream.write (out,0,2);
  fPortOutStream.flush ();
  try {
    String reply = receiveText ();
    if (!reply.equals ("PW OK!")) {
      fParent.println ("Password accepted");
      return true;
    }
    else {
      fParent.println ("Bad Password!");
      return false;

    }
  }
  catch (IOException e) {
    fParent.println ("Input or output error: " + e);
    return false;
  }
} // sendPW
```

```java
/**
 * Read text one byte at a time. Cast each byte to char
 * and then append to a Stringbuffer.
 */
public String receiveText () throws IOException {
  byte ch;
  fStrBuf.delete (0, fStrBuf.length ());
  while ((ch = (byte)fPortInStream.read ())!= -1) {
    fStrBuf.append ((char)ch);
    // Use \r as an end of text marker.
    if (ch =='\r') break;
  }
  return fStrBuf.toString ();
} // receiveText

/**
 * Receives 2 bytes from Javelin and converts them to
 * an int value using a ByteArrayInputStream.
 *
 * Javelin sends bytes in Big-Endian manner, which means
 * the high order byte arrives first. For example, if
 * the value 258=0x0102 were sent from the Javelin, the
 * byte 0x01 arrives first and then 0x02.
 *
 * The readInt () method in DataInputStream, which wraps
 * the ByteArrayInputStream, treats the four bytes in
 * the array as a four byte int value. So to give a
 * value 258, the values in the four byte array must
 * go as:
 *
 * fByteArray=    [0]         [1]         [2]         [3]
 *             [00000000] [00000000] [00000001] [00000010]
 *
 * We therefore place the first byte obtained from the
 * Javelin into element 2 of the array and the second
 * byte into element 3.
 */
public int receiveInt () throws IOException {
  int input =0;
  // Read high order byte
  if ((input = fPortInStream.read ()) == -1)
    throw new IOException ();
  // A cast to byte truncates 3 top bytes from int value
  fByteArray[2] = (byte)input;
```

```
    // Read low order byte
    if ((input = fPortInStream.read ()) == -1)
       throw new IOException ();
    fByteArray[3] = (byte)input;
    ByteArrayInputStream fByteIn =
       new ByteArrayInputStream (fByteArray);
    DataInputStream fDataIn =
       new DataInputStream (fByteIn);
    return fDataIn.readInt ();
  } // receiveInt
} // class GetJavalinData
```

The parameters in the constructor,

```
GetJavelinData (Outputable parent, int rate)
```

include the `rate` parameter that determines how often the data value from the device is read. The other parameter is a reference to the `Outputable parent` that is used for "callbacks" to display messages sent via the `print (String)` and `println (String)` methods in the text area in the `SerialToJavelin` user interface.

The data from the sensor arrive over the serial port as a sequence of bytes. Each pair of bytes must be converted to an `int` primitive value. We use the techniques described in Chapter 9 in which a byte array becomes the source for a `ByteArrayInputStream`, which in turn is wrapped with a `DataInputStream` class. The latter class offers the `readInt()` method that returns an `int` value with the data bytes in the lower 2 bytes of the 4-byte value.

Note that the code in `receiveInt()` in `GetJavelinData` expects the Java standard big endian format in which the bytes arrive in order of the highest-order byte first and the lowest-order byte last. So the Javelin must be programmed to follow this format.

The loop in the `run()` method first sends the password, which here consists of just a 2-byte value that matches a value set in the device program. If the password is not accepted, the thread processing stops. Otherwise, the raw data value is read. Then a "calibration" is done, which here simply consists of a slope and offset correction. The `Outputable` reference provides a callback to the user interface to print the temperature value.

Though many improvements and custom features can be added, this program illustrates the essentials of communicating over the serial line with a device to obtain data. In Chapter 24 we show how to set up the other end of the line with a device that illustrates an application of embedded Java processors.

23.5   Web Course materials

The Web Course Chapter 23 provides additional examples of serial port communications. In one demonstration program, a socket-based client/server system (see Chapters 14 and 15) involves a server program that uses the `GetJavelinData` class to provide temperature data to remote clients.

References

[1] Glen McCluskey, *Using Runtime.exec to Invoke Child Processes*, Core Java Technologies Tech Tips, March 4, 2003, `http://java.sun.com/developer/JDCTechTips/2003/tt0304.html`.

[2] `javax.comm` for Windows and Solaris platforms, `http://java.sun.com/products/javacomm/`.

[3] `javax.comm` for Linux, `www.rxtx.org`.

[4] jUSB – open source USB support for Linux, `http://jusb.sourceforge.net`.

[5] Elliotte Rusty Harold, *Java I/O*, O'Reilly, 1999.

[6] Ian F. Darwin, *Java Cookbook*, 2nd edition, O'Reilly, 2004.

[7] Christopher E. Strangio, *The RS-232 Standard: A Tutorial with Signal Names and Definitions*, CAMI Research, 2004, `www.camiresearch.com/Data_Com_Basics/RS232_standard.html`.

Resources

Craig Peacock, *Interfacing the Serial and RS-232 Port*, 2001, `www.beyondlogic.org/serial/serial.htm`.

`SerialPort` – Commercial software from `www.serialio.com`.

Al Williams, *Embedded Internet Design*, McGraw-Hill, 2003.

Chapter 24
Embedded Java

24.1 Introduction

Although Java first gained fame with applets in Web browsers and then became a popular tool for creating large enterprise services, the developers of Java originally intended it for *embedded* applications in consumer devices such as TV remote controls and Personal Data Assistants (see Chapter 1). The term "embedded" generally refers to encapsulating a processor into a device, along with programs stored in non-volatile memory, to provide services and features specific to that device. Microcontrollers, for example, are the most common type of embedded processors.

By *embedded Java* we refer to a device that contains either a conventional processor running a JVM or a special type of processor that either directly executes Java bytecodes or assists a conventional processor with executing bytecodes. The motivations for device designers to embed Java depend on the particular device, but, in general, Java provides flexibility, interactivity, networking, portability, and fast development of the software for embedded projects.

Today several types of commercial devices come with Java built into them. As mentioned in Chapter 1, over 600 million JavaCards have been sold around the world as of mid-2004, and several hundred cell phone models include Java.

Embedded applications typically must deal with very limited resources. A full-blown J2SE application on a desktop with a Swing graphical interface might require several megabytes of RAM. A typical embedded environment consisting of a basic processor and a minimal amount of memory presents a challenge to programmers who must fit codes within the limitations of the system while still providing the solid reliability required for a consumer device. A cell phone should never need rebooting!

In this chapter, we give a brief introduction to J2ME (Java 2 Micro Edition), which provides a systematic approach to choosing a subset of packages and classes that fit into small systems with memory resources ranging from a few hundred kilobytes on a basic cell phone to a few megabytes in a PDA or a high-end cell phone. We also look at the topic of real-time Java since embedded systems frequently involve hard real-time requirements.

JVM used in J2SE, it monitors a program's performance dynamically and it compiles frequently invoked methods (i.e. "hotspots") into native instructions.

CDC platforms provide sufficient resources to allow for a JVM that meets all of the official specifications [4]. Sun provides a reference JVM called the CDC Hotspot Implementation (formerly called CVM) that provides the full J2SE specs, but it doesn't implement all of the acceleration techniques of the J2SE Hotspot and is optimized for a limited resource environment.

24.3.2 J2ME profiles

A profile expands a configuration so that it works for a specialized application:

- **Mobile Information Device Profile (MIDP)** – this profile adds classes to the CLDC to provide for networking, graphical user interfaces, and local storage. It is aimed at wireless systems, especially cell phones, and allows for the downloading of MIDlets, which are similar to applets but load and run on wireless platforms that are limited to the CLDC capabilities. There are two versions of MIDP – version 1.0 and 2.0. The latter adds numerous enhancements including multimedia and game APIs and support for HTTPS. Early implementations of MIDP 2.0 were built on top of a CLDC 1.0 base. Newer implementations are built on a CLDC 1.1 base. Thus there are devices with CLDC 1.0/MIDP 1.0, devices with CLDC 1.0/MIDP 2.0, and devices with CLDC 1.1/MIDP 2.0.
- **Information Module Profile (IMP)** – this subset of the MIDP also applies to wireless systems but those with little or no graphical interface. It is intended for wireless access to remote devices such as alarm systems, meteorological stations, electric meters, and so forth.
- **Foundation profile** – this profile for the CDC applies to systems such as printers and embedded servers with no graphical interfaces. It only adds three packages dealing with security tasks.
- **Personal profile** – a CDC profile that provides for an AWT-based GUI. Intended for high-end PDAs, smart phones, and other resource limited systems as compared to desktops PCs.

Several other profiles are in development such as a PDA profile for CLDC that is specialized for those devices. There is also an optional package for the CDC that provides for a subset of the J2SE RMI classes to provide for distributed computing with micro-devices.

24.4 Real-time Java

Some embedded applications, such as controlling a cell phone or a heart-lung machine pump, require real-time programming so we provide a brief overview of real-time Java here [5,6]. Real-time essentially means providing both periodic services and responses to asynchronous demands within strictly enforced time deadlines. A real-time system is said to be *deterministic*, meaning that it can be relied upon to execute a given task in a predictable amount of time, every time that task is executed. A real-time system is not necessarily "real fast." What matters is that the time taken to complete a task is guaranteed to be within a known maximum allowed time, i.e. a *worst-case* delay.

Systems are often classified as "hard real-time" or "soft real-time." Hard real-time systems do not tolerate any missed deadlines at all. Soft real-time allows for some degree of delay and missed deadlines, i.e. performance degrades rather than fails. Often only a part of a program requires hard real-time execution, say the part that controls a device of some sort, and the rest of the program, such as a user interface, gets by with soft real-time performance.

With reasonable care, a program using J2SE or J2ME code can provide acceptable soft real-time performance such as responding quickly enough to inputs on a graphical user interface to satisfy usability requirements. However, for hard real-time tasks Java requires special techniques and/or extensions to the standard set of packages. For example, one technique has been to connect Java to C/C++ programs via JNI (see Chapter 22) to carry out the most time critical tasks.

The biggest obstacle to real-time processing with Java comes from the Garbage Collector (GC) [7,8]. The GC works as a threaded process to manage the allocation of memory in the "heap," i.e. the data buffer for a program. The GC provides memory for new objects and periodically reclaims memory from objects with no references to them. The GC relieves the programmer of the burden of memory management and is often cited as one of the primary advantages of Java programming.

However, for real-time applications, the uncertainty as to when and for how long the GC will run is unacceptable. A process responding to a critical request cannot stop what it is doing while the GC runs. The JVM specification does not require a particular type of GC algorithm, and many such algorithms are not deterministic. The GC can be turned off completely, as in one of the Java processors we discuss below, but then the user must carefully monitor the heap to avoid overflows. Another technique is to invoke `System.gc()` during free periods when critical processing isn't required. Though not required to do so, most JVMs run the GC immediately when this method is invoked. If the programmer also ensures that there is sufficient memory available for new objects (otherwise the GC will run to free up memory), then the GC will not interfere with the real-time activities.

There are several commercial real-time JVMs available. For example, NewMonics provides its PERC JVM, which is compatible with JDK 1.3 and works with several real-time operating systems [9]. The Jamaica VM from Aicas "provides hard real-time guarantees for all features of the languages together with high performance runtime efficiency" [10].

Incremental collection is a common technique used to create a real-time-compatible GC. Unlike many GC algorithms that must either fully complete their pass through the memory or start over from the beginning if they are interrupted, an incremental GC works in short steps and thus allows for interruptions in between the steps. The JVMs from NewMonics and Aicas, for example, use incremental GC.

The JVM specifications do not preclude enhancements needed to make a real-time implementation, and in 1998 work began on the *Real-Time Specification for Java* (RTSJ) by a group in the Java Community Process [11,12]. Their job was to detail a set of standard extensions for a real-time JVM. The RTSJ was released in 2003 and the company TimeSys provides the reference implementation [13]. The extension classes come via the `javax.realtime` API. Standard Java programs can run without modification in a real-time JVM.

The RTSJ does not specify a particular GC algorithm such as incremental collection. It does define new types of memory areas that allow for avoiding the GC altogether. There is *immortal memory* in which objects are never destroyed except when the program ends. *Scoped memory* is used only while a process works within a particular section, or scope, of the program such as a method. Objects there are automatically destroyed when the process leaves the scope. Neither immortal nor scoped memories are garbage collected, so using them avoids the problems of GC interference. Note, however, that the programmer must watch out for overflow of the immortal memory.

Another important aspect of the RTSJ was the addition of *real-time threads*, which provide for more precise scheduling than with standard threads. They have 28 levels of priority and their priority is strictly enforced. They are not subject to so-called *priority inversion* situations where a lower priority thread has a block on a resource needed by the higher priority thread and thus prevents the higher priority thread from running. In addition, the RTSJ includes "non-heap real-time threads" that cannot be interrupted by the GC.

The RTSJ also provides for *asynchronous event handlers* that deal with external events (or *happenings*, as they are called, to distinguish them from the events in the AWT). *Asynchronous transfer of control* allows one thread to interrupt another thread in a safe manner, unlike the deprecated `suspend()` and `stop()` methods for standard threads.

Timing is obviously important for real-time programming so the `javax.realtime` API includes the abstract class `HighResolutionTime` and its subclass `AbsoluteTime`, which represents a point in time, and `RelativeTime`, which represents a duration. The base class uses a `long` value for milliseconds and an `int` value for nanoseconds. Unlike standard Java, RTSJ requires that an implementation provide sub-millisecond precision. (Of course, the *accuracy* will vary according to the capability of the clock on a particular system.)

While still maintaining security protections, the RTSJ allows direct access to physical memory. This means that device drivers can be created with Java. Previously, Java had to link to native code to communicate directly with hardware.

Full implementations of the RTSJ specification can be too large for some embedded systems and not all of its capabilities are required for every real-time application. So some Java real-time systems use subsets of the RTSJ or they use independent extensions to obtain a smaller memory footprint.

Real-time programming in Java, and in general, involves a number of complex topics and techniques. See the book by Dibble [5] and the other real-time references for more information [6–13].

24.5 Java real machines

Until this chapter we have always assumed that Java program files provide byte-code instructions to Java *Virtual* Machine programs running on conventional processors. However, there is nothing to block the development of a Java *Real* Machine that executes bytecodes directly in hardware. Sun Microsystems proved this in the late 1990s with its *picoJava* and since then several independent Java hardware implementations have hit the market.

The JVM specification requires a stack-based processing scheme rather than the register approach common in conventional hardware (see the JVM discussions in the Web Course supplements section for more information about the JVM design). The language designers wanted Java to run on a wide range of processors, including simple embedded types with few registers, and so decided the stack approach was the most portable.

Java processors cover a wide range of designs and purposes. A *standalone* Java chip provides all the capabilities needed to act as a general-purpose computer. The processor executes the bytecodes directly. Most of the current processors actually execute only a subset of the full Java instruction set. For example, a chip might leave out the floating-point instructions since for many applications, such as for a micro-controller, floating-point instructions may or may not be needed.

Another approach is to add a Java *co-processor* to a conventional processor. The Java co-processor in some designs translates the bytecodes into the instruction set of the conventional processor and accelerates the running of Java programs. In a "companion processor" approach, the Java hardware takes over the execution of Java bytecodes completely whenever a Java program runs while the operating system and non-Java code run separately in the conventional processor.

A Java processor may refer to a *core*, which is a circuit that can be added to a FPGA (Field-Programmable Gate Array) or ASIC/SoC (Application Specific Integrated Circuit/System-on-a-Chip). In some cases a core acts as the primary processor while in others it acts as a co-processor to speed up Java programs. Such cores are not generally available on silicon but are sold as intellectual property, in the form of RTL (Register Transfer Language) descriptions of the circuits, to those who make the chips.

The JVM uses an 8-bit instruction word and currently about 200 of the possible 256 instructions are used in the class bytecodes [4]. Some instructions are

used far more often than others and some instructions carry out more complicated operations than others. So many of the hardware Java processors directly
implement only a subset of the instructions and emulate the others.

24.6   Benefits of hardware processors

Real Java machines offer a number of advantages for embedded applications
where the processor must typically work with limited memory and power
resources. As we mentioned in the J2ME section, for many of the small platforms
there may be insufficient resources for a JVM with Just-in-Time compilation or
other sophisticated acceleration capabilities. Furthermore, there may not even be
room for both a JVM and a program to run. A JVM can easily take up a megabyte
or more of RAM, while the entire memory available on a micro-platform like a
cell phone may consist of 500 kilobytes or less.

One option is to use AOT (ahead-of-time) compilers that interpret the Java
code in advance and transform it into machine code for a target platform. The
code then runs at full native speeds. This obviously eliminates portability, but
it works well for permanent, non-networked situations such as a controller in a
washing machine. For platforms that can download new Java programs, such as
a cell phone or PDA, an interpreter either as a JVM or in hardware is required.

Even when there is sufficient memory for a JVM, a hardware Java processor
could provide greater speed than a low power embedded conventional processor
running a JVM program. If a processor needs to handle both Java and non-Java
programs, an option is to add a Java accelerator core that assists a conventional
processor with Java programs. Multimedia operations in video, audio, and 2D/3D
graphics can benefit especially from the performance enhancements of Java acceleration hardware.

24.7   Java processors

We survey here a sample of commercially available systems to illustrate the
range of capabilities and designs of hardware Java processors. (We warn that
some of these products, and even some of the companies, will leave the scene in
the coming years as the marketplace determines which designs are viable.) We
consider Java processors that come as a complete chip and those provided as
cores to be implemented as part of other systems.

24.7.1   Java chips

The following Java processors are available as hardware chips and are usually
sold commercially as part of an electronics module for embedded applications:

- **aJile aJ-100, aJ-80** – the 32-bit aJ-100 derives from the JEM™ Java processor first
  developed at Rockwell Collins [14]. It executes the full bytecode instruction set,

including floating-point, with extended bytecodes for I/O and threading operations. It can also implement custom bytecodes for special operations. The processor core comes with 48 KB internal memory (16 KB for microcode and 32 KB for data) and can access up to 250 MB external RAM via an 8-bit, 16-bit, or 32-bit interface. It includes dual UARTs, five 8-bit I/O ports, and other I/O features.

It is suitable for real-time operations and supports the RTSJ. It offers a hard real-time multithreading kernel with synchronization and deterministic scheduling queues. The thread yield, wait, notify, and monitor enter/exit operations are implemented directly with extended bytecodes so a RTOS (Real-Time Operating System) is not needed to manage the threads. A thread will yield to another within 1 microsecond.

The system can implement the J2ME Connected Limited Device Configuration (CLDC) and, for wireless systems, can run MIDlets. Two independent applications can run simultaneously with no interference as if they were operating in two separate JVMs. A timer allocates time slices to each JVM and a separate timer is used by each JVM for thread time slices.

The aJ-80 provides essentially the same features as the aJ-100 but offers only an 8-bit memory interface with corresponding slower I/O.

- **Imsys Cjip processor** – the Cjip processor provides a complete Java instruction set, including floating-point operations [15]. The chip allows for reconfiguring of internal microcode (even in real-time), and most Java bytecodes are implemented in the microcode. Instruction sets for C/C++ and assembler are also included. The system offers *Virtual Peripherals* (VPs), which perform tasks normally requiring external circuitry and include timers, I/O, graphics processing and many other services. A VP is loaded at the microcode level and allows for fast, deterministic performance. J2ME in the CLDC configuration and MIDP profile is supported. (More about Imsys systems in Sections 24.8.2 and 24.10.)
- **Javelin Stamp Interpreter Chip (Parallax)** – this chip is sold separately or as part of the Javelin Stamp, which comes in a 24-pin DIP (Dual In-Line Package) module. The Javelin Stamp is discussed further in Sections 24.8.4 and 24.9 and shown in Figure 24.1. The Stamp module is nearly a complete computer system with only the need for power and a serial line to begin computing. The Java code runs in a version of the Ubicom SX48BD microcontroller chip. The chip translates a subset of Java bytecodes into the SX48 instructions and executes them. The interpreter chip is now available for those who want to use it separately from the Javelin Stamp [16,17].
- **JA108 (Nazomi)** – this processor comes in a standard 16-bit SRAM/FLASH memory package and plugs into the SRAM bus. It then acts as a co-processor to accelerate Java programs and multimedia applications [18].

Figure 24.1 The Javelin Stamp module comes as a 24-pin DIP package. See also Figure 24.3, which shows an evaluation card with the Javelin Stamp. (Photo courtesy of Parallax Corp.)

24.7.2 Java cores

Several companies offer Java processors for implementation on FPGAs and ASIC/SoC. As indicated above, some processors can act as the main processor for a Java-only program environment while others are intended to assist a conventional processor to accelerate the execution of Java programs (some cores can do both). Note that most of the processors listed below adhere to the J2ME CLDC and MIDP standards since they are aimed at the mobile device market.

- **Java Processor Cores (Aurora VLSI)** – this company offers several processor cores. The AU-J1000 executes all Java bytecodes (14 use "software assists") and also accelerates some common actions such as constant pool access and array bounds checking. It can operate as a single processor or as a co-processor. The company also offers several other Java cores. The AU-J2000 provides similar features but around 30% higher speed. The AU-J1100 and AU-J1200 cores are bilingual, meaning they include both a Java processor and a conventional processor [19].
- **Jazelle (ARM)** – the standard ARM processor, which is very popular for mobile applications, can run two different instruction sets – the standard ARM set and the compressed Thumb set. The Jazelle cores offer an extended version of the ARM that can run a third instruction set – Java bytecodes. The Jazelle directly executes most bytecodes while the rest are emulated [20].
- **JEMcore (aJile)** – the core of the aJ-100 processor (see above) can be licensed as a core for integration into other systems.
- **JVXtreme Accelerator (Synopsys)** – this co-processor directly executes 92 bytecodes and emulates the rest. It interfaces with many kinds of conventional processor via the system bus so it does not interfere with non-Java activities [21].
- **lavaCORE Configurable Java Processor Core (Xilinx)** – this 32-bit processor core from Xilinx directly executes Java bytecodes and fits onto Xilinx FPGAs. The core can be configured with all bytecode instructions included in hardware and firmware or with a subset of the instructions. Floating-point operations, a garbage collector, and encryption come as optional units. A simulator and other software development tools allow for determining what configuration is needed for a particular application and then generating the gate-level code needed to program a chip [22].
- **Lightfoot and Bigfoot Cores (DCT)** – the Lightfoot 32-bit processor core is compatible with J2ME and Java Card editions and executes native Java bytecodes directly in hardware. It is a stack-based processor with 128 bytecode instructions implemented in hardware and others implemented in software.

    The Bigfoot core is built around the 32-bit ARCtangent-A4 processor from ARC Inc. Java instructions are mapped one-to-one to ARCtangent extension instructions [23,24].
- **Moon 2 (Vulcan Machines)** – the Moon 2 core provides a 32-bit processor that directly executes bytecodes and uses a stack approach as in the JVM. A core set of the bytecode instructions execute in hardware. It can work as the primary processor in a Java only environment or as co-processor to a RISC core in a mixed code environment to accelerate Java programs [25].

24.8   Java boards

A number of board-level systems with the above Java processors are available. These include boards that provide more or less complete computer systems with features such as additional memory, I/O via serial lines and Ethernet, output digital-to-analog and input analog-to-digital converters. There are also boards that are intended to plug into the expansion slots of carrier cards that in turn provide these additional capabilities.

We discuss some examples of Java boards below. Note that many of the Java processors discussed above can be obtained in evaluation boards for easy experimentation and testing.

24.8.1   TINI^{TM} – Tiny InterNet Interface

The TINI^{TM} specification, developed by Dallas Semiconductor, aims to bring a network connection and high-level control and monitoring capabilities to devices ranging from industrial processing equipment to consumer appliances [26–28]. The TINI interface comes in a small, low-power package yet it can implement a complete TCI/IP node and an embedded Java server.

The TINI reference specification consists of a microcontroller on a 72-pin SIMM (Single Inline Memory Module) format card. Figure 24.2(a) shows a SIMM card from Imsys that follows the TINI format (but with the Imsys Cjip processor.) The card also includes an Ethernet controller, RS232, a 1-Wire Bus (a proprietary standard from Dallas Semiconductor), and SRAM. A Java runtime system is held in flash ROM. The TINI card plugs into a carrier board that provides power and the physical connectors for the I/O interfaces. Several carrier boards with a variety of features are available for the TINI module from different vendors.

An operating system called TINI OS provides basic services including a file system, memory and I/O management, and task switching. You can ftp or telnet into a Unix-like command shell (called *slush*) to load and run Java programs. The shell can be loaded via the serial port with an IDE called JavaKit available from Dallas Semiconductor. After development is finished, a standalone program can be loaded in place of the shell. The JVM takes up only 40 KB yet allows for all the basic Java capabilities including threading and the complete set of primitive types. A garbage collector runs as a native task while all other tasks run as Java applications. A round-robin task scheduler allocates processing in fixed 8-ms time slices.

Programs can be compiled elsewhere with the standard SDK compiler but with the TINI class packages instead of the standard ones. Your programs can use classes from the core packages – `java.lang`, `java.io`, `java.net`, `java.util` – and the set of custom packages (`com.dalsemi.*`) from Dallas

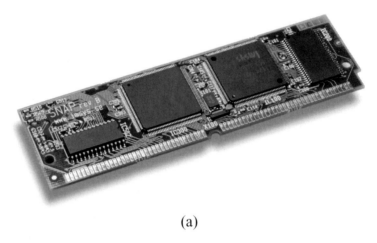

(a)

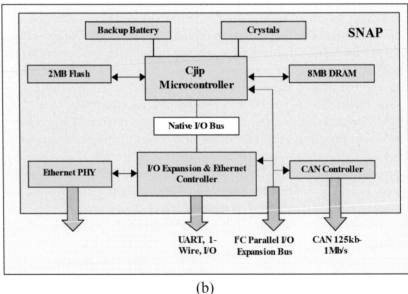

(b)

Figure 24.2 (a) Imsys SNAP card has the 72-pin SIMM format and is compatible with carrier boards that conform to the TINI™ specifications from Dallas Semiconductor. (b) Block diagram of the SNAP card showing the primary components [15]. (Photo courtesy of Imsys Technologies.)

Semiconductor. The latter packages include classes that provide access to system resources such as the 1-Wire bus. They also include useful classes like `HTTPServer` with which you can build custom servers. There is also a considerable amount of open-source software available for TINI, such as a server that runs servlets [29]. The book by Williams discusses servlet designs that work with this server and also provides several other programs for TINI [30].

Table 24.1 *SNAP vs. TINI timing comparison [15].*

Operation (10000 times)	SNAP	TINI
integer multiplications	32 ms	800 ms
integer division	82 ms	890 ms
float additions	52 ms	3390 ms
float multiplications	50 ms	3290 ms
float division	120 ms	69990 ms
double additions	78 ms	3550 ms
double multiplications	76 ms	3180 ms
double divisions	284 ms	49790 ms

By embedding Java and a network connection in a remote system, a TINI card opens up a wide range of new capabilities. As discussed in Chapters 14–15, a server can provide monitoring and control to a distant client. Potentially every device in a complex industrial facility or in a large scientific experiment could "go online" and provide access and control at a very fine-grained level.

In Section 24.10 we give an example of a customized server running on a TINI type system (with the SNAP card described next). The book by Loomis gives more examples of TINI applications including a remote data logger running on a TINI platform [31]. The code runs TCP/IP with either Ethernet or dial-up networking using a PPP network interface via serial I/O. This allows a remote sensor, for example, to contact a home base periodically by telephone to upload data, report on the status of the sensor, and so forth.

24.8.2   SNAP – Simple Network Application Platform

The SNAP card from Imsys Technologies comes as a TINI compatible SIMM card but instead of a conventional microcontroller running a JVM, it uses the company's Cjip Java hardware processor (see Section 24.7). This provides a significant improvement in processing speed (see Table 24.1). Figure 24.2(a) shows a photograph of the SNAP and Figure 24.2(b) shows a block diagram of the main components on the card.

The card provides expanded capabilities compared to the standard TINI reference system. It holds 2 MB of flash memory and 8 MB of DRAM. The system supports the J2ME CLDC 1.0 configuration. Figure 24.3(a) shows a development board from Systronix that is holding a SNAP in the SIMM connector. The card includes a serial port and an Ethernet connector [32].

24.8.3   aJile aJ-PC104 single board computer

The aJ-PC104 board from aJile Systems follows the popular PC/104 format and includes the company's aJ-100 Java processor (see Section 24.7). It can act as

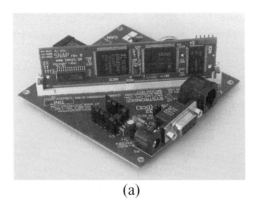

(a)

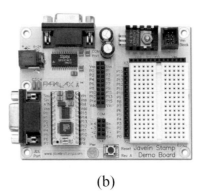

(b)

Figure 24.3   (a) The TILT (TINI Initial Learning Tool) board from Systronix includes a SIMM expansion slot. Here it is plugged with the Imsys SNAP card (the vertical card). The TILT includes an RS232 connector, Ethernet RJ45, and other interfaces. (b) The Javelin Stamp evaluation board for the Parallax Javelin Stamp includes a breadboard and serial ports – one for programming the module and the other for I/O with programs running in the module. (Tilt photo by Th. Lindblad, Javelin photo courtesy Parallax Corp.)

a standalone Java computer or fit into a PC/104 stack. Systronix offers a Java module called the JStamp (not to be confused with the Javelin Stamp mentioned below) that comes in a 40pin DIP module and uses the aJile aJ-80 Java processor chip [32].

24.8.4   Parallax Javelin Stamp

The Javelin is derived from the popular Basic Stamp series of microcontrollers but it runs programs written in Java instead of Basic [16,17]. The Javelin (see Figure 24.1) comes as a 24-pin DIP that holds an interpreter chip that runs a subset of the Java instruction set. The module includes 32 KB of non-volatile EEPROM and 32 KB of RAM. Programs are stored in the EEPROM and loaded into RAM for execution. The RAM holds the program's stack and heap. Programs can access both free RAM and EEPROM space. The processor executes up to 8000 instructions per second.

Four of the pins are used for power, ground, and reset. Another four pins are used for serial communications with the chip to load programs and debug them. Sixteen pins provide general-purpose I/O. These can handle digital-to-analog and analog-to-digital conversions, serial I/O, pulse modulated output, and pulse input.

The serial interface provides for programming the module. The programs are saved in the EEPROM and program execution begins after a reset. Virtual Peripheral (VP) objects run UARTs, pulse-width modulators, timers and other services, and use minimal system resources. The company offers a development system with an IDE for developing, downloading, and debugging programs on the module.

The Javelin Stamp runs a downsized version of Java with a small subset of the usual classes. In the next section we discuss how to program the Javelin and give a demonstration of a Javelin program that uses the general-purpose I/O pins to obtain data from a sensor and send it over a serial line.

24.9 Programming the Javelin Stamp

The Javelin Stamp provides an interesting example of a Java hardware platform. Of all the Java processors discussed above, it offers the most restricted set of bytecode instructions and the most limited number of classes. It does not conform to the CLDC framework and instead offers even fewer capabilities (but more than the JavaCard). However, it is low cost and fairly simple and straightforward to use. For many microcontroller applications, its capabilities suffice.

Parallax intended the Javelin to be as easy to use as its popular Basic Stamp modules [16,17]. The company provides a development board with a small breadboard (see Figure 24.3(b)) for experimentation and tests. To get to know the system you can place LEDs, buttons, and other circuits on the breadboard and connect them to the Javelin module's 16 general-purpose I/O pins, which are accessible to Java programs running in the module.

The Javelin Stamp's interpreter differs in several ways from the JVM that comes with your desktop SDK. For example, there is no multitasking or multithreading. Instead you can use a timer object to allocate times for tasks to run. As mentioned earlier, six commonly needed functions are provided by the VPs that come built into the Javelin's firmware. These VPs include a timer, UART for serial communications, PWM (Pulse Width Modulation) for pulse train generation, DAC and ADCs. You can install up to Six VPs at a time. They run in the background and so use a minimal amount of processing time and resources. Each VP requires one or more of the 16 available I/O pins.

The Javelin processor offers no garbage collection, so once you create objects they exist until the module is reset. This places tight restrictions on the number

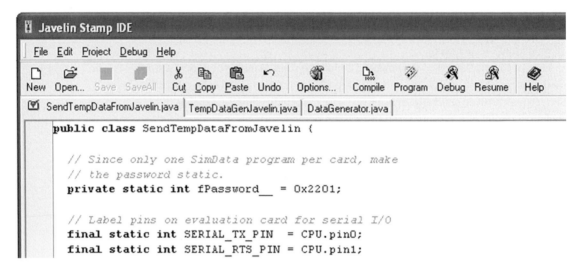

Figure 24.4 The Javelin Stamp IDE from Parallax provides a programming interface to the processor. It offers numerous features including a source editor and a java compiler. It can test the connection to the module and download programs to the module. Programs running on the Javelin Stamp send the output of `print ()` methods to a window on the IDE. You can also run programs with a debugger. [16, 17]

and size of objects that you can create without filling up the available memory. However, the absence of a GC means that the system can respond without the uncertainty as to when and for how long a GC might run.

Other limitations of the Javelin include a 16-bit maximum data width, even for the `int` type (there is no `long` type). The standard packages only allow integer math. However, an optional 16-bit FP package is now available [17]. You can create 1D arrays but not 2D. A number of core Java packages are either missing (no `java.net` for example) or truncated.

The IDE for the Javelin uses a serial port on the module to load a program and to start it. This port differs from the one used with the UART VP and the I/O pins. Figure 24.4 shows the IDE interface. You can edit programs, compile them, link (gather up all the classes needed by the processor), and then download them (via the "`Program`" command). While the serial line is connected, the `System.out.println()` methods send output to a window in the IDE. There is also a debugger for testing programs while they run on the processor.

Once the program is loaded into EEPROM, the serial port can be disconnected and the module runs independently. A reset pin restarts a program when brought to ground and then released. Sixteen of the 24 pins on the module are accessible from the Java program for input or output. For example, you can set a pin to 5 V, send out a pulse train, or read an analog value. You can use a set of pins to transmit and receive signals for a serial UART communications port.

24.9.1   Javelin demonstration program

The program `SendTempDataFromJavelin` shown below runs in the Javelin evaluation card and provides temperature readings obtained from a Dallas Semiconductor chip DS1620 that is installed on the breadboard. Via the serial line, the processor receives requests from the `SerialToJavelin` program discussed in Chapter 23 and transmits back a temperature value. The program creates a *receive* and a *transmit* UART virtual peripheral object to provide two-way serial communications. Each constructor assigns two I/O pins to the appropriate function and sets the serial mode for 9600 baud.

The reset signal starts the `main()` routine, which begins by creating an instance of `TempDataGenJavelin`. Since we might want to use this program as a template for obtaining other kinds of sensor data, we let `TempDataGenJavelin` extend `DataGenerator`, which is an abstract class (Javelin does not allow interfaces). Any class that extends `DataGenerator` must override `getData()` with a method that returns an `int` value.

The process in `SendTempDataFromJavelin` then goes into a loop and immediately invokes `checkPW()`. This method uses a simple protocol that requires that the request for data first include a short password number. The `checkPW()` method invokes `receiveInt()`, which waits for the requestor to transmit a 2-byte number that matches the password value. The method receives 2 bytes in big-endian format. That is, the most significant byte arrives first and the least significant byte last. Then an `int` value is made from the 2 bytes (remember that in the Javelin, the integers are a maximum of 2 bytes long.)

If the password value is valid, then a string confirmation is transmitted to the requestor over the serial line. Back in the main loop, the data `int` value is obtained from the `DataGenerator` via `getData()` and then transmitted as 2 bytes in big-endian format (since the Java specifications require a big-endian representation). The process then loops back to the `receiveInt()` method again and waits for the next data request.

```
import stamp.core.*;

/**
  * This program transmits temperature readings from the DS1620 chip when it
  * receives a request over the serial line. The request must include a password
  * number at the start. If the password is OK, then a temperature value is
  * obtained and transmitted. The "raw" temperature readings are obtained
  * via the TempDataGenJavelin class, which in turn uses the DS1620 class
  * provided in the stamp.peripheral.sensor.temperature package. The serial
  * communications code uses the stamp.core.Uart virtual peripheral class.
  **/
public class SendTempDataFromJavelin
{
```

```java
// Since only one program per card, make
// the password static.
private static int fPassword__ = 0x2201;

// Label pins on evaluation card for serial I/O
final static int SERIAL_TX_PIN = CPU.pin0;
final static int SERIAL_RTS_PIN = CPU.pin1;
final static int SERIAL_CTS_PIN = CPU.pin2;
final static int SERIAL_RX_PIN = CPU.pin3;

// Create the UART vp for transmission via COM serial port
static Uart fTxUart__ = new Uart (Uart.dirTransmit, SERIAL_TX_PIN,
                                  Uart.dontInvert, SERIAL_RTS_PIN,
                                  Uart.dontInvert, Uart.speed9600,
                                  Uart.stop1);
// Create the UART vp for reception via the COM serial port
static Uart fRxUart__ = new Uart (Uart.dirReceive, SERIAL_RX_PIN,
                                  Uart.dontInvert, SERIAL_CTS_PIN,
                                  Uart.dontInvert, Uart.speed9600,
                                  Uart.stop1);

/** Resetting Javelin will start the program here. **/
public static void main () {

  // Local variable
  int data=0;

  // Create a temperature data sensor.
  DataGenerator temp_sensor = new TempDataGenJavelin ();

  // Loop continuously, each time waiting for a request to arrive with
  // the password (PW). Then get the temperature reading and send it.
  do {
    // Go into receive mode and wait for PW
    checkPW ();

    // Get the data
    data = temp_sensor.getData ();

    // and then send it to the requestor.
    sendInt (data);

  } while (true);
} // main

 /**
  * Utility method to send int value as
```

```
    * two bytes. Use "big-endian" mode with most
    * significant byte sent first.
   **/
  static void sendInt (int data) {
     fTxUart__.sendByte (data >>> 8);
     fTxUart__.sendByte (data & 0x00FF);
  }

  /**
    * Utility method to receive 2 bytes and make an
    * int value (16 bits in Javelin) from them.
   **/
   static int receiveInt () {
     int byte1 = fRxUart__.receiveByte ();
     int byte2 = fRxUart__.receiveByte ();
     return byte2 | (byte1 << 8);
  }

   /**
    * Utility method to receive an int value and compare it to the password.
   **/
   static void checkPW () {
     do {
       int data = receiveInt ();
       if (data == fPassword__) {
         fTxUart__.sendString ("PW OK!\n\r");
         break;
       }
        fTxUart__.sendString ("Wrong PW!\n\r");
     } while (true);
   } // checkPW
} // class SendTempDataJavelin
```

The class `TempDataGenJavelin` creates an instance of the `DS1620` class from the package `stamp.peripheral.sensor.temperature` and passes the numbers of the pins that connect to the chip. The `getData()` method provides the temperature readings in units of 0.5 degrees Celsius. See Figure 23.1 for an example of temperature readings from the chip.

```
import stamp.core.*;
import stamp.peripheral.sensor.temperature.DS1620;

/**
  * Create a class that obtains the current temperature with the DS1620 chip.
  * It extends the DataGenerator class and overrides the getData() method to
```

```
 * return the temperature value in an int value.
 **/
class TempDataGenJavelin extends DataGenerator
{
      DS1620 fThermometer;

      /** Constructor creates an instance of DS1620. **/
      TempDataGenJavelin () {
      fThermometer = new DS1620 (CPU.pin4, CPU.pin5, CPU.pin6);
   }

   /** Return temperature in units of 0.5 degrees Celsius. **/
   int getData () {
      return fThermometer.getTempRaw ();
   }
} // class TempDataGenJavelin

/**
   * Javelin does not allow for interfaces so use an abstract class to represent
   * types of data generators.
   */
public abstract class DataGenerator
{
   abstract int getData ();
}
```

24.9.2   A Javelin Stamp for an unmanned aerial vehicle

Another demonstration project [32], discussed in more detail in the Web Course, involved the development of a system with the Javelin Stamp to provide location measurement and sensor data recording in a small unmanned aerial vehicle (UAV). The goal was for the plane to fly autonomously and measure ambient radioactivity. The entire payload, including batteries and the UAV, is less than 7 kg.

The position of the UAV in the horizontal plane comes from a GPS-module read via a serial line. The GPS module provides data in a standard protocol and code was written for the Javelin to translate the readings to longitude/latitude values. For the altitude coordinate the Javelin measured the analog voltage output from a pressure gauge. A virtual peripheral object provides for 8-bit analog-to-digital conversion (ADC) for voltages between 0 and 5 V on any of the I/O pins using just a few passive components (two resistors and a capacitor). For the radiation measurements, a standard Geiger counter with passive components and an operational amplifier gave an output proportional to the count rate, and this output was measured with another ADC on the Javelin Stamp. Garari and

Mansouri carried out bench tests of the GPS and radiation counter systems and found that it performed satisfactorily.

24.10 An embedded web server

We have frequently referred to the benefits of custom client/server programs for embedded processors. Here we finally install a basic web server on a SNAP board (see Section 24.8.2) that can send requested files as before but also present voltage readings taken from an analog-to-digital converter (ADC) on the board. Unlike the Javelin system, the SNAP implements the official J2ME standard – the CLDC 1.0 configuration. It provides an operating system with a command shell so you can `telnet` and `ftp` to it when the board is connected to the network via Ethernet.

24.10.1 Programming the SNAP

Creating and running programs on the SNAP are not fundamentally different from that for the PC desktop but do differ in some significant practical ways. First of all, there are limitations on the classes available for your programs:

- The core language classes include only those in the following packages: `java.lang`, `java.io`, `java.util`, and `javax.microedition.io`.
- Some of the core language classes available in the J2SE version are not included. For example, there is no `java.lang.StringTokenizer`.
- Some classes are available but with some methods missing. For example, `random()` is removed from `java.lang.Math`.

There are, however, additional packages available with classes that assist in developing programs for the device:

- `com.dalsemi.onewire.*` and `com.dalsemi.system` packages are available and are fully compatible with the TINI standard.
- `se.imsys.com`, `se.imsys.net`, `se.imsys.ppp`, `se.imsys.system`, and `se.imsys.util` packages provide lots of useful classes such as `HttpServer` for building custom servers.
- `org.xm.sax` and `uk.co.wilson` packages provide some tools for XML handling.

After creating class files compatible with the above libraries, you can use the `javac` compiler from the J2SE SDK on your desktop computer to compile them, but you must instruct the compiler to use the set of packages listed above. For example, if you install the SNAP software into the `c:\SNAP` directory, then you compile `HelloWorld.java` as follows:

```
c:\> javac -target 1.1 -bootclasspath c:\SNAP\classes
HelloWorld.java
```

The target option tells the compiler to produce bytecode compatible with a Java version 1.1. With the J2SE 5.0 compiler, you must also add the option "-source 1.3" to indicate that the code does not contain assertions and other post 1.3 additions. The bootclasspath option indicates that the core language classes should be taken from the SNAP set of packages rather than from the standard J2SE packages.

With J2SE the class files are loaded by the JVM, which first checks that the bytecode conforms to all the standard specifications and doesn't do anything illegal. The CLDC is intended for platforms with limited resources so it requires that some of this bytecode checking is carried out prior to loading the processor. The class files are run through a *preverification* program that checks the code and creates new class files with annotations that the processor uses to accelerate its own code checking.

So before the class files are moved to the SNAP, the following preverification step is required:

```
c:\SNAP\bin\preverify -classpath c:\SNAP\classes;.
-nofinalize -d o HelloWorld
```

The -nofinalize flag is necessary because CLDC platforms don't allow for invoking the finalize() method inherited from Object. The -d o flag sends the verified class file to the o\ subdirectory.

You can use ftp to place HelloWorld.class on the SNAP. To run the program you can telnet into the Unix style command shell and run the program with

```
> java -r HelloWorld &
```

which responds with the output

```
Hello World!
```

Here the -r flag tells the system to restart the Java processor in case it is still running an old version of a class file. The & flag is not necessary here since the program returns to the telnet prompt immediately after it prints the "Hello World!" string. However, for a long-running program like a server, this flag gives the console back to the shell in the normal Unix fashion so you can log off while the process continues to run.

24.10.2 The web server program

The program below is a modified version of the MicroServer program described in Chapter 14. The name is changed to SnapAdcServer since it reads an ADC value if the page requested is adc.html. Otherwise, it acts as in MicroServer and looks for files that are requested and returns them if found.

To obtain the ADC value the program uses the `DataPort` class in the `com.dalsemi.system` package. The system bus is mapped to memory so that by specifying a memory address a particular device can be accessed. This code in the `Worker` class:

```
DataPort p0 = new DataPort(0x380001);
    // read ADC
int adcValue = p0.read();
```

first creates the `DataPort` object for the address specified in the constructor and then reads the ADC value. The reading is combined with a hypertext header and footer code to create an HTML file with the data. The Web Course provides a link to a SNAP with its ADC connected to a device such as a solar panel voltage output.

```
import java.net.*;
import java.io.*;
import java.util.*;
import com.dalsemi.system.*;

/** Modified Version of MicroServer for the Imsys
  * SNAP card. **/
public class SnapAdcServer
{
    /** Start program with optional port number argument. **/
    public static void main (String args[]) throws
      IOException {

        int port; // port number

        // Get the port number from the command line.
        try {
          port = Integer.parseInt (args[0]);
        }
        catch (Exception e) {
           port = 2223; // Default
           System.out.println ("Use default port = 2223");
        }

        // Create a ServerSocket object to watch that
        // port for clients
        ServerSocket server_socket = new ServerSocket (port);
        System.out.println ("Server started");

        // Loop indefinitely while waiting for clients
        // to connect
```

```
        while (true) {

            // accept() does not return until a client
            // requests a connection
            Socket client_socket = server_socket.accept ();

            // Now that a client has arrived, create an
            // instance of our Worker thread subclass to
            // tend to it.
            Worker worker = new Worker (client_socket);
            worker.start ();
            System.out.println ("New client connected");
        }
    } // main
} // class SnapAdcServer
```

The SNAP I/O package does not include the `PrintWriter` class so we use `PrintStream` instead. We flush the stream after each set of print methods to ensure all data is moved from the internal buffers. We note that, like `PrintWriter`, the `PrintStream` methods don't throw `IOException` and instead the class offers the `checkError()` method, which returns true if an `IOException` occurred. (As we discussed in Chapter 14, you could modify this code to check for errors after every print invocation or put the print statements into utility methods that throw `IOException` as we did with the `DataWorker` class discussed in Chapter 15.)

Another difference with the `MicroServer` application in Chapter 14 is that the `split()` method in `String` class is not available in the SNAP `java.lang` package. So we created a `split()` in the `Worker` class to located substrings, i.e. tokens, separated by blank spaces.

```
import java.net.*;
import java.io.*;
import java.util.*;
import com.dalsemi.system.*;


/** Threaded process to serve the client connected to
  * the socket. **/
public class Worker extends Thread {
    Socket fClient;

    String fWebPageTop =
```

```
     "<html> <head> \n <TITLE>Solar Panel Voltage - ADC
                          Readout</TITLE>
       \n </HEAD>" ;

  String fWebPageBot = "</body> \n </html>";

  /** Pass the socket as a argument to the constructor **/
  Worker (Socket client) throws SocketException {
     fClient = client;

     // Set the thread priority down so that the
     // ServerSocket will be responsive to new clients.
     setPriority (NORM_PRIORITY - 1);
  } // ctor

  /**
    * This thread receives a request from the client for a
    * a web page file. The file name is found relative to
    * the directory of this code.
    **/
  public void run () {
    boolean read_adc = false;
    try {
       BufferedReader client_in = new BufferedReader (
         new InputStreamReader(fClient.getInputStream()));

       // Now get an output stream to the client.
       OutputStream client_out =
         fClient.getOutputStream ();

       // Use PrintStream for SNAP output
       PrintStream pw_client_out =
         new PrintStream (client_out);

       // First read the message from the client
       String client_str = client_in.readLine ();
       System.out.println ("Client message: "+client_str);

       // Split the message into substrings.
       String [] tokens = split (client_str);

       // Check that the message has a minimun number of
       // words and that the first word is the GET command.
       if ((tokens.length >= 2) &&
            tokens[0].equals ("GET")) {
         String file_name = tokens[1];
```

```java
    // Ignore the leading "/" on the file name.
    if (file_name.startsWith ("/"))
        file_name = file_name.substring (1);

    // If no file name is there, use index.html
    // default.
    if (file_name.endsWith ("/") ||
        file_name.equals (""))
        file_name = file_name + "index.html";

    // Check for a request for the ADC reading.
    if (file_name.endsWith ("adc.html")) read_adc =
        true;

    // Check if the file is hypertext or plain text
    String content_type;
    if (file_name.endsWith (".html") ||
        file_name.endsWith (".htm")) {
            content_type = "text/html";
}
else {
    content_type = "text/plain";
}

    // Now either read a file from the disk and
    // write it to the output stream to the client
    // or send the ADC reading if that is requested
    try {
        // Send the header.
        pw_client_out.print ("HTTP/1.0 200 OK\r\n");

        if (read_adc){
            pw_client_out.print ("Server: ReadADC
                                   1.0\r\n");
            pw_client_out.print ("Content-length: " +
                                   500 + "\r\n");
            pw_client_out.print ("Content-type: " +
                                   content_type +
                                   "\r\n\r\n");
            pw_client_out.print (fWebPageTop);
            // Connect to ADC.
            DataPort p0 = new DataPort (0x380001);
                        //DataPort.CE3);
            // read ADC
            int adc_value = p0.read ();
```

```
            pw_client_out.println ("ADC = " + adc_value);
            pw_client_out.println (fWebPageBot);
            pw_client_out.flush ();


          }
        else {// Or just send a requested file

            // Open a stream to the file. Remember that
            // by this point all the text except for
            // the file name has been stripped from the
            // "request" string.
            FileInputStream file_in =
              new FileInputStream (file_name);

            // Send the header.

            File file = new File (file_name);
            Date date = new Date (file.lastModified ());
            pw_client_out.print ("Date: " + date +
                                "\r\n");
            pw_client_out.print ("Server: MicroServer
                                1.0\r\n");
            pw_client_out.print ("Content-length: " +
                                file_in.available () +
                                "\r\n");
            pw_client_out.print ("Content-type: " +
                                content_type +
                                "\r\n\r\n");
            pw_client_out.flush ();
            // For PrintStream with SNAP

            // Creat a byte array to hold the file.
            byte [] data =
              new byte [file_in.available ()];

            file_in.read (data);
              // Read file into the byte array
            client_out.write (data);
              // Write it to client output stream
            client_out.flush ();
              // Remember to flush output buffer
            file_in.close (); // Close file input stream
          }
        }
      catch (IllegalAddressException err) {
```

```
                    // If no such file, then send the famous 404
                    // message.
                    pw_client_out.println(
                       "ADC readout failure - Illegal Address
                          Exception");
                    pw_client_out.println(fWebPageBot);
                    pw_client_out.flush ();
                       // For PrintStream with SNAP

                } catch (FileNotFoundException e) {
                    // If no such file, then send the famous 404
                    // message.
                    pw_client_out.println ("404 Object Not Found");
                    pw_client_out.flush ();
                       // For PrintStream with SNAP
                }


        }
        else {
          pw_client_out.println ("400 Bad Request");
          pw_client_out.flush ();
             // For PrintStream with SNAP
        }
      }
    catch (IOException e) {
        System.out.println ("I/O error " + e);
    }

      // Close client socket.
      try {
        fClient.close ();
      }
      catch (IOException e) {
        System.out.println ("I/O error " + e);
      }
  } // run

/**
  * Since the platform may not include the J2SE 1.4
  * String class with the split() method, we provide
  * a substitute. This method returns an array with
  * the tokens in the string parameter that are
  * separated by blank spaces.
  **/
```

```
String [] split(String source_string) {

    // First get rid of whitespace at start and end of the
    // string
    String string = source_string.trim();
    // If string contains no tokens, return a zero length
    // array.
    if(string.length () == 0) return (new String [0]);

    // Use a Vector to collect the unknown number of
    // tokens.
    Vector token_vector = new Vector();
    String token;
    int index_a = 0;
    int index_b = 0;

    // Then scan through the string for the tokens.
    while(true){
        index_b = string.indexOf (' ', index_a);
          if (index_b == -1) {
                token = string.substring (index_a);
                token_vector.addElement (token);
                break;
          }
        token = string.substring (index_a, index_b);
        token.trim ();
        if (token.length () >= 1)
            token_vector.addElement (token);
        index_a = index_b + 1;
      }

    // Copy elements into a string array.
    String [] str_array =
      new String[token_vector.size ()];
    for(int i=0; i < str_array.length; i++)
        str_array[i] =
            (String) (token_vector.elementAt (i));
    return str_array;

} // split

} // class Worker
```

24.11   Java processor performance

Java processors, whether virtual or in hardware, vary greatly in their performance capabilities. Many embedded applications, such as controlling an appliance, do not need tremendous speed. On the other hand, some applications, such as creating detailed animations for a PDA screen, do require high performance.

There is no universally accepted set of benchmarks for measuring Java processing performance (though at least one organization is trying to develop them [33]). One popular measure of Java processing speed, however, is the Caffeine-Mark developed by Pendragon Software [34] in the mid-1990s. It consists of a suite of tests such as a prime number search, recursive method invocations, drawing images, and so forth. Another measure is the VolanoMark, developed by Jeff Neffenger, which emphasizes server performance [35].

Benchmarks are never perfect, and you can usually find some bias that favors one system over another. In general, it is best to focus on the particular platform and the features desired and compare only the systems that could fulfill those requirements. There is no point in comparing, say, the Javelin Stamp with a JVM running on a modern desktop machine. The more advanced chips can perform at higher speeds and offer floating-point, but many do not offer the performance of even a modest desktop-level machine. For example, in a series of tests involving tasks such as sorting and pattern recognition, a 1.6-GHz Pentium was roughly 100 times faster than an aJile evaluation board with the aJ-100 chip [36].

More of an *apples versus apples* comparison is given in Table 24.1. It shows a comparison of speeds for several types of operations for the SNAP board from Imsys Technologies, which uses a hardware Java processor, and a standard TINI^{TM} board that uses a JVM running in a conventional microcontroller. For these tests, the hardware approach provides one to two orders of magnitude faster speed [15].

24.12   Web Course materials

In addition to more details about the chips, boards, and programs discussed here, the Web Course Chapter 24 offers more examples and demonstrations of embedded Java. Code listings, diagrams, and other resources are included. As new Java hardware is introduced, these will be added to the resources section.

References

[1] Java Card at Sun Microsystems, http://java.sun.com/products/javacard/.
[2] Kim Topley, *J2ME in a Nutshell*, 2002, O'Reilly.
[3] *KVM Porting Guide, CLDC, Version 1.1, Java^{TM} 2 Platform, Micro Edition*, Sun Microsystems, Inc., March 2003.
[4] Tim Lindholm and Frank Yellin, *The Java^{TM} Virtual Machine Specification,* 2nd Edn, Sun Microsystems, Inc. 1999, http://java.sun.com/docs/books/vmspec/.

[5] Peter C. Dibble, *Real-Time Java Platform Programming*, Sun Microsystems, Inc., 2002.

[6] Greg Bollella, *Greg Bollella on Controlling Physical Systems with Real-time Specification for Java: Combining Real-time and Non-real-time Operations on a Single Processor*, System News, Vol. 75, Issue 4, May 24, 2004, `http://sun.systemnews.com/articles/75/4/feature/13057`.

[7] Brian Goetz, *Java Theory and Practice: A Brief History of Garbage Collection*, IBM Developer Works, 2003, `www-106.ibm.com/developerworks/java/library/j-jtp10283/`.

[8] *Tuning Garbage Collection with the 1.4.2 Java[tm] Virtual Machine*, Sun Microsystems, Inc. 2003, `http://java.sun.com/docs/hotspot/gc1.4.2/`.

[9] NewMonics, Inc., Tucson, AZ, `www.newmonics.com`.

[10] aicas GmbH, Karlsruhe, Germany, `www.aicas.com`.

[11] *JSR 1: Real-time Specifications*, Java Community Process, `www.jcp.org/en/jsr/detail?id=1`.

[12] Sione Palu. Real-time Specification for Java (RTSJ), `www.developer.com/java/article.php/1367671`.

[13] TimeSys Corp., Pittsburgh, PA, `www.timesys.com`.

[14] aJile Systems, Inc, San Jose, CA, `www.ajile.com`.

[15] Imsys Technologies, Upplands Väsby, Sweden, `www.imsystech.com`.

[16] Parallax, Inc., Rocklin, CA, `www.parallax.com`.

[17] Javelin Stamp resources available at `www.parallax.com/javelin/`.

[18] Nazomi Communications, Santa Clara, CA, `www.nazomi.com`.

[19] Aurora VLSI Inc., Santa Clara, CA, `www.auroravlsi.com`.

[20] ARM Computer, Cambridge, UK, `www.arm.com`.

[21] Synopsys, Mountain View, CA, `www.synopsys.com`.

[22] Xilinx, San Josa, CA, `www.xilinx.com`.

[23] Digital Communication Technologies Ltd. (DCT), Middlesex, UK, `www.dctl.com`. (DCT has merged with Velocity Semiconductor, `www.velocitysemi.com`.)

[24] Peter Clarke, *Startup builds Java processor with ARC core*, EE Times, September 17, 2001, `www.commsdesign.com/news/product_news/showArticle.jhtml?articleID=16503516`.

[25] Vulcan Machines Ltd, Royston, UK, `www.vulcanasic.com`.

[26] Dallas Semiconductor is owned by Maxim Integrated Products, Sunnyvale, CA, `www.maxim-ic.com/TINIplatform.cfm`.

[27] TINI™ official web site, `www.ibutton.com/TINI/`.

[28] Dan Eisenreich and Brian DeMuth, *Designing Embedded Internet Devices*, Newnes, 2003.

[29] Smart Software Consulting, unofficial TINI information site, `www.smartsc.com/tini/`.

[30] Al Williams, *Embedded Internet Design*, McGraw-Hill Education, 2003.

[31] Don Loomis, *The TINI™ Specification and Developer's Guide*, Addison-Wesley, 2001. Available on line at: `www.maxim-ic.com/products/tini/pdfs/tinispec.pdf`.

[32] Systronix, Salt Lake City, UT, `www.systronix.com`.

[33] Embedded Microprocessor Benchmark Consortium (EEMBC), `www.eembc.hotdesk.com`.

[34] Pendragon Software, Libertyville, IL, `www.pendragon-software.com`.

[35] The Volano report and benchmark tests, `www.volano.com/benchmarks.html`.

[36] Hosseyn Karimi Garari and Daryoush Mansouri, *Radiation Surveillance by Unmanned Aerial Vehicle (UAV)*, Diploma Work, Physics Dept., Royal Institute of Technology, 2003.

[37] David Sikter, *Benchmarking and Feasibility Study of the aJ100tm Java Processor in a Real-Time Image Processing Application*, Diploma Work, Physics Dept., Royal Institute of Technology, 2002.

Appendix 1
Language elements

Table A.1.1 *Keywords.*

abstract	else	interface	switch
assert	enum	long	synchronized
boolean	extends	native	this
break	false	new	throw
byte	final	null	throws
case	finally	package	transient
catch	float	private	true
char	for	protected	try
class	goto	public	void
const	if	return	volatile
continue	implements	short	while
default	import	static	
do	instanceof	strictfp	
double	int	super	

Notes on keywords:

- the `goto` keyword is reserved but not used (in modern language design, such jumping between lines of code is considered very bad programming practice)
- `const` is also reserved but not used
- `assert` is a new keyword added with Java 1.4
- `enum` is a new keyword added with Java 5.0

Java is case sensitive so in principle you could use these words as identifiers if you change any character to upper case. But that's not recommended!

Table A.1.2 *Reserved symbols in Java.*

Symbol	Description
;	Semi-colon indicates the end of a statement.
( )	Parentheses are used in several places. They override the default precedence in an expression that contains multiple operators.
	They indicate a method and also a casting operator. They surround the logic test in an `if` statement, e.g.
	`if (test) doSomething();`
[ ]	Array declaration and an array element specification.
{ }	Curly braces enclose the fields and methods of a class, the code for a method, the code body for an `if` statement, a `for` loop statement, a synchronized block, and initial values for an array declaration.
//	Indicates a single line comment.
/*. .*/	Bracket a set of comments that can span more than one line.
or	The `/** */` version is the same as `/* */` except the double asterisks tells `javadoc` to
/**. .*/	use the comments in its output.
:	Colon is used in switch statements and the conditional operator. Also, it is used in the enhanced for-loop.
"xx"	Double quotes surround a string literal.
`x`	Single quotes surround a character literal.
+, -, etc	Operator symbols. See Appendix 2 for a listing of operators.
@	Used with annotation. (J2SE 5.0)
< >	Indicates generics. (J2SE 5.0)
%	Used with the `Formatter` class and `printf()` to specify output formats. (J2SE 5.0)
?	Used with the conditional operator:
	`x = boolean ? y: z;`
	As of Java 5.0, '?' acts also as a type wildcard in generics.

Notes:

- identifiers (names of data, methods and classes) cannot begin with a number
- whitespace (space, line return) is ignored in Java code
- non-printing ASCII characters use backslash, e.g. `\t` = tab, `\n` = return.
- Unicode character specified with `\u + 4 hex values`, e.g. `\u03c0` = π.

Appendix 2
Operators

Table A.2.1 *Assignment operators. x and y must be numeric or char types except for "=", which allows x and y also to be object references. In this case, x must be of the same type of class or interface as y. If mixed floating-point and integer types, the rules for mixed types in expressions apply.*

Operator	Description
=	Assignment operator:
	`x = y;`
	y is evaluated and x set to this value.
	The value of x is then returned.
+=, -=, *=, /=, %=	Arithmetic operation and then assignment, e.g.
	`x += y;`
	is equivalent to
	`x = x + y;`
&=, |=, ^=	Bitwise operation and then assignment, e.g.
	`x &= y;`
	is equivalent to
	`x = x & y;`
<<=, >>=, >>>=	Shift operations and then assignment, e.g.
	`x <<= n;`
	is equivalent to
	`x = x << n;`

Table A.2.2   *Arithmetic operators. x and y are numeric or char types. If mixed floating-point and integer types, then floating-point arithmetic is used and a floating-point value returned. If mixed integer types, the wider type is returned. If double and float mixed, double is returned.*

Operator	Description
x + y	Addition.
x - y	Subtraction.
x * y	Multiplication.
x / y	Division. If FP arithmetic and y = 0.0, then infinity returned if x is not zero, NaN if x is zero. `ArthmeticException` thrown if x and y are integer types and y is zero.
x % y	Modulo – remainder of x/y returned. If floating-point arithmetic and y = 0.0 or infinity, then NaN returned. `ArthmeticException` thrown if x and y are integer types and y is zero.
-x	Unary minus. Negation of x value.

Table A.2.3   *Increment and decrement operators. x and y are numeric*
(floating-point and integer) or char types.

Operator	Description

x++	Post-increment: add 1 to the value.
	The value is returned *before* the increment is made, e.g.
	
	```x = 1;```
	```y = x++;```
	
	Then y will hold 1 and x will hold 2

x--	Post-decrement: subtract 1 from the value.
	The value is returned *before* the decrement is made, e.g.
	
	```x = 1;```
	```y = x--;```
	
	Then y will hold 1 and x will hold 0.

++x	Pre-increment: add 1 to the value.
	The value is returned *after* the increment is made, e.g.
	
	```x = 1;```
	```y = ++x;```
	
	Then y will hold 2 and x will hold 2.

--x	Pre-decrement: subtract 1 from the value.
	The value is returned *after* the decrement is made, e.g.
	
	```x = 1;```
	```y = --x;```
	
	Then y will hold 0 and x will hold 0.

Table A.2.4   *Boolean operators. x and y are Boolean types. x and y can be expressions that result in a Boolean value. Result is a Boolean* `true` *or* `false` *value.*

Operator	Name	Description
x && y	Conditional AND	If both x and y are true, result is true. If either x or y are false, the result is false If x is false, y is not evaluated.
x & y	Boolean AND	If both x and y are true, the result is true. If either x or y are false, the result is false Both x and y are evaluated before the test.
x || y	Conditional OR	If either x or y are true, the result is true. If x is true, y is not evaluated.
x | y	Boolean OR	If either x or y are true, the result is true. Both x and y are evaluated before the test.
!x	Boolean NOT	If x is true, the result is false. If x is false, the result is true.
x ^ y	Boolean XOR	If x is true and y is false, the result is true. If x is false and y is true, the result is true. Otherwise, the result is false. Both x and y are evaluated before the test.

Table A.2.5   *Comparison operators. x and y are numeric or char types only except for* `==` *and* `!=` *operators, which can also compare references. If mixed types, then the narrower type converted to wider type. Returned value is Boolean* `true` *or* `false`.

Operator	Description
x < y	Is x less than y?
x <= y	Is x less than or equal to y?
x > y	Is x greater than y?
x >= y	Is x greater than or equal to y?
x == y	Is x equal to y?
x != y	Is x not equal to y?

Table A.2.6 *Bitwise operators. x and y are integers. If mixed integer types, the result will be of the wider type.*

Operator	Name	Description
~x	Compliment	Flip each bit, ones to zeros, zeros to ones.
x & y	AND	AND each bit a with corresponding bit in b.
x | y	OR	OR each bit in a with corresponding bit in b.
x ^ y	XOR	XOR each bit in x with corresponding bit in y.
x << y	Shift left	Shift x to the left by y bits. High-order bits lost. Zero bits fill in right bits.
x >> y	Shift right –	Shift x to the right by y bits. Low-order bits lost.
	Signed	Same bit value as sign (0 for positive numbers, 1 for negative) fills in the left bits.
x >>> y	Shift right –	Shift x to the right by y bits. Low-order bits lost.
	Unsigned	Zeros fill in left bits regardless of sign.

Table A.2.7   *Class and object operators.*

Operator	Name	Description
`x instanceof` `c`	Class test operator	The first operand must be an object reference. c is the name of a class or interface. If x is an instance of type c or a sub-class of c, then `true` returned. If x is an instance of interface type c or a sub-interface, then `true` is returned. Otherwise, `false` is returned.
`new c (args)`	Class instantiation	Create an instance of class c using constructor `c (args)`
"`.`"	Class member access	Access a method or field of a class or object: `o.f` – field `f` access for object o `o.m()` – method `m()` access for object o

`()`	Method
	invocation
		Parentheses after a method name invokes (i.e. calls) the code for the method, e.g.
		`o.m()`
		`o.m(x,y)`

`(c)`	Object cast	Treat an object as the type of class or interface c: `c x =(c)y;` Treat y as an instance of class or interface c

`+`	String
	concatenation
		This binary operator will concatenate one string to another, e.g.
		`String str1 = "abc";`
		`String str2 = "def";`
		`String str3 = str1 + str2`
		results in `str3` holding "abcdef".
		For mixed operands, if either a or b in (a + b) is a string, concatenation to a string will occur. Primitives will be converted to strings and the `toString()` method of objects will be called.
		(This is the only case of operator *overloading* in Java.)
		Note that the operator "+=" will also perform string concatenation.

`[]`	Array element
	access
		In Java, arrays are classes. However, the bracket operators work essentially the same as in C/C++. To access a given element of an array, place the number of the element as an `int` value (`long` values cannot be used in Java arrays) into the brackets, e.g.
		`float a = b[3];`
		`int n = 5;`
		`char c=c[n];`
		where b is a `float` array and c is a `char` array.

Table A.2.8 *Other operators.*

Operator	Name	Description
`x=boolean?y:x`	Conditional Operator	The first operand – *boolean* – is a Boolean variable or expression. First this Boolean operand is evaluated. If it is `true` then the second operator is evaluated and x is set to that value. If the Boolean operator is `false`, then the third operand is evaluated and x is set to that value.

`(primitive type)`	Type Cast	To assign a value of one primitive numeric type to a more narrow type, e.g. `long` to `int`, an explicit cast operation is required, e.g.
		
		`long a = 5;`
		`int b = (int)a;`

Table A.2.9 *Operator precedence. The larger the number, the higher the precedence.*

1	2	3	4	5	6	7	8	9	10	11	12	13	14	15
`=`	`?:`	`||`	`&&`	`|`	`^`	`&`	`==`	`<`	`<<`	`+`	`*`	`new`	`++x`	`.`
`*=`							`!=`	`<=`	`>>`	`-`	`/`	`(type)`	`--x`	`[]`
`/=`								`<`	`>>>`		`%`		`+x`	`(args)`
`%=`								`<=`					`-x`	`x++`
`+=`													`~`	`x--`
`-=`													`!`	
`<<=`														
`>>=`														
`>>>=`														
`&=`														
`^=`														
`|=`														

Notes:

- `(type)` refers to the casting operator.
- "`.`" is the object member access operator.
- `[]` is the array access operator.
- `(args)` indicates the invocation of a method.
- In column 11, the + and `--` refer to binary addition and subtraction. Also, the + refers to the string concatenation operator. In column 14, the + and `--` refer to the unary operations +x and -x and specify the sign of the value.
- `|`, `^`, and `&` refer to both the bitwise and Boolean operators.

Table A.2.10 *Operator associativity. The following operators have right-to-left associativity. All other operators (see precedence table above) are evaluated left to right.*

`=`	`<<=`	`?:`	`-x`
`*=`	`>>=`	`new`	`~`
`/=`	`>>>=`	`(type cast)`	`!`
`%=`	`&=`	`++x`	
`+=`	`^=`	`--x`	
`-=`	`|=`	`+x`	

Appendix 3
Java floating-point

Floating point values in Java are represented by two types: `float` and `double`. Java follows most of the standard IEEE 754 floating-point specifications but not all. In Chapter 2 we discussed floating-point and here we provide some additional information.

A.3.1 Minimum/maximum values

Table 2.3 gave the bit allocations for the two floating-point types and Section 2.11.2 described the values that can be taken for the exponents and significands. Below we show the minimum and maximum values in binary and decimal representations for the two types for both the normalized and denormalized cases.

A.3.1.1 `float`

- Normalized

```
-126 ≤ exponent ≤ +127
min = 2⁻¹²⁶ * 1.00000000000000000000000 = 1.17549435E-38
max = 2⁺¹²⁷ * 1.11111111111111111111111 = 3.4028235E38
```

- Denormalized

```
exponent = -126
min = 2⁻¹²⁶ * 0.00000000000000000000001 = 1.4012985E-45
max = 2⁻¹²⁶ * 0.11111111111111111111111 = 1.1754942E-38
```

A.3.1.2 `double`

- Normalized

```
-1022 ≤ exponent ≤ +1023
min = 2⁻¹⁰²² *
      1.0000000000000000000000000000000000000000000000000000
    = 2.2250738585072014E-308
```

```
max = 2^+1023 *
        1.1111111111111111111111111111111111111111111111111111
    = 1.7976931348623157E308
```

- Denormalized

```
exponent = -1022
min = 2^-1022 *
        0.0000000000000000000000000000000000000000000000000001
    = 4.9E-324
max = 2^-1022 *
        0.1111111111111111111111111111111111111111111111111111
    = 2.225073858507201E-308
```

A.3.2   Special values

As discussed in Chapter 2, operations with floating-point never result in an exception thrown. (Exceptions are Java error conditions. See Section 3.9.) For example, even if an operation results in a *divide by zero* there is no exception thrown. (An integer divided by zero does throw an exception.)

Instead of error messages for abnormal operations, the floating-point result is filled with one of several special floating-point values (see code in Section 2.12.2):

- Floating-point special values:
  - `Float.POSITIVE_INFINITY`: overflow of a positive value
  - `Float.NEGATIVE_INFINITY`: overflow of a negative value
  - `Float.NaN` – Not-a-Number: zero divided by zero, square root of -1
  - `Positive zero`: underflow from positive direction, e.g.

    ```
    x = 2.0e-45 * 1.0e-10
    ```

  - `Negative zero`: underflow from negative direction, e.g.

    ```
    x = -2.0e-45 * 1.0e-10
    ```

- Finite floating-point numbers and the special values are ordered from smallest to largest as follows:
  1. `NEGATIVE_INFINITY`
  2. Negative finite values
  3. `Negative zero` and `Positive zero`
  4. Positive finite values
  5. `POSITIVE_INFINITY`
- The positive and negative zero values act as follows:
  - `Positive zero` and `negative zero` compare as equal
  - `1.0 / (positive zero)` → `POSITIVE_INFINITY`
  - `1.0 / (negative zero)` → `NEGATIVE_INFINITY`

Table A.3.1 *Floating-Point Type Specifications*

Parameter	float	float-extended-exponent	double	double-extended-exponent
N	24	24	53	53
K	8	> 10	11	> 14
E_{max}	+127	> +1022	+1023	> +16382
E_{min}	−126	< −1021	−1022	< −16381

- NaN values are unordered. This means that:
  - Numerical comparisons and tests for numerical equality result in `false` if either or both operands are NaN.
  - A test for numerical equality of a value against itself results in `false` if and only if the value is NaN.
  - A test for numerical inequality results in `true` if either operand is NaN.

A.3.3 Extended exponents

The JVM Specifications after version 1.1 allow for a JVM implementation to include extended exponent versions of either or both the float and double types during intermediate calculations so as to avoid over/under flows.

Table A.3.1 maps the floating point specifications for the four types with the symbols defined as follows:

- N = number of bits in significand
- K = number of bits in exponent
- E_{max} = maximum value of exponent
- E_{min} = minimum size of exponent

The final accessible floating-point results will be in `float` or `double` types but intermediate floating-point values can use the larger extended exponent representations if the platform processor allows it. There is no access for the Java programmer to the extended exponent types.

The JVM does not support either the official IEEE 754 single extended or double extended format since these extended formats require extended precision, i.e. longer significand, in addition to the extended exponent ranges shown in the above table. The documentation for a particular JVM should indicate whether it uses the extended exponent options.

The modifier `strictfp` in front of a method will force the precision to remain at 64-bit for all calculations within that method. This is useful if one wants to ensure exactly the same results regardless of the platform or JVM implementation.

A.3.4   More about floating-point

Additional notes of interest about Java floating-point include:

- Literals default to `double` unless appended with f or F:

```
float  x = 1.0;  // compile time error
float  x = 1.0f; // OK
double x = 1.0;  // OK
```

- Floating-point rounding:
  - The JVM uses IEEE 754 *round-to-nearest* mode: inexact results are rounded to the nearest representable value, with ties going to the value with a zero least-significant bit.
  - Instructions that convert values of floating-point types to integer values will round towards zero.
- See Section 10.14.3 for a discussion of bit operations on floating-point numbers

Resources

Joseph D. Darcy, *What Everybody Using the Java*™ *Programming Language Should Know About Floating-Point Arithmetic*, Sun Microsystems, JavaOne Conference, 2002, `http://servlet.java.sun.com/javaone/sf2002/conf/sessions/display-1079.en.jsp`

David Flanagan, *Java in a Nutshell*, 4th edn, O'Reilly, 2002.

David Goldberg, *What Every Computer Scientist Should Know About Floating-Point Arithmetic*, Computing Surveys, March 1991, `http://docs.sun.com/source/806-3568/ncg_goldberg.html`

James Gosling, Bill Joy, Guy Steele, and Gilad Bracha, *The Java Language Specification*, 2nd edn, Addison-Wesley, 2000. Available online at: `http://java.sun.com/docs/books/jls/second_edition/html/jTOC.doc.html`.

Ronald Mak, *Java Number Cruncher: The Java Programmer's Guide to Numerical Computing*, Prentice Hall, 2003.

Glen McCluskey, *Some Things You Should Know about Floating-Point Arithmetic*, Java Tech Tips, February 4, 2003, `http://java.sun.com/developer/JDCTechTips/2003/tt0204.html#2`

Index

Classes are listed by name, followed by the package in which they belong in parentheses, e.g. `HashMap (java.util)`. Methods are indicated by name followed by parentheses with no argument list, e.g. `println()`.

697